Fluorescence Techniques in Cell Biology

Edited by
A. A. Thaer and M. Sernetz

With 303 Figures

Springer-Verlag Berlin · Heidelberg · New York 1973

Dr. Andreas A. Thaer · Dr. Manfred Sernetz
Battelle-Institut e. V.
Biosciences Department
Frankfurt a. M./W. Germany

Proceedings of the Conference on
"Quantitative Fluorescence Techniques as Applied to Cell Biology"
held at
Battelle Seattle Research Center
4000 N.E. 41st Street
Seattle, Washington
March 27–31, 1972

In accord with that part of the charge of its founder, Gordon Battelle, to assist in the further education of men, it is the commitment of Battelle to encourage the distribution of information. This is done in part by supporting conferences and meetings and by encouraging the publication of reports and proceedings. Towards that objective this publication, while protected by copyright from plagiarism or unfair infringement, is available for the making of single copies for scholarship and research or to promote the progress of science and the useful arts.

ISBN-13: 978-3-642-49206-8   e-ISBN-13: 978-3-642-49204-4
DOI: 10.1007/978-3-642-49204-4

Preface

.For there to be progress in science, there must first be communication between experts of different disciplines. This is particularly true of modern biology which is becoming more and more of an interdisciplinary field. The present situation in cell biology clearly reflects this development and demonstrates that the application of physical techniques was necessary before this field of biological research could be developed on an objective and quantitative basis. The utilization of optical phenomena as measuring parameters at the microscopic level has provided the basis for the development of quantitative cytochemistry. This rapidly growing extension of conventional cytochemistry and histochemistry is based on the visual oberservation of qualitative chemical criteria in correlation with the microscopically resolved structure of cells and tissues. Furthermore, the introduction into cytochemistry of such optical measuring techniques as absorption photometry, interferometry, and fluorometry, as well as the measurement of optical anisotropy, diffraction and scattered light, has provided the methodological bridge for the exchange of knowledge between cell biology on the one hand and biochemistry, or molecular biology, on the other.

Travelling in one direction across this bridge, the cell biologist in general and the cytochemist in particular profit by the accumulated experience of the biochemist concerning the theoretical basis and the application of these optical measuring techniques and the interpretation of results at the molecular level. The traffic in the opposite direction involves an essential part of the results obtained by biochemistry: unless these data are applied and interpreted in terms of the functional structure and organization of the cell, they will lack most of the desired biological significance. Thus, what cell biologists have learned in this field should be of equal interest and value to the biochemist. There is probably no better example of this fruitful interaction than the application of quantitative fluorescence techniques in both biochemistry and cell biology.·

More than any other method, fluorescence techniques are not just a tool for applying quantitative cytochemistry to cellular composition, but must also be seen as providing perhaps the most important basis for "biochemistry at the level of the single intact cell and its structure" and may be expected to make an essential contribution to the understanding of the molecular organization of the cell.

The subject and scope of the conference the proceedings of which are published in this volume should be seen against this background. The organizers of the conference habe tried to construct a scientific program which reflects not only the present stage in the application of quantitative fluorescence techniques in biochemistry and cytochemistry, but also the increasing interaction between these two disciplines. Moreover, the program was intended to pinpoint crucial problems which probably cannot be solved except by such an interaction. They confidently hope they have gone some way toward meeting this goal, despite the sometimes highly divergent views with regard to the material to be investigated, the application of fluorescent techniques for obtaining quantitative information, and the interpretation of the results. There is still much to be done in overcoming "language problems".

The sequence selected for sessions I to V is certainly not the only possible one. However, the methodological aspect of the conference was the dominant consideration. Thus, it seemed reasonable to start with two reviews (Part I) on the utilization of two fluorescence parameters – fluorescence polarization and decay time – that have been succesfully used for the biochemical investigation of solutions, suspensions and homogenates, but have not yet been fully utilized for measuring single cells at the microscopic level.

The papers in Part II deal with the instrumental and standardization requirements for microscope fluorometry and spectrofluorometry on single cells. Not only do they reflect the present state of the art in this field, they also open the way for the systematic introduction of spectrofluorometry on solutions into ultramicroanalysis at the microscopic level. This in turn permits the direct comparison of information obtained for cells and solutions under identical microscopic conditions, which is most welcome because the data can be used for calibration and standardization and also for interpreting the results obtained for single cells. These contributions thus form a bridge to Part III and the papers dedicated to the central subject of quantitative cytochemistry: the chemical composition of cells and cellular compartments. In addition to the progress achieved in the cytofluorometric determination of DNA, RNA, proteins and mucopolysaccharides microscope fluorometry has successfully penetrated new areas: chromosomal analysis, study of biogenic amines, and immunology.

The new techniques of automated cytofluorometry permit the rapid assessment of cytochemical parameters in cell populations and their recording in histogram form. Such techniques open up new horizons for studies of cell population kinetics, for instance in pharmacology and for clinical diagnostics.

The same is true, at least in part, of the application of microscope fluorometry to the measurement of cellular activity at the level of the single cell and its compartments (Part IV). For instance, it has been used to investigate the intracellular turnover of fluorogenic substrates of different cell populations under various influences. A further step towards the ultimate goal of revealing the functional structure of the living intact cell is to apply techniques for the rapid microfluorometric recording of fluorescence intensity and spectrum to the study of intracellular enzyme reactions and transport phenomena and to correlate the findings with the microscopically resolved cellular structure.

Finally, the various papers in Part V deal with the use of fluorescent substances as molecular probes to study the configuration, conformation and reaction sites of biological macromolecules. This session illustrates particularly well the complementary nature of studies on known solutions on the one hand, and those on organized biological material like fractionated cellular compartments and even intact cells on the other.

Models like artificial membranes are becoming increasingly important in the study of such properties of well-defined macromolecular structures. Models can aid in the interpretation of the results obtained by measuring the fluorescence parameters of fluorescent molecular probes bound to cellular structures. There is no doubt that during the next few years this very promising application of fluorescence techniques in cell biology will considerably increase in importance, for it correlates biochemical information and cell structure, i. e. the morphology and chemical topology of the intact cell.

The editors – who were also responsible for the organization of the conference – gratefully acknowledge the cooperation of all authors who contributed to this volume and their understanding of the editors' problems. They also wish to thank the session chairmen, Dr. H NEURATH (I), Dr. S. S. WEST (II), Dr. F. RUCH (III), Dr. B. THORELL (IV) and Dr. G. WEBER (V), for their effective coordination and their determination to make each session a real success.

The organizers' and editors' heartfelt thanks are due to Dr. T. CASPERSSON for taking on the conference chairmanship, and for his valuable advice concerning the preparation of the scientific program, his introductory remarks to the conference, and his own outstanding contribution.

The conference was made possible by the generous support provided by the Battelle Institute, Life Sciences Program. The organizers would like to express their sincere thanks to Dr. H. NEURATH, University of Washington, at that time coordinator of the life sciences activities sponsored by the Battelle Institute, for his deep and permanent interest in the organization of the conference and for his suggestions concerning the preparation of the scientific program.

The help of the staff of the Battelle Seattle Research Center, in particular of Mr. L. BONNEFOND and his staff, before and during the conference is also gratefully acknowledged.

V

The essential part of the work concerning the preparation of all manuscripts for the delivery to the publisher was performed by Mr. A. W. ROECKER, Librarian of the Battelle Seattle Research Center, and by his staff, to whom the editors' thanks are due.

September, 1973   The Editors

Contents

OPENING REMARKS

by

T. Caspersson

I am very pleased to have been asked to give some opening remarks on this occasion, for I believe that this will be an unusually productive conference, and that it is being held at precisely the right time. Work in the field of biological fluorescence is in a period of almost explosive development, but it is proceeding along so many different lines — biophysical, methodological, organochemical, cell physiological, genetical, etc. — that it is impossible for one man to follow more than part of it. Thus, just now, there exist real information gaps, and this conference is very ably planned to begin bridging these gaps. One of the crucial reasons why development has accelerated so quickly during the last few years is that, at long last, really quantitative methods have been developed and used. It is thus very proper to use the quantitative approach as the common denominator in our discussions this week.

The field of microscopy was the first field in biology to benefit from fluorescence approaches. The great microscopy pioneer, August Köhler, gave the first impetus as early as in 1904, and in 1911 Lehmann built a good instrument and produced the first photographs in 1913. I believe it is correct to say that above all, it was the primary fluorescence of porphyrins that lay behind the great widening of the work in the late twenties and the thirties, even though interesting work was performed using secondary fluorescence on different cell constituents. Acridine orange was even then a very popular substance.

At that time the fluorescence microscopes used carbon arcs or sparks between metal electrodes as light sources and functioned quite well even if their stability was not very good. Optics and filters were of a very satisfactory standard, although the choice was not very wide. Moreover, soon after came the gas discharge lamps which replaced all other light sources. Many different lines developed and many types of biological materials were studied.

Little or nothing was then heard of microscope fluorometry. We tried twice in the early forties to build ultramicrofluorometers at our laboratory. In applying them to biological objects on a microscale, however, fluorescence decay became such a problem that we had to abandon the project.

While optics, filters, and light sources were satisfactory, the photoelectric aspects offered almost insurmountable difficulties until photomultipliers came along after the war. Then suddenly, everything improved. At the same time, however, there were so many other fields for the new sensitive light detection systems that, to my knowledge, little was done on fluorescence until in this last decade, when work was taken up in many places, and really quantitative work on cells was started.

Fluorescence analysis is a complex field. Many physical and chemical factors influence the intensity and the character of the fluorescence of an object. It is most encouraging that during these last years these fields have grown so rapidly and that therefore the amount of information available has also greatly increased. We will be hearing a great deal on this subject during our meeting.

Furthermore, physicochemical studies have shown that fluorescence analysis can offer much more than a means for quantitative chemical determinations in biological materials. Methods such as measurement of fluorescence polarization and depolarization and work on relaxation kinetics have opened new vistas in the study of the macromolecular organization of cellular compounds.

During the last two decades we have received hardly any essentially new basic tools in the field. However, different types of biological problems make very different demands on the instrumentation to be used, both in routine fluorescence measurements and in measurements of more sophisticated kinds. This field is well covered in this symposium.

All in all, this syposium presents surveys of a number of fields, often of very different kinds, but all aimed at really quantitative work on the composition and structure of cellular constituents. Many of these ostensibly rather unrelated fields

are difficult for workers in other areas to penetrate. Thus this symposium is an unusually meaningful convocation of research workers who share a common interest but
who are following different lines.

There will also be a series of papers giving examples of applied biological work
in several fields, from enzyme kinetics to nerve function, and these, together with
the presentation of physical and chemical background information, will, I am sure,
stimulate all of us to a further widening of the realm of biological fluorometry.

We are all very grateful to the Battelle organization for this opportunity.

Part I
Introductory Papers

POLARIZED FLUORESCENCE

Gregorio Weber

Roger Adams Laboratory, Department of Biochemistry
University of Illinois, Urbana, Illinois

ORIGIN OF FLUORESCENCE POLARIZATION

The emission from a fluorescent solution is always polarized to some degree. By this we understand that an observer who sees the light emitted in a direction at right angles to the direction of the excitation through a polarizer observes changes in the intensity of the emission as he rotates the polarizer (Figure 1a). This polarization, first discovered by Weigert (1920) has received considerable attention; its origin and significance are well understood in the case of fluorescent solutions. Briefly, the polarization results from the existence of a preferential orientation in the molecules at the time of the emission. The exciting light produces a selection of the orientations because the molecules preferentially excited are those in which the transition moment in absorption is parallel or at a small angle to the electric vector of the polarized excitation. This original orientation is preserved if the molecules undergo negligible motion between excitation and emission, in which case the polarization is a maximum, called the fundamental or limiting polarization P_0.

EFFECT OF MOLECULAR MOTIONS

On the other hand, if the brownian motion of the molecules is very lively, the initial orientation is all but lost in the few nanoseconds elapsing between excitation and emission; in this case only a residual polarization of the order of a few parts per thousand to a few parts per hundred is observable. It follows that in dilute solutions the polarization of the fluorescence is a good indicator of the motion of the particles responsible for the fluorescence. In fact, the polarization observed is determined by only three factors: the limiting polarization P_0 which would obtain if no disorientation followed the excitation, the rate of rotation R of the particles, and the interval between excitation and emission, τ. In 1926 Francis Perrin showed that these quantities were linked by the relation:

$$\frac{1}{P} - \frac{1}{3} = (\frac{1}{P_0} - \frac{1}{3})(1 + 6R\tau)$$

(1)

The simplicity of equation 1 is deceptive. While P, P_0 and τ have a simple physical significance and can be subjected to direct experimental measurement, R can have a very complex origin. If the fluorescent elements are attached to rigid spherical particles, R is the rate of rotation of a sphere, which Einstein in 1906 showed to be equal to:

$$R = \frac{1}{6} \frac{kT}{\eta V}$$

(2)

Here k is the Boltzman constant, T the absolute temperature, η the viscosity of the solvent in which the particle moves, and V the volume of the particles. While some simple molecules of weight of a few hundred daltons may be considered to come close to "rigid spheres," the fluorescent molecules of the biological specimens cannot always be assumed to have both of these characteristics. Some particles will be irregular, so that their motion is not characterized by a single rotational rate but by three rates R_1, R_2, and R_3 about a set of cartesian axes attached to the particles. The effects that the rotations about these different axes would produce as regards the polarization of the fluorescence will depend upon the angles that the three axes make with the transition moments in absorption and emission. Both a phenomenological (Weber 1971)and a rigorous diffusion theory(Perrin 1936 ;Belford et al.1972;Ehrenberg and Rigler1972) have been developed to account for the rotational motions and the ensuing depolarization of the fluorescence of these irregular particles. We shall not dis-

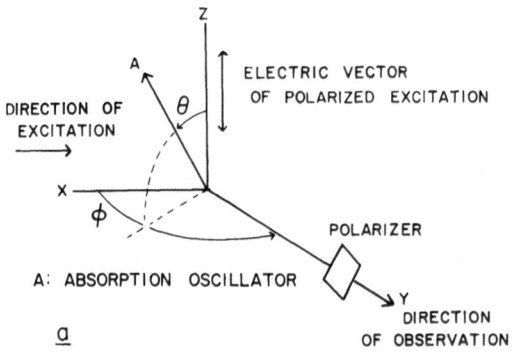

Figure 1a. *Coordinates showing direction of excitation and observation*

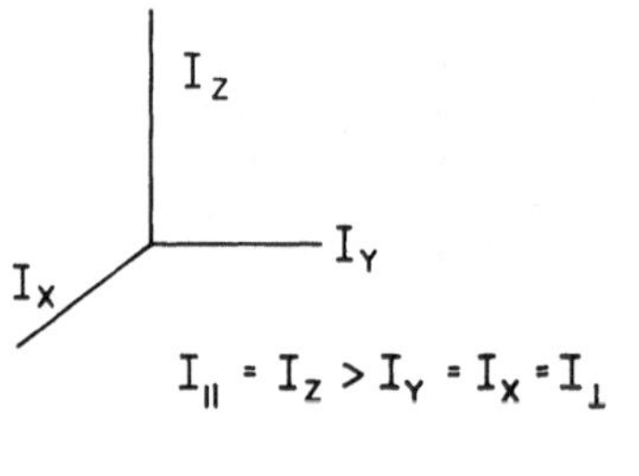

$$I_{\parallel} = I_z > I_y = I_x = I_{\perp}$$

Figure 1b. *Intensity of the polarized components* I_z, I_y, I_x, *present when the solution is excited with light polarized along OZ*

cuss these here, but will refer the interested readers to reviews and original papers on this subject.Some general conclusions,however,may be stated. If we consider the conditions of the experimental observations,we see (Figure 1)that the excited molecules form a system with an axis of symmetry determined by the electric ve ctor of the polarized excitation.As a result of this symmetry,some molecular rotatory motions do not produce depolarizing effects.It follows that the depolarization reflects the rotations very much like a two-dimensional picture reflects a three-dimensional object. In the latter case, the three-dimensional object may be seen in its true shape by taking pictures from various angles. The analogy here is the observation of the depolarization under excitation by several different wavelengths. The limiting polarization P_0 can be shown in fact to be a function of the exciting wavelength. According to an equation originally due to Levshin (Perrin 1936; Weber 1971)

$$Cos^2\theta = \frac{1 + 3P_0}{3 - P_0} \qquad (3)$$

where θ is the angle between the transition moments in absorption and emission. The emission oscillator is placed in a fixed direction in the molecule, while the absorption oscillator has a different direction for each electronic absorption band.

POLARIZATION IN AROMATIC FLUOROPHORES

Most, if not all, of the fluorophores extensively employed in biochemistry and biology are aromatic molecules in which the absorption transitions as well as the unique fluorescence transition are of typical $\pi \rightarrow \pi^*$ nature. There is general agreement (Platt 1949) — if not extensive experimental proof — to the effect that the transition moments of $\pi \rightarrow \pi^*$ transitions in aromatic molecules are all contained in the plane of the aromatic rings. Consider then an irregular particle, typically a macromolecule carrying an aromatic fluorophore rigidly attached to it in a unique orientation (Figure 2). The rotational motion of the particles takes place about three axes of which one, R_3, is normal to the plane of the aromatic ring. As the absorption falls upon different directions contained in the plane of the ring, different

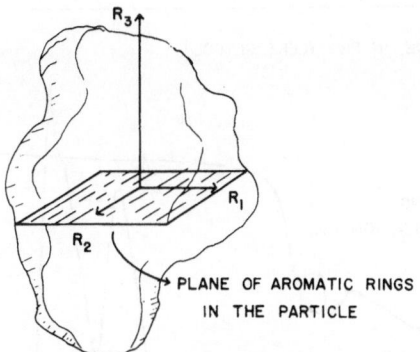

PLANE OF AROMATIC RINGS
IN THE PARTICLE

R_3 = ROTATIONAL RATE ABOUT AN AXIS NORMAL
TO THE RING PLANE DETERMINES IN-PLANE
ROTATIONS

R_1, R_2 = ROTATIONAL RATES ABOUT TWO AXES
CONTAINED IN THE RING PLANE DETERMINE
RATE OF OUT-OF-PLANE ROTATIONS

Figure 2. *Phenomenological axes of rotation, when an aromatic fluorophore is rigidly attached to a macromolecule*

weights are assigned to the effects of the rotational axes. When the absorption and emission oscillators are at 45° to each other, that is when P_0, according to equation 3, equals 1/7, the effect of R_3 upon the depolarization becomes null. This results from the fact that under these conditions all orientations obtained by rotation about R_3 are equally represented at the time of excitation. Further rotation about R_3 can only exchange positions; it cannot destroy the original random orientation. In this case therefore, the only active rotations are those that bring the aromatic rings out of the plane that they originally occupied at the time of the excitation (out-of-plane rotations). On the other hand, if the absorption and emission are coincident $(P_0 = 1/2)$ the rotations that bring the aromatic ring out of its own plane, and those about R_3 that keep the rings in it (in-plane rotations) have equal weight. This generalization is valid only if the rotations are small, that is, if the values of P observed are not very much smaller than those of P_0. In these cases Perrin's Equation 1 will apply, with R being some weighted average, $\bar{R}$, of R_1, R_2 and R_3. From our description of the motions we can write for this average value $\bar{R}$,

$$\bar{R} = (Rip + Rop)/2 \quad if\ P_0 = 1/2$$
$$\bar{R} = Rop \qquad\qquad if\ P_0 = 1/7$$

(4)

where Rip = rate of in-plane rotations, Rop = rate of out-of-plane rotations. Therefore, simple observations of the dependence of the polarization at these values of P_0 can tell us whether the in-plane and out-of-plane rotations of the ring are equivalent or otherwise. The figures below show observations of this type for two different fluorophores: perylene (Figure 3 and 4) and 1-naphthylamine (Figures 5 and 6). In both cases, but particularly so in perylene, the out-of-plane rotations are markedly more sluggish than the in-plane rotations.

POLARIZATION OF FLUORESCENCE FROM PROTEIN CONJUGATES

In the case of a protein carrying a fluorophore rigidly attached in a fixed orientation as depicted in Figure 2, the same considerations apply. Witholt and Brand (1970) have studied the effect of changing the wavelength of excitation upon the observed value of $\bar{R}$ in complexes of anilino naphthalene sulfonate (ANS) and bovine serum albumin (BSA). They found that as P_0 decreases the value of $\bar{R}$ also decreases, which, from the

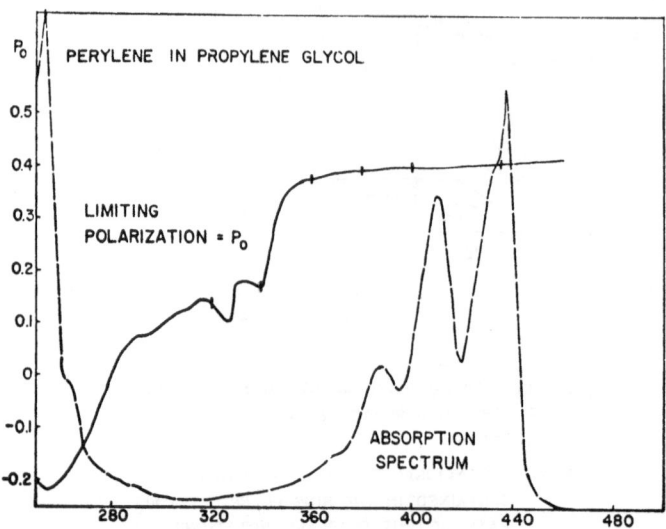

Figure 3. *Absorption and polarization spectra of perylene. The fluorescence polarization is that observed when the whole of the emission at wavelength longer than 450 nm is measured as function of excitation wavelength*

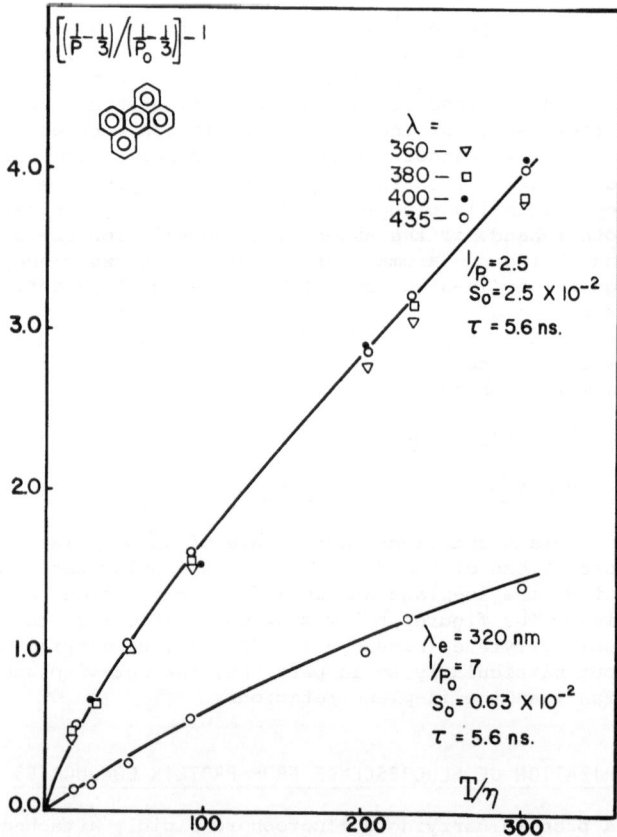

Figure 4. *Plots of relative anisotropy against temperature/viscosity for perylene in propylene glycol. Temperatures varied from -20°C to 25°C. S_0 is the initial slope in the plots. The in-plane rotations are some seven times faster than the out-of-plane rotations*

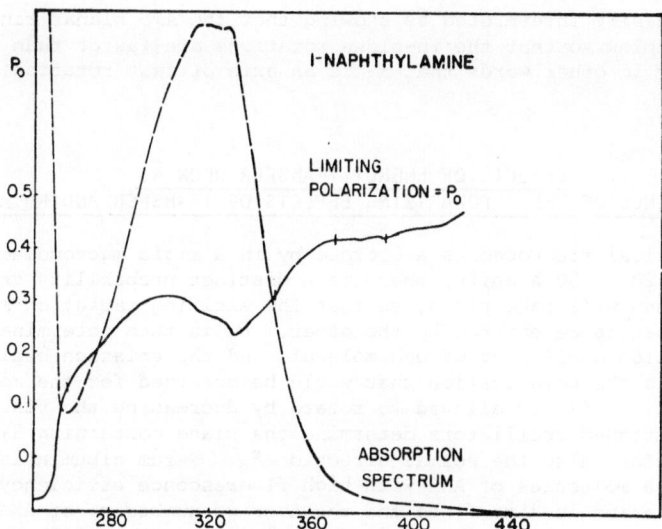

Figure 5. *Absorption and polarization spectra of 1-naphthylamine. The fluorescence at wavelengths longer than 425 nm was observed*

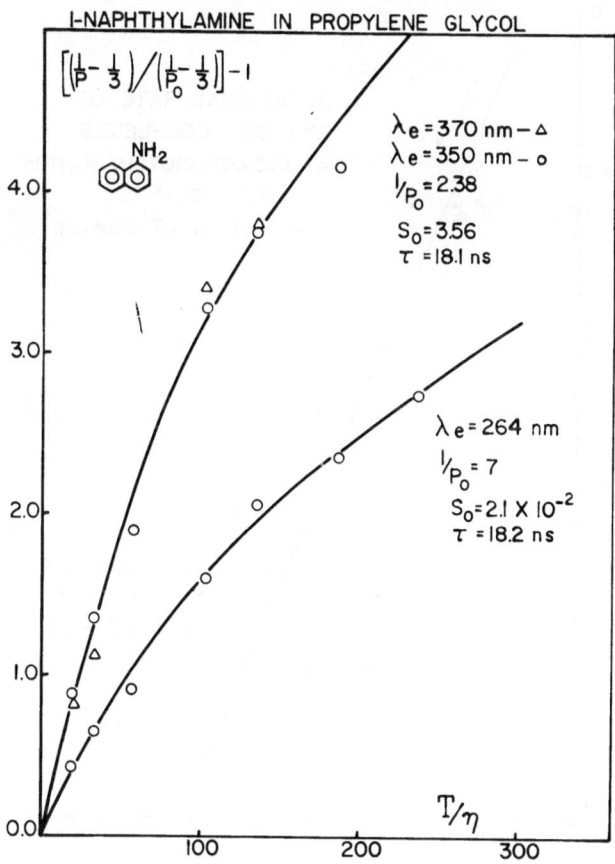

Figure 6. *Plots of relative anisotropy against T/η, as in Figure 4. From the initial slopes it follows that in-plane rotations are twice as fast as out-of-plane rotations*

above remarks, is easily interpreted as showing that the *ANS* planar ring system is oriented in the complex so that the in-plane rotations are faster than the out-of-plane rotations, or in other words that R_3 is an axis of fast rotation as compared with R_1 or R_2.

EFFECTS OF ENERGY TRANSFER UPON $\bar{R}$.
EQUIVALENCE OF THE DEPOLARIZING EFFECTS OF TRANSFER AND ROTATION

When two identical fluorophores are close by in a rigid macromolecule, and by "close by" we mean 20 or 30 Å units, there is a distinct probability that energy transfer between them will take place, so that the exciting radiation may be absorbed by one and the fluorescence emitted by the other. P_0 is then determined by the angle between the absorption oscillator of one molecule and the emission oscillator of the other, since this is the polarization that would be observed for the motionless particle. If the particle is now allowed to rotate by decreasing the viscosity of the medium, the two mentioned oscillators determine the plane containing R_1 and R_2 in Figure 2, and therefore also the normal direction R_3. Serum albumin is capable of binding four or five molecules of ANS with high fluorescence efficiency. If the wavelength of excitation is kept fixed and the average number $\bar{n}$ of ANS molecules bound per BSA is allowed to increase, $\bar{R}$ is determined increasingly by the absorption and emission oscillators belonging to different ANS molecules. If there were no pre-

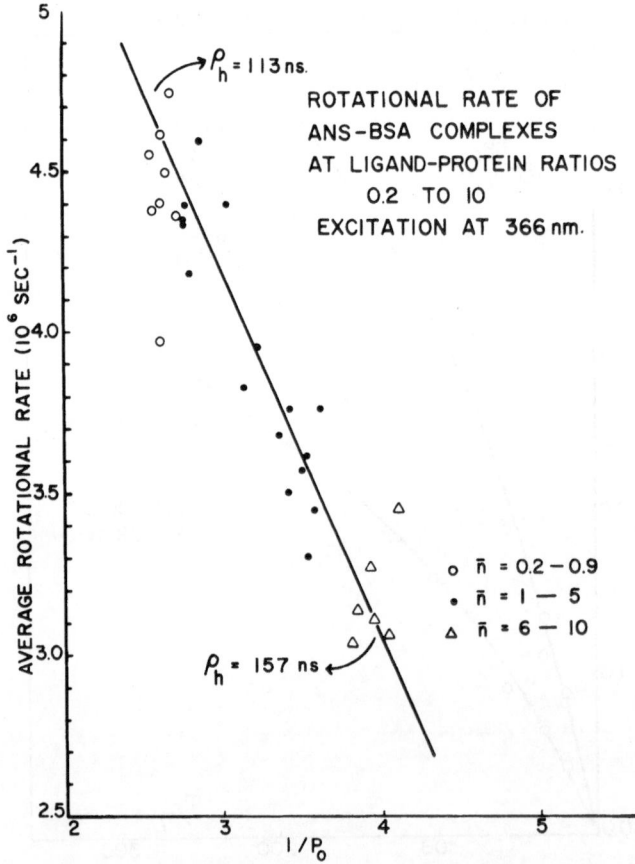

Figure 7. *Data of Weber and Anderson replotted to show the dependence of the apparent rotational rate upon limiting polarization when this varies due to energy transfer among ANS molecules*

ferential orientation of these molecules with respect to each other, $\bar{R}$ would be constant; but if they have a mutual preferential orientation, as well as preferential orientation within the protein, $\bar{R}$ will be found to vary with $\bar{n}$. It is easy to see that if R_3 is an axis of fast rotation as compared with R_2 and R_1, $\bar{R}$ will decrease as $\bar{n}$ increases; but if R_3 is an axis of slow rotation $\bar{R}$ will increase with $\bar{n}$. Figure 7 shows a study by Anderson and Weber (1969) of the average rate of rotation of BSA-ANS complexes as a function of $\bar{n}$ for excitation with the 366 nm Hg line. $\bar{R}$ decreases systematically with $\bar{n}$ as P_0 decreases. Evidently the ANS fluorophores must align themselves so that the plane determined by the oscillator of absorption for 366 nm excitation of one molecule and the emission oscillator of the molecule to which the excitation is preferentially transferred is normal to an axis of fast rotation. The results are then in agreement with those already quoted of Witholt and Brand; in their study, they kept $\bar{n}$ constant and obtained the variation of P_0 by changing the excitation wavelength. These two studies furnish good examples of the equivalence of the effects of brownian rotations and energy transfer as regards the depolarization.

DEPOLARIZATION MEASUREMENTS IN REAL TIME

The static measurements of the rate of rotation of the molecules described above have one serious limitation: they require observations at a series of temperatures and viscosities. Thus, although they are practicable for the physical chemist or biochemist, who examines the behavior of molecules in solution, they do not appeal as much to the biologist, who does not dare disturb his more complex and fragile material by alterations of either variable. An alternative method, less accurate and convenient but without the mentioned drawbacks, is available however. It is based on an initial theory and observations of Jablonski in 1935(also Wahl 1966) and consists of the measurement of the rate of rotation in real time. Suppose that the conditions of excitation are those of Figure 1, but that instead of a steady light, a flash of light of a short duration, as compared with the fluorescence lifetime, is used for excitation. At the instant of the excitation, the intensities of the polarized components of the emission I_z, I_x, and I_y are as shown in the scheme of Figure 1b. I_z parallel to the electric vector of the exciting light is greater than I_x or I_y; these are equal to each other because of symmetry. The molecular rotations effectively transport the excitation from one component to another during the fluorescence lifetime. If an observer determines the intensity I_z following the exciting flash by looking at it through a polarizer, he will see the intensity decrease as a result of two processes; emission and rotation. The latter decreases the number of excited molecules emitting light polarized along Z, since a rotation of 90° makes their emission unobservable through the polarizer. It is true that molecules excited along I_x or I_y will, through their own rotation, become visible to the observer looking for polarization along Z, but because I_z is originally greater than I_x or I_y, it loses to them more than it gains from them. The net result is that the decay of the emission polarized along Z is faster than the exponential decay that would be observed if no rotation took place. For the very same reasons I_x and I_y decay more slowly than in simple exponential fashion. The decay of the polarization with time, $P(t)$, brought about by the equalization of the intensities as described above, is a function of the rates of rotation and emission of the same form as the steady-state polarization P, which is in fact nothing more than an average value of $P(t)$ weighted over all times of possible emission. For a spherical molecule it is found that the real-time polarization decays according to the simple relation,

$$A(t) = A(0)\ e^{-6Rt} \tag{5}$$

when A is the emission anisotropy defined by Jablonski (1960) as

$$A = (2/3)(1/P - 1/3)^{-1} \tag{6}$$

According to the last equation, if log $A(t)$ is plotted against t, a single rate of decay will be observed if the molecule is close to spherical. If the molecule is of irregular shape, it should be possible to observe more than one rate of rotational decay, although it must be acknowledged that up to the present no case has been reported in which the rotational decay can be clearly attributed to the existence of different rates of rotation in a *rigid* molecule.* As in the steady-state measurements, it is possible to recognize the existence of different rates of rotation by measuring the average rate $\bar{R}$ of rotations over a small angle as a function of P_0 (or $A(0)$). This method of determining the existence of different rates of rotation is far more promising than the resolution of the anisotropy decay into components of different rates. This is demonstrated in Figure 8 which shows a computer plot of log $A(t)$ as a function of time for excitation at two wavelengths, for which $P_0 = 1/2$ and $1/7$ respectively, in the case of an irregular molecule with rotational rates in the ratios of 4:2.5:1. It is seen that the change in slope over the decay amounts to no more than 20% if the observations are carried out at values of $A(t)$ close to $A(0)$ and $A(0)/10$. When it is

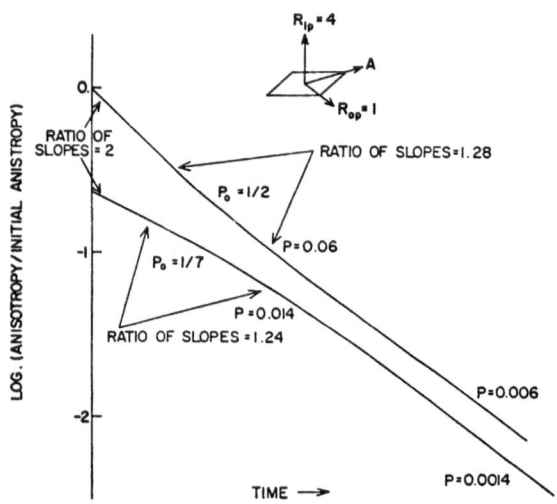

Figure 8. *The decay of the logarithm of the anisotropies for an irregular mole-cule (axial ratio = 4:1) excited at wavelengths for which $P_0 = 1/2$ and $P_0 = 1/7$. The demonstration of two rates of rotations by observations of the two initial slopes is relatively simple, as these are in the ratio 2:1. The same demonstration by the change in slope of log A vs. time in either plot appears much more difficult, since the change in slope is only 30% and requires measurements at very small value of P for its demonstration*

remembered that $A(t)$ cannot be directly measured, but must be constructed from separate observations of $I_\parallel(t)$ and $I_\perp(t)$, it becomes obvious that under present conditions of observation we shall not be able in this case to assert that more than one rate of rotation is active in the depolarization. On the other hand, comparison of the initial slopes for $P_0 = 1/2$ and $P_0 = 1/7$ shows that they differ by 100%, and this difference may be detected with certainty.

We can conclude that the observation of the fluorescence polarization of solutions is capable of furnishing the overall rate of rotation of the particle, and that the state of the art is such that the demonstration of more than one rate of rotation in the case of irregular molecules is a practical possibility.

FLUORESCENCE POLARIZATION OF ORGANIZED SYSTEMS

The fluorescence of solutions forms an indispensable background for the understanding of the phenomena, but it requires some extension before it can be applied to

---

* *Yguerabide, et al., (1970), have observed two rotational rates in gamma globulin coupled to DNS. The two rates do not correspond, however, to those of a rigid body of the dimensions of the gamma globulin molecule, but must be attributed to the existence of a limited internal rotation involving only a portion of the molecule. This portion must comprise the fluorescent DNS residue and a fraction of the protein structure, which according to Yguerabide et al., is as large as the F_{ab} fragment itself.*

the more complex case of organized systems, which are the main interest of the biolog-ist. The two most important differences between systems with random orientation of the components and the organized systems are the preferential orientation of the fluorophores at the time of the excitation; and the possibility of anisotropic mo-tion, owing not to the shape of the particles, but to the anisotropic resistence pre-sented by the medium in which the particles move. In these cases it is no longer possible to assume the existence of symmetry about the electric vector of the ex-citing light as is done for solutions. It would be tedious, and not very useful, to try to enumerate the type and number of observations required to define completely an arbitrary anisotropic system. Instead we shall discuss a simple test to determine that preferential orientation of the emitting elements exists in the system.

If the emission is observed in the direction of the electric vector of the ex-citing light, the absence of preferential orientation about this direction is shown by the absence of polarization. If a polarizer is rotated, no change in intensity is observed in such case. Conversely, the presence of polarization demonstrates the existence of preferential orientation in the emitting elements. If the excitation is directed along the different local coordinates of the illuminated object, it should be possible to draw certain conclusions about the preferential orientations of the emit-ting elements present. Already a number of observations have been made in which — by the method described above or by what are, in fact, equivalent measurements — the ex-istence of orientation of fluorophores in bilayer bubbles (Yguerabide and Stryer 1971), muscle (Aronson and Morales 1969; Dos Remedios et al.1972 b) and nerve (Tasaki et al.1971)has been demonstrated. In opposition, no case has been demonstrated of the influence of the medium in the production of anisotropic rotations. Although there is at present no theory to describe these possible effects, the observation of the fluo-rescence polarization of biological specimens is certain to provide both the pheno-mena and the interest that will lead to its development.

REFERENCES

ANDERSON, S.R., and Weber, G., (1969), *Biochemistry*, **8**, 371.

ARONSON, J. F., and Morales, M. F., (1969), *Biochemistry*, **8**, 4517.

BELFORD, G. C., Belford, R. L., and Weber, G., (1972), *Proc. Natl. Acad. Sci. USA*, **69**, 1392.

DOS REMEDIOS, C. G., Millikan, R. G. C., and Morales, M. F., (1972a), *J. Gen. Physiol.*, **59**, 103.

DOS REMEDIOS, C. G., Yount, R. G., and Morales, M. F., (1972b), *Proc. Natl. Acad. Sci. USA*, **69**, 2542.

EHRENBERG, M., and Rigler, R., (1972), *Chem. Phys. Letters.*, **14**, 539.

JABLONSKI, A. Z., (1935), *Physik.*, **95**, 53.

JABLONSKI, A., (1960), *Bull. Pol. Acad. Sci.*, (Phys. Ser)., **8**, 259.

PERRIN, F., (1926), *J. Phys. Rad.*, **7**, 930.

PERRIN, F., (1936), *J. Phys. Rad.*, **7**, 1.

PLATT, J. R., (1949), *J. Chem. Phys.*, **17**, 484.

TASAKI, I., Watanabe, A., and Hallet, M., (1971), *Proc. Natl. Acad. Sci. USA.*, **68**, 938.

WAHL, P., C.R. (1966), *Acad. Sci. Paris.*, **263**, 1525.

WEBER, G., (1971), *J. Chem. Phys.*, **55**, 2399.

WEIGERT, F., (1920), *Verh. d. Deutsch. Phys. Ges.*, **23**, 100.

WITHOLT, B., and Brand, L., (1970), *Biochemistry*, **9**, 1948.

YGUERABIDE, J., Epstein, H.F., and Stryer, L., (1970), *J. Mol. Biol.*, **51**, 573.

YGUERABIDE, J., and Stryer, L., (1971), *Proc. Natl. Acad. Sci. USA.*, **68**, 1217.

TECHNIQUES FOR FLUORESCENCE LIFETIME MEASUREMENTS AND TIME-RESOLVED EMISSION SPECTROSCOPY

William R. Ware

Department of Chemistry, University of Western Ontario, Canada

Transients on a nanosecond time scale imply frequencies in the range of 500 to 2000 mHz: the measurement of such transients has traditionally been regarded as difficult. Such transients, for example, occur subsequent to the flash excitation of fluorescence. The first successful approach to the problem of fluorescence lifetime measurements involved the measurement of the phase lag between excitation and fluorescence emission when the fluorescence was excited by sinusoidally modulated light. (Gaviola 1926). This approach avoids the problem of ultrahigh frequencies completely. Between 1945 and the mid-1960's the phase-shift technique was most popular for fluorescence lifetime measurements, although pulse sampling and photocurrent sampling techniques also came into use during this period (Ware 1971). In addition, a very fast method based on a gated image converter was developed (Yguerabide 1965). In the late 1950's the single-photon time correlation method was developed and applied first to scintillation and atomic fluorescence decay experiments and later to fluorescence lifetime measurements of molecules (Bollinger and Thomas 1961;Koechlin 1961; Laustriat et al.1963; Pfeffer et al.1962). However, the widespread use of the single-photon technique did not occur until about a decade later; this technique is now by far the most popular. Since the single-photon technique is inherently very sensitive and can be used not only for fluorescent decay measurements, but also for time-resolved emission spectroscopy (Ware 1971; Ware et al.1971), on a nanosecond time scale and for the measurement of time dependence of polarization anisotropy, it has great appeal to those interested in the application of fluorescence techniques to biological systems. In this discussion, techniques for lifetime measurements will be reviewed briefly with emphasis on the single-photon method. Several applications will also be described.

It must be emphasized that the fluorescence decay time of a molecule which is in dilute solution or bound to a macromolecule is a composite of several rate parameters. Consider the simple competitive mechanism for fluorescence deactivation:

$$A \xrightarrow{h\nu} A* \qquad \text{(Excitation)}$$

$$A* \xrightarrow{k_F} A + h\nu_F \qquad \text{(Fluorescence)}$$

$$A* \xrightarrow{k_{NR}} \text{Products of nonradiative processes}$$

This yields exponential decay, i.e.,

$$[A^*] = [A^*]_0 e^{-t/\tau_F} \tag{1}$$

where $1/\tau_F = k_F + k_{NR}$.

However, $k_{NR} = \Sigma k_i$ where k_i are the individual rate constants for the various nonradiative processes in operation. If biomolecular quenching is present, the following occurs

$$A* + Q \xrightarrow{k_Q} Products$$

and

$$1/\tau_F = k_F + k_{NR} + k_Q [Q] . \tag{2}$$

A further complication arises when the quenching step involves electronic energy transfer or when the efficiency of the quenching process is so high that the rate is controlled by diffusion. In both cases, k_Q may exhibit time dependence, and the fluorescence decay in general is expected to be nonexponential (Ware 1971). Aside from these complications, the important point is that the observed changes in the fluorescence lifetime can result from changes in either k_F or k_{NR} and that great caution must be exercised in the interpretation of such changes. Both can simultaneously change and both may be temperature or medium sensitive. If the quantum yield of fluorescence is measured, K_F and K_{NR} can be calculated since

$$\phi_F = k_F / (k_F + k_{NR}) \ . \tag{3}$$

Quantum yield measurements are routine in dilute solution, although only fair agreement generally exists among laboratories as to absolute yields (Demas 1971). A vastly more difficult problem appears to be the measurement of accurate absolute or relative quantum yields for molecules bound to macromolecules, and especially for fluorescent probe molecules within cells. In these cases, the difficulty in estimating the fraction of the incident light absorbed by the fluorophore makes even relative measurement difficult. This is an area that merits attention, since the complete understanding of changes of fluorescence properties as one goes from the free molecule in solution to the bound molecule, inside or outside the cell, ultimately depends on a knowledge of changes in individual transition probabilities.

In discussing fluorescence lifetime measurements it is convenient to divide the methods into those based upon phase detection and those that depend upon the measurement of a fluorescence transient in a more direct fashion. The latter are generally called flash methods because they employ a repetitive flash lamp of nanosecond duration for excitation.

In the phase-shift technique one measures the phase angle and the percent modulation of the fluorescence signal relative to the modulation of the excitation. For sinusoidal modulation, an exponential fluorescence decay law results in the sinusoidal modulation of the fluorescence (Ware 1971, Birks 1967), that is;

$$A \xrightarrow{\quad I_a(t) \quad} A^* \ . \tag{4}$$

If $I_a(t) = a + b \, sin\omega t$

then $[A^*] = c + d \, sin(\omega t - \phi)$

where $\phi = tan^{-1} \, (\omega \tau)$.

Also $m = cos \, (\phi)$

where m is the percent modulation of the light. It is to be emphasized that these simple relationships hold only for exponential decay; if the decay is nonexponential, one obtains disagreement between the lifetimes calculated from ϕ and m. In addition, the fluorescence lifetime will vary with frequency if the decay is nonexponential. Methods are available to deal with decay laws corresponding to the sum or difference

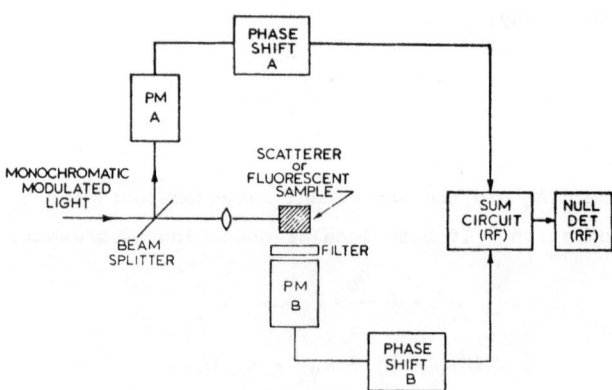

Figure 1. *Illustration of the basic idea of phase-shift fluorometry*

of two exponentials. If one simply measures ϕ and calculates the lifetime, then there is always the risk that an incorrect and misleading result has been obtained, since no test has in fact been made for the exponential character of the decay associated with the calculated lifetime value.

Phase-shift instruments vary from relatively simple to highly sophisticated machines, and a number have been described in the literature (Ware 1971, Birks 1967, Spencer 1970). Figure 1 shows a block diagram of one approach to the phase-lag measurement commonly employed in these instruments. The modulated light has been split into two beams to provide a reference signal for phase comparison with the fluorescence signal. If a cell containing a Rayleigh scatterer is used in place of the fluorescent sample, only instrumental phase shifts should be present, and these can be reduced to zero with the phase-shift circuits. In the presence of a fluorescer rather than a scatterer in the cell compartment, the fluorescent signal will be found to lag the exciting signal, and this phase shift can now be compensated either by calibrated phase-shift circuits or by a calibrated delay line. The balance in phase, which frequently must be accompanied by a balance in amplitude, is detected with a null detector operating at RF frequencies, although many instruments use single - or multiple - frequency conversion to reduce the null signal ultimately to a low frequency. A number of variations on this basic scheme can be found in the literature (Ware 1971, Birks 1967, Spencer 1970). Light is generally modulated for phase-shift measurements either by an ultrasonic Debye-Sears water tank modulator or a Pockels cell (solid state electrooptic) light modulator. The former results in a frequency doubling, a decided advantage when RF detection is used, since the power oscillator driving the ultrasonic crystal operates at one half the primary RF detector frequency. The water tank also offers a very large aperture and minimum requirements as to the collimation of the light beam. The Pockels cell, on the other hand, requires a highly collimated light beam and is driven at the light modulation frequency; thus introduces undesired RF signals into the area where a sensitive RF detector must be operated.

At present, most phase-shift instruments operate at only one or two frequencies, both generally in the range of 5-25 mHz. The ideal instrument would continuously sweep frequency, giving phase as a function of the modulation frequency. A Fourier transform would then yield a decay curve. While a variable frequency instrument has great appeal to those who are enthusiastic about the phase-shift methods, such an instrument does not appear to have actually been put into operation. However, its construction is contemplated by several groups. With the phase-shift technique one can, of course, employ any suitable DC light source.

Today, the single-photon methods enjoys the greatest popularity of any technique, and it will therefore be described in somewhat more detail than the phase-shift technique. It employs a nanosecond flash lamp, either free running or gated, and individual fluorescence photons are detected and timed relative to the flash-lamp pulse. A time-to-amplitude converter (TAC) is used, and the data are stored in either a computer or a multichannel pulse-height analyzer. A block diagram of such an instrument is shown in Figure 2. The lamp flash is generally detected with a 1P28 and the resultant analog pulse is used, after discrimination, to start the TAC. Fluorescence photons are detected by a fast photomultiplier such as the 56DUVP, and the individual

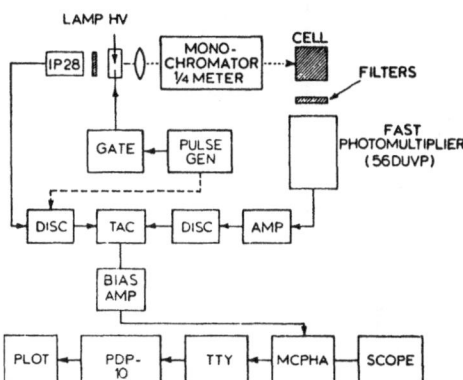

Figure 2. *A block diagram of a typical single-photon instrument. Components labeled DISC are discriminators and MCPHA is the multichannel pulse-height analyzer. The pulse generator operating the lamp gate can also be used to start the time-to-amplitude converter (TAC), if desired*

single photon pulses are used to stop the TAC. The TAC puts out a pulse, whose amplitude is proportional to the time interval between start and stop, and thus produces data for a probability distribution histogram that is in fact a decay curve. This is illustrated in Figure 3. For this method to produce true decay curves, the rate of detecting photons must be less than 10% of the lamp-repetition rate (Ware 1971). Faster collection of fluorescence photons results in what is called "pulse pile-up,"

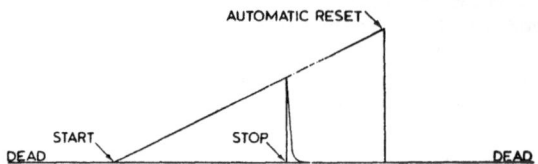

Figure 3. *The basic idea of time-to-amplitude conversion. Note the analogy to an oscilloscope sweep. The output pulse amplitude is proportional to $t_{stop} - t_{start}$)*

biasing of the decay curve in favor of early events. However, pulse pile-up corrections can be made if it is considered desirable to collect curves at a high repetition rate (Ware 1971). The output of the TAC, which is generally stored in a multichannel pulse-height analyzer, can then be read out onto paper or magnetic tape, a teletype, an oscilloscope, or plotter or can be fed directly into a computer. Most analyzers provide live display of the decay curve as it is collected.

In the single-photon technique, there is no base line such as one encounters in analog techniques. Instead there is simply a background count rate which is uncorrelated in time and which can be subtracted from the decay curve. Its effect is negligible unless the count rate is very low. In addition, one can improve the signal-to-noise ratio through the use of cooled photomultiplier tubes that have been selected for low noise. Thus, it is possible with the single-photon technique to obtain fluorescence decay curves covering 3-4 decades with 10^5 to 10^6 counts in the channel corresponding to the maximum in the curve. The standard deviation of each point is given by Poisson statistics; i.e., $\sigma = \sqrt{N}$ where N is the number of counts in the channel in question. A typical decay curve is illustrated in Figure 4.

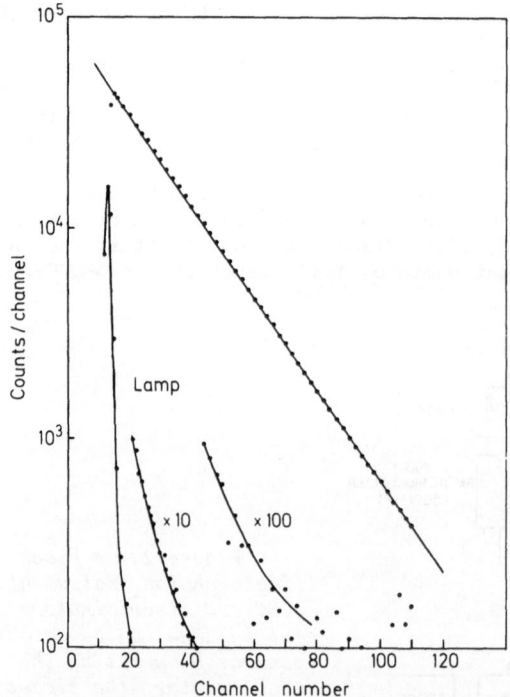

Figure 4. *A typical decay curve with good counting statistics. Decay time 85 nsec*

The time resolution of the single-photon technique depends upon the uncertainty of timing the detection of single-photon pulses relative to the flash-lamp pulse. The principal time jitter arises from the fact that the amplitude of single-photon anode pulses is not constant but covers a broad range in pulse heights. The jitter associated with timing by single-level crossing - e.g., with a simple discriminator - is a result of different amplitude pulses crossing the level at different times. This is illustrated in Figure 5. "Low-level timing" or "constant-fraction timing" discriminators help to minimize this effect. It appears that the photomultiplier resolution of commonly used fast multipliers (such as the Amperex 56DUVP or the RCA 8575) is on the order of 800 picoseconds. This is the width at half maximum of counts as a function of time of the distribution curve under the conditions where the excitation is a delta pulse of light, i.e., instantaneous. It can be shown experimentally that the 1P28 anode pulse, when detected by a simple fast discriminator, times the occurrence of the lamp flash with a filter of less than 100 psec.

Both free-running and gated lamps have been employed with a single-photon technique (Ware 1971; Birks 1967). The principal of the free-running lamp is illustrated in Figure 6. It is essentially a relaxation oscillator with C charge through R until the breakdown voltage of the lamp is reached, whereupon the lamp discharges and the cycle is repeated. To achieve fast lamps, C must be in the range of a few micromicro farads and R is adjusted to give repetition rates in the range of 5 to 40 kHz. The

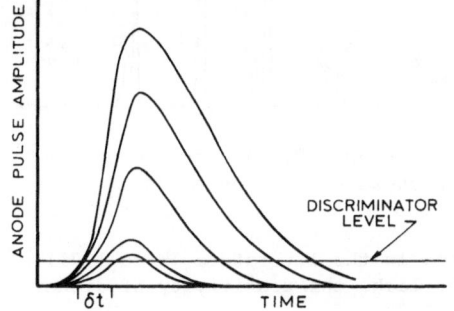

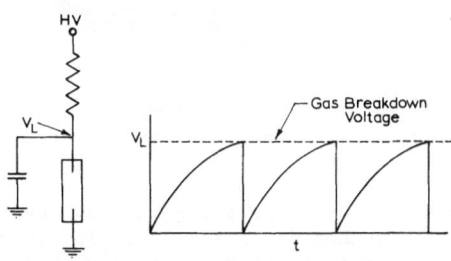

Figure 5. *The origin of some of the time jitter in the photomultiplier anode pulses derived from single-photon events. The variable amplitude introduces an uncertainty δt in the measurement of the arrival of a photon at the photocathode*

Figure 6. *The operation of a free-running flash lamp*

voltage at which the lamp breakdown occurs is controlled by the pressure inside the lamp. Free-running lamps are operated at pressures all the way from less than one atmosphere to 20 or 25 atmospheres. The gated lamp, on the other hand, employs a Thyratron which acts as an ultrafast switch and allows the lamp to discharge when the thyratron grid is triggered with a pulse. Discharge voltages considerably above the breakdown voltage of the gas are possible with this type of lamp. A typical gated lamp circuit is illustrated in Figure 7. Both free-running and gated lamps have been described and reviewed recently (Ware 1971, Birks 1967).

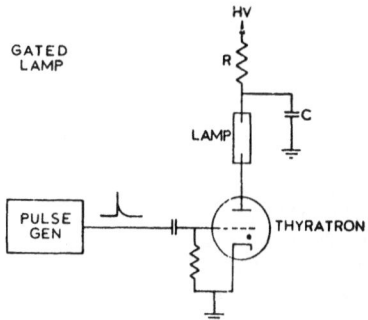

Figure 7. *A typical flash-lamp gate circuit. The thyratron switches the lamp to ground, discharging* C

A method for obtaining fluorescence lifetimes simpler than either of the two described above is called the pulse sampling method, because it employs a pulse sampling oscilloscope. Pulse sampling oscilloscopes have rise times that are faster than the response times of most photomultipliers currently used for fluorescence decay work. Such an instrument is diagrammed in Figure 8. Several variations that use digital data

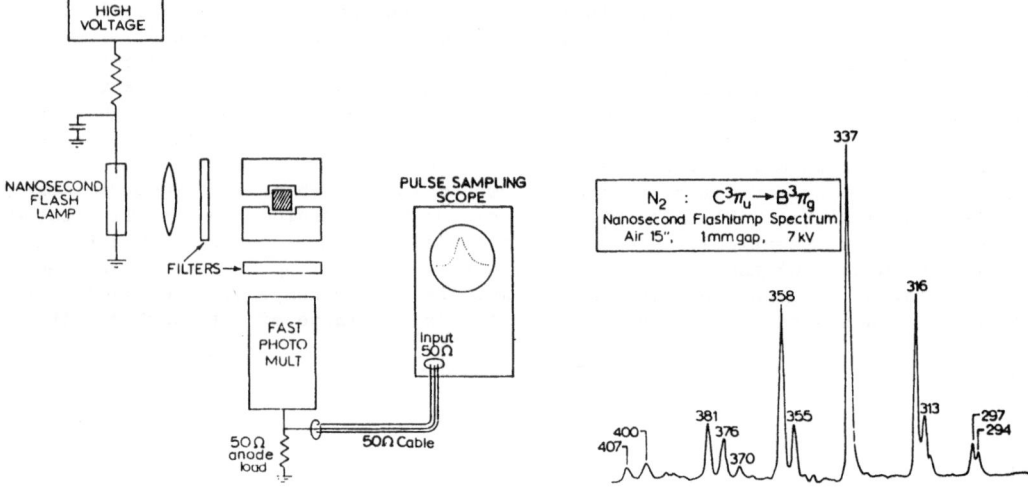

Figure 8. *The pulse sampling method for fluorescence lifetime measurements*

Figure 9. *A nitrogen or air flash-lamp emission spectrum*

processing and signal averaging have been described in the literature (Ware 1971). Since the method is basically analog in nature, it lacks the sensitivity of either the phase-shift or single-photon methods.

TABLE I

USEFUL SPECTRAL RANGE OF NANOSECOND FLASHLAMPS

Gas			
		294 (vs)	358 (vs)
		297 (w)	376 (s)
N_2	290-400nm	313 (s)	381 (s)
(Air)	Bands	316 (vs)	
	FWHM - 1-2nm	337 (vs)	
H_2	VAC-UV to 400nm continuum plus structure in VAC-UV.		
O_2	250-600nm Some structure - mostly continuum		

In Table I are presented the useful emission regions of nanosecond flash lamps commonly used for fluorescence lifetime measurements. If air or nitrogen is used, one obtains a banded spectrum as indicated in Figure 9. This yields a series of semi-monochromatic sources, provided the vibrational bands are isolated; the band width is about 2nm. Oxygen is useful in the visible and red regions of the spectrum, and deuterium is satisfactory as a flashing gas for operation in the ultraviolet. Lamp design has recently been reviewed and will not be discussed further here   (Ware 1971).

The minimum lifetime that can be measured with the single-photon technique is also limited by the finite width and decay of the lamp used to excite fluorescence.

It can be shown by elementary Laplace transform techniques that the instrument output $I_k^{obs}(t)$ is given by the convolution integral

$$I_k^{obs}(t) = \int_0^{t_k} G(\lambda) F(t_k - \lambda) d\lambda \tag{5}$$

where $G(t)$ is the time response of the fluorescence system when excited with the delta-pulse of light, and $F(t)$ is the time dependence of the lamp *distorted by the detection system*. To obtain the desired function $G(t)$, one must solve this integral equation, a process called deconvolution. $F(t)$ is of necessity empirical, a fact which further complicates the problem. Various techniques have been proposed for deconvolution; some require the assumption of the decay law $G(t)$; others do not. The principal methods are as follows:

    a) convolution of an assumed $G(t)$, varying parameters in $G(t)$ to obtain a fit, (Selinger 1971)

    b) the method of moments, (Isenberg and Dyson 1969)

    c) Fourier transform,

    d) Laplace transform (Ware 1971)

    e) multiple-series method.

The first four methods have been reviewed recently (Ware 1971) and will not be discussed further here. The last method (Ware and Nemzek) involves representing $G(t)$ as a set of 10 to 30 exponentials. The lifetimes used in the exponential functions are fixed, and least-squares techniques are used to obtain the best weighting factors for each exponential in order to minimize the sum of the squares of the deviations between the observed decay curve and that calculated from the lamp curve and this empirical function. The equation is

$$I_i^{calc} = \int_0^{t_i} \sum_n a_n e^{-k_n \lambda} F(t_i - \lambda) d\lambda \tag{6}$$

The residuals are

$$\rho_i = \sum_n a_n \int_0^{t_i} e^{-k_n \lambda} F(t_i - \lambda) d\lambda - I_i^{obs} \tag{7}$$

and the problem reduces to finding a_n such that $\sum_i \rho_i \partial \rho_i / \partial a_j = 0$; $j = 1, 2 \ldots$.

    Since the k_n are not varied, a set of linear equations results; i.e.,

$$\sum_n (\sum_i F_{ij}^2) a_n = \sum_i F_{ij} I_i^{obs}; \quad j = 1, 2 \ldots \tag{8}$$

where

$$F_{ij} = \int_0^{t_i} e^{-k_j \lambda} F(t_i - \lambda) d\lambda .$$

Examples of deconvolution by this method for sums and differences of exponential functions are illustrated in Figures 10 through 13. The input data contain typical Poisson noise and the decay curves analyzed were obtained by convoluting a lamp curve (Figure 10) with an assumed $G(t)$ which was either the sum or difference of two exponentials. The resultant curve for the sum of two exponentials is shown in Figure 11. This was submitted for analysis with no assumption as to the weighting coefficients for the series representation of $G(t)$; therefore, no assumption was made during the deconvolution process as to the nature of the decay function. The ultimate weighting factors for the assumed sum of exponentials, which in this case numbered 24, were such as to reproduce the input function exactly over several decades. This can be seen in Figures 12 and 13. Furthermore, it was found that from a lamp having a decay time of 0.9 nanoseconds it was possible to recover a single exponential curve with a lifetime of 0.5 nanoseconds by using a sum of 24 exponentials. That is to say that $G(t)$ expressed as a sum of 24 exponentials was found in fact to give a straight line on a semilog plot over 4 decades with a slope equivalent to a lifetime of 0.5 nanoseconds, exactly the lifetime introduced when the lamp was convoluted to produce an artificial decay curve. While these decay curves are artificial, the noise they contain is realistic; therefore, they do not differ in any significant respect from actual decay curves. The great advantage of this method is in not having to assume a specific functional form for $G(t)$ other than a very general functional form encompassed by a sum of 10–30 exponential functions. Since the sum of even a small number of exponential functions is generally successful in fitting practically any smooth decay curve, it is an obvious choice. The philosophy, then, in using the multiple-series method is that one starts out with the raw data and obtains a representation of the undistorted decay curve in terms of the weighting coefficients for the exponentials. This can then be plotted out and constitutes the starting point for analysis in terms of a physical model. Artifacts such as scattered light and filter fluorescence are immediately apparent and can be corrected for. It is emphasized that no physical significance is attached to the a_n coefficients or the k_n values used in synthesizing $G(t)$. In general, it has proved satisfactory to make up $G(t)$ in terms of 10–30 exponential functions with lifetimes in the range of 0.1 to 15 nanoseconds to fit fluorescence decays in the range of 0.5 to 10 nanoseconds.

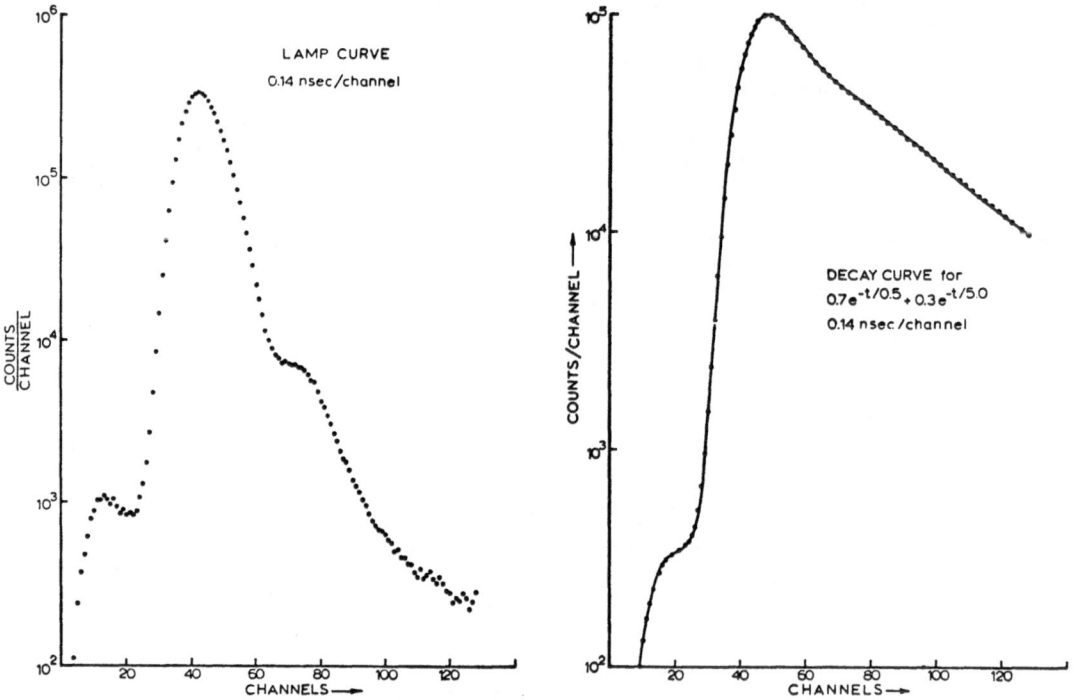

Figure 10. *Decay curve of the flash lamp used for curves in Figures 11, 12 and 13*   Figure 11. *Decay curve of the sum of two exponentials*

It is perhaps useful at this point to summarize the advantages and disadvantages of the single-photon method as compared with the phase-shift technique. The advantages of the single-photon technique are that:

a) one obtains directly a representation of the decay curve, though a distorted one if the lamp and fluorophore have similar decay times;

b) the sensitivity is high and signal averaging over long periods (24 hrs.) is practical;

c) the shape of the histogram obtained, that is to say the decay curve, is independent of the lamp intensity, an important consideration if long counting periods are employed (Lamp intensity merely controls the rate of data collection!), and

d) the instrumentation is commercially available, modular, compact, extra-ordinarily stable and reliable, and the multichannel analyzer lends itself to computer data processing. In-lab plotting from a time-sharing computer is a valuable additional capability. The disadvantages of the single-photon technique are:

a) long counting times may be required (overnight, for example);

b) for short lifetimes relative to the lamp decay, numerical deconvolution is an absolute necessity.

The advantages of the phase-shift technique may be listed as follows:

a) any light source can be used which possesses the required intensity;

b) sensitivity is high with intense sources; and

c) high-precision lifetimes can be determined for systems with an exactly exponential decay law, and measurements below one nanosecond are practical. The disadvantages of the phase-shift technique appear to be as follows:

a) the water tank modulator is not commercially available;

b) a knowledge of RF circuitry (1-25 mHz) and techniques is desirable;

c) samples exhibiting nonexponential decay curves present difficulties that are not easily overcome; and

d) it is difficult and perhaps impossible to do time-resolved emission spectroscopy and time-resolved polarization measurements unless variable-frequency instruments come into use.

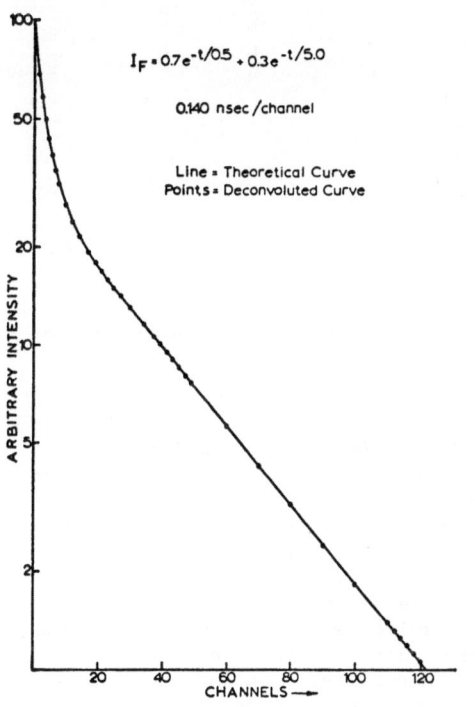

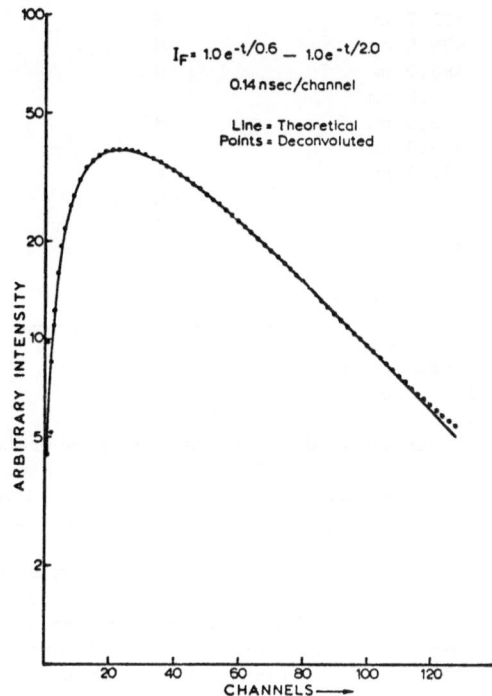

Figure 12. *Results of deconvoluting the curve in Figure 11*

Figure 13. *Results of deconvoluting a decay curve made up of the difference in two exponentials*

It is relatively safe to predict that within a few years, if not sooner, lasers will replace both conventional light sources and nanosecond flash lamps. In particular, the CW ion lasers show great promise in this direction. They can be mode-locked to produce sub-nanosecond pulses; they can be "cavity-dumped" to produce pulses with widths of several nanoseconds. The output can be doubled in frequency to yield ultraviolet light, and the CW laser can be used to pump dye lasers, which can be tuned and doubled. In addition, pulse-repetition rates up to several mHz are possible with the mode-locked or cavity-dumped CW laser, a capability which makes them ideal sources to replace the repetitive nanosecond flash lamps now used with the single-photon technique. Although there will no doubt be problems due to the high degree of polarization, the high intensity, and in some cases the high repetition rate of these sources, they should nevertheless prove useful. The wavelengths currently available are listed in Table 2.

TABLE 2

WAVELENGTHS AVAILABLE FROM COMMERCIAL
CW ION LASERS

Wavelength	Gas Fill		
	Argon	Krypton	Argon/Krypton
799.3 nm		L	
793.1 nm		L	
752.5 nm		M	
676.4 nm		M	L
647.1 nm		H	M
568.2 nm		M	L
530.9 nm		M	L
520.8 nm		L	L
514.5 nm	H		M
501.7 nm	M		L
496.5 nm	M		L
488.0 nm	H		M
482.5 nm		L	L
476.5 nm	M		L
476.2 nm		L	
472.7 nm	L		
465.8 nm	L		
457.9 nm	M		L
454.5 nm			
351.1 nm + 363.8 nm		L	
350.7 nm + 356.4 nm		L	

H - high power
M - medium power
L - low power

There is good agreement among laboratories as to the fluorescence lifetime of selected compounds. In Table 3 are listed eight compounds that have been examined by several laboratories using either the pulse sampling, photocurrent sampling, or phase-shift methods. While the agreement is quite satisfactory, better methods of cali-bration (Ware 1971) and the high sensitivity of the single-photon method would give even closer agreement if the experiments were repeated today.

An area of considerable interest at present involves time-resolved emission spec-troscopy. Two methods are used for obtaining time-resolved emission spectra with the single-photon technique. In one, (Brand) fluorescence decay curves are obtained at a series of emission wavelengths. If these curves are normalized to a constant number of TAC start pulses, they can be used to construct fluorescence spectra at any given time subsequent to the flashing of the lamp. An alternative approach (Ware et al., 1971) involves the measurement of the emission spectrum through a time window created by a single-channel pulse-height analyzer placed after the TAC. The single-channel pulse-height analyzer allows only pulses between E and E+δE to pass, and thus an emis-sion spectrum is generated for events occurring only between t and t+δt after excita-tion. A block diagram of such an apparatus is shown in Figure 14. In practice, the instrument is first operated in the single-photon lifetime mode, and the upper and lower discriminator levels of the multichannel pulse-height analyzer or a single-channel analyzer is adjusted in order to obtain the desired time window. This can be seen visually on the oscilloscope associated with the analyzer during the collection of the decay curve, since the upper and lower level discriminators will block out the leading and trailing edges of the decay curve and can be used to create a window with any time size. The instrument is then switched to multichannel scaling and a spectrum is obtained through the time window created by the analyzer discriminators.

The decay-curve method has the advantage of yielding spectra which are undistorted by the lamp or instrument, provided the decay curves are first deconvoluted. The mul-tiple-series deconvolution method should prove very useful here, since in general one does not know $G(t)$ for decay distorted by spectral shifts. This method also uses all of the fluorescence decay curve.

Emission spectra which vary with time are anticipated under the following circumstances (Ware et al. 1971; Ware and Chakrabarti 1971):

a) when reorientation relaxation of solvent molecules around a polar excited molecule causes energy-level shifts;

b) when one species decays and an excited fluorescent product species grows in; and

c) when intramolecular motion influences the energy of an excited state; for example, the twisting of a phenyl ring may shift the level of an excited state by a measureable amount. Relaxation from an initially formed conformation to a fully relaxed conformation in the excited state can produce time-dependent spectral shifts.

TABLE 3

FLUORESCENCE STANDARDS

FLUORESCENCE LIFETIMES OF SEVERAL COMPOUNDS
IN DILUTE OXYGEN-FREE SOLUTION AT 25°C

Compound	Solvent	Av. τ	% Av. dev.*	No. of Labs.
Anthracene	Benzene	4.08	2.9	5
9(10H)Acridone	H_2O Saturated	15.5	1.5	4
9(10H)Acridone	95% EtOH	13.5	4.5	3
Quinine	1 NH_2SO_4	19.7	1.5	2
Bisulfate	$10^{-2}M$ HNO_3	20.5	0.8	3
Anthracene	EtOH	5.50	2.7	4
Perylene	Benzene	5.02	1.6	4
9,10-Diphenyl anthracene	Benzene	7.35	0.7	2
Biphenyl	Cyclohexane	16.2	2.3	3

* % deviation between two or more laboratories. Both flash and phase-shift

measurements are included. (Ware 1971).

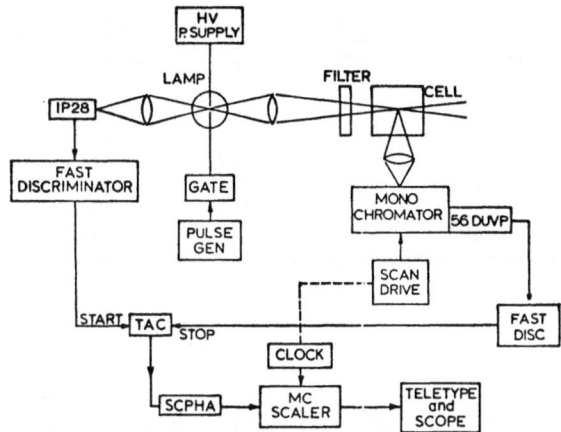

Figure 14. *Block diagram of a typical single-photon time-resolved emission spectrometer. The component labeled SCPHA is a single-channel pulse-height analyzer*

Molecules which experience large changes in dipole moment upon excitation generally yield time-resolvable emission spectra in polar solvents at low temperatures (Ware et al. 1971). Alcohols such as propylene glycol and glycerol at reduced temperatures have dielectric relaxation times comparable with fluorescence decay times, and one can observe significant shifts in fluorescence spectra due to solvent reorientation about the excited molecule during its fluorescence lifetime. An example of such a system is given in Figure 15, in which the time-dependent emission of 4-aminophthalimide is shown. ANS also exhibits similar time-dependent spectral shifts (Ware and Chakrabarti, 1971), presumably for the same reasons. In addition, the ANS fluorescence lifetime and fluorescence quantum yields are extremely sensitive to solvent polarity.

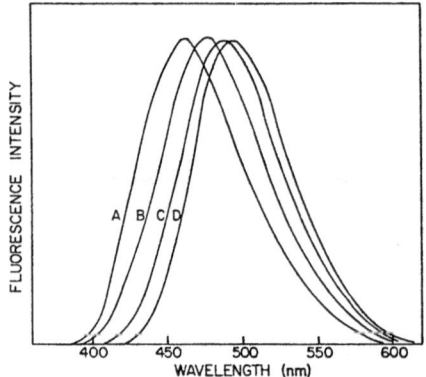

Figure 15. *Time-dependent spectra from 4-aminophthalimide in normal propyl alcohol at -95°C (Ware et al., 1971). Time after flash: (a) 4 nsec, (b) 8 nsec, (c) 15 nsec, (d) 23 nsec*

ANS provides a good example of the care that must be exercised in interpreting changes in fluorescence quantum yields and lifetimes. It is well known that the quantum yield (Stryer 1965) decreases sharply as one goes from a solvent of low polarity, such as butyl alcohol, to a more polar solvent such as ethylene glycol. The lifetime also decreases with increasing solvent polarity. If these two parameters are used to calculate K_F and k_{NR}, it is found that k_F decreases and k_{NR} increases with solvent polarity. The results are summarized in Table 4. The effect of solvent on

TABLE 4

FLUORESCENCE LIFETIMES AND SPECTRAL SHIFTS FOR ANS AT 25°C IN SEVERAL SOLVENTS

Solvent	ε^b	λ_{max}^F (nm)	ϕ_F^c	τ_F^a (nsec)	k_F $\times10^{-7}$ sec^{-1}	k_{NR} $\times10^{-7}$ sec^{-1}
n-Butyl alcohol	19.2	477	0.56	21.6	2.6	2.1
n-Propyl alcohol	22.0	479	0.46	21.0	2.2	2.5
Ethanol	25.8	485	0.37	18.0	2.1	3.5
Methanol	31.2	490	0.22	13.6	1.6	5.7
Ethylene glycol	37.7	495	0.15	11.8	1.3	7.2
Glycerol	42.5	501		9.0		

[a]Taken from Ware and Chakrabarti 1971

[b]Dielectric constant

[c]Taken from Stryer 1965

k_F presumably arises from the intramolecular charge transfer nature of the transition to the first excited singlet state. The effect of polarity on k_{NR}, however, is far from clear. While ANS is without a doubt a very useful probe for polarity, the details of the photophysics of this molecule when bound to a macromolecule must be considered open to question since the reorientation relaxation as well as a variation in k_F and k_{NR} are presumably taking place. In addition, one must consider the possibility of physical constraints on the molecular conformation, something which has not as yet been studied.

The single-photon technique also lends itself to the study of time-dependent fluorescence polarization (Stryer 1968). Polarization anisotropy, $A(t)$, defind as

$$A(t) = \frac{F_y(t) - F_x(t)}{F_y(t) + 2F_x(t)} \qquad (9)$$

where F_x and F_y are the parallel and perpendicular components of the polarized emission. $A(t)$ can be measured with a single-photon instrument suitably modified by the introduction of polarizers. This technique has many applications for the study of macromolecular systems of biological significance, as will be clear from the papers in this volume.

REFERENCES

BIRKS, J. B., In: *Progress in Reaction Kinetics*, (Porter, G., ed.), Vol. 4, Pergamon New York.

BOLLINGER, L. M., and Thomas, G. E., (1961), *Rev. Sci. Instru.*, <u>32</u>, 1044.

BRAND, L., Private Communication.

DEMAS, J. N., and Crosby, G. A., (1971), *J. Phys. Chem.*, <u>75</u>, 991.

GAVIOLA, E., (1926), *Z. Physik.*, <u>35</u>, 748.

ISENBERG, I., and Dyson, R. D., (1969), *Biophysical J.*, <u>9</u>, 1337.

KOECHLIN, Y., (1961), Doctoral Thesis, Paris.

LAUSTRIAT, G., Coche, A., Lami, H., and Pfeffer, G., (1963), *Compt. Rend.*, <u>257</u>, 434.

PFEFFER, G., Lami, H., Laustriat, G., and Coche, A., (1962), *Compt. Rend.*, <u>254</u>, 1035.

SELINGER, B. K., (1971), *Spectrochim Acta*, <u>27A</u>, 1223.

SPENCER, R. D., (1970), Thesis, University of Illinois.

STRYER, L., (1965), *J. Mol. Biol.*, <u>13</u>, 482.

STRYER, L., (1968), *Science*, <u>162</u>, 526.

WARE, W. R. (1971), In: *Creation and Detection of the Excited State*, (A. Lamolla, ed) Dekka, New York.

WARE, W. R., and Chakrabarti, S. K., (1971), *J. Chem. Phys.*, <u>55</u>, 5494.

WARE, W. R., Lee, S. K., Chow, P., and Brant, G., (1971), *J. Chem. Phys.*, <u>54</u>, 4729.

WARE, W.R. and Nemzek, T., *J.Phys. Chem.*, in press (Oct.1973), see also:

WARE, W.R., Lewis, C.,Doemeny, L. and Nemzek, T.,(1973), *Rev. Sci. Inst.* <u>44</u>, 107

YGUERABIDE, J., (1965), *Rev.Sci. Inst.*, <u>36</u>, 1734

Part II
Instrumentation and Standardization in Fluorometry and
Microscope Fluorometry

SOME ASPECTS OF INSTRUMENTATION AND METHODS AS APPLIED TO FLUOROMETRY AT THE MICROSCALE

G. von Sengbusch and A. Thaer

Battelle Institut e.V., Frankfurt, Germany

INTRODUCTION

During the last ten years fluorometry has become a well-known and very important tool in the field of quantitative cytochemistry, whereas biochemists had already developed this technique into an extremely valuable methodology ten to twenty years before. With regard to the quantitative staining reactions involved, cytofluorometry has evolved predominantly from the fluorescence microscopy of cells and tissues. Of course, the experience gained in the biochemists' development of fluorometric methodology must also be utilized for the quantitation of fluorescence microscopy; no doubt, the use of fluorescence parameters to derive quantitative information in cytochemistry requires the same careful consideration of the physicochemical laws of fluorescence phenomena as that needed in chemistry and biochemistry. Thus, fluorometry of single cells and cellular compartments should profit by the considerable amount of experience that is available in the field of macrofluorometry with regard to instrumentation, methodology, and interpretation of results.

However, accepting this highly desirable flux of information from the biochemical to the cytochemical field of application of fluorometric techniques, one must keep in mind the specific situation and requirements in cytochemistry: the microscopic dimensions of the objects to be measured and, in most cases, the necessity of correlating the fluorometric data with a cellular structure or morphological information obtained visually by means of the light microscope. It is just this correlation of biochemical information with the single cell and its structure that makes the intensive interaction between the two disciplines or fields of application of fluorescence techniques so profitable.

LOWER LIMITS OF DETECTABILITY OF MACRO- AND MICROSCOPE FLUOROMETRY

In order to compare the absolute mass of fluorescent substance available for microscope fluorometry of cytochemical objects with that available for macrofluorometric analysis of solutions, homogenates, and suspensions (for instance of cells), one has to take into consideration that there is a difference in volume of roughly 10^9 between milliliter cuvettes of normal fluorometers and cells. This is clearly demonstrated by the data represented in Table I. A volume of 10^{-12} l, corresponds to

TABLE I

COMPARISON OF SAMPLE VOLUMES AND LOWER LIMITS OF DETECTABILITY IN
MACROFLUOROMETRY AND MICROSCOPE FLUOROMETRY

Range	Volume (liters)	Lower Limits of Detectability		
		Concentration	Mass	Number of Molecules
Macro	10^{-3} l	$10^{-8} - 10^{-12}$ M	$10^{-11} - 10^{-15}$ moles	$10^{12} - 10^{8}$
Micro	10^{-12} to 10^{-14} l	$10^{-5} - 10^{-6}$ M	$10^{-17} - 10^{-18}$ moles	$10^{6} - 10^{5}$

the volume of a relatively big cell like the ciliate tetrahymena pyriformis, a single mitochondrion represents a volume of about 10^{-14} l, and is close to the lower limit of optical resolution. Microscope fluorometry can detect 10^{-17} to 10^{-18} moles of a fluorescent compound, for instance DPNH (Chance 1962; Kohen 1964; Kohen et al.1973) or acridine orange (Loeser and West 1962; West 1969) in single cells or quinacrine mustard (Caspersson et al. 1969) in chromosomal regions; for instance 10^{-15} g DNA is clearly detectable by microscope fluorometry (Rigler 1969). On the other hand, the lowest concentration of a fluorescent substance detectable by fluorometry with ml cuvettes is of the order of magnitude of 10^{-12} M (Blyum and Shcherbov 1967; Spencer and Weber 1972) provided that special measures are taken to improve the optical signal-to-noise ratio. This corresponds to a mass of 10^{-15} moles or roughly 10^8 molecules in one milliliter, whereas by use of microfluorometry 10^5 molecules may still be detectable, however, this is true only at concentrations higher by a factor of 10^6. The same lower limit of detectability could be achieved by measuring solutions of fluorescent substances in capillaries used as microscope cuvettes (Sernetz and Thaer 1972).

Compared with a macrofluorometer, the microfluorometer is able to collect for measurement a much higher percentage of the fluorescence emitted by the object. This is because of the high numerical aperture of the microscope objective; for the same reason, much higher intensities of exciting radiation per excitable molecule are also reached. These factors contribute substantially to the obviously higher sensitivity of microfluorometry with regard to the smallest detectable amount of a fluorescent compound. There is also a corresponding reduction of the influence of the optical background noise that is a limiting factor to the detectability of very low concentrations of fluorescent compounds in macrofluorometry (Spencer and Weber 1972).

Thus, microfluorometry and macrofluorometry may be considered as being complementary with respect to the accessible concentration ranges and detectable number of fluorescent molecules (see also the following chapter).

Apart from these purely methodical considerations, a meaningful combination of macro- and microfluorometry is also urgently needed in order to correlate biochemical information gained from fluorescent molecules in solution with that obtained from those molecules in the single cell and its compartments. Moreover, the interpretation of results obtained by microscope fluorometry often needs support from, or confirmation by parallel experiments with macrofluorometry on model solutions or cell suspensions. This is necessary because the optical conditions for achieving high resolution of the fluorescence microscope in some cases are not optimum conditions for measuring fluorescence parameters. It is well known that there is sometimes rapid fading of fluorescence due to the high excitation energy concentrated on the object.

Another very interesting problem for study, is the measurement of fluorescence polarization. A system is here described which permits the alternative recording of fluorescence polarization spectra for single cells and picoliter volumes of solutions and for solutions or cell suspensions in a 0.3 ml cuvette using the same instrumentation. Thus, it is possible to compare not only the results obtained on picoliter volumes with those measured on fractions of milliliters of the same solution but also the quantitative information derived from cells with that obtained from model solutions, both in the ultramicro- and the macrorange.

A SYSTEM FOR COMPARATIVE MEASUREMENTS OF FLUORESCENCE POLARIZATION IN THE MACRO- AND MICROSCOPIC RANGE

Figure 1 illustrates the light paths usually applied in macrofluorometry and those available for microscope fluorometry. Incident-light excitation using the microscope objective itself as condenser (bright-field illumination in incident light; see Figure 1) has become the optical arrangement widely used for fluorescence microscopy and microfluorometry. It is chosen because of its simplicity and the fact that it permits one to achieve stable correlation of the focal planes of the exciting beam with the object plane simply by focusing the fluorescent object in the microscope. It also offers excellent possibilities for combination with transmitted light illumination (for instance, phase contrast). Moreover, microscope objectives provide optimum conditions for focusing the exciting radiation onto the object and for forming a highly demagnified chromatically and spherically corrected image of a field diaphragm in the object plane. The loss of light intensity by reflection at the beam-

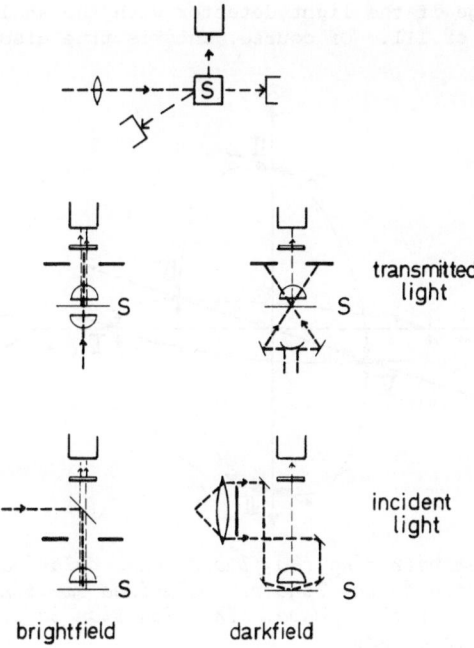

transmitted
light

incident
light

brightfield   darkfield

Figure 1. *Optical arrangements (schematic) for macrofluorometry and microscope fluorometry (S = sample or object plane)*

splitting plate deflecting the exciting light through the objective onto the object can be reduced by using dichromatic mirrors instead of beam-splitting plates semi-transparent for light of any wavelength. For fluorescence polarization measurements, however, beam-splitting plates with neutral spectral characteristics of reflectance and transmission have to be used, since dichromatic mirrors have different ratios of reflectance-to-transmission as a function of wavelength.

Thus, the most important difference between macrofluorometric and microfluorometric optical arrangements is not the various possibilities of orienting the exciting beam and the measuring direction with respect to each other but the large numerical aperture of the optics needed for focusing the excitation energy on the microscopic sample. It is this aperture that permits one to collect as much of the emitted fluorescence light as possible and thus to achieve the microscopic resolution required.

What does this mean for fluorescence polarization measurements on microscopic objects? Figure 2 illustrates schematically the measuring of the fluorescence polarization of molecules with their emission and absorption oscillators parallel to each other in a direction parallel or perpendicular to the direction of incidence of the horizontally polarized exciting beam. Provided that the opening angle of the linearly polarized exciting beam is negligibly small and the fluorescence polarization is measured only parallel to the microscope axis, the situation is comparable with that in macrofluorometric measurements of the fluorescence polarization at zero angle to the exciting beam; there is no optical resolution and — using microscope optics — very low light energy is available for the measurement. The situation for fluorescence polarization measurements with increasing numerical aperture is also illustrated in Figure 2:

The measured ratio
$$P = \frac{I_{\|} - I_{\perp}}{I_{\|} + I_{\perp}}$$

$I_{\|}$ = fluorescence intensity measured with parallel analyzer and polarizer.

$I_{\perp}$ = fluorescence intensity measured with crossed polarizer and analyzer.

reflecting the amount of depolarization due to rotation of the absorption (emission) oscillators does not change if the light detector with the analyzer is moved from position I to position II or III. Of course, this is true also for the direction of

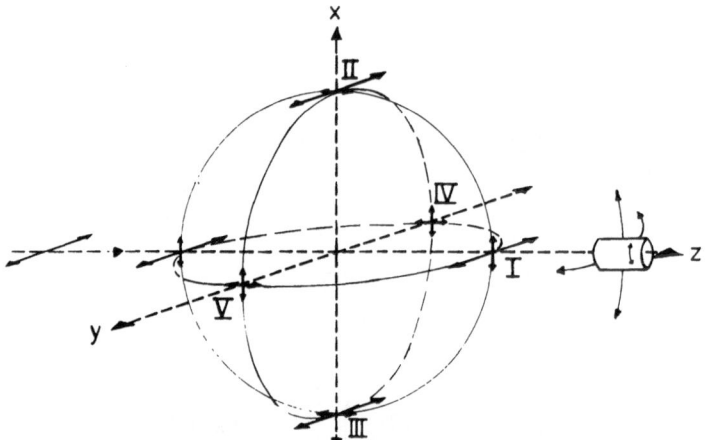

Figure 2. *Diagram demonstrating the fluorescence polarization measured in different directions relative to the linearly polarized exciting beam. The object is assumed to be in the center of the sphere. The situation is illustrative of positive fluorescence polarization (see text)*

incidence of the exciting beam if the light source together with the polarizer were moved from position 0 to position II or III.

On the other hand, entirely different fluorescence polarization values are measured in positions IV and V. This is because the fluorescent light emitted by the excited molecules oriented with their absorption (emission) oscillators parallel to the electric vector of the polarized exciting beam would not be measured if the oscillators were to maintain their orientation during the lifetime of the excited state. With increasing freedom of rotation, however, they emit fluorescence also in directions IV and V, and thus contribute equal amounts to the intensities measured in the two vibration directions of the analyzer in these positions.

In principle, the same applies if measurements are made of molecules with their absorption oscillator vertically oriented to the emission oscillator (see also Udenfriend, 1969), P does not change if the detector is moved from position I to positions II or III, but varies between positions I, IV and V. The situation differs only in cases where the emission and the absorption oscillator are parallel to each other, as the emission oscillators of the excited molecules are now oriented parallel to the plane through I, II, and III, but are oriented in random distribution relative to the different vibration directions of the analyzer in position IV and V. The rotation of the excited molecules during the lifetime of the excited state causes a measurable decrease of fluorescence intensity which is identical at all vibration directions of the analyzer in these positions.

If we apply these conditions to the optical arrangement preferred for microscope fluorometry, we may utilize the entire numerical aperture of the objective — especially of a high-power objective — only in a plane through positions I, II and III. This applies both to the exciting beam and to the fluorescence light collected by the objective (see Figure 2). With the effective numerical aperture increasing perpendicular to this plane, however, the actual fluorescence polarization (P_0) is increasingly underdetermined to the extent that the measured values have to be corrected (see Figure 4).

Thus, in practice, satisfactory optical conditions for measuring fluorescence polarization on microscopic objects are obtained if, for instance, the exit slit of the entrance monochromator of a microspectrofluorometer is imaged into the rear focal plane of the microscope objective, thereby limiting the numerical aperture effective for the exciting beam to less than, say, 0.2 in one direction by the slit width and fully utilizing the numerical aperture of the objective in the other. With regard to the emitted fluorescence collected by the objective, the effective numerical aperture should be reduced in the same direction as for the exciting beam. A slit-shaped adjustable diaphragm in the rear focal plane of the objective or in a conjugated plane of the microscope optical pathway may be used for this purpose, as is shown in Figure 3. The arrangement depicted in the diagram of this figure would also permit the observation of the microscopic object with the full aperture of the objective — for instance in transmitted light — by adjusting the rectangular diaphragm accordingly.

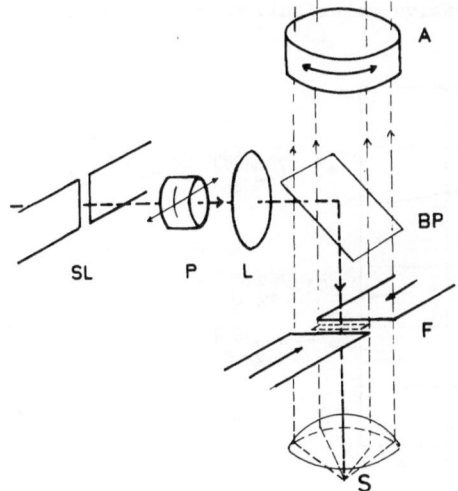

Figure 3. *Optical arrangement (schematic) for reducing the numerical aperture of the microscope objective for excitation and uptake of emission in incident light microscope fluorometry. The aperture is reduced perpendicular to the plane through positions I, II, and III in Figure 2 (see text). SL = slit, P = polarizer, F = rear focal plane of the objective, L = Lens for imaging SL into F, BP = beam-splitting plate, S = sample or object plane, A = analyzer*

In general, however, the loss of by far the greatest part of fluorescence light available for measurement is not acceptable. It is by all means preferable, therefore, to utilize the full aperture of the objective for the collection and photometry of the fluorescence light emitted by the object and to introduce a correction function where called for.

The diagram in Figure 4 shows the calculated values for the fluorescence polarization $P(\alpha)$ to be measured as a function of the aperture angle α effective for the uptake of fluorescence light in a plane perpendicular to the plane through I, II, and III in Figure 2 for three different values of the actual fluorescence poarization P_0 that would be obtained by measuring $I_{\parallel}$ and $I_{\perp}$ at $\alpha = 0$. The calculation is based on the assumption that the aperture angle of the exciting light focused by the objective onto the object is zero for the same plane for which the aperture angle α effective for the uptake of fluorescent light by the objective is varied. The values for $I_{\parallel}(\alpha)$ and $I_{\perp}(\alpha)$ can then be calculated as a function of α, $I_{\parallel 0}$ and $I_{\perp 0}$. In Figure 4 the fluorescence polarization $P(\alpha)$ described by the equation

$$P(\alpha) = \frac{I_{\parallel}(\alpha) - I_{\perp}(\alpha)}{I_{\parallel}(\alpha) + I_{\perp}(\alpha)}$$

is plotted versus α for three different values of

$$P_0 = \frac{I_{\parallel 0} - I_{\perp 0}}{I_{\parallel 0} + I_{\perp 0}}$$

We thus obtain the factors for correction of $P(\alpha)$ as a function of α for given P_0. Since the correction factor is also a function of P_0, the calculation results in a number of correction curves. The effective aperture angle α can be calculated from the numerical aperture A_n of the microscope objective used:

$$\sin \alpha = \frac{A_n}{n} \qquad (n = \text{refractive index of the medium between front lens of the objective and specimen})$$

For instance, an oil immersion objective with a maximum numerical aperture A_n = 1.40 gives an aperture angle α = 68°. Thus, the values for aperture angles >68° shown in Figure 4 are of theoretical interest only. As can be taken from the graph in Figure 4, the values to be measured for $P(68°)$ may be expected to be about 20 percent below P_O for P_O = 0.2.

Figure 4 depicts only the situation for positive polarization; i.e., for fluorescent molecules with the emission and absorption oscillator parallel to each other. For negative polarization, i.e., for molecules with the emission and absorption oscillator approximately vertical to each other similar correction values can be calculated considering the fact that the maximum value for negative polarization reaches 0.33 compared with 0.5 for positive polarization.

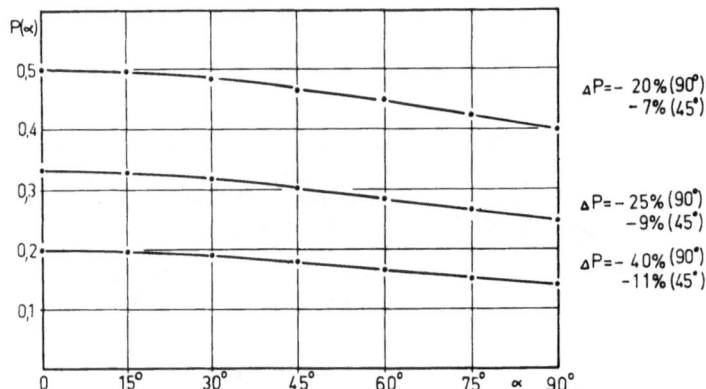

Figure 4. *Fluorescence polarization* P *(α) measured by means of microscope fluorometry as a function of the aperture angle* α *for the uptake of emission by the microscope objective. The values for* P(α) *are calculated for three different values of* P_O. *for correction* ΔP *can be taken from this graph for a given aperture angle and measured* P(α)

COMPARATIVE MEASUREMENTS OF FLUORESCENCE POLARIZATION USING SOLUTIONS AT THE MACRO- AND MICROSCALE

Preliminary results of a current experimental study being carried out at our laboratory confirm these theoretical calculations and demonstrate the general feasibility of fluorescence polarization measurements by use of a microscope fluorometer. This study is based on the measurement of milliliter and picoliter volumes of identical solutions. Figure 5 schematically illustrates the system used: in a microscope spectrofluorometer a 0.3 ml cuvette with quartz glass windows can be placed in the light path instead of the neutral beam-splitting plate and the objective. For measuring the same solution in picoliter volumes under microscope conditions, quartz glass capillaries (inside diameters between 3 and 20 μm, embedded in glycerol between slide and coverglass) are used as microscope cuvettes Sernetz 1970; 1972). These are described in detail in the following article of this volume.

This system not only permits us to compare the results obtained with milliliter and picoliter volumes of solutions, but also enables alternative measurements on cell suspensions and single cells under identical spectral conditions of excitation and fluorometric measurement. The equipment used for this type of microspectrofluorometry is an extended and slightly modified version of the Leitz Microspectrograph (Ruch 1960). A similar version with different attachments for microscope spectrofluorometry is described by Dr. F. W. D. Rost in this volume. Photon-counting equipment is used for measuring the fluorescence intensity either behind the exit monochromator for recording polarization spectra of emission or behind the measuring diaphragm for measuring polarization spectra of excitation.

Figure 6 shows the polarization spectra of excitation measured on 0.3 ml (●●●) and picoliter volumes (ooo) of different fluorescent dye solutions. The polarization resulting from fluorescein in glycerol (Figure 6a) in the 0.3 ml cuvette (P = 0.33) is

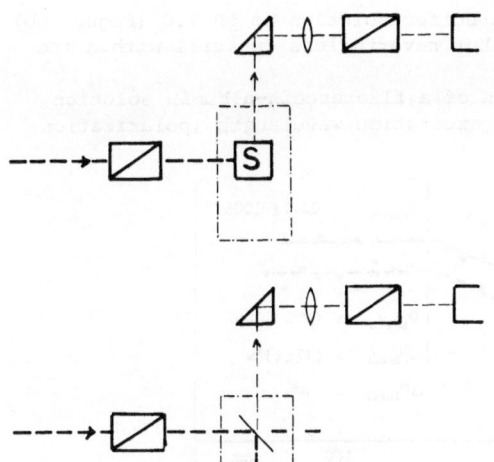

Figure 5. *Optical arrangement (schematic) for alternative measurement of fluorescence polarization on either solutions and cell suspensions in a 0.3 ml cuvette or microcapillary samples of solutions and single cells. The diagram illustrates the central part of a microscope spectrofluorometer; the incident illuminator with the objective can be exchanged for the 0.3 ml cuvette in a preadjusted holder*

in satisfactory agreement with that previously reported (Chen 1967), whereas the measurement on picoliter volumes of the same solution in microcapillaries by use of microscope optics (objective Zeiss Ultrafluar 100/1.25 Glyc., numerical aperture effective for excitation <0.1) resulted in values for P lower by 11 ± 1% compared with $\Delta P = 12\%$ expected from the graph in Figure 4. ΔP between the 0.3 ml and the pl sample measured

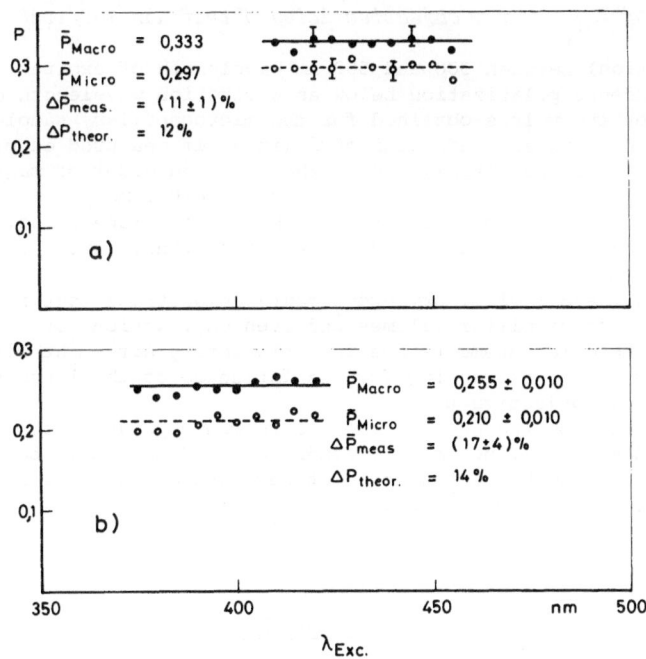

Figure 6. *Fluorescence polarization of fluorescent dye solutions measured as a function of the excitation wavelength in the visible spectrum on 0.3 ml (P_{macro}) and on picoliter volumes in microcapillaries (P_{micro}) (objective Zeiss Ultrafluar 100/1.25 glyc).*

a) 3.75×10^{-4} M fluorescein -Na_2 in Glycerol (13% H_2O), $\lambda_{Em} > 530$ nm

b) 2.5×10^{-4} M albumin + 2.5×10^{-4} ANS in 0.125 phosphate buffer solution (pH 7.0), $\lambda_{Em} > 460$ nm) in the case of ANS bound to albumin in phosphate buffer solution at pH 7.0 (Figure 6b) is slightly higher than expected theoretically, but nevertheless is still within the limits of error.

Figure 7 shows the fluorescence polarization of a fluorescein-albumin solution in 0.125 M phosphate buffer as a function of the excitation wavelength (polarization

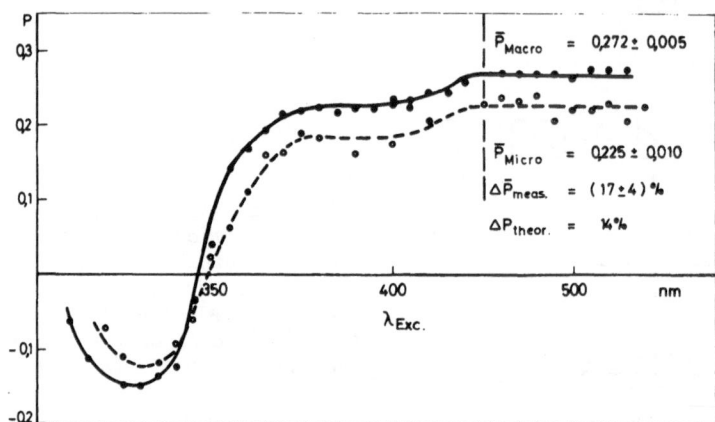

Figure 7. *Fluorescence polarization of a fluorescein-albumin solution in 0.125 M phosphate buffer, pH 7.0 (1 × 10⁻⁴ M fluorescein-Na₂ + 4 × 10⁻⁴ M albumin) measured as a function of the excitation wavelength (polarization spectrum of excitation) on 0.3 ml (P_macro) and on picoliter volumes in microcapillaries (P_micro), λ_Em > 530 nm. For determination of P_micro the objective Zeiss Ultrafluar 100/1.25 Glyc. was used*

spectrum of excitation) between 300 and 500 nm wavelength of excitation. The change to negative fluorescence polarization below an excitation wavelength of 300 nm is demonstrated also by the values obtained for the microcapillary sample. The values for ΔP, i.e., for the apparent reduction of P (in %) if measured on microcapillary samples by use of microscope optics, are of the expected order of magnitude throughout the spectral region of excitation covered by the experiment.

The orientation of the capillary axis in the object plane relative to the vibration direction of the exciting light did not exert a significant influence on the polarization values measured.

The results of these preliminary experiments demonstrate that fluorescence polarization measurements on picoliter volumes and even on fractions of a picoliter are feasible and that these measurements are in satisfactory agreement with those carried out on macrovolumes, after correcting for the influence of the large numerical aperture of the microscope objective used.

Investigation based on the comparative fluorescence polarization measurement on cell suspensions under macrofluorometric conditions and on single cells in a microscope spectrofluorometer are now being conducted at our laboratory. The results of these experiments will be reported elsewhere.

ACKNOWLEDGEMENT

This work was supported by the Battelle Memorial Institute, Columbus, Ohio.

REFERENCES

BLYUM, I.A., and Shcherbov, D. P., (1967), *Zh. Anal. Khim.*, <u>22</u>, 670.

CASPERSSON, T., Zech, L., Modest, E. J., Foley, G. E., Wagh, U., and Simonsson, E., (1969), *Exp. Cell Res.*, <u>58</u>, 141.

CHANCE, B., (1962), *Ann. NY. Acad. Sci.*, <u>97</u>, 431.

CHEN, R. F., (1967), *Anal. Biochem.*, <u>19</u>, 374.

KOHEN, E., (1964), *Exp. Cell Res.*, <u>35</u>, 303.

KOHEN, E., Kohen, C., Thorell, B., and Wagener, G., (1973), this volume.
LOESER, C. N., and West, S. S., (1962), *Ann. N. Y. Acad. Sci.*, <u>97</u>, 346.

ROST, F. W. D., (1973), this volume.

RIGLER, R., (1969), *Exp. Cell Res.*, <u>58</u>, 460.

RUCH, F. (1960), *Z. wiss. Mikrosk.*, <u>64</u>, 453.
SERNETZ, M., and Thaer, A., (1970), *J. Microscopy.*, <u>91</u>, 43.

SERNETZ, M., and Thaer, A., (1972), *Anal. Biochem.*, <u>50</u>, 98.

UDENFRIEND, S., (1969), *Fluorescence Assay in Biology and Medicine, Vol. II*, Acad. Press, New York.

WEST, S. S., (1969), In: *Physical Techniques in Biological Research*, (A. W. Pollister, ed.) vol. 3C, Acad. Press, New York. 253-321.

SPENCER, R. D., and Weber, G., (1972), this conference, unpublished.

ADDENDUM

Notes Added in Proof

A microfluorometer for the investigation of fluorescence polarization has also been developed by Russian scientists (Barsky et al. 1971, Shiffers et al. 1971). It was successfully applied to the study of the polarized ultraviolet fluorescence of giant muscle fibers of *Balanus rostratus* (Borovikov et al. 1972). In this study the fluorescence anisotropy was determined, as caused by two factors; by the orientation of fluorescent molecules with respect to the fiber axis, and by the polarization of the spectrum of excitation of the tryptophan residues in the anisotropic bands. Both these anisotropics could be determined separately by measuring the polarization spectra at different vibration angles of exciting polarized light with reference to the fiber axis.

BARSKY, I. Ya., Rosanov, Yu. M., and Shiffers, L. A., (1967), *Zitologija*, <u>9</u>, 1026.

SHIFFERS, L. A., Rosanov, Yu. M., Barsky, I. Ya., Brumberg, E. M., Ioffe, V. A., and Klyuchnikov, V. V., (1971), *Zitologija*, <u>73</u>, 1314.

BOROVIKOV, Yu. S., Rosanov, M., Barsky, I. Ya., Shudel, M. S., and Chernogryadskaya, N. A., (1972), *Zitologija*, <u>14</u>, 953.

MICROCAPILLARY FLUOROMETRY AND STANDARDIZATION FOR MICROSCOPE FLUOROMETRY

M. Sernetz and A. Thaer

Battelle-Institut e.V., Frankfurt, Germany

INTRODUCTION

Standardization in fluorometry is still connected with difficulties. The theoretical and practical reasons for this have recently been discussed during the National Bureau of Standards conference in Washington (Conference on Accuracy). In the field of microscope fluorometry standardization is even more difficult because of the tiny volumes and amounts of substances involved and their correlation to microscopic objects to be measured. Previous attempts at the calibration of instruments and standardization of fluorescence intensities were mainly based on, (a) fluorescent glass plates or crystals for calibration of instruments, (b) biological standards such as spermatozoa (Ruch 1961), and (c) microdroplets (Rotman 1966, Rigler 1966, Jongsma et al. 1971) or microspheres. All these attempts claim to produce reproducible fluorescence signals and permit the measured intensities to be correlated with the amounts of fluorescent substances in biological objects.

Since 1969, we have used microcapillaries as optical cuvettes for microfluorometric investigations (Sernetz and Thaer, 1970, 1972). They permit not only the exact calibration of the microscope fluorometer but also the highly accurate correlation of measured fluorescence intensities with the mass of the excited fluorochrome.

METHODS

The capillary cuvettes are quartz glass capillaries with an inside diameter of 3 to 20 μm. Up to 20 of these capillaries can be arranged parallel to each other and embedded like microscopic specimens between slide and cover glass; this process uses glycerol, which has about the same refractive index as quartz glass. The inside diameters of the empty capillaries are measured microinterferometrically (Figure 1, Plate I) according to

$$d[\mu m] = \frac{\Gamma[\mu m]}{n_Q - n_A} \text{,} \tag{1}$$

where:

Γ = path difference determined by measuring displacement of interference fringes in wavelengths [μm]; n_Q = refractive index of quartz glass; n_A = refractive index of air (= 1.0).

The capillaries are then filled, for instance, with a defined fluorochrome solution in linear or logarithmic dilution. Figure 2, Plate I shows a photomicrograph of a capillary section filled with a fluorescein solution superimposed by the image of the measuring aperture of the microscope fluorometer.

Under exact "Koehler" illumination conditions, the image of the rectangular, adjustable, and variable field diaphragm frames an illuminated area of homogeneous radiant flux distribution in the object plane. Another variable rectangular diaphragm in the image plane of the microscope photometer limits the measuring field (Figure 2, Plate I).

The capillary volume v within the dimensions of the measuring field (i.e., the length ℓ and the inner diameter d together with the concentration c) defines the mass m of the fluorochrome measured:

$$m = \frac{\pi}{4} d^2 \cdot \ell \cdot c \tag{2}$$

Thus, in contrast to the normal situation in fluorometry, we obtain primarily a mass-proportional signal rather than a concentration-proportional signal.

PLATE I

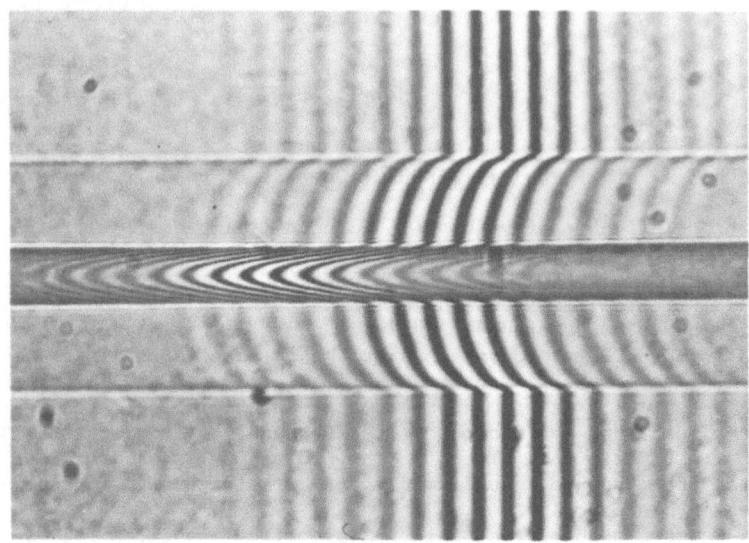

Figure 1. *Empty capillary cuvette, displacement of white light interference fringes due to the path difference between the quartz glass tube and the inner capillary space.*
Leitz Interference Microscope (Mach-Zehnder principle) objective pair F1 Oil 100/1.32, magnification 2000 x (Sernetz and Thaer 1970)

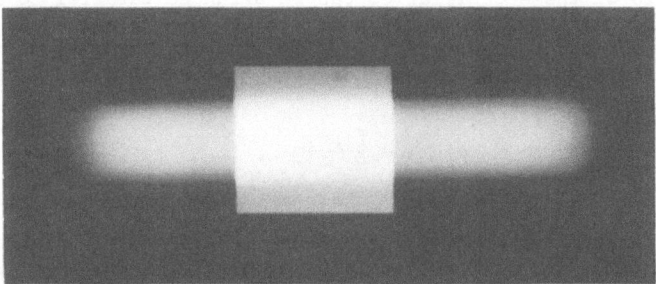

Figure 2. *Fluorescence photomicrograph from a section of a capillary cuvette filled with fluorochrome solution superimposed by the image of the aperture of the measuring diaphragm. Magnification 2000 x (Sernetz and Thaer 1970)*

Using fluorescence excitation in incident light, the following relation is valid:

$$I_F = I_0 \cdot \Phi (1 - e^{-kcd}) \qquad (3)$$

where: I_F = fluorescence intensity emitted by the sample

I_0 = intensity of exciting light entering the sample

Φ = quantum efficiency

k = molar extinction coefficient

d = thickness of the fluorescent sample in incident light direction.

As the intensities measured in the capillaries are emitted from defined sections, i.e., from defined volumes, the concentration c may be replaced by mass per volume;

$$I_F = I_0 \cdot \Phi \cdot (1 - e^{\frac{-kdm}{\vartheta}}) \qquad . \qquad (4)$$

As long as the exponent is small compared with unity, a linear relation between mass and fluorescence intensity is obtained:

$$I_F = I_0 \cdot \Phi \cdot \frac{k\,d}{v} \cdot m \qquad (5)$$

Figure 3 shows results of fluorescence intensity measurements as a function of concentration (or mass), using capillaries in a microscope fluorometer (b) and normal cuvettes in a fluorometer (a). The fluorochrome in both cases is fluorescein in 0.06 M phosphate buffer of pH 7.0.

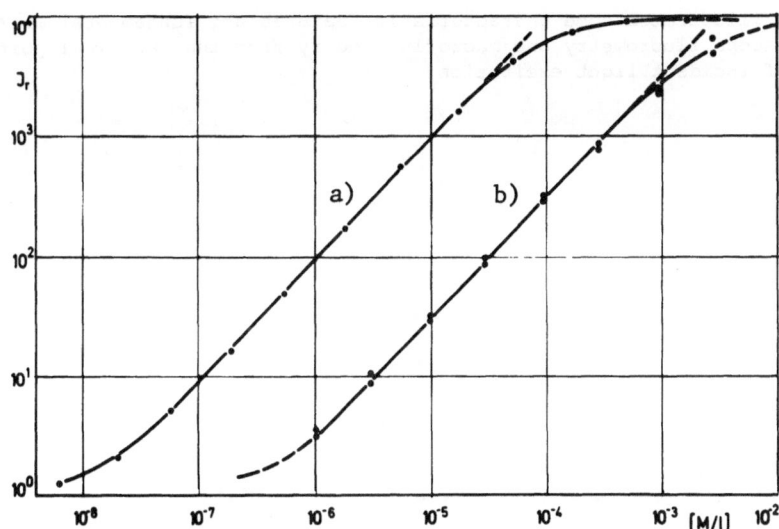

Figure 3. *Comparison of the measuring ranges of macrofluorometry* (a) *and microscope fluorometry by means of capillary cuvettes* (b). *Relative fluorescence intensities corrected for identical signal-to-noise ratio as a function of the fluorescein concentration (Sernetz and Thaer 1972)*

RESULTS AND DISCUSSION

The intensity versus concentration (or mass) plots (a) and (b) in Figure 3 are corrected for identical signal-to-noise ratio. Fluorometry and microscope fluorometry differ in the useful concentration range by about two decades. In view of the capillary thickness of $d < 10^{-3}$ cm the exponent kcd will remain very small with respect to unity even at concentrations up to $10^{-3}M$ and extinction coefficients up to $10^4M^{-1}cm^{-1}$. We thus obtain a linear measuring range between fluorescence signal and concentration or mass, unless other influences such as concentration quenching have to be taken into account.

At the lower end of the standardization curve for microscope fluorometry, the limit of detectability seems to be reached at fluorescein concentrations of about $10^{-6}M$. From our experience, however, this limitation is mainly due to the increasing portion of background fluorescence of the embedding medium, quartz glass, microscope optics, and barrier filters. Thus, at the moment, the lower level of detectability seems to be rather a technical limit.

However, it should be borne in mind that, in terms of mass of excited fluoro-chrome, a concentration of $10^{-6}M$ and an average volume of the excited capillary in the measuring field of about 10^{-12} l correspond to 10^{-18} moles of fluorescein. Moreover, capillary fluorometry extends the measuring range to higher concentrations so that the two methods may complement one another.

Curve b in Figure 3 also shows that the accuracy of measurements is satisfactory, all the more as they involve not only the fluorometric measurement but also the inter-ferometric determination of the inside diameter d of the capillary. The error of the diameter determination enters the calculation with the second power, whereas the determination of the length l is relatively uncritical. This is important to note when comparing the capillary technique with the microdroplet methods, where the error of diameter determination enters the volume determination of the sphere with the third power. Another important difference between these techniques is the fact that in the microdroplet methods, apart from the necessary statistical evaluation of a mean cali-bration value from randomly distributed sphere volumes, the entire droplets are ex-cited for measurement and therefore may undergo fluorescence decomposition. In the capillary, however, only a small fraction of the total volume is excited, and it is permanently exchanged by diffusion and convection so that a constant signal can be achieved.

Figure 4 illustrates some characteristic features and fundamental differences between microscope fluorometry and macrofluorometry from the technical point of view in the case of incident light excitation.

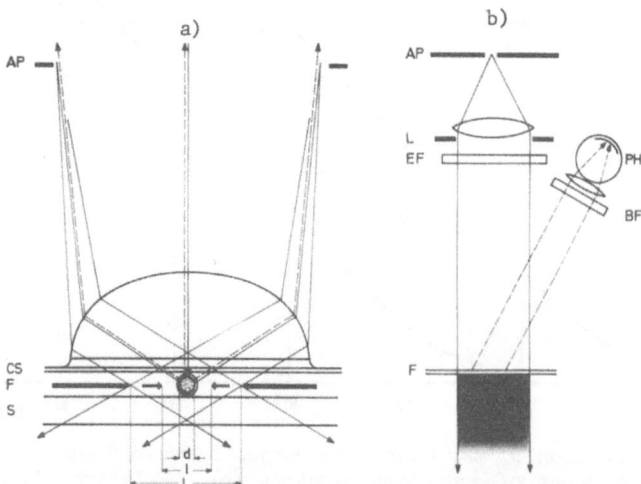

Figure 4. *Comparison of incident light fluorometry with respect to the differ-ent optical arrangement* (a) *microscope fluorometer,* (b) *macrofluorometer. The dia-grams are arranged for identical positions of the light source* (AP, b) *or its image in the rear focal plane of the objective* (AP, a) *and the fluorescence cuvettes.* CS = *cover slide,* F = *fluorescent capillary* (a) *or front of cuvette* (b), S = *slide,* d = *diameter of capillary,* l = *diameter of measuring field* (= *length of capillary section measured*), L = *diameter of illumination field diaphragm* (b) *or its image in the object plane* (b) *respectively,* EF = *excitation filter,* BF = *barrier filter,* PH = *photometer (Sernetz and Thaer 1972)*

First this applies to the relationship between the dimensions of the optical beam and those of the cuvette. In microscope fluorometry (Figure 4a) the measuring cuvette (microcapillary, droplet, or cell) is smaller than the diameter of the exciting beam. Thus, primarily a mass m of the fluorochrome is defined as being excited in a volume v. In the case of macrofluorometry (Figure 4b), however, the diameter of the exciting beam is smaller than the dimensions of the measuring cuvette, and the fluorescence intensity is thus a measure of the concentration c.

Secondly, the differences also relate to the optical arrangement used for illumination. Since it is necessary to achieve homogeneous light energy distribution for excitation in the object plane in order to obtain a mass proportional signal in microfluorometry irrespective of the position of the cuvette in the object plane, the lightfield diaphragm has to be imaged into the object plane, and the light source of the exit slit of the monochromator must be imaged into the rear focal plane of the objective)("Koehler" illumination) (Figure 4a). In macrofluorometry, usually the exit slit of the monochromator is imaged into the center of the cuvette (Figure 4b); this method of illumination corresponds to the "critical" illumination in microscopy.

Thirdly, to achieve the desired optical resolution and energy gain, the solid angles (numerical apertures of microscope objectives) applied for excitation and uptake of fluorescence in microscope fluorometry normally tend to be as large as possible, whereas in macrofluorometry they are kept very small.

Figure 5 shows that the energy distribution in the object plane is sufficiently homogeneous under appropriate Koehler illumination. This has been established by

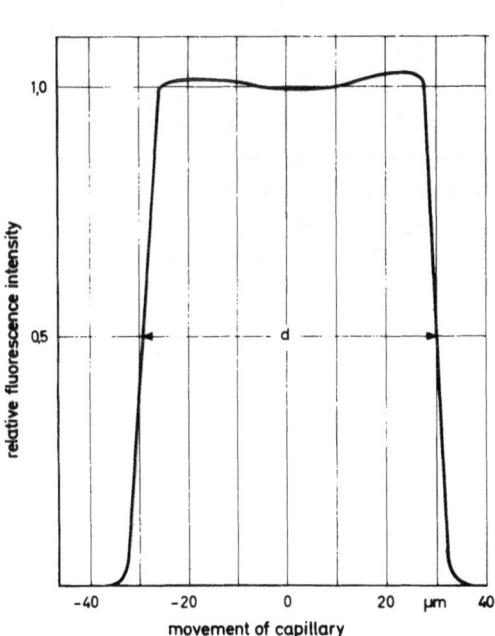

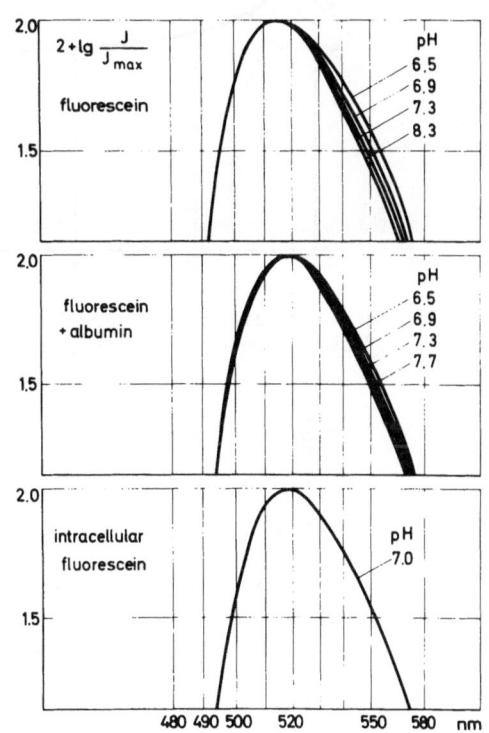

Figure 5. *Homogeneous energy distribution of the exciting light in the illuminated area in the object plane under Köhler illumination conditions, as demonstrated by movement of a fluorescent capillary across the rectangular illuminated area with diameter* d *(Sernetz and Thaer 1970)*

Figure 6. *Comparison of fluorescence spectra of fluorescein (a) and fluorescein albumin solutions (b) at different pH, measured in capillaries with the fluorescence spectrum of intracellular fluorescein (c) liberated by enzymatic hydrolysis of fluorescein diacetate in a single living cell (Sernetz 1973)*

measuring the fluorescence intensity during movement of a capillary across the illuminated area in the object plane limited by the light-field diaphragm.

We have used the capillary technique to compare fluorescence spectra of intracellular fluorochromes with those of authentic fluorochromes in solution under identical optical and fluorometric conditions. This applies to the spectral characteristics of light sources, monochromators, filters, mirrors, and photomultipliers.
The results are shown in Figure 6 for fluorescein, intracellularly liberated from
fluorescein-diacetate by enzymatic hydrolysis (Sernetz 1973). They suggest an interaction of intracellular fluorescein with, for instance, cytoplasmic proteins, since
the spectral shift observed for intracellular fluorescein is identical with that of
the system fluorescein-albumin used as a model.

Stimulated by these first applications of the microcapillary technique for
investigating interactions between fluorochromes and macromolecules in picoliter
volumes, the capillaries have been extensively used in our laboratory for studies of
analytical binding between fluorescein and albumin in order to compare the results
of microscope fluorometry and macrofluorometry.

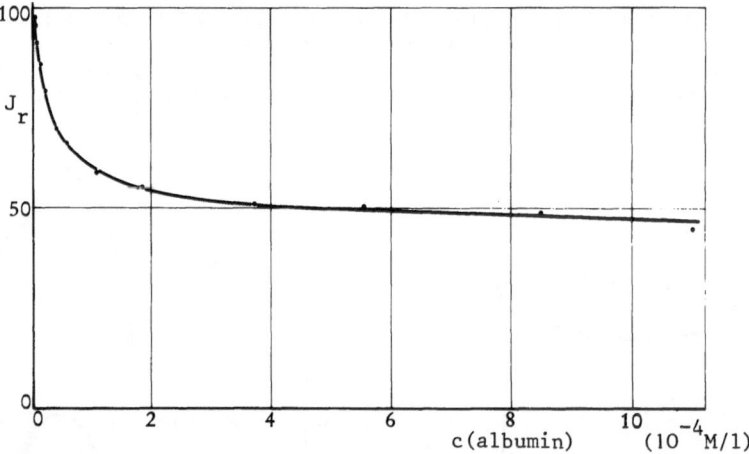

Figure 7. *Relative fluorescence intensity of 5.29 × 10⁻⁵M fluorescein as a
function of the albumin concentration measured fluorometrically (Sernetz and Thaer
1972)*

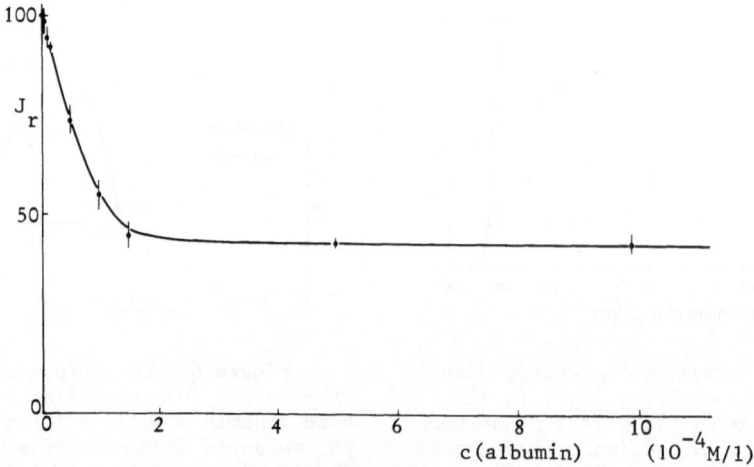

Figure 8. *Relative fluorescence intensity of 3.47 × 10⁻⁴M fluorescein measured
by means of microscope fluorometry as a function of the albumin concentration (each
value was measured on two capillary samples) (Sernetz and Thaer 1972)*

Figure 7 depicts the titration curve obtained for fluorescence quenching measurements in the presence of increasing albumin concentration at a constant fluorescein content using a macrofluorometer. Figure 8 shows the results of an analogous experiment using a microcapillary preparation; each value represents the mean of two capillaries.

We converted these fluorometric values first according to Blake and Peacocke (1968), to get the adsorption isotherm of fluorescein for albumin as a function of the free fluorescein concentration; and secondly, according to Klotz (1946) to determine the reciprocal specific binding capacity as a function of the reciprocal free fluorescein concentration (Figure 9). From this it may be concluded that the capillary method not only represents a microtechnique for fluorometric protein-binding studies, but also permits the measuring range to be shifted to higher saturation with respect to macrofluorometry by about one decade. This reduces the extrapolation for determining the binding parameters by means of a Klotz-plot, especially with respect to the error characteristics of the transformations. We found a single set of $n = 4$ binding sites per albumin molecule and a binding constant of 4.4×10^4 M^{-1}, in satisfactory agreement with earlier findings of Laurence (1952).

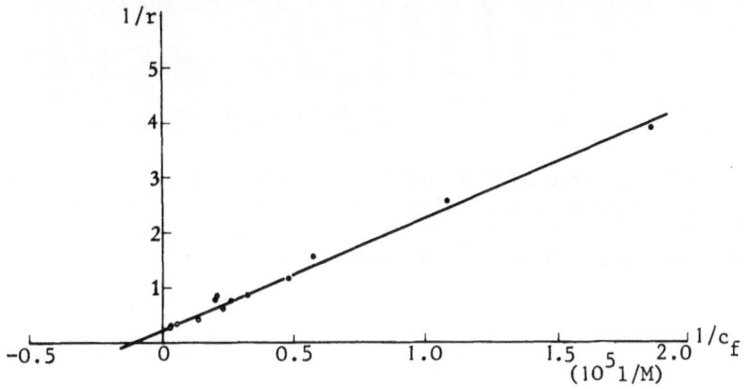

Figure 9. *Reciprocal specific binding capacity as a function of the reciprocal free fluorescein concentration - (●) macrofluorometric and (o) microscope fluorometric determinations (Sernetz and Thaer 1972)*

These experiments demonstrate the applicability of microcapillaries as optical cuvettes for fluorometry. Of course they also offer new prospects for absorption spectrophotometry and refractometry as demonstrated by the following example. The unknown refractive index n_L of a solution in the capillary may be determined by microscope interferometry (Figure 10, Plate II) from the relation of the path differences Γ_{QL} and Γ_{QA} and the known refractive index of quartz glass n_Q and air n_A according to the equation

$$n_L = n_Q + \frac{\Gamma_{QL}}{\Gamma_{QA}} (n_Q - n_A) \tag{6}$$

(L = *liquid*, Q = *quartz glass*, A = *air*).

For instance, the unknown albumin concentration of a solution could be determined by this method in a volume as little as 10^{-12} liter.

ACKNOWLEDGEMENT

This work was supported by the Battelle Memorial Institute, Columbus, Ohio.

PLATE II

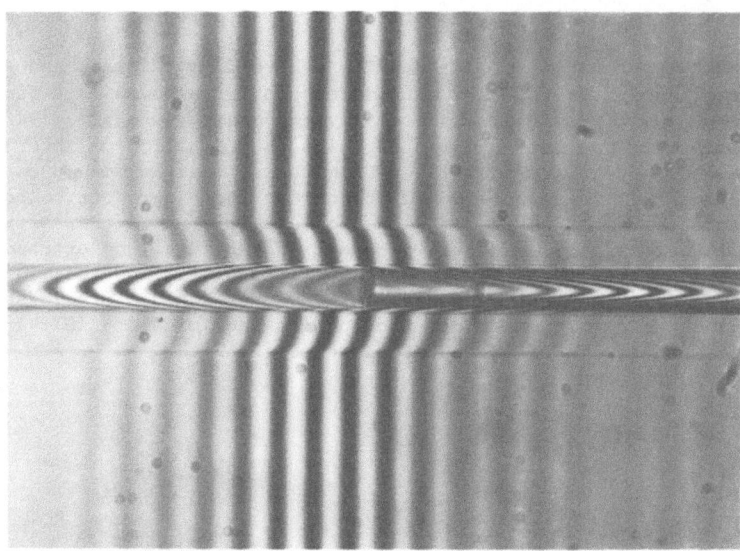

Figure 10. *Refractometry of a solution (di-iodomethane, $n_D^{25} = 1,764$) in capillary cuvettes by microscope interferometry. Displacement of the interference fringes according to the path differences Γ_{QL} between quartz glass Q and solution L (right) and Γ_{QA} between quartz glass Q and air A (left)*

REFERENCES

BLAKE, A., and Peacocke, A. R., (1968), *Biopolymers, 6,* 1225.

CONFERENCE ON ACCURACY, In: *Spectrophotometry and Luminescence Measurements,* March 22-24, 1972. National Bureau of Standards, Gaithersburg, Md. (in Press).

JONGSMA, P. M., Hijmans, W., and Ploem, J. S., (1971), *Histochemistry, 25,* 329.

KLOTZ, J. M., (1946), *J. Amer. Chem. Soc., 68,* 2299.

LAURENCE, D. J. R., (1952), *Biochem. J., 51,* 168.

RIGLER, R., (1966), *Acta Physiol. Scand. Suppl., 67,* 267.

ROTMAN, B., (1961), *Proc. Natl. Acad. Sci. U.S., 47,* 1981.

RUCH, F., (1961), In: *Introduction to Quantitative Cytochemistry* (G.L. Wied, ed.) Academic Press, New York, P. 281.

SERNETZ, M., (1973), This volume.

SERNETZ, M., and Thaer, A., (1970), *J. Microscopy., 91,* 43.

SERNETZ, M., and Thaer, A., (1972), *Anal. Biochem., 50,* 98.

THE USE OF HUMAN LEUKOCYTES AS A STANDARD FOR THE CYTOFLUOROMETRIC DETERMINATION OF PROTEIN AND DNA

F. Ruch

Department of General Botany, Swiss Federal Institute of Technology, Zurich, Switzerland

INTRODUCTION

Since cytofluorometric methods yield only relative results, standard cellular material of known chemical composition is necessary for calibration. Furthermore, such material may be of importance for methodical investigations such as testing new staining reactions or cell preparations.

Contrary to the situation in the field of cytochemical DNA determination, little is known about the reliability and reproducibility of protein reactions. Highly differentiated cells such as sperms suitable for DNA standardization, may give results in the case of protein determinations that are not comparable with those obtained with other cells (Rosselet and Ruch, 1968). The protein content of mitotically and metabolically active cells varies too much for standard purposes. The usefulness of isolated nuclei is somewhat limited due to the possibility of loss of substance in the isolation process. We have made investigations on human leukocytes (Ruch and Leemann 1972) using different cytochemical reactions; the studies were made with special regard to an evaluation of DNA and proteins in absolute terms and to the reproducibility of the reactions.

MATERIALS AND METHODS

Leukocytes were isolated from venous blood of healthy adults by a standard method (Ruch and Leemann, 1973). The cells were fixed with different liquids. The slides were prepared by centrifugation of the cells onto coverslips, a method described elsewhere in this volume (Ruch 1973). The following cytochemical determinations have been tested:
DNA was determined by the BAO fluorescence reaction (Ruch,1970). Basic protein (histone) was stained, after hydrolysis with TCA, with sulfaflavine at pH 8.0 (Leemann 1972), total protein with sulfaflavine at pH2.8 (Leemann, 1972). Arginine determination was carried out by the ninhydrin reaction (Rosselet, 1967). Lysin was determined by the dansylchloride reaction (Rosselet and Ruch, 1968). Sulfhydryl groups were stained with mercury orange (Bennet 1958; Burns 1967). In the PAS reaction for polysaccharides BAO was used instead of pararosaniline. The primary fluorescence of tyrosine/tryptophan in proteins was measured on unstained preparations (Saurer 1966). Dry weight measurements were carried out with the interference microscope.

Tests on the influence of different fixations (composition, time and temperature) proved that best results are obtained with 4% Formalin (buffered at pH 7.0) and ethanol acetic acid 99:1 for 20 hours at 4°C. Leukocytes may be stored in 70% ethanol at 4°C for 3 to 4 weeks without there being significant changes in dry weight, DNA, and protein content.

Special attention was paid to proper measuring conditions for cytofluorometry. The new Zeiss cytofluorometer (Figure 1. Plate I) used for our measurements was especially developed by Ruch and Trapp, 1973 for routine measurements on cells, and offers the following advantages:

a) Each cell is measured within 20 milliseconds and is exposed to exciting radiation during that time only. Thus, the influence of photodecomposition is practically eliminated.

b) After the cell has been selected and focused by hand in phase-contrast, all other steps required for a single measurement are automatically executed according to a preselected program. This permits the necessary number of measurements to be carried out within a relatively short time.

c) Recording and evaluation of the cytofluorometric data are accomplished by means of a printer, a desk-top calculator, or a computer. Figure 2 shows a histogram

PLATE I

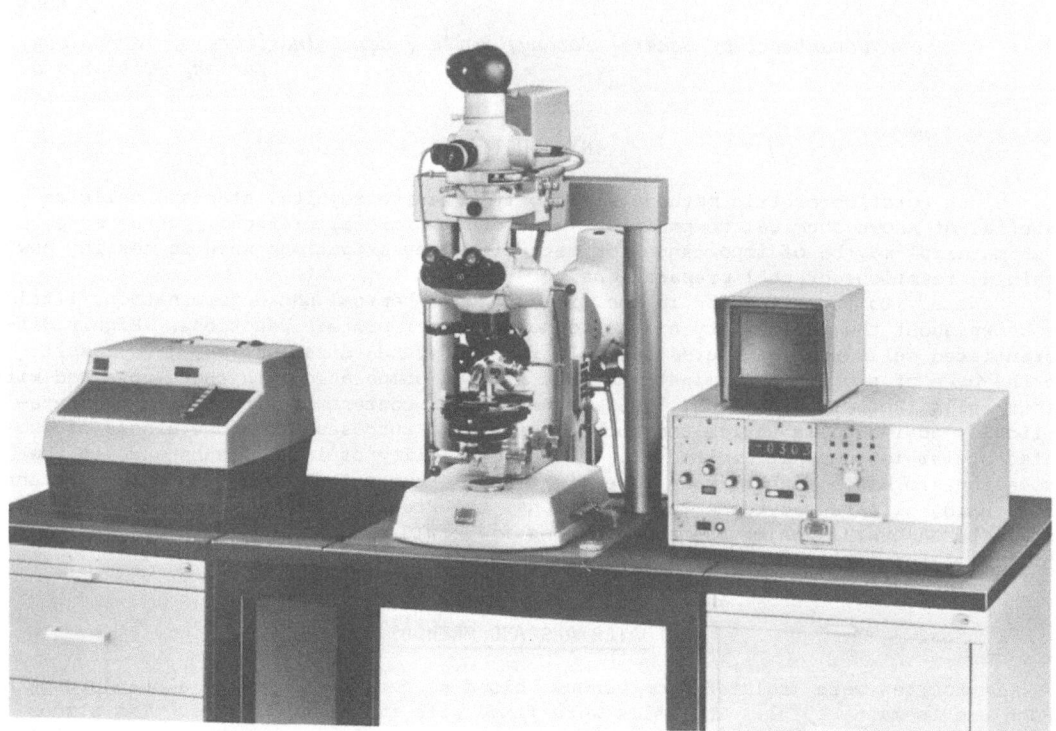

Figure 1. *Zeiss Cytofluorometer, based on the large Zeiss fluorescence micro-scope with vertical illuminator for incident light excitation with television and printer. The object is focused in transmitted light phase-contrast*

resulting from measurements carried out by use of this instrumentation.

```
        KE(FO)          DNS

 A      N           10        20        30        40        50        60

        X
 91     XX
        XXX
        XXXXXX
        XXXXXXXXXXX
        XXXX
100     XXXXXXXXX
        X
        XXXXXX
        XXXXXXXXXXXX
        XXXXXXXXXXX
140     XXXXXXXXXXXXXX
        XXXXXXXXXXXXXXXXXXXXXXXXX
        XXXXXXXXXXXXXXXXXXXXXXXXXXXXXXXXXXXXXX
        XXXXXXXXXXXXXXXXXXXXXXXXXXXXXXXXXXXX
        XXXXXXXXXXXXXXXXXXXXXXXXXXXXXXXXXXXXXXXXXXXXXXXXX
200     XXXXXXXXXXXXXXXXXXXXXXXXXXXXXXXXXX
        XXXXXXXXXXXXXXXXXX
        XXXXXX
        X
        X
280
        XX

        XXXXXX
        X
400
```

Figure 2. *Histogram printed by a pdp 8E computer.* Abscissa: *number of cells measured (DNA).* Ordinate: *fluorescence intensity in arbitrary units*

RESULTS AND DISCUSSION

Coefficients of Variation between Individual Cells

Lymphocytes: (Table I) DNA measurements are in accordance with Feulgen values in the literature and the coefficient of variation does not greatly exceed that of sperms. Protein reactions show coefficients of variation between 10% and 20%. According to our experience, the most reliable protein determinations (lysine, sulfhydryl, primary fluorescence, dry weight) yield coefficients of variation of 10% to 13%. This indicates biological variations of up to 10%.

Polymorphonuclear Cells: The coefficients of polymorphonuclear cells are generally high (12% to 23%). In addition, Table I demonstrates the ratios between the values of two selected cell types, namely small lymphocytes (LC) and polymorphonuclear leukocytes (PC). Even though the selection of these cell types is somewhat arbitrary, the ratio is remarkably constant for a given fixation technique.

Variations Among Leukocytes From Different Samples

Dry weight and total protein measurements made on leukocytes from different batches at different times, show a considerable variation (Table II). The fact that this variation is larger than the one found between slides from the same batch points to an actual difference in mass among the cells from different samples. The fact that the values of dry weight and total protein are correlated confirms this result. It is interesting to note, that the ratio PC/LC is more constant than are the means of dry weight and total protein. The figures given in Table II are relatively high, since staining and measurements were made on different days.

TABLE I
Human Leukocytes: DNA and Protein Determinations

| | Lymphocytes | | | Ratio PC/LC | |
| | | | | Fixative | |
Substance	Average Amount 10^{-12}gr	Coefficient of variation n: 40-50	Fixative	EA	FO
DNA	5	6	EA	1.30	1.10
Histone	7	10	FO		1.35
Protein-bound lysine	3	10	EA	1.80	1.35
Total protein	29	15	EA	1.80	1.50
Dry weight	38	13	EA	1.45	1.35
	54	14	FO		

EA: ethanol acetic acid, FO: Formalin

TABLE II
Human Leukocytes: Dry weight (DW) and Total Protein (TP) Determination

| | Coefficient of Variation between the Means | | | Correlation between | n |
	DW	TP	Ratio PC/LC	DW-TP	
1 Batch of Leukocytes	8	14			8
Different batches	21	24	10	0.83	10
Estimated variation between batches	15-20	15-20			

n: independently measured series of 10-30 (DW) and 50 (TP) measurements.
PC: polymorphonuclear cells; CL: lymphocytes.

Absolute Amounts

Table I shows absolute amounts of dry weight, DNA, histone and total protein. The histone and total protein values are obtained directly from comparison with model experiments (Ruch and Leemann, 1972). RNA and polysaccharides have been omitted from the calculations because of the difficulties encountered in measuring their absolute amounts cytochemically. They constitute too small a percentage of the dry weight to seriously affect the results given in Table I (glycogen 0.5% - 2% for PC, 0% for LC; acid mucopolysaccharides 0.1%; RNA 10% or less). The deviation in the amount of total protein determined independently by interference microscopy and from that determined fluorometrically after staining with sulfaflavine is reasonably low (10%-20%).

CONCLUSIONS

For standardization purposes, it makes sense to focus on small lymphocytes as these can easily be selected and have relatively small standard deviations. For testing of cytochemical reactions, as well as for measuring procedures, the existence both of different cell types and of the characteristic protein distribution is advantageous: any method determining the same substance under similar conditions should result in the same distribution. Compared with DNA determinations, the deviations of the protein measurements among individual cells, as well as those among different batches of leukocytes, are relatively large. This is, at least to some extent, due to the fact that the biological variation of protein content is higher than that of DNA content. Nevertheless, further refinement of cytochemical techniques is necessary, especially with regard to fixation and preparation. Despite all the limitations discussed above, leukocytes should still prove to be a useful material both for standardization and for further methodical studies.

REFERENCES

BENNET, H. S., and Watts, R. M., (1958), In: General Cytochemical Methods., Vol I (J. F. Danielli, Ed.), Academic Press, New York, p.317.

BURNS, J., (1967), *Histochemie* , 10, 293.

LEEMANN, U., and Ruch, F., (1972), *J. Histochem. Cytochem.*, 20, 659.

ROSSELET, A., (1967), *Z. wiss.Mikr.*, 68, 22.

ROSSELET, A., and Ruch, F., (1968), *J. Histochem.Cytochem.*, 16, 459.

RUCH, F., (1970), In: Introduction to Quantitative Cytochemistry , Vol. II, (G. L. Wied and G. F. Bahr, Eds.), Academic Press, New York, p.431.

RUCH, F., (1973), This Volume.

RUCH, F., and Leemann, U., (1972), *Acta Cytol.*, 16,342.

RUCH, F., and Trapp, L., (1973), *Zeiss Inform.*, in press.

SAURER, W., (1966), Thesis No 3834, ETH, Zurich.

A MICROSPECTROFLUOROMETER FOR MEASURING SPECTRA OF EXCITATION, EMISSION AND ABSORPTION IN CELLS AND TISSUES

F. W. D. Rost

Department of Histochemistry, Royal Postgraduate Medical School
London, England

INTRODUCTION

This paper describes an instrument for microspectrofluorometry, based on the Leitz microspectrograph. A recent description of this instrument has been published by Rost and Pearse (1971); there is also an earlier report on the subject by the same authors (Pearse and Rost 1969). Important features of the present instrument are monochromatic epiillumination, a reference channel, and a measuring system using photon counting, with automatic recording of data in digital form. To minimize fading, measurements of fluorescence intensity can be made within as little as 200 msec from the commencement of irradiation.

The instrument was designed primarily for the measurement of fluorescence and excitation and emission spectra of substances in tissue sections and similar preparations. It can also be used to measure transmission, in both visible and ultraviolet light, and reflection. In designing and developing the microfluorometric apparatus, the following instrumental requirements and considerations particularly were taken into account.

a) The system should be based on a good microscope, one that permits observation and photography of the object being investigated.

b) Substantially monochromatic light should be available for excitation, with wavelength continuously variable from ultraviolet to blue, approximately in the range 250-500 nm.

c) Excitation by epiillumination should be possible (i.e., illumination from above through the microscope objective).

d) Fluorescence should be observable and measurable at any desired wavelength from the near ultraviolet to the red (approximately 360-700 nm).

e) Operation should be sufficiently rapid to allow measurement of rapidly fading fluorescence.

f) Sensitivity of the photometric system should be adequate to allow measurements of quite weakly fluorescent objects and/or of small objects such as lysosomes.

g) Provision should be made for recording the rate of fluorescence fading.

h) The instrument should be as versatile as possible, to maximize its usefulness.

At the time of commencement of this project (1966), a number of microspectrofluorometers had been described in the literature (Nordén 1953; Rousseau 1957; Olson 1960; West et al. 1960; Caspersson et al. 1965; West 1965), but no proper microspectrofluorometer was commercially available. It seemed that the Leitz microspectrograph (Ruch 1960) offered best hope of being adaptable to our needs, because:

a) it was designed for use over a wide spectral range, including ultraviolet,

b) a monochromator which could be used as the emission monochromator was included in the instrument, and

c) the manufacturers could fit a monochromator on the excitation side.

The Leitz microspectrograph (Ruch 1960) was originally designed for measurement of ultraviolet absorption; this was accomplished by photographing the object in transmitted light of appropriate wavelength, and then measuring the density of the photograph in a separate instrument, such as the Joyce-Loebel scanning densitometer. The instrument is now commercially available in a modified form suitable for microfluorometry; however, each instrument appears to have been modified individually to suit the laboratory in which it has been installed (Rost and Pearse 1971; Björklund 1968; Sprenger 1971).

PLATE I

Figure 1. *General view of the microspectrofluorometer, showing (left to right) the electronic system (photon counting and data logging), the Teletype, the optical system (microspectrograph), and a chart recorder*

DESCRIPTION OF APPARATUS

A general view of the apparatus is presented in Figure 1 (see Plate I); the optical portion is shown on the right, the major part of the electronic system is contained in the console on the left, and the data is typed either on the Teletype machine (center) or drawn on the chart recorder (far right).

A block diagram of the optical system is presented in Figure 2. The most usual arrangement of the optical system for fluorometry is as follows. The light source is

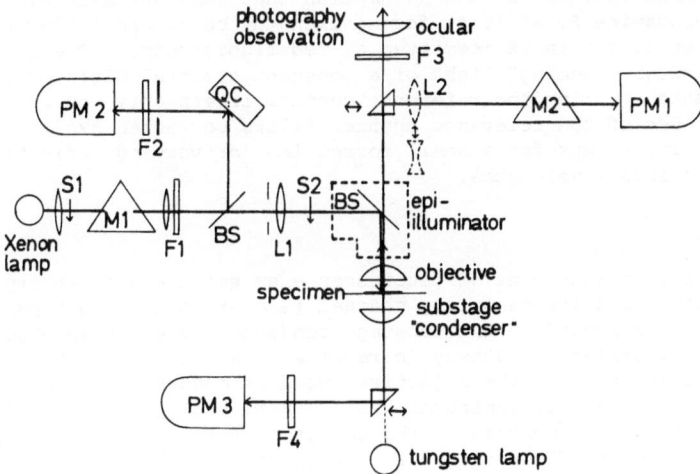

Figure 2. *Block diagram of the optical system of the microspectrofluorometer.*
Key to abbreviations: BS, beamsplitters (reference channel and epiilluminator);
F1, supplementary filter for monochromator; F2, F3, barrier filters; F4, UV-pass filter;
L1, Köhler field lens; L2, auxiliary lens for forming an image of the rear focal plane
of the objective in the plane of the exit slit of M2; M1, M2, monochromators; PM1, PM2,
PM3, photomultipliers of main, reference and third channels; QC, quantum converter; S1,
S2, shutters

a xenon arc lamp (Osram XBO 150 W/2) with a stabilized DC supply. The light is monochromated with the now obsolete Leitz prism monochromator, with separate prisms for the ranges 210-440 nm and 370-1100 nm. A supplementary glass filter is required, placed behind the monochromator to reduce stray light: during the measurement of emission spectra, UG1 glass and an excitation wavelength of about 360 nm are usually employed; for measuring excitation spectra, UG5 or BG12 is employed with the quartz and glass prisms respectively. A quartz lens with field diaphragm is used to focus the exit slit of the monochromator into the rear focal plane of the microscope objective, thereby giving approximate Köhler-type epiillumination. A nondichromatic beamsplitter is normally employed in the epiilluminator to avoid distortion of spectra. A portion of the light passing through the field diaphragm is split off by a quartz glass beamsplitter for the reference channel; the remaining light passes through a leaf shutter and enters the epiilluminator. Fluorescence from the specimen is collected by the objective and passes back up through the epiillumination beamsplitter. Fluorescence can be examined or photographed through a conventional barrier filter and ocular. A sliding prism is provided to divert the fluorescence to the measuring system. A lens (L2) images the exit pupil of the objective into the entrance slit of the second monochromator (M2), which is an integral part of the original microspectrograph. Alternatively, a negative lens can be used, in which case the object field is imaged into the entrance slit of M2; this is suitable for very small or linear objects.

Reference Channel

The beamsplitter consists of a quartz glass plate at 45° to the axis of the main beam. The proportion of light reflected by this beamsplitter at various wavelengths was computed from known refractive indices (Rost and Pearse 1971). In the reference channel, the light passes first through a diaphragm which corresponds in diameter and position with the entrance pupil of the epiilluminator. The light then impinges on a quantum converter, consisting of a quartz glass cuvette containing a concentrated solution of Rhodamine B, 3 g/l in ethylene glycol (Ritzén 1967). The cuvette, which is rectangular with light paths of 2 cm and 1 cm, is orientated so that the reference beam strikes one of the broad faces at an angle of 45°, the fluorescence of the Rhodamine B is then observed from the same face by a photomultiplier. A red barrier filter (3 mm RG 620) and an iris diaphragm are placed between the cuvette and the photomultiplier. The reference channel operates on the assumption that substantially all the light is absorbed by the Rhodamine B, at least for wavelengths below about 530 nm, and that a constant proportion of quanta is reemitted as red fluorescence. The reference photomultiplier is therefore "seeing" light of a constant spectral distribution, and measures the number of quanta entering the reference channel irrespective of their wavelength.

Accordingly, use of the reference channel allows corrected excitation spectra to be measured directly, except for a small correction for varying reflectivity of the beamsplitter at various wavelengths.

The Third Channel

For simultaneous measurement of fluorescence emission and of absorption at the exciting wavelengths, a third measuring channel is available. Light passing through the specimen can be collected by the substage condenser (used as an objective) and passed along a diaillumination pathway in reverse to a third photomultiplier. In this case the effective aperture of the objective, used as condenser, is limited by an aperture diaphragm in the epiilluminator, this limits the aperture of the illumination beam without affecting the aperture of the objective for picking up fluorescence emission. In order to prevent fluorescence from reaching the third-channel photomultiplier, a glass filter is placed in the third channel; this is usually BG12 or UG1, depending on the wavelength used for excitation. This third channel makes it possible to follow changes of quantum efficiency during kinetic studies, *e.g.*, of fading.

Electronic System

The photomultipliers in the main and reference channels are EMI type 9558QA, having a S-20 cathode and a quartz glass window (the latter is not necessary for the reference channel) with a usable spectral range of approximately 200-750 nm.

In earlier microspectrofluorometers, photomultiplier output was determined by measuring the DC current, either directly with a galvanometer (Nordén 1953) or chart recorder (Thieme 1966), or indirectly with a mechanical light chopper and a tuned amplifier (Caspersson et al.1965).In the present instrument, photomultiplier output is measured by photon counting. The electronic requirements of this technique in fluorometry have been described by Kemplay (1962). Each time an electron is emitted from the photomultiplier cathode as a result of a photon striking the cathode, a brief pulse is produced in the anode circuit. These pulses are amplified and counted for a preset time, using conventional equipment (Nuclear Enterprises (GB) Ltd., Edinburgh). Each channel consists of a preamplifier (type NE 5281) situated within the photomultiplier housing, a pulse amplifier (NE 5259) and a scaler (NE 5079). Both scalers are triggered simultaneously by a timer unit (NE 5058, modified), which gives counting times from 0.2 to 10^4 sec. In addition to the scalers, each pulse amplifier is connected to a ratemeter (NE 5456, modified), which produces an analogue signal approximately proportional to the count rate integrated with time constants variable from 12 msec to 100 sec. These analogue signals are fed to a simple computing device which produces an analogue signal proportional to the main channel reading divided by that of the reference channel, after subtraction of dark currents. This analogue signal is read on a meter and can also be fed to a chart recorder or the Y axis of an oscilloscope. Wavelength analogue signals from either of the monochromators can be fed to the X axis of the oscilloscope, so that excitation or uncorrected emission spectra can be plotted directly on the oscilloscope.

A print control unit transfers the digital data on command to a Teletype machine, which gives both a typewritten record and a punched paper tape for input to a computer, which in turn makes the necessary corrections. The data to be printed by the Teletype is selected by a switch, and may include not only counts from the main and reference channels, and (optionally) the time indicated by a digital clock, but also a digital voltmeter reading representing either the third-channel reading or a wavelength analogue.

The advantages of photon counting over DC measurements are well known in other fields, particularly astronomy. The main advantages for the present purposes are:

a) Very great sensitivity can be achieved, since in order to measure very dim light, one virtually need only increase the counting time.

b) A direct digital output is obtained, instead of one that has to be read from a scale; this is quicker, more convenient, and gives greater precision; and also makes it easy for automatic recording equipment to print the results and punch them on a paper tape.

c) Low-frequency noise, such as mains hum, does not affect the system.

d) This method gives a much truer picture of what is happening, since the light is in fact quantized. It is possible to calculate the probability of errors to be expected, in relation to a particular count, from the statistical distribution of photons.

From the above it will be evident that individual measurements of fluorescence intensity can be made with great speed and efficiency, since the photon-counting system has a "counting time" rather than a "response time and integration time". If the object is located by phase contrast, and the photon counting system is triggered by the opening of the shutter in the excitation beam, an intensity measurement can be made with only 22 msec of irradiation (the shortest counting time available with this equipment), during which all the pulses are counted and not wasted. For measurements of rapidly fading fluorescence, a photon counting system is greatly preferable to any analogue system, however rapid the response time of the measuring device. For recording fading, measurements can be made and recorded automatically at intervals of about 4 sec or longer.

Alternatively, the analogue output from the ratio computer can be fed to an oscilloscope, for recording extremely rapid fading.

Corrections

The quantum efficiencies of the photomultiplier cathodes were determined at various wavelengths by the National Physical Laboratory. The band width of the emission monochromator at various wavelengths was calculated from the dispersion curve for which data were supplied by the manufacturers. Corrections for the wavelength dependence of reflectivity of the reference-channel beamsplitter were calculated from known refractive indices. The linearity of the pulse-counting system was tested with neutral-density glasses, as follows. Using epiillumination, a small piece of front-aluminized mirror was used as an "object," reflecting light from the source into the microscope and thence through the second monochromator (set to the same wavelength) to the main channel photomultiplier. A piece of neutral-density glass (NG4, approximately 5% transmission) was divided into pieces, and the count rate measured with zero, one, and two thicknesses of the glass in the light path. From these data, the effective "dead time" of the counting system was calculated. (The "dead time" is the period after receipt of one pulse when another pulse will not be registered.) A value of 1.7 µsec was obtained, and used in subsequent correction of observed count rates. Although the expected nonlinearity has been presumed to be due to dead time, other causes of non linearity may contribute. Nonetheless, the method substantially corrects for non-linearity from all causes.

Data Processing

It has already been indicated that corrections have to be made on each reading for counter dead time reflectivity of the reference beamsplitter (excitation spectra only), and bandwidth of the emission monochromator (emission spectra only); it is also necessary to subtract dark-current readings and to divide the readings of the main channel by those of the reference channel. Finally, it is usually necessary to subtract a control reading (or, in the case of absorption measurements, to calculate percent transmission and optical density). For these calculations, the Elliott 4100 computer at the Royal Postgraduate Medical School Computer Center was utilized.

APPLICATIONS

As already indicated, the apparatus was intended primarily for the determination of excitation and emission spectra. This technique has been applied in a number of investigations, including the study of formaldehyde-induced fluorescence in melanomas (Rost and Polak 1969), confirmation of the uptake of L-dopa and 5-hydroxytryptophan by cells of the APUD series (Polak 1970); the study of autofluorescence of certain types of amyloid (Pearse, Ewen et al., 1972). It has also been utilized in the development of methods for masked metachromasia (Bussolati et al. 1969; Rost and Maunder 1971; Maunder and Rost 1972), and of new methods for the histochemical demonstration of catecholamines, tryptamines, and other arylethylamines by acid- and aldehyde-induced fluorescence (Rost and Ewen 1971; Ewen and Rost 1972).

Dynamic studies of changes in fluorescence with time have been applied to studies of fluorescence fading (Rost and Pearse 1968), and to the microfluorometric characterization of alkaline phosphatase in cryostat tissue sections, using α-naphthylphosphate as a fluorogenic substrate (Rost et al., 1970).

ACKNOWLEDGEMENTS

Construction and further development have been made possible by grants from the British Medical Research Council. Additional assistance from the Muscular Dystrophy Group of Great Britain, and the Wellcome Trust, is gratefully acknowledged. It is a pleasure to acknowledge also the cooperation of Ernst Leitz G.m.b.H. (Wetzlar) and E. Leitz (Instruments) Limited (London). Many modifications and additions to the original equipment were carried out by members of the staff of the Department of Biophysics and of the Workshop in the Royal Postgraduate Medical School.

REFERENCES

BJÖRKLUND, A., Ehinger, B., and Falck, B., (1968), *J. Histochem. Cytochem.*, <u>16</u>, 263-70.

BUSSOLATI, G., Rost, F. W. D., and Pearse, A. G. E., (1969), *Histochem. J.*, <u>1</u>, 517-30.

CASPERSSON, T., Lomakka, G., and Rigler, R., (1965), *Acta Histochem. Suppl.*, <u>6</u>, 123-26.

EWEN, S. W. B., and Rost, F. W. D., (1972), *Histochem. J.*, <u>4</u>, 59-69.

KEMPLAY, J. R., (1962), *Electron. Engng.*, <u>34</u>, 820.

MAUNDER, C., and Rost, F. W. D., (1972), *Histochem. J.*, <u>4</u>, 145-53.

NORDÉN, G., (1953), *Acta Path. Microbiol. Scand. Suppl.*, <u>96</u>.

OLSON, R. A., (1960), *Rev. Sci. Instrum.* <u>31</u>, 944-49.

PEARSE, A. G. E., Ewen, S. W. B., and Polak, J. M., (1972), *Virchows Arch. B. Zellpath.*, <u>10</u>, 93-107.

PEARSE, A. G. E., and Rost, F. W. D., (1969), *J. Microsc.*, <u>89</u>, 321-28.

POLAK, J. M., and Pearse, A. G. E., (1970), *Experientia*, <u>26</u>, 288-89.

RITZEN, M., (1967), *Cytochemical Identification and Quantitation of Biogenic Amines.*, M.D. Thesis. Stockholm.

ROST, F. W. D., and Ewen, S. W. B., (1971), *Histochem. J.*, <u>4</u>, 59-69.

ROST, F. W. D., and Maunder, C., (1971), *Proc. R. Microsc. Soc.*, <u>6</u>, 20.

ROST, F. W. D., Nägel, L. C. A., and Moss, D. W., (1970), *Proc. R. Microsc. Soc.* <u>5</u>, 76-7.

ROST, F. W. D., and Pearse, A. G. E., (1968), *3rd Int. Congr. Histochem. Cytochem. Reports.*, Springer-Verlag, New York., p.226.

ROST, F. W. D., and Pearse, A. G. E., (1971), *J. Microscopy.*, <u>94</u>, 93-105.

ROST, F. W. D., and Polak, J. M., (1969), *Virchows Arch. Path. Anat. Physiol.*, <u>347</u>, 321-26.

ROUSSEAU, M., (1957), *Bull. Micr. Appl.*, 7, 92-4.

RUCH, R., (1960), *Z. wiss. Mikrosk.*, 64, 453-68.

SPRENGER, E., and Böhm, N., (1971), *Histochemie* , 25, 163-76.

THIEME, G. A., (1966), *A cta Physiol. Scand.*, 67, 514-20.

WEST, S. S., (1965), *Acta Histochem. Suppl.*, 6, 135-53.

WEST, S. S., Loeser, C. N., and Schoenberger, M. D., (1960), *IRE Trans. Med. Electron.*, ME-7, 138-42.

Part III
Cytofluorometric Determination of Cellular Substances

ABSORBANCE AND FLUORESCENCE CYTOPHOTOMETRY OF NUCLEAR FEULGEN DNA. A COMPARATIVE STUDY

Norbert Böhm, Ernst Sprenger, and Walter Sandritter

Institute of Pathology, University of Freiburg,
West Germany

1. INTRODUCTION

Fluorescence cytophotometric measurements of intracellular substances, either those with native fluorescence or those treated with appropriate fluorescence staining are based upon physical laws that are different from those of absorbance cytophotometry.

Nevertheless, there is an obvious relation between the phenomenon of fluorescence emission and the absorption of light, since, at least under ordinary conditions, fluorescence can only occur when light has been absorbed by the fluorescent molecule. Because of some radiationless loss of energy from the excited molecule, the wavelength of the emitted radiation is usually longer than that of the absorbed light (Stokes' law).

2. THEORECTICAL CONSIDERATIONS

Provided that both reabsorption and fluorescence quenching of the emitted fluorescence light are negligible, there is a linear proportionality between the fluorescence intensity F and the radiant energy absorbed. This proportionality is equivalent to the difference between the radiant energy of the incident I_0 and the transmitted part I of the excitation beam.

$$F = (I_0 - I) \, Q \qquad (1)$$

The proportionality coefficient Q resembles the fluorescence quantum yield. It may be expressed as the quotient of the light quanta emitted per light quanta absorbed.

From Lambert-Beer's absorption law I may be written as

$$I = I_0 \, e^{-cd\varepsilon} \qquad (2)$$

so that Eq. (1) will read

$$F = (1 - e^{cd\varepsilon}) \, I_0 \, Q \qquad (3)$$

where c is the concentration of the fluorescent substance, d the thickness and ε the molar absorbance coefficient. The product of these three factors resembles the absorbance $E \, (= cd\varepsilon)$.

In Eq. (3) an exponential relation is shown between fluorescence intensity F and concentration of the fluorescent substance c which is presented diagrammatically in Figure 1.

If, on the other hand, the fluorescence intensity F is expressed in terms of the transmission $T \, (= e^{-cd\varepsilon})$ or absorption $A \, (= 1 - T)$, a linear proportionality is achieved;

$$F = (1 - T) \, I_0 \, Q \qquad (4)$$

$$F = A \, I_0 \, Q \; . \qquad (5)$$

This, however, has no meaning for fluorescence cytophotometry, since *not absorption but absorbance is directly related to the concentration* of the fluorescent substance. In an inhomogeneous structure, such as a Feulgen stained nucleus, the densities, and hence the fluorescence intensities, are different at various sites of the nucleus,

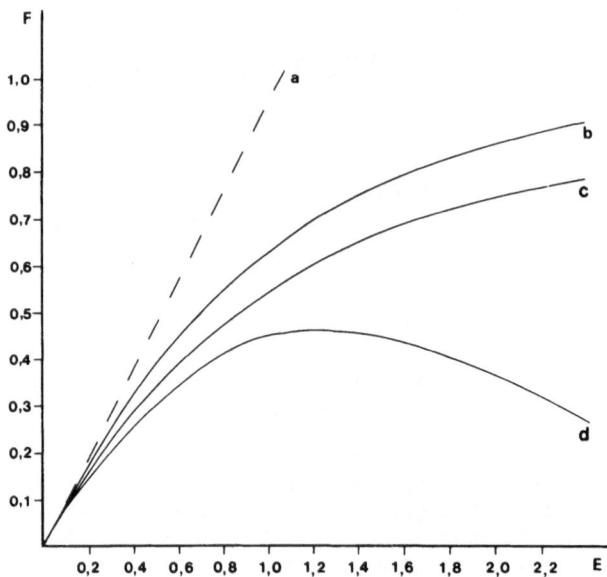

Figure 1. *Schematic presentation of the relation between concentration of the fluorescent substance (in terms of absorbance E) and the relative fluorescence intensity F. Curve a shows an "ideal" linear relationship that is not achievable. Curve b represents the exponential function according to Eq. (3) without reabsorption and fluorescence quenching. The curve was calculated under the arbitrary assumption of $I_0 \cdot Q = 1$. Curve c (excitation by epi-illumination) and curve d (excitation by substage illumination) indicate the condition under which reabsorption has to be considered*

which have to be measured at the same time. The sum total of the different fluorescence intensities within the measuring field a is recorded as "total fluorescence," expressed in fluorescence units *FU* (in analogy to absorbance or arbitrary units *AU* standing for "total absorbance" in absorbance cytophotometry)

$$FU = \sum_{x=0}^{a} F\, x \quad .$$ (6)

This, however, requires linearity between fluorescence intensity and concentration, otherwise *FU* will not be proportional to the amount of fluorescent material that should be determined.

Therefore, either we have to integrate the logarithms of the different fluorescence intensities at various sites of the inhomogeneous object (which is exactly as time consuming and expensive as integrating microdensitometry), or we have to limit the fluorescence measurements to an absorbance range, in which the exponential curve yields an approximate linearity.

3. EXPERIMENTAL DATA

We did some experiments to determine the upper limits of absorbance in the excitation beam; in order to avoid errors in quantitative fluorescence cytophotometry these limits should not be exceeded. After acriflavine-Feulgen staining, the fluorescence intensities F, obtained with epi- and substage illumination, and absorbance E were measured in pairs from various nuclear plugs (field size about $2\mu m^2$) with different densities. The experimentally established curves (Figure 2 a) reveal the exponential relationship to be expected from Eq. (3) and curve b in Figure 1, while the plot of fluorescence versus absorption A, calculated from the same measurements, results in a linear function (Figure 2 b) in accordance with Eq. (5). Thus, the experimental data confirm the previous theoretical considerations. In addition, an approximately linear proportionality between concentration (in terms of absorbance E) and fluorescence intensity F up to absorbance values of 0.4 is discernible in Figure 2 a. Similar results were obtained from coriphosphine-Feulgen stained specimens (Böhm 1970 and 1972).

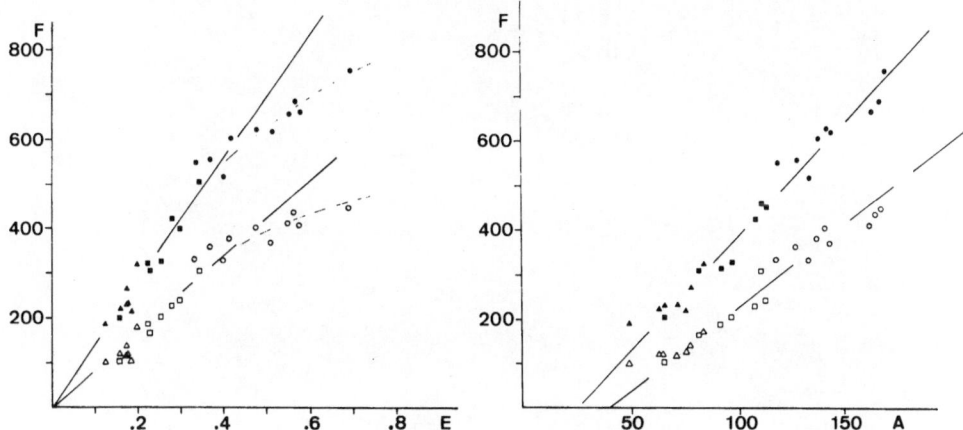

Figure 2. *Acriflavine-Feulgen staining: Comparison of plug measurements*
(a) Fluorescence F versus absorbance E
(b) Fluorescence F versus absorption A
• *Ascites tumor cells,* ■ *liver nuclei and* ▲ *sperm excited with epi-illumination and substage illumination* (o □ Δ). *Light source for fluorescence excitation: XBO 75 W (Osram). Excitation filter BG 12/BG 38*

This absorbance limit of 0.4 is surprisingly high when compared with the 5% absorption ($E = 0.02$) allowed for fluorometric determinations of solutions. It may, be explained however, by lack of dynamic quenching, which plays an important role in fluorescence measurements of solutions, but does not in those of solids or under conditions in which the fluorescent molecule is rigidly and covalently bound to a macromolecular tissue component.

In order to omit absorbance values higher than 0.4, we kept the dye concentrations as low as 0.01% in the fluorescence Schiff-type reagents prepared with acriflavine, coriphosphine O, and acridine yellow. These acridine dyes were found to be excellent substitutes for pararosaniline in the Feulgen reaction for nuclear DNA (Böhm and Sprenger, 1968). Acriflavine-Schiff had already been used by Culling and Vassar (1961) to increase the sensitivity of the Feulgen reaction and by Prenna (1964) for quantitative fluorescence measurements of nuclear DNA.

When the low dye concentrations of 0.01% were used, even with blue excitation near the absorption maximum of the acridine dyes at λ460 nm, only a few sites were found with extinction values higher than 0.4; and these were usually measured from pyknotic or clumped nuclei. We may, therefore, conclude that our fluorometric data are obtained under safe conditions. This is also evident from the histograms and the regression lines presented subsequently.

4. SPECIFICITY OF THE FLUORESCENCE FEULGEN REACTION

The specificity of the Feulgen staining with acriflavine- (Figure 3, see Plate I), coriphosphine-, and acridine yellow-Schiff is as good as with the conventional pararosaniline-Feulgen, when approporiate barrier filters with the transmission beyond λ590 nm are applied that extinguish the faint green nonspecific fluorescence of the cytoplasm (Böhm, 1972). Furthermore, the introduction of the barrier filter limits the fluorescence measuring range to the long wavelength part of the spectrum, so that no reabsorption of the fluorescence radiation can occur by the nonexcited molecules (Figure 4).

PLATE I

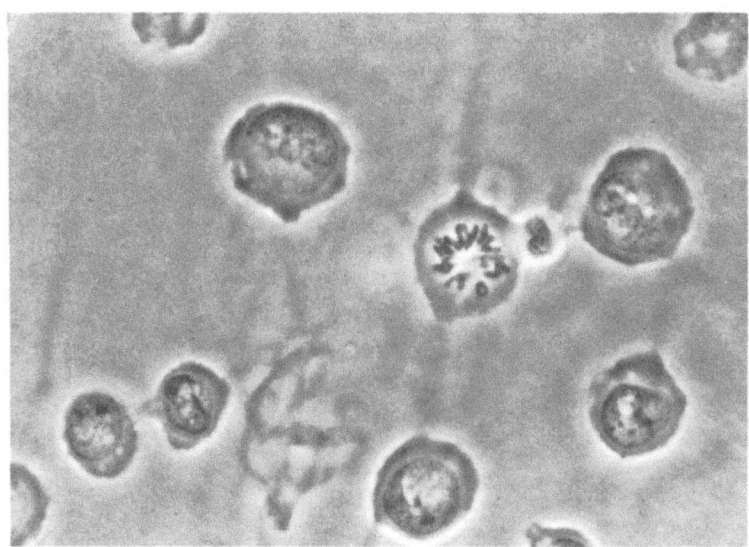

Figure 3. *Acriflavine-Feulgen stained ascites tumor cells. Leitz-MPV.*
(a) Phase contrast photo micrography (900 X) showing the nuclear chromatin as well
as a rim of cytoplasm around each nucleus. Next to the center a metaphase plate

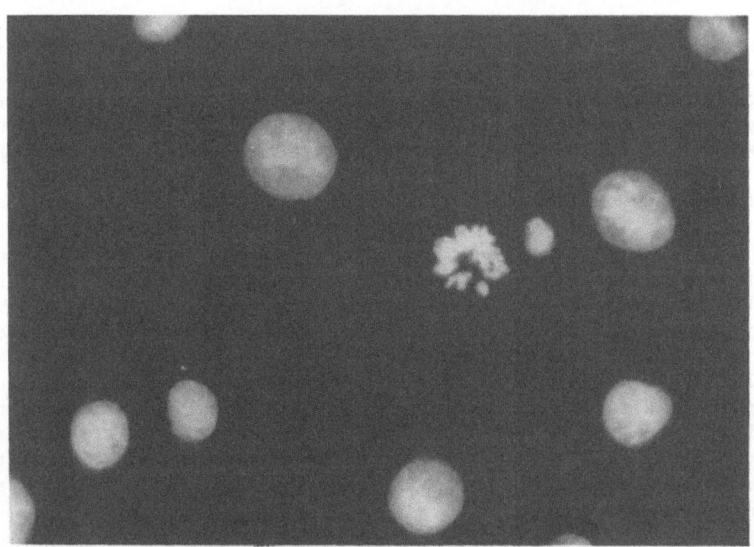

(b) Same area as (a). Bright red specific fluorescence of the chromatin. Blue
excitation with BG 12/BG 38. Barrier filter λ590 nm

Besides these important measuring conditions another essential requirement has to be fulfilled for quantitative fluorescence cytophotometry: the elimination of fluorescence fading, a phenomenon inherent to virtually all organic fluorescent compounds. Because of the rapid photo-fading, the time of exposure to short-wavelength excitation must be kept as brief as possible. This may be achieved by a photoshutter in connection with a fully automated short-time measuring and recording unit such as we have developed in our laboratory (Sprenger and Böhm, 1971).

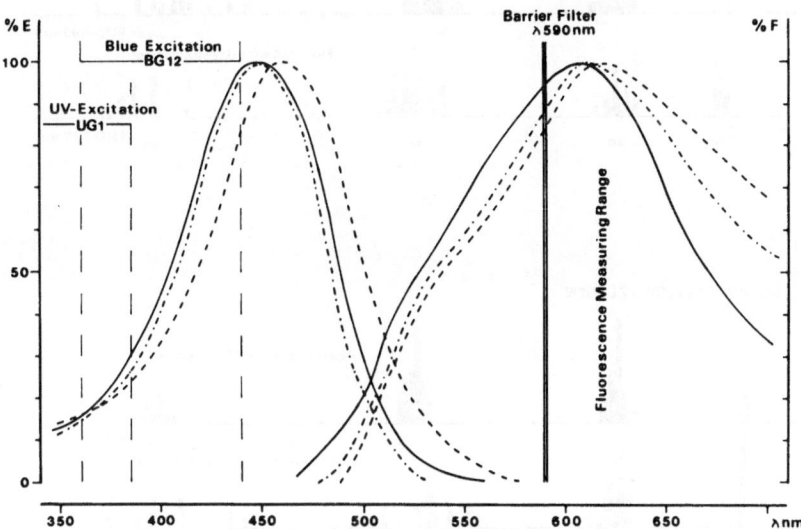

Figure 4. *Spectral absorption curves (left, not corrected for fluorescence) and emission curves (right) from acridine-yellow (———), coriphosphine- (-•-•-) and acriflavine-Feulgen (-----) stained liver nuclei. % E relative absorbance. % F relative fluorescence intensity. The spectral transmission range of the barrier filter λ590 nm is entirely outside the spectral absorption range of the fluorochromized nuclei. Thus, no reabsorption can influence the results. From Böhm (1972)*

5. COMPARISON OF HISTOGRAMS

With strict adherence to all the previously mentioned measuring conditions, the following comparative measurements were performed with acriflavine-, coriphosphine-, and acridine yellow-Feulgen stained liver smears and guinea pig sperm. The fluorescence determinations were done first with blue excitation (Schott filters BG 12/BG 38) as well as UV-excitation (UG 1/BG 38), followed by the absorbance measurements using the integrating microdensitometer (Deeley, 1955). The histograms in Figure 5 reveal corresponding DNA distribution patterns with each of the different staining and measuring conditions; the haploid sperms are grouped around the 1 c-line, the tetraploid liver nuclei around 4 c, and the octoploid nuclei are scattered around 8 c; the diploid (2 c) liver nuclei were chosen as reference values for normalization of the histograms (Böhm and Sprenger, 1968).

6. REGRESSION LINES

In another more detailed approach we compared, two by two, the fluorescence and absorbance readings obtained from the same nucleus - a comparison that was achieved by mapping the smears. Ascites tumor cells were chosen as the objects for measurement, because of the wide range of their DNA values. The results obtained from an acridine yellow-Feulgen stained specimen are presented in Figure 6, showing straight regression lines with correlation coefficients r greater than 0.98 when fluorescence measurements

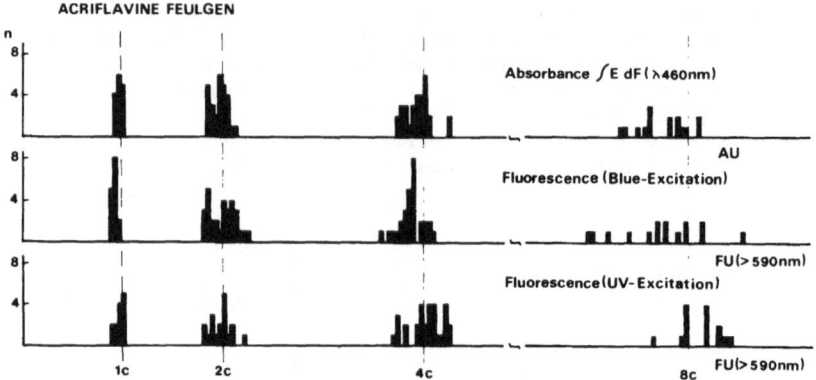

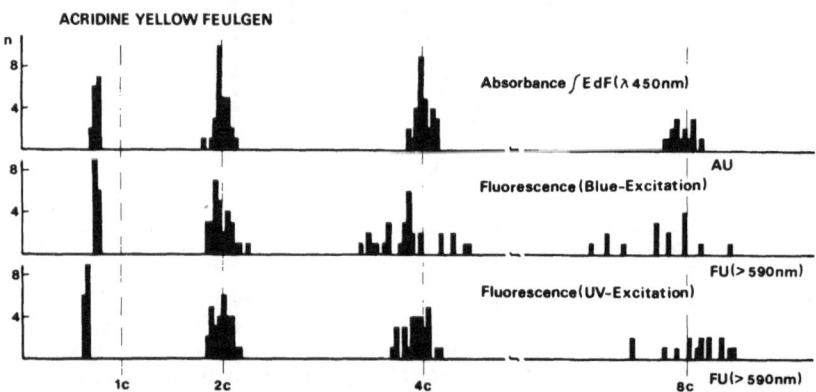

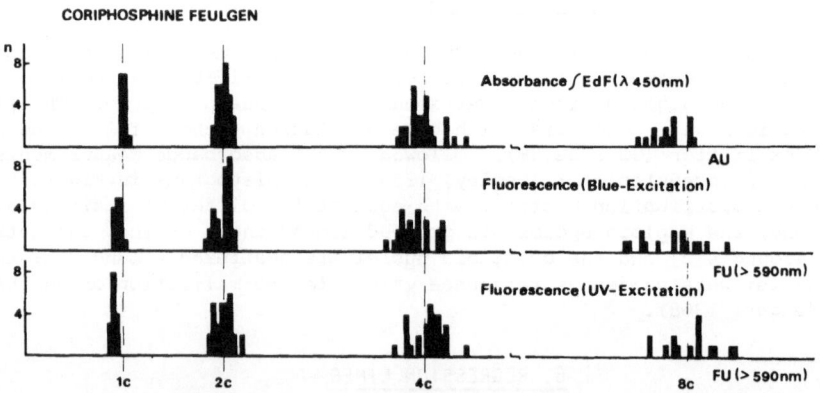

Figure 5. *Acriflavine-, acridine yellow- and coriphosphine-Feulgen stained sperm, di-, tetra- and octoploid liver nuclei. Histograms of the normalized absorbance values (AU) and fluorescence values (FU) obtained from one slide under different measuring conditions. The mean of the diploid population (2 c = 100%)was chosen as reference value for normalisation. n represents the number of nuclei (relative frequency). From Böhm and Sprenger (1968)*

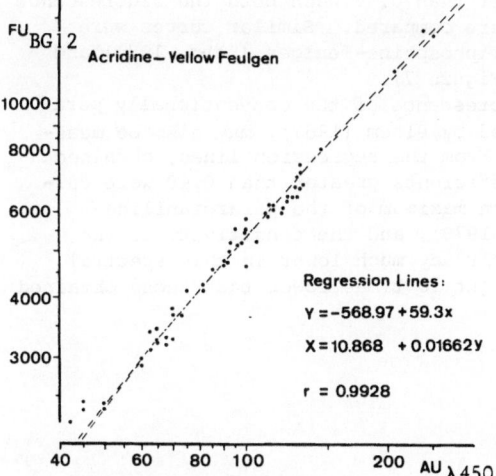

Figure 6. *Acridine yellow-Feulgen stained ascites tumor cells.*

(a)   Fluorescence blue excitation ($FU_{BG\ 12}$) versus integrated extinction at $\lambda 450$ nm ($AU_{\lambda 450}$)

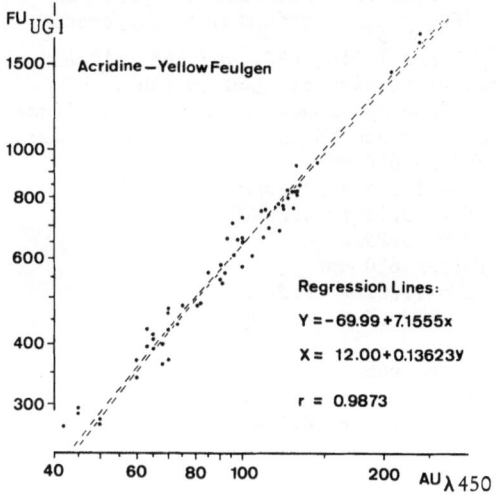

(b)   Fluorescence UV excitation ($FU_{UG\ 1}$) versus integrated extinction at $\lambda 450$ nm ($AU_{\lambda 450}$)

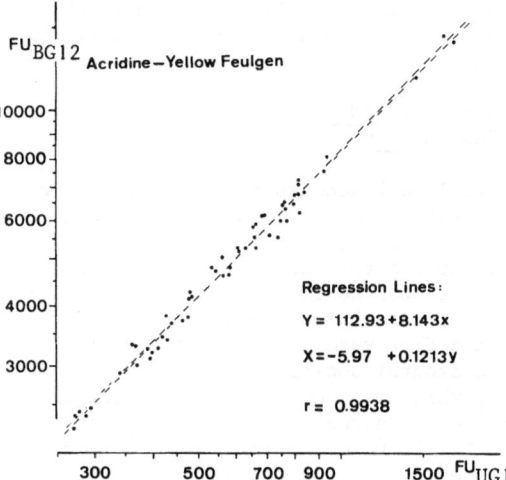

(c)   Fluorescence blue excitation ($FU_{BG\ 12}$) versus fluorescence UV excitation ($FU_{UG\ 1}$)

were compared with absorbance values, and greater than 0.99 when both the fluorescence readings obtained with blue and UV excitation were compared. Similar curves were found with acriflavine-Feulgen (Böhm, 1972), coriphosphine-Feulgen (Böhm, 1970) and with pararosaniline-Feulgen stained specimens (Figure 7).

We were able to show that the deep red fluorescence of the conventionally pararosaniline-Feulgen stained nuclei, first observed by Ploem (1967), may also be measured quantitatively (Böhm and Sprenger 1968). From the regression lines, obtained with different barrier filters, correlation coefficients greater than 0.98 were calculated (Figure 7). Since, however, the emission maximum of the pararosaniline-Feulgen stained nuclei is beyond $\lambda 700$ nm (Böhm 1970), and the sensitivity of the photocathode of the photomultiplier applied is already much lower in this spectral range, the absolute fluorescence readings are about 10 times lower than those obtained with the schiff-type acridine dyes.

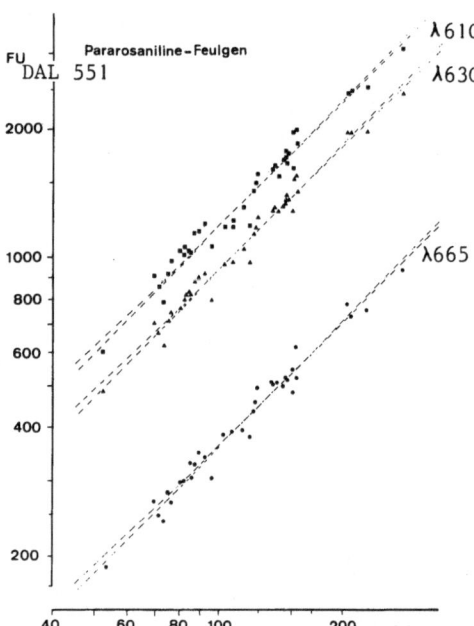

Figure 7. *Pararosaniline-Feulgen stained ascites tumor cells. Fluorescence green excitation (FU$_{DAL\ 551}$)obtained with different barrier filters ($\lambda 610$, 630, and 665 nm)versus integrated extinction at $\lambda 560$ nm (AU).*
From the measuring values the regression lines and correlation coefficients were calculated:
Barrier filter 610 nm:
$$Y = 51.22 + 11.250x$$
$$X = 0.13 + 0.0866$$
$$r = 0.9813$$
Barrier filter 630 nm:
$$Y = 41.21 + 8.867x$$
$$X = 0.65 + 0.109y$$
$$r = 0.9841$$
Barrier filter 665 nm:
$$Y = 15.55 + 3.430x$$
$$X = 1.06 + 0.280y$$
$$r = 0.9839$$

These experimental results demonstrate and prove the comparability and reliability of both the measuring methods, absorbance and fluorescence cytophotometry, for the quantitative determination of nuclear Feulgen DNA. What after all, we may now ask, is the decisive advantage of fluorescence cytophotometry over absorbance measurements?

7. APPLICATION OF FEULGEN FLUORESCENCE FOR "FLOW-THROUGH CYTOPHOTOMETRY"

Integrating absorbance readings require about 1 to 2 seconds per nucleus at the Barr and Strought microdensitometer, whereas fluorescence measurements may be obtained in an extremely short measuring time, amounting to "split seconds." This made possible the development of an automated flow-through cytofluorometer (scheme is presented in Figure 8), in which the fluorochromized nuclei are suspended in water and fed through the focal plane of the objective by means of a longitudinal fluid stream carrying the stained cells along the axis of the microscope at a flow rate of 100 to 1000 cells per second. Finally they are flushed off by a horizontal suction current. While passing through the focal plane of the objective, which is focused on the upper rim of platinum-iridium diaphragm that terminates the cell channel at its opening into the suction channel, the nuclei are excited and the fluorescence impulse height of each single nucleus is recorded and classified by an impulse-height discriminator.

From the discriminator the data may be recalled either digitally, or analogically as a histogram that is finally automatically drawn on a paper chart (Sprenger et al. 1971 a; Sprenger 1971).

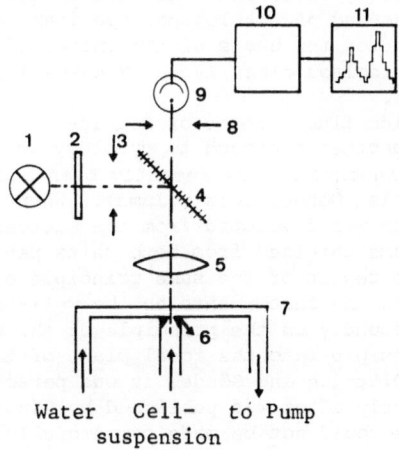

Figure 8. *Technical layout of the flow-through cytofluorometer. 1 light source, 2 excitation filter, 3 field diaphragm, 4 dichromatic beam splitter, 5 objective, 6 platinum- iridium diaphragm, 7 flow-through chamber, 8 measuring diaphragm, 9 photo-multiplier, 10 impulse-height discriminator, 11 analogue tracer or digital printer. From Sprenger et al. (1972)*

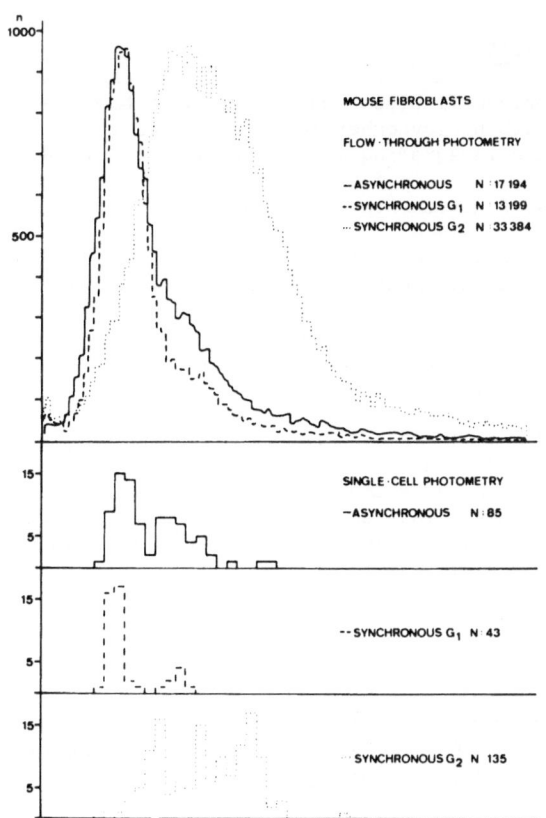

Figure 9. *Acriflavine-Feulgen DNA-histograms of mouse fibroblast cultures growing asynchronously, and synchronously in G1 or G2 phase of the cell cycle. The histograms obtained with flow-through cyto-fluorometry correspond well to the respective DNA distribution patterns measured by static single-cell cyto-fluorometry of slide specimens on which the tissue culture cells were directly grown*

A typical DNA distribution pattern obtained from our flow-through cytofluorometer is presented in Figure 9. The static fluorescence measurements of the same
material (fibroblast cultures) reveal corresponding histograms in each of the different growth conditions.

Flow-through cytofluorometry of acriflavine-Feulgen stained material has been
used successfully for prescreening of cytological specimens obtained by cervical
scrapes with the Ayre-spatula, on the basis of the increased nuclear DNA content of
atypical cells found in Papanicolaou class IV and V cases (Sprenger et al. 1971 b
and c; Sprenger et al. 1972).

Based upon ethidiumbromide fluorescence of double-stranded DNA, Dittrich and
Göhde (1969) have described another approach to cytology automation by means of
"Impulsfluorometrie". Their instrument has recently been made commercially available by
Phywe-Comp., Göttingen (see also Göhde, this volume). When compared with our flow-
through cytofluorometer, which was developed from the microscope photometer MPV (Leitz-
Comp., Wetzlar), the histograms obtained from both units reveal a very similar shape.
This similarity is probably a result of the same principle of construction of the flow
channels, which depend in both the instruments on the well-known principle of the
aspirator, or still more profoundly on the principle of the water suction pump. The
idea to incorporate this principle into the focal plane of the objective of a microscope photometer belongs to Dittrich and Göhde, it was personally communicated to our
group (Dr. Sprenger) and shortly after was published by Göhde (1971). Since the flow
chamber of Dittrich and Göhde could not be made commercially available to us at that
time, we were forced to develop our own. Our chamber was successfully achieved by the
introduction of a platinum-iridium diaphragm with an extremely even opening of 100 μm
(Sprenger 1971; Sprenger et al. 1971 a), as is used in electronmicroscopy, instead
of the quartz capillaries handmade by Göhde and Dittrich.

Designs for highly sophisticated and very useful devices for automated microfluorometry of biological cells have also been published by Kamentsky (1965, 1970) and
van Dilla et al. (1969).

8. CONCLUSIONS

Quantitative determinations of DNA by Feulgen fluorescence cytophotometric measurements have proved to be as reliable as absorbance cytophotometric estimations of
the nuclear Feulgen dye content by means of the integrating microdensitometer. Moreover, fluorescence cytophotometric measurements are much faster and hence superior to
conventional absorbance cytophotometry. Thus, fluorescence cytophotometry has gained
scope in two important areas of application:

(a) *Measurements of single cells and nuclei in smears or imprints* of any tissue
or cytological materials, which exhibit an excellent preservation of cellular micro-
phology. This allows Feulgen DNA determinations from cells and nuclei that have been
selectively identified under phase-contrast observation. For example, dyscariotic
and malignant cells in cervical smears, may even be stained and classified according
to Papanicolaou, before the Feulgen staining and measuring of the same slide is performed in a subsequent step (Böhm et al. 1971).

(b) *Flow-through cytofluorometry for rapid analysis of large cell populations*
amounting to several thousand measurements of individual cells, including also simultaneous recording of different cellular parameters as described by Kamentsky and Melamed
(1969), opens the tentative possibility of automated mass screening for cervical cancer.
However, the applicability of this procedure has to be carefully tested in field trials,
before the practical usefulness of the method can be judged conclusively.

REFERENCES

BÖHM, N., (1970), *Fluoreszenzcytophotometrie,* Professorial Thesis, Freiburg.

BÖHM, N., (1972), In *Techniques of Biochemical and Biophysical Morphology,* D. Glick and
R.M. Rosenbaum, (eds)., Wiley-Interscience, New York, *89-141.*

BÖHM, N., and Sprenger E., (1968), *Histochemie,* 16, 100-18.

BÖHM. N., Sandritter, W. and Sprenger E., (1970), In *Cytology Automation. Proc. Sec. Tenovus Symp., Cardiff, Sept. 24-25, 1968*, D.M.D. Evans, (ed), Livingstone, Edinburgh and London, 48-54.

BÖHM, N., Roka, S., Sprenger, E., and Wagner, D., (1971), *Acta Histochem., Suppl. X*, 233-42.

CULLING, C., and Vassar, P., (1961), *Arch. Pathol.*, 71, 88/76-92/80.

DEELEY, E., (1955), *J. Sci. Instrum.*, 32, 263-67.

van DILLA, M. A., Trujillo, T. T., Mullaney, P.F., and Coulter, J. R., (1969), *Science*. 163, 1213-14.

DITTRICH, W., and Göhde, W., (1969), *Z.f. Naturforschung*, 24 b, 360-61.

GÖHDE, W., and Dittrich, W., (1971), *Acta Histochem., Suppl. X*, 429-37.

KAMENTSKY, L. A., (1965), *Science*, 150, 630-31.

KAMENTSKY, L. A., (1970), *In Cytology Automation. Proc. Sec. Tenovus. Symp., Cardiff, Sept. 24-25, 1968* D.M.D. Evans, (ed.), Livingstone, Edinburgh and London. 177-85.

KAMENTSKY, L. A., and Melamed M. R., (1969), *Ann. N.Y. Acad. Sci.* 157, 310-23.

PLOEM, J. S., (1967), Z. *wiss. Mikr.* 68, 129-42.

PRENNA, G., and Bianchi, U. A., (1964), *Riv. Istoch. norm. pat.* 10, 667-76.

SPRENGER, E., (1971), *Fluoreszenzzytophotometrische DNA-Messungen in der zytologischen Diagnostik*. Professorial Thesis, Freiburg.

SPRENGER, E., and Böhm N., (1971), *Acta Histochem., Suppl. X.* 243-46.

SPRENGER, E., Böhm, N., and Sandritter, W., (1971a), *Histochemie*, 26, 238-57.

SPRENGER, E., Böhm, N., and Sandritter, W. W., (1971b), *German Medicine*, 2, 47-48.

SPRENGER, E., et al.,(1971c), *Beitr. Path.* 143, 323-44.

SPRENGER, E., et al, (1972), *Acta Cytologica*, 16, 297.

AUTOMATION OF CYTOFLUOROMETRY BY USE OF THE IMPULSMICROPHOTOMETER

Wolfgang Göhde

Institute of Radiation Biology
University of Münster, Münster/Westfalen, Germany

INTRODUCTION

For the automation of microscope fluorometry, two different principles can be applied. First, it is possible to measure fluorescent particles deposited on slides using a scanning microscope photometer. Secondly, chemical and physicochemical properties of fluorescent particles can be measured by means of an automatic flow-through device. For studying the proliferation kinetics of cells exposed to physical (e.g., ionizing or nonionizing radiation) or chemical treatment (e.g., cytostatica), flow-through methods are very suitable because of their high measuring speed. The same is true for cytopathological problems.

Two different flow-through principles in microscope fluorometry are possible for relating flow direction to the optical axis of the microscope:

a) The suspension is led perpendicularly to the optical axis of the microscope photometer.

b) The suspension moves parallel to the optical axis through the focal plane of the objective.

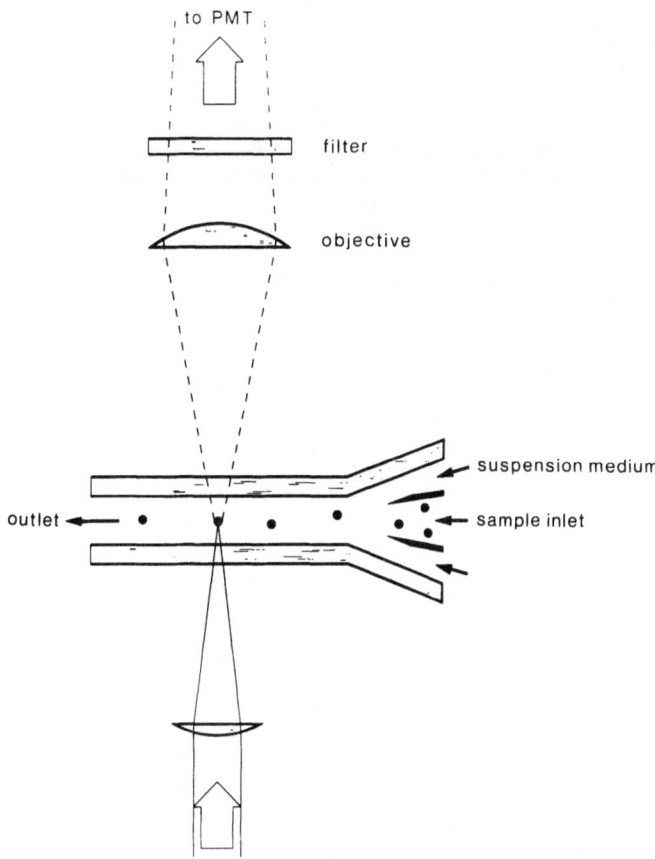

Figure 1. *Flow-through system introduced by Crosland-Taylor (1953) as laminar flow system. The suspension is led perpendicular to the optical axis of the microscope photometer. The particles do not necessarily pass the microscope axis in the plane focused by the objective*

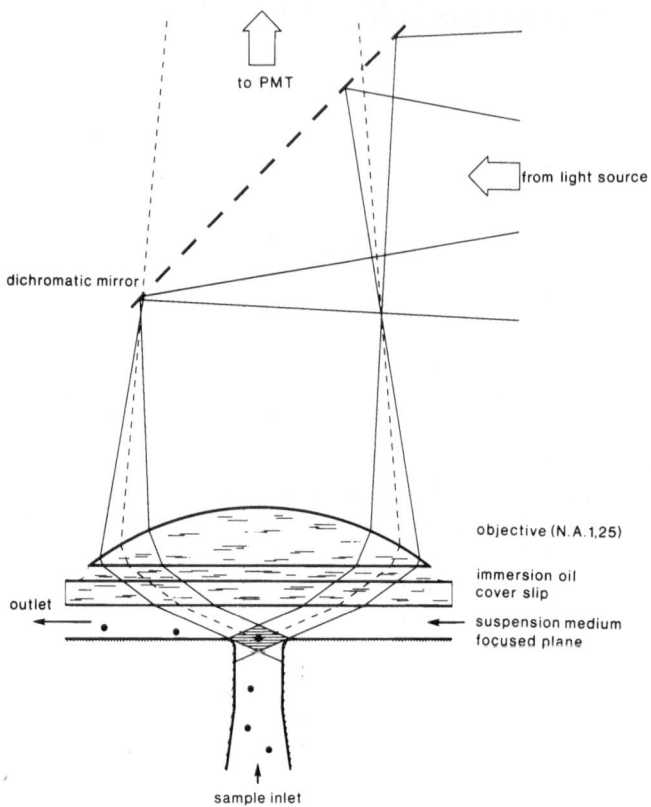

Figure 2. *In this flow-through system the suspension runs in the microscope axis and is fed into a particle-free laminar stream in the plane focused by the object- ive. Thus, each particle must pass the homogeneously illuminated zone (//////) and the focused plane before it is swept along by the particle-free laminar stream (for instance, suspension medium). Microscope objectives with the highest available numerical aperture may be used. For this system a flow-through chamber with inter- changeable cover slip was developed (Göhde 1972, 1973)*

In Figures 1 and 2 both technical principles are illustrated. Figure 1 shows the arrangement introduced as laminar flow system by Crosland-Taylor (1953) for cell counting, and recently applied to fluorescence measurement (van Dilla 1969; Traganos 1972). The particles move perpendicularly to the optical axis, but not necessarily in the focused plane if the diameter of the particle suspension stream in the central part of the laminar flow significantly exceeds the focal depth of the microscope ob- jective. Passage of particles above or below the focused plane causes measuring er- rors. It is possible to reduce these errors by using objectives with lower numerical apertures and thus larger focal depth. However, this reduces the energy flux per unit area of exciting radiation in the object plane and the percentage of emitted fluores- cence available for measurement. This loss of fluorescence yield can be at least partially be compensated for by using a laser as light source (van Dilla et al. 1969; Traganos et al. 1972).

Figure 2 shows a diagram of the flow-through system developed in our laboratory (Dittrich and Göhde 1970; Göhde and Dittrich 1972). Using the Köhler illumination principle, this system provides for a homogeneously illuminated area filling the fo- cused exit of the suspension capillary. The main advantage of this method is that no focusing problems are encountered: each cell passes the homogeneously illuminated measuring area and thus also the focused plane. Furthermore, in applying this flow- through principle, the same objective has to be used for incident light excitation and for measuring the intensity of emitted fluorescence. This requires objectives free of chromatic focus difference between the spectral excitation and emission ranges, but simplifies the adjustment of focal planes of the illuminating and measuring system

PLATE I

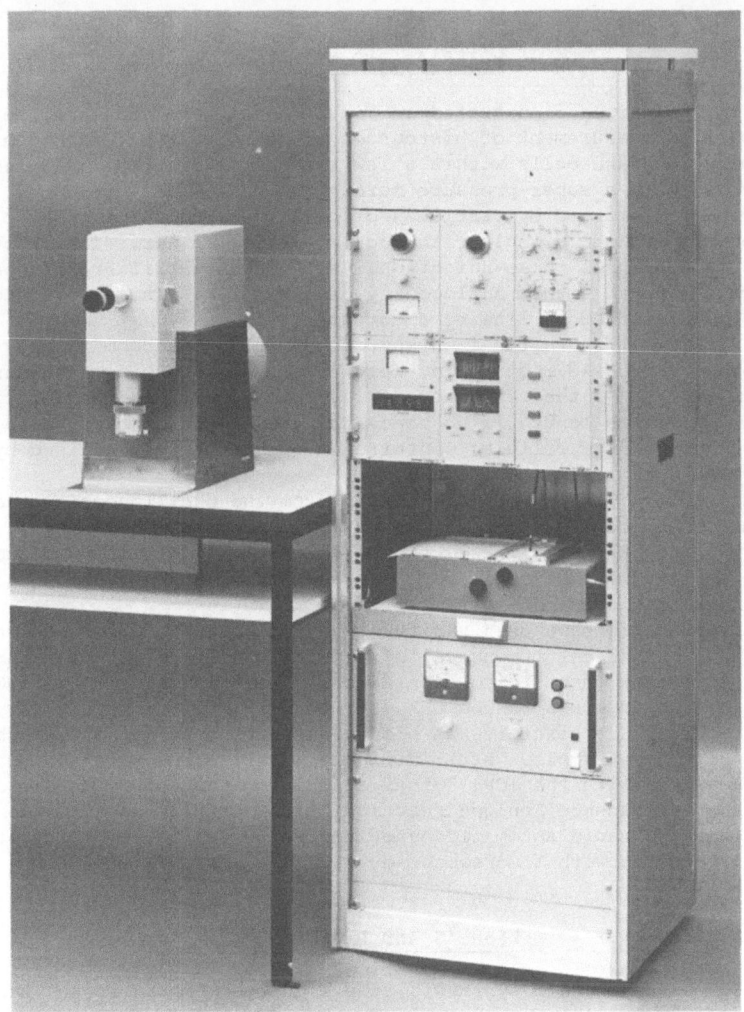

Figure 3. *Impulscytophotometer (ICP II Phywe AG, Göttingen). The optical/mech-anical part consists of the flow-through system, the super-pressure mercury lamp (HBO 100 W/2, Osram), the optical components for incident-light illumination - including the microscope objective - one or two photomultiplier tubes (for simultaneous measure-ment of two different parameters), an ocular for controlling the measuring area, and a suction pump for the flow-through chamber. The electronic part includes the power supply for the super-pressure mercury lamp, the high-voltage supply for the photo-multiplier tube, an amplifier with variable threshold, a peak detector, the multi-channel analyzer, an X-Y-recorder, an oscilloscope for displaying the data stored by the multichannel analyzer, an oscilloscope for monitoring the measuring signals during operation, a scaler, a rate-meter and attachments for digital data read-out or for on-line connection to a computer*

to each other. Since the numerical aperture of the objective is about 1.3, the solid
angle for excitation and uptake of fluorescence is larger than 130°. For these
reasons a conventional super-pressure mercury lamp is sufficient in most cases.

MATERIAL AND METHODS

Applying the principle illustrated in Figure 2, we have developed an automated
system which permits measurement of histograms of quantitatively fluorochromed cellular
compounds of many thousand cells within a few seconds or minutes. Fluorescence excitation is provided by a super-pressure mercury lamp.

The fluorescence emitted by particles or cells passing the focused plane reaches
the photocathode of a photomultiplier through a barrier filter. The electrical pulses
are fed into a multichannel pulse-height analyser via an amplifier and peak detector.
The data can be presented either analogously or digitally. The impulscytophotometer
(ICP) can also be connected on line to a computer.

Figure 3 (Plate I), shows the impulscytophotometer as a commercially available
piece of equipment (PHYWE AG, Göttingen, West Germany). By using a measuring area
with a diameter of 120 μm the cells move through the measuring chamber with a speed of
about 3 m/sec. The pulse length is 40 μsec. and the height of each pulse is proportional to the amount of the cellular constituent quantitatively stained by an appropriate fluorescent dye.

In Figure 4 (Plate II), a photograph taken from the oscilloscope screen shows
signals from human leucocytes stained with ethidium bromide (see below). Since most
of the leucocytes are in the G_1 phase, the majority of signals are expected to give
about the same amplitude. The oscilloscope traces in Figure 4 justify this expectation.
The noise-to-signal ratio is insignificant.

For our investigations we used various types of cells (human leucocytes, bone
marrow, skin, Ehrlich ascites tumour cells). Cells are fixed in 96% ethanol, treated
with RNase and stained with ethidium bromide (EB) (Dittrich and Göhde 1969). EB
specifically binds to double-stranded DNA (Le Pecq 1967). With only very few exceptions we obtained very satisfactory DNA histograms with this staining technique. The
staining with EB is very simple. After mixing the cell suspension and EB solution,
the cells are measured with the ICP. Other staining methods for DNA are also possible;
for example, the fluorescence Feulgen reaction using acriflavine (Prenna 1964) as
fluorochrome. For the rapid automated cytofluorometry, of course, many other cellular
substances are stainable with fluorescent dyes.

RESULTS AND DISCUSSION

In contrast to the protein synthesis, the amount of DNA increases discontinuously
in mammalian cells during the exponential growth. Therefore, the fractions of cells
in G_1, S, and G_2+M phases can be determined from DNA histograms of normal cells. The
ratio of cells in G_1, S, and G_2+M phase immediately gives information on the proliferation kinetics of the cell population.

In Figure 5 the DNA histogram is given for Ehrlich ascites tumor cells, harvested
eight days after intraperitoneal inoculation into an adult femal NMRI mouse. The relative amount of cells in G_1, S, and G_2+M can be obtained from the histogram by plan-
imetry or simple calculation. Furthermore, these histograms may be used for estimating
the relative duration of the different phases of the cell cycle.

Various possible applications of rapid automated cytofluorometry for biological
and medical research are:

Application of the ICP in Radiation Biology

For both radiobiological investigations and radiation therapy as well it is important to know the composition of the irradiated cell material with regard to the
different phases of cell cycle, because of their very different response to irradiation. In radiobiological experiments, the ICP permits rapid determination of this
composition of cell populations under investigation. An example is shown in Figure 6.
Ehrlich ascites tumor cells irradiated with neutrons *in vivo* were harvested after

PLATE II

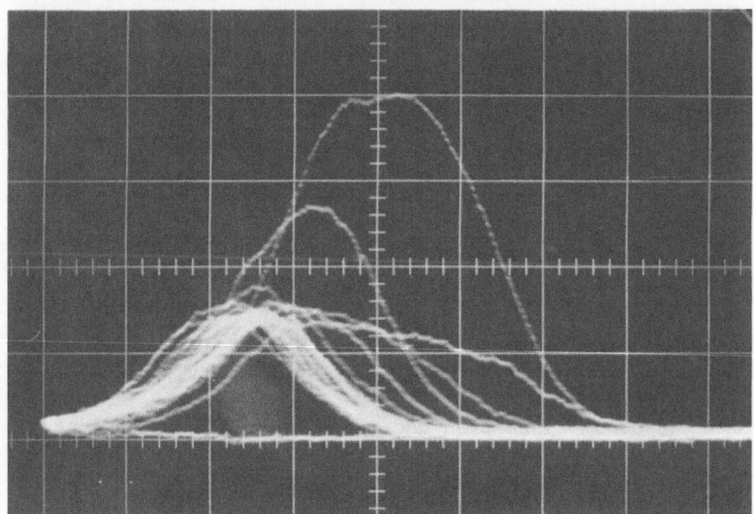

Figure 4. *Photograph taken from the monitoring oscilloscope shows the fluorescence signals of human leucocytes stained with Ethidiumbromide. Most of the signals have about the same amplitude. 1 unit of time base = 10 μ sec*

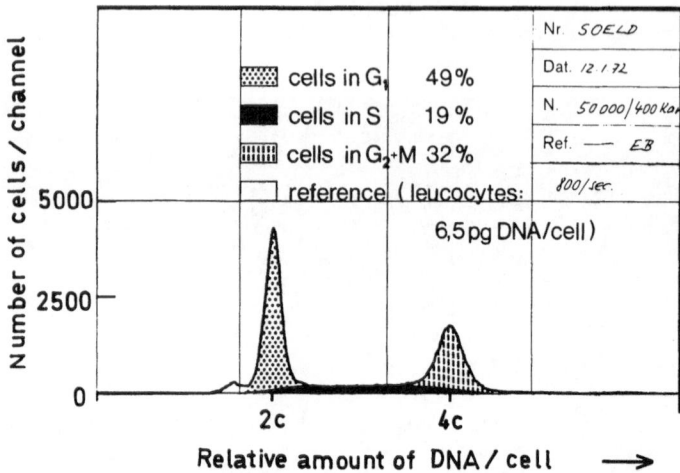

Figure 5. *DNA histogram of Ehrlich ascites tumor cells as presented by the X-Y-recorder. The measured values are stored in a 400-channel analyzer. The X-axis can be calibrated by cells with known DNA content (for instance; leucocytes). About 50,000 cells per minute are measured. The standard deviation for the G_1 peak is about 4%*

irradiation with a neutron dose of 900 rad shifted almost completely from G_1 and S phase to G_2+M (histograms at 24, 48, 72 and 96 h). The inhibition of mitosis continued for several days, as shown in the right part of Figure 6. At lower doeses as demonstrated by the results obtained with a neutron dose of 300 rad, the inhibition was of shorter duration, and the cells shifted by dividing from G_2+M to G_1 (histogram at 72 and 96h). Thereafter, the cells can even synthesize DNA again (histogram at 96 h). Such results may be important for the interpretation of repair processes in irradiated cells. Such investigations are also of theoretical value for radiation therapy.

The split-dose method is usually applied in studying the repair capacity of cells. In these experiments, the effect of radiation caused by a certain dose given at once is compared with the effect caused by fractionated irradiation. The different survival rates are usually explained by repair mechanisms which become effective between the dose fractions. The radiation damage may already be partially repaired before the second dose is applied. Therefore, the survival after split doses is larger in case of repair. It is possible, however, that this interpretation needs supplementation, since the composition of the cell population changes during the experiment. For example, the first dose blocks many cells at mitosis. Thus, the second dose is applied to a completely different cell pool having a different response to radiation.

Application of the ICP for Examination of Cellular Response to Chemical Noxes

The ICP is very useful in elucidating the mechanism of the effect of cytostatica on tumor cells. So far, investigations into the effect of cytostatica on the proliferation of cells were carried out by use of the time-consuming autoradiomicrography or by determination of the mitotic index. The ^{3}H-thymidine method, however, may give misleading results since the incorporation of radioactivity in treated cells is not necessarily due to new DNA synthesis; it may be also due to unscheduled (repair) synthesis (Stich et al. 1970). Moreover, it was shown that treated cells occasionally do not incorporate ^{3}H-thymidine although DNA synthesis was not inhibited (Grunicke et al. 1973). By use of the ICP, these difficulties can be avoided. We can demonstrate the effect of daunomycine on DNA synthesis and proliferation of Ehrlich ascites tumor cells (Göhde and Dittrich 1971). In contrast to all results hitherto reported by other authors (Wang et al. 1972) we found that daunomycine does not inhibit DNA synthesis at the primary stage (Göhde and Dittrich 1971). Our results clearly indicate that daunomycine blocks the cells in G_2 phase. Even at high concentrations of this cytostatic compound the cells shift through the S-phase to G_2.

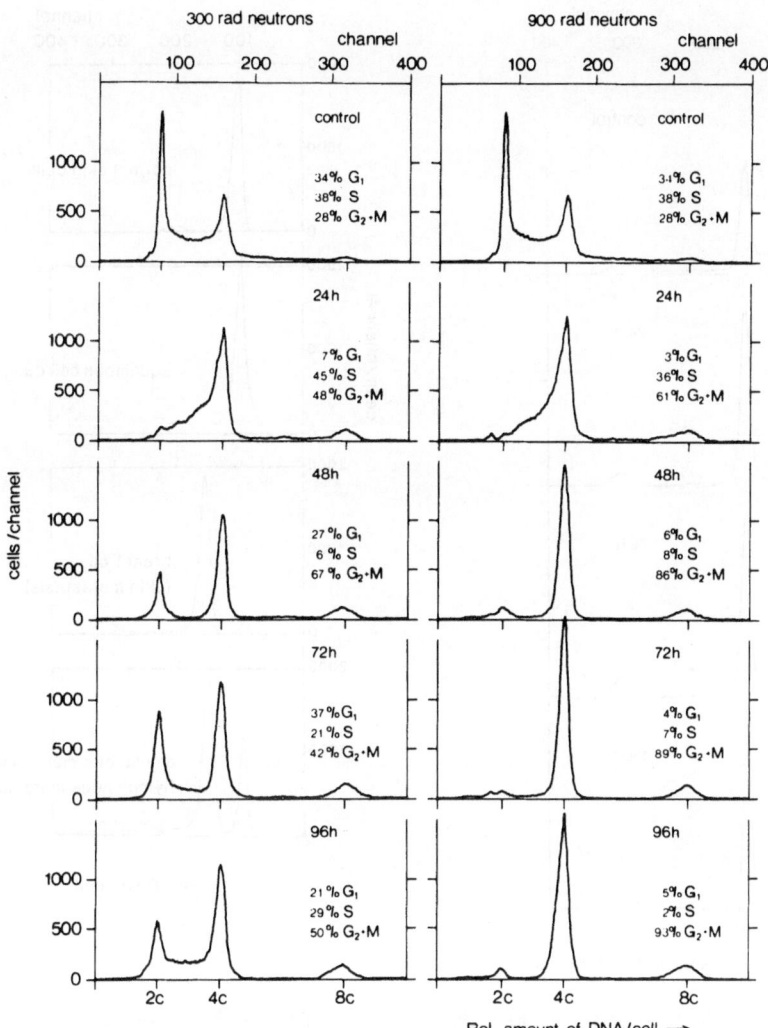

Figure 6. *DNA histograms of Ehrlich ascites tumor cells irradiated with a neutron dose of 300 rad (left) and 900 rad (right). These were collected at different times after irradiation, and were fixed, stained, and recorded by use of the ICP. Each histogram represents 50,000 cells measured within about one min.*

Similar results are shown for Ehrlich ascites tumor cells after treatment with adriamycine *in vivo* (Figure 7). Again, cells in G_1 and S phases shifted completely to G_2. Because there is no mitosis after treatment, no cells entered the G_1 phase. Thus an accumulation of cells in G_2 was observed. With sufficiently high doses, the blocking of cells in G_2 is maintained for several days (Figure 7, 120 h).

These investigations became practically feasible only after development of instrumentation for automated flow-through cytofluorometry. The application of conventional cytofluorometric methods would be extremely time consuming (up to 10,000 to 100,000 times) for reaching the same statistical accuracy. Using the ICP these measurements, including the recording of histograms, require only about 10 minutes.

These few applications indicate the usefulness of rapid automated cytofluorometry not only as a screening technique for studying the effects of substances on the proliferation kinetics of cells, but also as a method for investigating the mechanism underlying these effects.

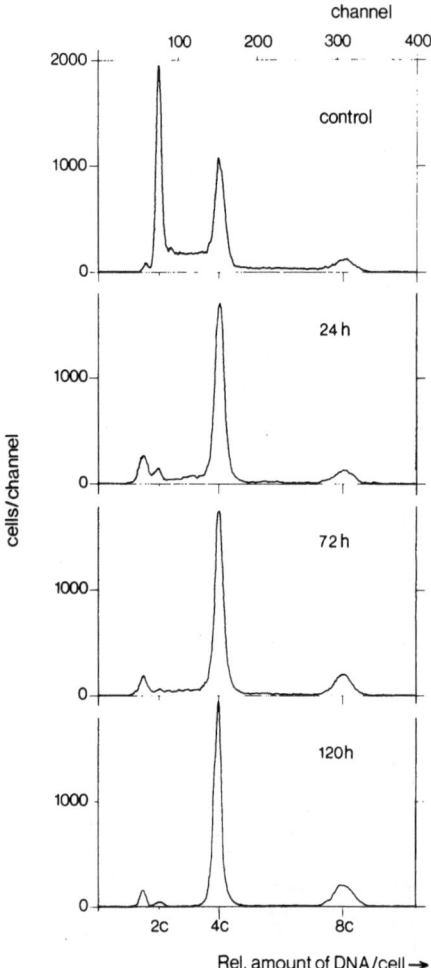

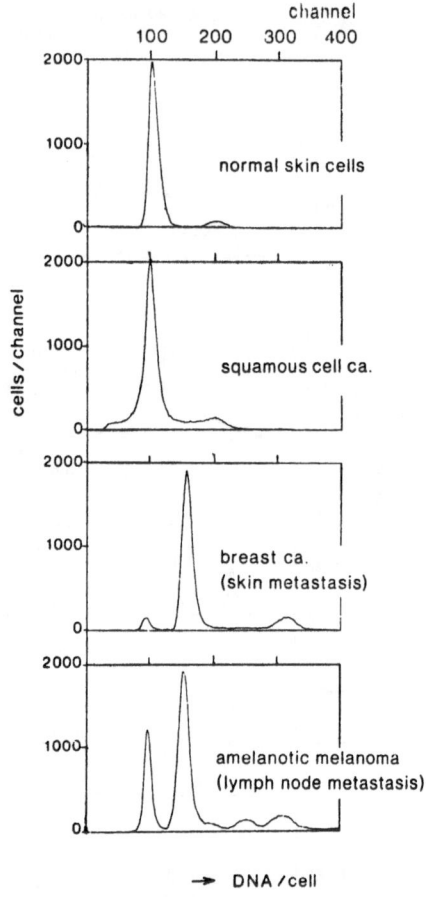

Figure 7. *DNA histograms of Ehrlich ascites tumor cells treated with adriamycin in vivo (exponentially growing cells were treated with 40 mg. adriamycin/kg body weight five days after inoculation in adult mice). The cells were sampled at different times fixed, stained and processed with the ICP immediately. Each histogram represents 50,000 cells, measured within about one minute. (Some leucocytes are recorded left of 2c.)*

Figure 8. *DNA histograms of human skin cells. Cells of normal skin are shown in the top histogram. Malignant samples show characteristic differences compared with the control. The histograms represent 40,000 – 85,000 cells and were obtained within 1-2 minutes*

Applications of the ICP in Cytopathology

Applications of this technique for prescreening in gynecological cytology have been reported elsewhere (Göhde and Dittrich 1970a; Göhde et al. 1973; Dittrich et al. 1971; Göhde et al. 1972; Reiffenstuhl et al. 1971). In addition, we used this method for detecting differences between normal and malignant cells of various materials (Schuman et al. 1971; Büchner et al. 1971). Original DNA histograms as printed out from the ICP and obtained from different tissues (trypsinated bioptic material) are shown in Figure 8. Each histogram represents 40,000 to 85,000 cells and was obtained within roughly one minute.

The histograms of the malignant samples differ characteristically from the control sample (histogram at the top of Figure 8.) In abnormal samples the percentage of cells with increased DNA content (S_- and G_2+M phase cells) is higher (see histogram 2). In other DNA histograms of malignant cells we observed besides the normal G_1 peak other peaks representing aneuploid tumor cell lines. These anomalous peaks representing cells with high DNA content were recorded at high channel numbers (above channel 100 in the multichannel analyzer) as demonstrated by histograms 3 and 4.

CONCLUSION

The described system for rapid automated cytofluorometry based on the flow-through principle described above may be considered as an important step towards automation of microscope fluorometry. The main advantage of this flow-through system is that each cell must pass through the plane focused by the objective. Thus, differences in optical conditions of excitation and collection of fluorescence by the microscope objective between cells moving in axis and those moving more or less out of axis can be kept small (Figure 5), provided that homogeneous illumination across the capillary exit is achieved and that the photocathode response is independent of the light distribution in the object plane. A super-pressure mercury lamp can be applied in the same manner as it is used for conventional fluorescence microscopy in connection with an objective with a numerical aperture 1.3. When a laser is used as light source a substantial increase of sensitivity may be expected. Therefore, the assumption seems to be justified that the use of the ICP technique developed in our laboratory can be extended to other problems such as measurements of microorganisms and the fluorometric assay of cellular enzymes. Moreover, histograms of two parameters can be obtained automatically and simultaneously. This is done by measuring two different cell substances quantitatively stained with fluorochromes having different emission spectra. For instance it is possible to measure simultaneously the DNA content and the protein content of the same cell. This is of great importance both in investigations of cyto-pathological problems and for research on cytostatic compounds. It is obvious that by measuring the two parameters of a cell population simultaneously on each single cell much more information is obtained than by proceeding one by one on separate histograms. The simultaneous measurement of DNA and protein content per cell in populations of Ehrlich ascites tumor cells after treatment with vincaleucoblastine *in vivo* (Göhde and Dittrich 1970) is an example.

Of course, principles of illumination and light measurement other than fluorometry may also be applied. For instance, measurement of forward light scattering or absorption is possible using a different type of flow-through chamber (Dittrich and Göhde 1969) which permits illumination of the particles passing the measuring area in transmitted light. Measurement of back scattering may be made by using incident-light illumination as in the flow-through chamber described above. A cell-separation technique is also being developed in our laboratory.

The use of the impulsmicrophotometer has proved very suitable for investigations on proliferation kinetics of normal and tumor cells. Exact information on the behaviour of cell populations after their exposure to physical (radiation) or chemical treatment (for instance cytostatica) may be obtained by determining their composition with respect to G_1, S and G_2+M cells.

ACKNOWLEDGEMENTS

The author is very indebted to Dr. F. Otto, Institut für Aerobiologie der Fraunhofer Gesellschaft in Grafschaft, and Dr. J. Schumann, Fachklinik Hornheide in Handorf bei Münster for their help and to Miss J. Spies and Mrs. H. Schäfer for their excellent technical assistance. The assistance during the development of the ICP of Mr. S. Kiegler (electronics) and Mr. H. Mertens (mechanics and optics) is gratefully acknowledged.

REFERENCES

BÜCHNER, T., Dittrich, W., Göhde, W., (1971), *Verh. Deut. Ges. Inn. Med.*, **77**, 416-18.

CROSLAND-TAYLOR, P.J., (1969), *Nature*, **171**, 37-38.

DILLA, M.A., van, Trujillo, T. T., Mullaney, P. F., and Coulter, J. R., (1969), *Science*, **163**, 1213-14.

DITTRICH, W., and Göhde, W., (1969), *Z. Naturforsch*, **24b**, 360-61.

DITTRICH, W., and Göhde, W., (1970), *Atomkernenergie*, **15**, 174-76.

DITTRICH, W., Göhde, W., Severin, E., and Reiffenstuhl, G., (1971), Die Kern-Plasma-Relation in der Impulscytophotometrie des Cervicalsmears. *IV Intern. Congr. of Cytology*, London.

GÖHDE, W., (1972), *Mitteilungsdienst der Ges. zur Bekämpfung der Krebskrankheiten im Lande NRW*, **6**, 255-76.

GÖHDE, W., (1973), *Habilitationsschrift, Med. Fakultät der Universität Münster*.

GÖHDE, W., and Dittrich, W., (1970a), *Vortrag. Symp. Prim. Fed. Soc. Cytolog. Europ.*, Prag.

GÖHDE, W., and Dittrich W., (1970b), *Z. Anal. Chem.*, **252**, 328-30.

GÖHDE, W., and Dittrich, W., (1971), *Arzneimittelfschg.*, **21**, 1656-58.

GÖHDE, W., and Dittrich, W., (1972), *Acta Histochem. Suppl*, **10**, 429-37.

GÖHDE, W., Dittrich, W., Severin, E., Reiffenstuhl, G., (1973), *Rev. Cytol. Clin.*, (Paris), **6**, 9-15.

GÖHDE, W., Zinser, H. K., Dittrich, W., and Prieshof, J., (1972), *Geburtsh. u. Frauenheilk.*, **32**, 382-93.

GRUNICKE, H., Hirsch, F., Wolf, H., Bauer, U., and Kiefer, G., (1973), In: *Aktuelle Probleme der Therapie Maligner Tumoren*, (G.P. Wust, ed), Springer-Verlag, Heidelberg, . in press.

Le PECQ, J. B., and Paoletti, C., (1967), *J. Molec. Biol.*, **27**, 87-106.

PRENNA, G., (1964), *Riv. Istoch. Norm. Pat.*, **10**, 469-74.

REIFFENSTUHL, G., Severin, E., Dittrich, W., and Göhde, W., (1971), *Arch. Gynäk.*, **211**, 595-616.

SCHUMANN, J., Ehring, R., Göhde, W., and Dittrich, W., (1971), *Arch. Klin. Exp. Derm.*, **239**, 377-89.

STICH, H. F., San. R. H., and Kawazoe, Y., (1970), *Nature*, **229**, 416-19.

TRAGANOS, F., Adams. R. R., Kamentsky, L. A., and Melamed, M. R., (1972), *Acta. Cytol.*, **16**, 281-83.

WANG, J. J., Chervinsky, D. S., and Rosen, J. M., (1972), *Cancer Res.*, **32**, 511-15.

QUANTITATIVE DETERMINATION OF DNA AND PROTEIN IN SINGLE CELLS

Fritz Ruch

Department of General Botany, Swiss Federal Institute of Technology, Zurich, Switzerland

CELL PREPARATION

Uneven staining of slides and poor reproducibility can make quantitative work difficult or even impossible. These conditions often occur with protein stainings in smears, especially in the case of the histone reactions using fast green or sulfaflavine. Below is a brief discussion of experiments on human leukocytes stained with sulfaflavine at pH 8.0.

Smears fixed after air drying either do not react or react only insufficiently. Smears fixed before the drying process is complete yield very irregular preparations. Cells or isolated nuclei fixed in suspension and then air dried on slides also yield irregular staining, but to a lesser degree than smears of unfixed cells. Thus, the protein determination led to the conclusion that only those cells that are fixed in a wet state and never allowed to dry until completion of the staining process give consistent and reproducible results.

Similar observations were made regarding dry weight measurements with the interference microscope; in these cases, the embedding media should penetrate homogeneously into all the cells. Therefore methods were investigated for the preparation of slides from suspensions of cells or cell particles without air drying them (Leeman 1971). The following technique has proved to be adequate and usable for fixed or unfixed cells. A cover slip is placed on a carrier fitting into a centrifugation cup. Then a Teflon ring is sealed onto the cover slip. A few drops of cell suspension are put into the opening of the ring. Centrifugation is carried out for 5 min at 1000-1800 g. In the case of unfixed cells, fixative is then added and the cover slip removed and placed in a cuvette for washing and staining. After centrifugation of fixed cells, the cover slips are brought directly to washing and staining media. The result of this technique is shown in Figure 1 in comparison with the results of smear techniques. The measurements made at random over the whole area of the preparation clearly demonstrate the detrimental influence of air drying and the superiority of the centrifugation technique.

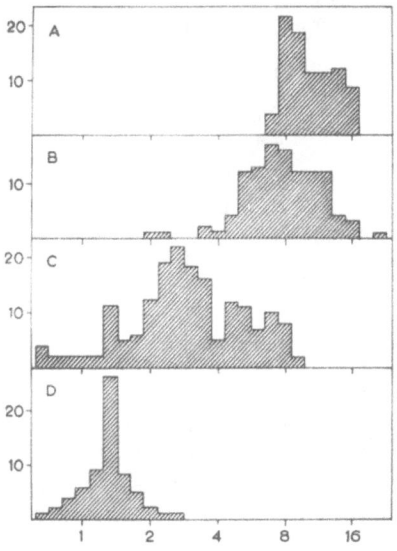

Figure 1. *Cytofluorometric determination of basic protein (histone) in human leukocyte preparations. Staining: sulfaflavine pH 8.0.*

A) cells fixed in suspension, attached to cover slips by centrifugation;

B) cells fixed in suspension, air dried on slides;

C) blood smear, semidried, fixed;

D) blood smear, air dried, fixed.

Abscissa: fluorescence intensity, arbitrary units; Ordinate: number of cells. [from Leeman, 1972]

PROPERTIES OF DYES

It is well known that samples of the same dye from different sources may stain quite differently. Moreover, even the same sample may give different results when used after different storage periods. Following is a discussion of this important point in the case of two dyes used in the Feulgen reagent. In our earlier studies we used auramin O (Bosshard 1964). Since this dye is not very stable under UV irradiation, and since the results varied according to supplier, we also tested many other dyes. We found the so-called BAO from CIBA* most suitable (Ruch 1970). As this dye is a pure chemical compound, it is constant in composition and thus has an advantage over most other stains. For some years we used several samples of this dye with good results. Occasionally, however, we observed unexpectedly low fluorescence intensities and great deviations among different cells. The same phenomenon was observed in other laboratories and sometimes led to misconceptions regarding the usefulness of the dye. Tests then showed that BAO is very sensitive to oxygen, and changes its properties when exposed to air. Therefore the dye must be stored in small, airtight bottles in a cool place for not longer than about one year. When these precautions have been taken, BAO gives very constant results.

SULFAFLAVINE FOR PROTEIN DETERMINATION IN CELLS

Acid dyes, such as fast green or bromphenol blue at pH 8.0 (Alfert, 1953; Ringertz, 1966) and naphthol yellow at pH 2.8 (Deitch, 1955), are widely used for histone and total protein determinations by absorption photometry. Both staining reactions depend on the positive charge of the basic amino acid residues. For some years we have been using one fluorochrome, namely sulfaflavine, for these determinations at both pH values (Leeman, 1972). This procedure also allows successive staining and determination of total protein and histone in the same cell.

Sulfaflavine was chosen from a series of acid fluorochromes both because of its identical absorption and emission spectra for the two pH values (figure 2) and for

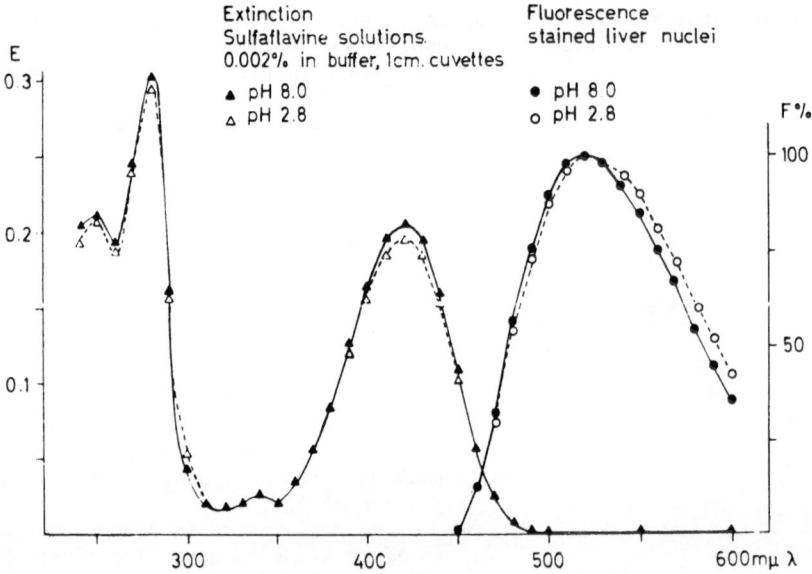

Figure 2. *Absorption and fluorescence spectra of sulfaflavine. The fluorescence values are not corrected for the spectral sensitivity of the apparatus. Irradiation at 366 nm. [From Leeman, 1972]*

---

* *Obtainable from FLUKA AG, Buchs, Switzerland.*

its brilliant fluorescence. Regardless of the excitation wavelength (360 to 450 nm), the maximum fluorescence intensity is emitted at 520 nm. The dye shows considerable photodecomposition under prolonged irradiation, but in case of short-term excitation (1/50 sec), this effect is less than 1%. DNA must be extracted prior to histone staining, e.g. with 5% TCA (3 hours at 60°C), in which case Formalin fixation is necessary. Alcohol fixation does not stabilize the proteins enough to resist hot acid hydrolysis. Staining patterns of different types of cells in smears and sections, as well as model experiments with filter paper and blocking reactions (Leeman, 1972), strongly suggest that sulfaflavine at pH 8.0 stains histones, and at pH 2.8, stains all proteins with the exception of those which are strongly acidic (pepsin). To test the proportionality between fluorescence intensity and protein mass, protein smears and nuclei of known protein content (determined by independent methods) were used. As discussed above, these experiments showed that proper initial treatment and fixation of the cells are of outstanding importance in obtaining satisfactory results. Standard deviations among uniform cells are in the same range as those observed with other protein reactions (with bull sperms 9-11% at both pH values; with small lymphocytes 10-11% for pH 8.0; 15% for pH 2.8).

Reproducibility

Table 1 shows the coefficients of variation between the means of different series of measurements. Between different locations on the same slide there are no statistically significant differences. Slides stained together whether measured at the same time or at time intervals within 24 days, show only slightly higher coefficients of variation. According to expections, the preparations stained and measured independently of one another show much higher coefficients of variation.

TABLE 1

REPRODUCIBILITY OF SULFAFLAVINE STAINING
HUMAN LEUKOCYTES

Set up of experiment	Coefficient of variation between the means of n series of approx. 40 measurements			
	pH 8.0	n	pH 2.8	n
1) 1 slide measured at different locations, on the same day	4.4	7	4.1	7
2) 1 slide measured at different locations, at time intervals, up to 24 days	8.5	7	6.2	7
3) n slides, stained together in the same staining solution measured on the same day	6.0	6	6.4	8
4) n slides, stained in the same staining solution; stained and measured at different times (up to 24 days)	22	8	15	8
5) n slides, stained and measured independently, at time intervals up to 24 days	20	5	12	5

Absolute Amounts

Table 2 shows that with most objects the quantity of total protein as determined by sulfaflavine agrees well with dry weight minus DNA; of course losses of substance with EA fixation must be considered. In the case of mastocytoma cells the deviation can be explained by the high content of RNA (ca. 15×10^{-12}g). The validity of the histone values can be judged on the basis of the histone/DNA ratio. This corresponds well with the values found in the biochemical literature (Leemann 1972).

TABLE 2

TOTAL PROTEIN, HISTONE, DNA AND DRY WEIGHT DETERMINATION ON VARIOUS OBJECTS

		fix.	calf thymus nuclei	human leuko-cytes	rat liver nuclei 2n	chicken erythro-cyte nuclei	bull sperm heads	masto-cytoma cells 2n
A.	total protein Sulfaflavine 2.8	10^{-12}gr	EA 1)	1) 17.5	1) 38	31	5.7	4.0	59
B.	histone Sulfaflavine 8.0	10^{-12}gr	FO	5.8	8.6	18.5	2.3	2.8	12
C.	nonhistone protein A - B	10^{-12}gr	EA	11.7	29.4	12.5	3.4	1.2	47
D.	dry mass	10^{-12}gr	FO / EA	22 / 22	63 / 46	40 / 37	6.5	8.6 / 7.6	85 / 85
E.	DNA 2) 3) 4)	10^{-12}gr	FO,EA / FO / FO,EA	6.0 / 7.1 / 6.0	6.0	7.1 / 6.9	2.7 / 1.8	3.5	7.1
F.	D - E	10^{-12}gr	EA	16	40	30	5) 4.7	4.1	78
G.	error; mass not accounted for F - A	10^{-12}gr	EA	-1.5	2	-1	-1	0.1	19
H.	ratio histone/DNA B/E			0.97	1.4	2.7	1.3	0.80	1.7
I.	ratio histone/non histone protein B/C		EA	0.50	0.29	1.5	0.68	2.3	0.25
K.	ratio histone/DNA, literature			0.48 / 1.0-1.2		2.3-2.64	1.35		

1) absolute values, by comparison with protein smears.
2) UV.
3) dry weight loss after TCA extraction.
4) BAO, relative to thymus 6.0 10^{-12}gr.
5) FO

EA: ethanol acetic acid
FO: Formalin

Successive measurements of total protein and histone are possible on Formalin fixed cells if corrections for the following points are made: Formalin decreases the results of total protein staining, whereas TCA extraction increases them. Figure 3 shows these measurements on thymus nuclei and leukocytes. For thymus nuclei, the ratio of histone to total protein is fairly constant (correlation 0.91). In case of leukocytes, however, the different histone/total protein ratio of lymphocytes and poly-morphonuclear cells is evident (correlation 0.27).

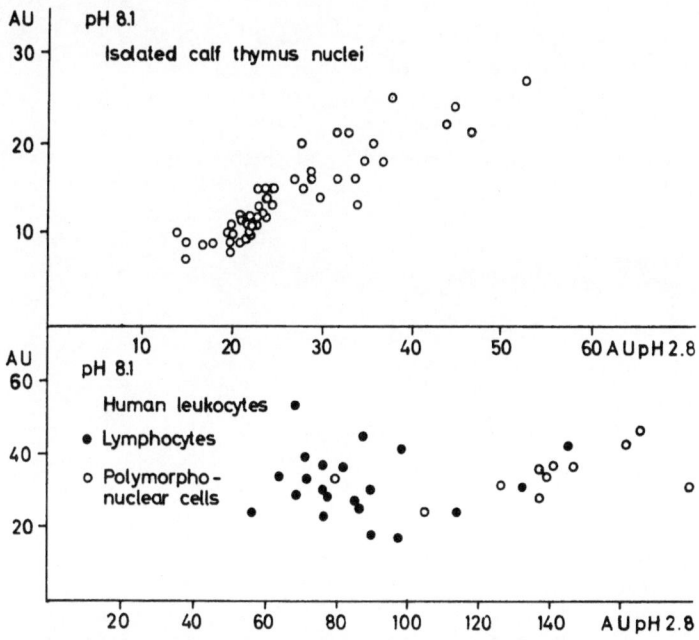

Figure 3. *Relations between the fluorescence intensities (AU: arbitrary units) of individual cells stained successively with sulfaflavine at pH 2.8 and 8.0. [From Leeman and Ruch, 1970]*

REFERENCES

ALFERT, M., and Geschwind, J.I., (1953), *Proc. Natl. Acad. Sci. US.* **39**, 991.

BOSSHARD, U., (1964), *Z. wiss. Mikr.*, **65**, 391.

DEITCH, A.D., (1955), *Lab. Invest.* **4**, 324.

LEEMANN, U., and Ruch, F., (1972), *J. Histochem. Cytochem.* **20**, 659.

RINGERTZ, N.R., and Zetterberg, A., (1966), *Exptl. Cell Res.* **42**, 243.

RUCH, F., (1970), *Introduction to Quantitative Cytochemistry II*, G.L. Wied and G.F. Bahr, eds. Academic Press, New York. p.431.

FLUORESCENCE PROPERTIES OF THE MONOAMINOACRIDINES AND SOME 2-AMINOACRIDINE-DERIVATIVES

D. Wittekind

Institute of Anatomy, University of Freiburg i.Br., Germany

INTRODUCTION

Investigations into the detection and identification of nucleic acids in living and fixed cells are largely based on the use of 3,6-diaminoacridine (proflavine) (Allison and Young 1964; Robbins 1960; and others) — and its tetramethylated derivative Acridine Orange (AO) (Schümmelfeder 1950, 1957; Rigler 1966).

In these investigations, carried out over the last ten years, quantitative determinations of AO binding groups in nucleo-proteins were based both on the theoretical considerations of the stoichiometric staining reactions involved (Zanker, 1952 a, b: 1954) and on theories concerning the specific interactions of 3,6-diaminoacridines with DNA and RNA macromolecules (Peacocke 1956; Steiner and Beers 1957; Stone and Bradley 1961; Lerman 1961; 1963; 1964).

After the detection of AO as a valuable biological fluorochrome (Bukatsch and Haitinger 1940; Strugger 1940; 1949) the "golden age" of this compound began when Gössner (1949), Schümmelfeder (1950; 1957; 1958), Armstrong (1956), and Von Bertalanffy (1956; 1963) detected its capacity to differentiate DNA from RNA, the former fluorescing green, the latter red in fixed and also in living cells (Weissmann 1953; Weissmann and Gilgen 1956; Stockinger 1958; Zeiger and Schmidt 1957; Wittekind 1958). Pioneer studies of the quantitative aspects of fluorescence microscopy are also based on the use of AO (West 1963; Rigler 1966). Red cytoplasmic fluorescence is of metachromatic nature, based on the high stacking effect exerted by RNA on closely associated dye molecules. Heparine with its anionic $-SO_3H$ groups is much superior to RNA in this respect (DNA 1.25; RNA 3.3; Heparine 787; (Rigler 1966; Appel and Zanker 1958). Analysis of absorption, excitation and fluorescence spectra clearly show AO bound to helix structures, like DNA, in a monomer molecular form - one acridine molecule intercalated between normally neighboring base pairs in a plane perpendicular to the helix axis (Lerman 1961).

Emission of fluorescence in nuclei of living cells, a property so far reserved for 3,6-diaminoacridine and AO (De Bruyn, Robertson et al. 1951; De Bruyn, Farr et al. 1953), was mentioned in the case of 3-Aminoacridine as well (Morthland et al. 1954), and could be shown to occur in living macrophages and fibroblasts (Wittekind and Kunze 1969). Since 3-Aminoacridine shares some biologically important features with AO, it seemed interesting to compare the fluorescence properties of that monoaminoacridine with its four isomerides. Some results have been published in a preliminary paper (Wittekind and Kunze 1969). Essential physico-chemical and biological data of the monoaminoacridines are given in Figure 1. These were collected mainly from the work of Albert and his collaborators (Albert and Linnell 1936a; b; Albert and Ritchie 1941; 1963; Albert 1966).

In the following, consideration is given to the fluorescence properties of 2-Amino-acridine (2-AA) and some of its derivatives (Figure 2) as compared with 3-AA, 9-AA, and AO. 4-AA is a very weak base, owing to the formation of an internal hydrogen bond between the ring nitrogen atom and that of the neighboring amino group (Wittwer and Zanker 1959). It is an excellent fluorochrome for lipids. The effect probably results from a very high oil/water partition coefficient (Figure 1). The 2-AA group deserves further consideration, since some of its compounds were reported to exhibit an orange-red fluorescence in the nuclear area of living and fixed cells (Kunze and Wittekind 1970).

MATERIAL AND METHODS

The compounds of the 2-AA group investigated are listed in Figure 2.* They were used in chromatographically pure form. Elementary analysis has been carried out for

---

* 2-Aminoacridine has been generously supplied by Prof. A. Albert (Canberra), the derivatives of 2-AA by Bayer-Werke, Wuppertal-Elberfeld (Prof. R. Gönnert), Germany.

the dimethyl and the methoxyderivatives**. There was satisfactory agreement between the quantities measured and those calculated for C, N, and O.

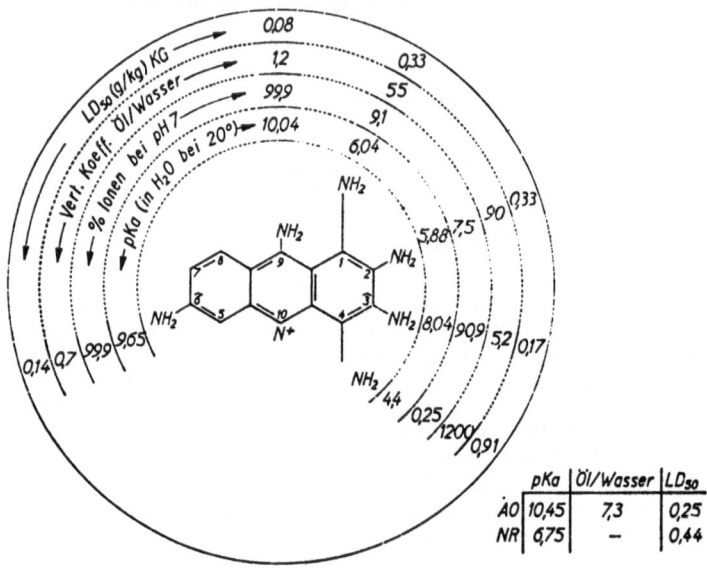

Figure 1. *Some physico-chemical and biological properties of the monoamino-acridines. Circles, from periphery to center: LD_{50}(Dosis letalis$_{50}$); oil/water partition coefficient; percent ionized at pH 7; pka; inset at right: Acridin Orange (AO) and Neutral Red (NR) for comparison (Wittekind and Kunze, 1969)*

Spectrophotometry

Each compound was dissolved in 0.1 M citric acid-Na_2HPO_4 buffer at pH 1; 2; 4; 6; 7; 8.75; 11. With regard to the poor solubility of 2-ĀA and its derivatives, concentrated solutions were prepared by dissolving the substance in 50% ethanol in aqua bidest. at 37°C. Lower concentrations of the dyes ($5 \times 10^{-4}M$; $1 \times 10^{-4}M$; $5 \times 10^{-5}M$; $1 \times 10^{-6}M$) were obtained by adding corresponding amounts of buffer solution. As expected from the low specific absorption coefficient (log ε = 3,65) of 2-AA at 462 nm (λ_{max} in the long wave-length range), pH 4.1, no reliable measurements could be made

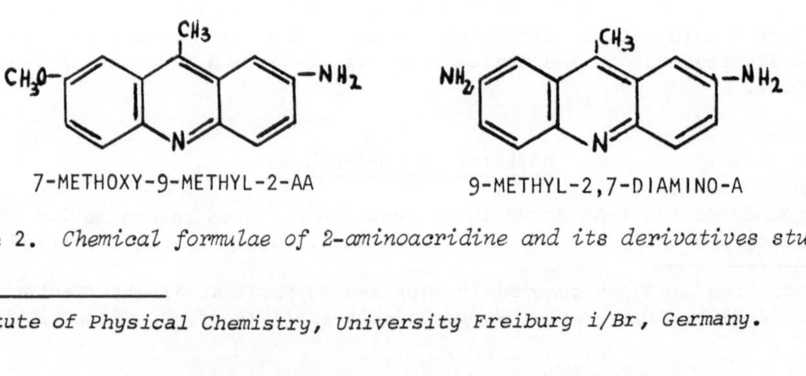

2-AMINO ACRIDINE

7-9-DIMETHYL-2-AA

7-METHOXY-9-METHYL-2-AA

9-METHYL-2,7-DIAMINO-A

Figure 2. *Chemical formulae of 2-aminoacridine and its derivatives studied in this paper*

** Institute of Physical Chemistry, University Freiburg i/Br, Germany.

at a concentration of $1 \times 10^{-6} M$.

The spectra were recorded with a Zeiss spectrophotometer PM Q II at room temperature over the wavelength range 320-550 nm. DNA (Serva, Heidelberg, "reinst";), and RNA (Serva, yeast-RNA, pract.) were added to solutions of 2-amino-7-methoxy-9-methylacridine, at a concentration of $1 \times 10^{-4} M$. All compounds were used in chromatographically pure form.

Cytological Studies

Cells. a) LM-fibroblasts grown in monolayer in Leighton tubes on cover slip 9 x 22 mm. Medium: Lactalbumin (Difco, Detroit; No. 5996-2) 0.5% in Hanks' balanced salt solution with 5% fetal calf serum (Flow Lab., Ayreshire, Scotland).
b) Mast cells and macrophages were obtained by rinsing the peritoneal cavities of rats with Hank's solution.

Vital Staining. The dyes indicated in Figure 2 were added to the cell cultures to arrive in the medium at concentrations $2 \times 10^{-5} M$. Cells were exposed to dyes for 24 hours and then examined under a Zeiss Universal Microscope equipped with combined transmitted light phase contrast and epi-illumination for fluorescence excitation using the exciting filter BG 12 or UG 1 and the barrier filter 41, 47 or 50. Further methodological details are given elsewhere (Wittekind and Kretschmer 1972).

Autophagic Vacuoles. These were induced by applying 9-amino-1,2,3,4-tetrahydroacridine (Tacrine, Ward, Blenkinsop Ltd., conc. $2 \times 10^{-4} M$) to the cell cultures according to a procedure described in more detail in another paper (Wittekind et al. 1970). The living cells, showing no fluorescence when exposed to this compound, were poststained with the substances listed in Figure 2.

Fixation and Staining

Cultures were fixed for 15 min. in ethanol:acetone = 3:1. Fluorochromation of substrate with the dyes dissolved in citric acid-Na_2HPO_4 buffer, at a concentration of $1 \times 10^{-4} M$, at pH 4.1, was carried out according to a procedure recommended by Rigler (1966). After rediffusion of unbound dye the specimens were sealed with buffer pH 4.1 on slides; cells fluorochromed with AO at the same concentration served as standard for comparison. By adapting the composition of the buffer, pH was varied from 1 to 10. Nucleic acids were extracted with 5% trichloracetic acid for 15 min. at 90°C (Schneider 1945). 5% cold trichloracetic acid was used for 4 and 6 hours to obtain a change in nuclear fluorescence from green to yellow and orange - a change comparable with the effect observed by Kasten (1965,1967) with 5% cold perchloric acid. Depolymerization (probably with simultanous denaturation) was carried out by an abrupt change from hot to cold water (Doty et al. 1960; Marmur et al. 1963).

RESULTS

Spectrophotometric Measurements

In Figures 3 and 4, spectral absorption curves within the long-wave UV and visible ranges of 2-AA and 9-methoxy-2,7-dimethylaminoacridine are shown. The spectrum of 2-amino-7methoxy-9-aminoacridine does not deviate significantly from that of dimethyl-2-AA derivative.

It is apparent that distinct absorption maxima are observed with the monocations at 460 and 454 nm at pH 4 respectively and, to a smaller degree, at pH 2. (Curves at pH 1, showing grossly abnormal maxima, have been omitted from the graphs). 2-AA has a pka of 5.88 (Albert 1966); pka of the derivatives have not yet been determined. They can be safely assumed to be similar. At pH 8.75 and certainly at pH 11, 2-AA and the derivatives under consideration here can be expected to exist in the neutral form. With a pH above 7, absorption in the visible range is markedly reduced, an observation that is essentially in agreement with findings by Craig and Short (1945), and Albert (1966).

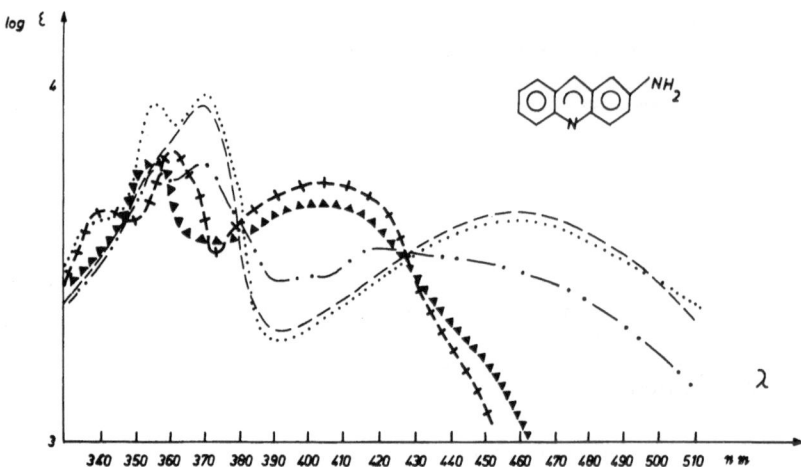

Figure 3. *Absorption curves of 2-aminoacridine at different pH.*
pH 2: • • • ; pH 4: – – – – ; pH 6: – •• – •• – ; pH 7 ▲▲▲▲ ; pH 8.75: + + + ;
abscissa: log ε ; ordinate: wavelengths [nm]

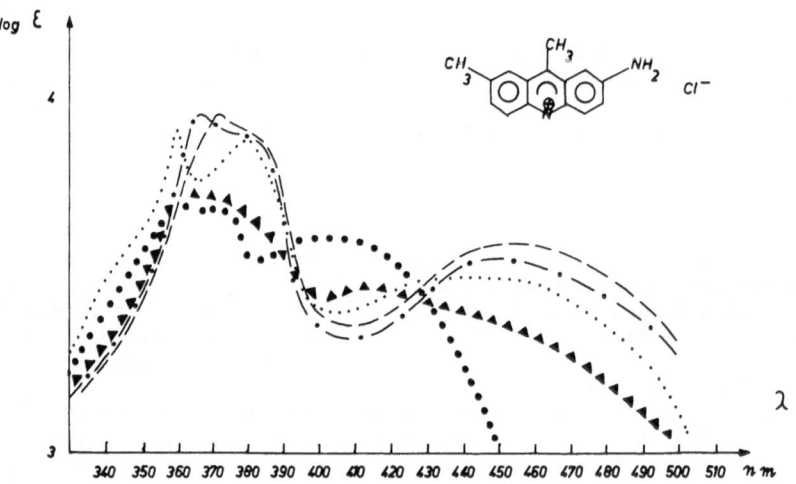

Figure 4. *Absorption curves of 7,9-dimethyl-2-aminoacridine pH 2: •••• ;*
pH 4: – – – ; pH 7: ▲ ▲ ▲ ▲ ; pH 11 : ● ● ● ●

Changes in Absorption after Addition of DNA and RNA. Both nucleic acids were added
in concentrations of 0.1g% to buffer solutions containing the aminoacridines in
$1 \times 10^{-4}M$ concentration. With regard to the cytological results (see below) and in
order to give a better indication of the influence of the nucleic acids pH7 was selected for the solvent (Figures 5 and 6). No attempt was made to calculate the phosphorus content of the DNA and RNA preparations used.

In comparing Figures 5 and 6 with the absorption curves of the monocations at
pH4 in figures 2 and 3, it can be seen that the flat maxima in the visible range around
450-460 nm have shifted to longer wavelengths in the case of the dye-DNA complex, but
scarcely so with RNA. Comparison of Figures 5 and 6, moreover, points to a stronger
effect of DNA on absorption by 7-methoxy-9-methyl-2-AA than by 2-AA itself. Absorption
by 2-AA in the longwave ultraviolet spectral range remains unchanged by DNA, again in
contrast to the 2-AA derivative.

The situation in the longwave ultraviolet spectral region is different in the
case of 7-methoxy-9-methyl-2-AA group: the spectral absorption peak at 377 (pH 7) disappears, and new maxima are observed at 407 (DNA) and 399 (RNA). Furthermore, in

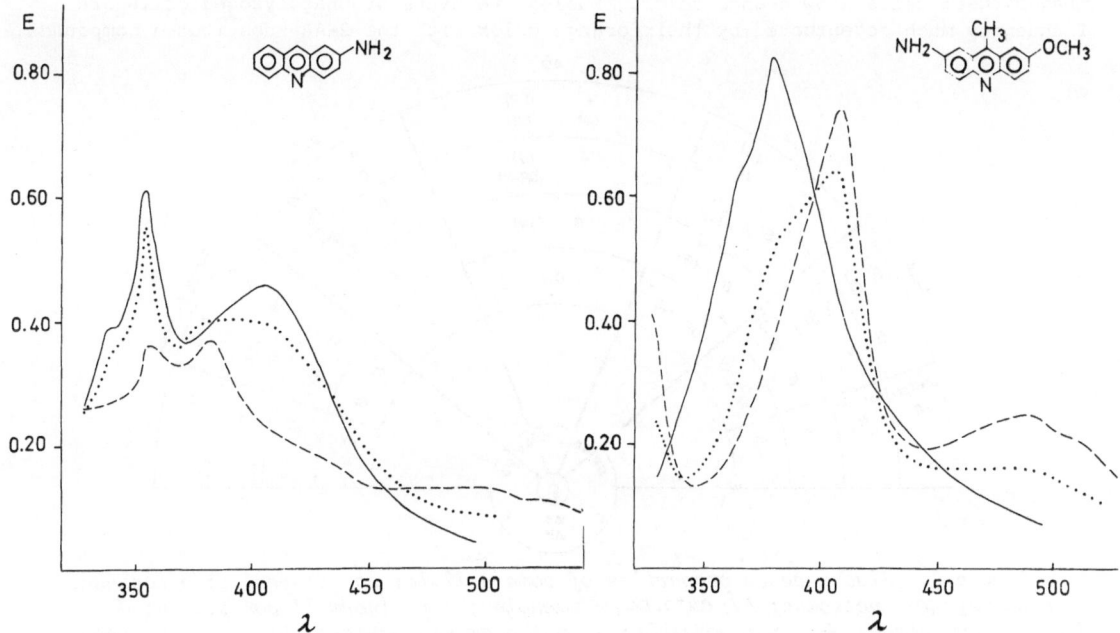

Figure 5. *2-aminoacridine; absorption spectrum at pH 7:——; absorption spectrum at pH 7 with DNA 0.1%: - - - ; with RNA 0.1% • • • • •; ordinate: E = absorbance; abscissa: wavelengths [nm]*

Figure 6. *7-methoxy-9-methyl-2-aminoacridine; absorption spectrum. Explanation: see Figure 5*

contrast to 2-AA, slight but distinct maxima at 490 (DNA) and 485 (RNA) occur. Changing the relation of either DNA or RNA to the aminoacridines does not result in measurable further shifts of absorption maxima in the visible range.

Fluorescence. In buffered solutions 2-AA and its derivatives exhibit three types of fluorescence when excited at 354 mn: green at pH 1, red at pH 4, yellow at pH 11. There are minor variations in intensity of fluorescence between the 2-aminoacridines.

Cytological Studies

Cells were fluorochromed with aminoacridines with special regard to acidic poly-anionic cellular structures: nuclei, nucleoli and cytoplasm for DNA and RNA, granules of mast cells for sulfated mucopolysaccharides; dye-induced autophagic vacuoles for acid phospholipids (Dingle and Barrett 1969).

Fibroblast Cultures - Living Cells. These large cells with ample cytoplasm lend themselves readily to the study of cytological detail. Results obtained by fluoro-chroming fibroblasts with diverse aminoacridines are summarized schematically in Figure 7. Without commenting on every detail the main points to be mentioned are the following.

(a) AO as the 'standard' dye stains nuclei and nucleoli green (or, rarely, slightly yellow). It has this property in common with 3-aa (and 3,6-diaminoacridine, as already mentioned).

(b) In the nuclear region quenching occurs with 9-AA. Its strong affinity for DNA is well known (Riva 1966; Löber and Achtert 1967).

(c) 2-AA fluorochromes the cytoplasm green. No staining of nuclei.

(d) 7-methoxy-9-methyl-2-AA stains cytoplasm green, nuclei and nucleoli orange. The dimethyl-2-AA derivative behaves similarly, although the orange fluorescence of nuclei and nucleoli is not equally strong. With 9-methyl-2-AA only nuclei of

phagocytosed cells show orange color. Nuclear remnants of phagocytosed cells are
frequently much accentuated by their orange color with the 2-AA substituted compounds.

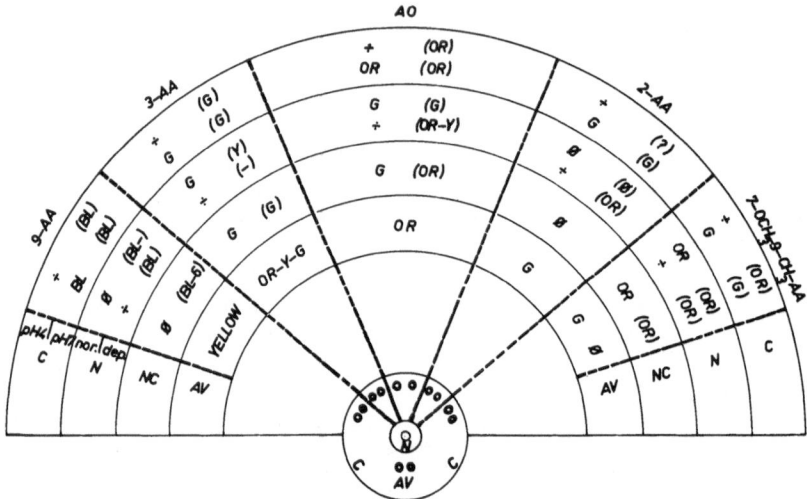

Figure 7. *Fluorescence properties of some cellular structures   C: cytoplasm;
N: nucleus; NC: nucleolus; AV: autophagic vacuoles; nor: 'normal' conditions;
dep: depolymerized; ÷ : not possible; -: not examined; colors; BL: blue; G: green;
BL-G: Blue-green; Y: yellow; OR: orange; symbols in parenthesis: fixed cells; without
paranthesis: living cells*

(e) Autophagic vacuoles are induced by the not fluorescing (in case of excitation
at 360 nm) compound 1,2,3,4-tetrahydo-9-AA (tacrine). These autolysosomes may attain
diameters up to 4 and 5 µm (Figure 8 see plate I.) After staining of autophagic
vacuoles in living fibroblasts, these structures show metachromatic reactions with AO,
3-AA, 9-AA. All of these compounds have high pka's (10.45; 8.04; 10.04 respectively)
and are equally capable of inducing autophagic vacuoles themselves (Kunze et al. 1970).

2-AA, both its methylated derivatives and its other derivatives not mentioned here,
have stained autolysosomes green, with varying intensity. A 2-AA compound with Cl in
position 6 was the most efficient. This compound also showed marked affinity for
cytoplasmic lipid inclusions. This quality is shared to a minor extent by other substances of the 2-AA group; 2-AA itself fluorochromed phagocytosed droplets of paraf-
finum liquidum an intense blue.

Fibroblast Cells Fixed in Ethanol-acetone. (a) Influence of changing pH: Nothing
new can be contributed concerning the staining effects of the three strongly cationic
aminoacridines (center and left of Figure 7.). Changing pH from 7 to 4.1 has an
especially profound effect on cells stained with the substituted 2-aminoacridines.
The orange color of nuclear regions is maintained, but the cytoplasm changes from
green to orange. Living cells and those stained after fixation at pH 7 are nearly
identical in appearance. Capacity to induce autophagic vacuoles is very weak or
missing in the 2-AA group, with the exception of 9-methyl-2,7-diaminoacridine. 2-AA
shows distinct green cytoplasmic fluorescence above pH 6.

(b) Desoxyribonuclease: Nuclear fluorescence is missing in all specimens observed.

(c) Ribonuclease: Cytoplasm shows loss of red fluorescence in the case of AO; at
pH 4.1 cytoplasmic fluorescence changes from orange to green under the 2-AA derivatives.

(d) Depolymerization of DNA by rapid change from hot to cold changes nuclear
fluorescence from green to orange-yellow in nuclei stained with AO, and from orange to
yellow with both 2-AA derivatives. Remarkably enough under these circumstances, 2-AA
stained cells exhibit a weak, but all the same distinct, orange fluorescence in the
nuclear region. "Partial" depolymerization by cold TCA for 4 hours has a similar
effect.

(e) Total extraction of nucleic acids with hot perchloric acid and tricheloracetic
acid results in a rather uniform picture with all aminoacridines: nuclei empty, a
small perinuclear rim fluorescing in a weak, greenish color.

PLATE I

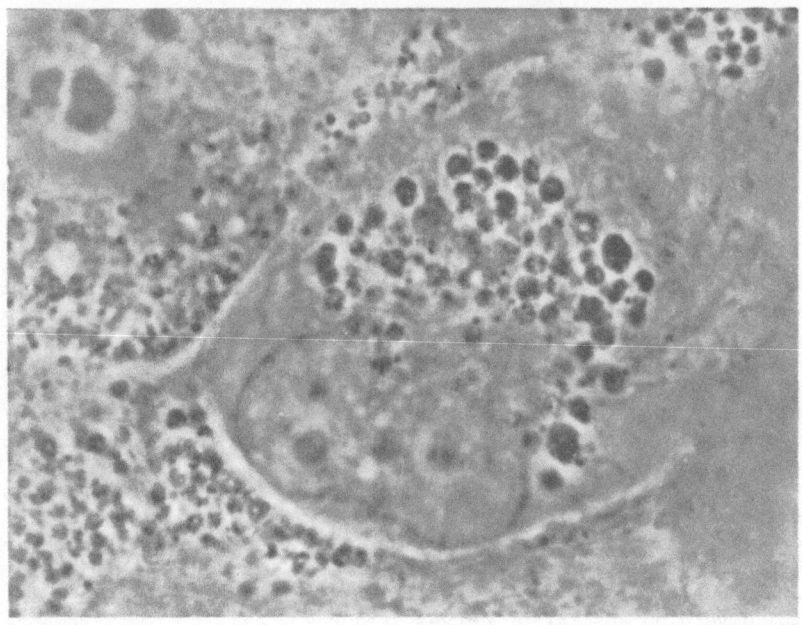

Figure 8. *Cultured cell after 24 hours exposure to 1, 2, 3, 4-tetrahydro-9-aminoacridine: cytoplasm studded with dark polymorphous autolysosomes. Objective: 100/1.32 oil, phasecontrast*

Mast Cells. The granules containing heparine fluoresce red with AO, green with 3-AA and 9-AA; fluorescence is missing in the 2-AA group.

DISCUSSION

Discussion of all physico-chemical and cytological aspects involved in this study would certainly be beyond the scope of the present paper. Only some points of immediate interest will be mentioned here.

(a) It is reasonably well established that nuclear orange-to-red fluorescence after application of the 2-AA derivatives points to DNA as the cellular anionic partner of these compounds.

(b) Cytoplasmic green fluorescence at pH 7 in living and fixed cells with the aminoacridines mentioned under a) may reflect interaction between the neutral form of the dyes and cellular proteins. Model experiments with fibrin and human serum albumin (fluorescing green) favor this assumption. Furthermore, at this pH, not more than a few percent of the compounds exist as monocations (e.g., at pH 7.3; 2-AA, 2% cation; 9-CH_3-2-AA, 3%; 2,7-diaminoacridin 4%; (Albert, 1966). Finally, at pH 4.1 (all-monocationic condition) nuclei and cytoplasm fluoresce orange.

It remains to be explained why only nuclear structures remain orange at neutral pH. The considerable power of both nucleic acids to maintain the acid form of dyes in more basic media has been emphasized by Morthland et al. (1954). Their results point to the existence of free anionic (phosphate-) binding sites specific for DNA. Nuclear orange fluorescence is demonstrable even at pH 9.

Like other weak cations (Förster 1946) the 2-aminoacridines exhibit marked changes in color with changing pH, and also demonstrate the well known (Linser 1932) pH-dependent fluorescence color. They have definite indicator properties. This is partly reflected by the diverging absorption curves (Figures 2 and 3). Measurements of emission spectra, especially by microfluorometric methods, seem clearly desirable and are being prepared.

(c) Another important point to be left for further investigation is the profound difference in fluorescence within the nuclear area between 2-AA and its methyl- and methoxy-substituted derivatives. The contribution of the auxochromic substitute CH_3O to the molecule's π-orbitals, and thus to absorption, is likely to occur, and will be certainly inferior to a second amino group. Supplementary investigations with 2,7-diamino-9-methyl-acridine have shown, that this substance also fluorochromes nuclei red, with about the same, and at pH7, even somewhat more intensity than does the methoxy-methyl derivative. CH_3 in position 9 is even less effective on the reactive molecular π-electrons, and the 7,9-dimethyl derivative is slightly inferior in marking the nuclear DNA than are members of the 2-AA group with substituents -NH_2 and -OCH_3 in position 7. The statement by Albert (1966) that C-substituents modify some property already inherent in the corresponding monoaminoacridine holds true for predictions made on absorption spectra, but apparently additional factors come into play when interactions of substituted aminoacridines with helically structured DNA become effective.

Certain conditions (e.g., sterical or hydrogen bonds) of the DNA helix may favor the orange fluorescence, which very probably is based on monomeric cations of the substituted 2-aminoacridines. Still unanswered, however, is the question as to why depolymerization (and denaturation) of DNA favors nuclear staining by 2-AA, which apparently shows no affinity for RNA where one expects a similar decreased order of nucleic acid chains.

(d) Nuclear orange fluorescence is certainly not of metachromatic nature. The minimal concentration of 2-AA solutions showing deviation from Lambert-Beer's law is 9.4 x 10^{-5} M (Albert 1966). A study of the absorption spectra of the 2-AA group did not provide unequivocal evidence either for reduction of α-band, or for new maxima appearing at shorter wavelengths with increasing concentrations up to 5 x 10^{-4} M. Scheibe (1939) and Zanker (1952 a, b; 1954) carried out extensive studies on metachromasy based either on the association of monomeric cations (rarely anions) to dimers (and polymers) in aqueous solutions or on certain polyanions. To be clearly ascertained in absorption curves it has to meet requirements with regard to molecular structure similar to those realized with such well known dyes as thionine, methylenblue, and acridine orange. They all have two NH_2 substituents, with the N adding to the aromatic cationic charge resonance, and accordingly with log ε above 4 in the visible range

(Zanker 1971). Thus the 2-AA compounds are scarcely entitled to their designation as dyes. According to Förster (1951), increased mutual contact of dye molecules leads to altered transition probabilities caused by a shift in the equilibrium of the excited electronic state. It may be interesting in this context to mention that 9-AA shows metachromatic properties only when bound to acidic phospholipids in autophagic vacuoles. To our knowledge, only in this exceptional case does this strongly basic monoaminoacridine exist in a dimeric form.

Since, in the binding of aminoacridines and their alkylated derivatives to DNA, the free energy increases linearly with increasing pka (Löber 1967), the 2-AA compounds cannot be expected to be tightly bound to DNA. Albert and his co-workers have convincingly demonstrated correlations between cationic strength of aminoacridines and their bacteriostatic power. Of immediate concern in this discussion is the statement by Albert (1966) about the slightly suppressing effect of 9-methyl-2,7-diamino-acridine on bacterial activity, and the loss of this antiseptic property by replacing NH_2- in position 7 by CH_3-. However, the resulting compound 7,9-dimethyl-2-AA has been shown, by observing its fluorescent properties at various cell constituents, to interact with DNA similarly to 2,7-diaminoacridine but differently from 2-AA, with both being almost equally ineffective against bacterial growth.

The assumption that some specific factors may be involved in binding members of the 2-AA group to the helical structure of DNA is supported by the intense green fluorescence exhibited by the 2-AA compounds at autophagic vacuoles. Their anionic partner in these structures is a strongly acidic phospholipid (Dingle and Barrett 1969), which allows the more basic aminoacridines to display metachromatic fluorescence (Figure 7).

The quenching of fluorescence is obvious when substituted 2-AA is bound to the strongly anionic $-SO_3H$ groups of heparine in mast cells. This is possibly an indication of molecular association in which the existence of metastabile triplet conditions may contribute to the loss of fluorescence. Investigating an eventual phosphorescence seems desirable in this case. Quenching of fluorescence at sulfated polysaccharides seems to demonstrate the significance of the distance between the dye molecules as influenced by the respective anionic cellular substrates providing ligands for these dye molecules.

The question of how far the substituted 2-aminoacridines lend themselves to quantitative microspectrofluorometric measurements remains unanswered at the moment. There is evidence, however, that in the nuclear region 7-methoxy-9-methyl-2-AA does show a red fluorescence which is stable over at least 5 seconds of exposure to ultra-violet and blue light excitation. Distinct differences in fluorescence intensity have been observed in interphase nuclei, fibroblasts, lymphocytes, and metaphase chromosomes.

ACKNOWLEDGEMENTS

The author wishes to express his gratitude to Professor V. Zanker (Munich) for valuable discussion of the results; to Miss V. Kretschmer, Miss I. Sohmer, and Mr. Unger for skillful technical assistance; to Hoffmann-La Roche Gmbh and Co. (Basle/Switzerland) for financial support of this study.

REFERENCES

ALBERT, A., and Linnell, W. H., (1936), *J. Chem. Soc.*, 88.

ALBERT, A., and Linnell, W. H., (1936b), *J. Chem. Soc.*, 1614.

ALBERT, A., and Ritchie, B., (1941), *J. Soc. Chem. Indust.*, 60, 120.

ALBERT, A., and Ritchie, B., (1963), *J. Chem Soc.*, 458.

ALBERT, A., (1966), *The Acridines*. 2nd Ed. E. Arnold Publ., London.

ALLISON, A. C., and Young, M. R., (1964), *Life Sci.* 3, 1407.

APPEL, W., and Zanker, V., (1958), *Z. Naturforsch.* 13b, 126

ARMSTRONG, J. A., (1956), *Exp. Cell Res.* 11, 640.

BERTALANFFY, L. Von., and Bickis, J., (1956), *J. Histochem. Cytochem.* $\underline{4}$, 481.

BERTALANFFY, L. Von., (1963), *Protoplasma.* $\underline{57}$, 51.

BUKATSCH, F., and Haitinger, M., (1940), *Protoplasma.* $\underline{34}$, 515.

CRAIG, D. P., and Short, L. N., (1945), *J. Chem. Soc.* 419.

DINGLE, J. T. and Barrett, A. J., (1969), *Proc. Roy. Soc.*, $\underline{169}$, 85.

DOTY, P., Marmur, J., Eigner, J., and Schildkraut, C. L. (1960), *Proc. N.A.S.* $\underline{46}$, 461.

FÖRSTER, TH., (1951), *Fluoreszenz Organischer Verbindungen.* Vandenhoeck und Rupprecht, Göttingen.

GÖSSNER, W., (1949), *Verh. Dtsch. Ges. Path.* $\underline{33}$, 102.

KASTEN, F., (1965), *Stain Technol.* $\underline{40}$, 127.

KASTEN, F., (1967), *Interm. Rev. Cytol.* $\underline{21}$, 141.

KUNZE, J., Staubesand, J., and Wittekind, D., (1968), *Acta Histochemica,* $\underline{\text{Suppl. IX}}$.693.

KUNZE, J., Staubesand J., and Wittekind, D., (1970), *64. Verh. Anat. Ges. (Erg. H.) Anat. Anz.* $\underline{126}$, 511.

LERMAN, L.S., (1961), *J. Mol. Biol.* $\underline{3}$, 18.

LERMAN, L. S., (1963), *Proc. N.A.S.* $\underline{49}$, 94.

LERMAN, L. S., (1964), *J. Mol. Biol.* $\underline{10}$, 367.

LINSER, H., (1932), *Biochem. Z.* $\underline{244}$, 157.

LÖBER, G., (1967), *Studia Biophys.* $\underline{2}$, 299.

LÖBER, G., and Achtert, G., (1967), *Studia Biophys.* $\underline{3}$, 113.

MARMUR, J., Rownd, R., and Schildkraut, C. L., (1963), *Progr. Nucleic Acid Res.* $\underline{1}$, 231.

MORTHLAND, F., de Bruyn, P. P., and Smith, N., (1954), *Exp. Cell Res.* $\underline{7}$, 201.

PEACOCKE, A. R., and Skerrett, J. N. H., (1956), *Trans. Faraday Soc.* $\underline{52}$, 261.

RIGLER, R., (1966), *Acta Physiol. Scand.* $\underline{67}$, Suppl. 267.

RIVA, S. C., (1966), *Biochem. Biophys. Res. Comm.* $\underline{23}$, 606.

ROBBINS, E., (1960), *J. Gen. Physiol.* $\underline{431}$, 853.

SCHEIBE, G., (1939), *Z. Angew. Chem.* $\underline{52}$, 631.

SCHNEIDER, W., (1945), *J. Biol. Chem.* $\underline{161}$, 293.

SCHÜMMELFEDER, N., (1950), *Virch. Arch.* $\underline{318}$, 119.

SCHÜMMELFEDER, N., Ebschner, K. J., and Krogh, R. E., (1957), *Naturwiss.* $\underline{44}$, 467.

SCHÜMMELFEDER, N., Krogh, R. E., and Ebschner, K. J., (1958), *Histochemie,* $\underline{1}$, 1.

STEINER, R. F., and Beers, R. F., (1962), *Polynucleotides.* Elsevier, Amsterdam.

STONE, A. L., and Bradley, D. F., (1961), *J. Amer. Chem. Soc.* $\underline{83}$, 3627.

STOCKINGER, L., (1964), In: *Protoplasmatologia, Hndb. Protoplasmaforsch.* $\underline{\text{II/Dl}}$, Springer, Vienna.

STRUGGER, S., (1940), *Jena. Z. Med. Naturw.* $\underline{73}$, 97.

STRUGGER, S., (1949), *Fluoreszenzmikroskopie und Mikrobiologie.* M. u. H. Schaper, Hannover.

WEISSMANN, C., (1953), *Z. Zellforsch.* $\underline{38}$, 374.

WEISSMANN, C., and Gilgen, A., (1956), *Z. Zellforsch.* $\underline{44}$, 292.

WEST, S. S., (1965), In: *Methoden und Ergbnisse der Cytophotometrie, Verh. Ges. Histochem. 1963; Acta Histochem.* $\underline{\text{Suppl. VI}}$.

WITTEKIND, D., (1958), *Z. Zellforsch.* <u>49</u>, 58.

WITTEKIND, D., and Kunze, J., (1969), 63. *Verh. Anat. Ges. (Erg.H.) Anat. Anz.* <u>125</u>, 241.

WITTEKIND, D., Staubesand, J., Kretschmer, V., and Möhring, E., (1970), *Cytobiologie.* <u>2</u>, 275.

WITTEKIND, D., and Kretschmer, V., (1972), *Z. Zellforsch.* <u>126</u>, 518.

WITTWER, A., and Zanker, V., (1959), *Z. Physik. Chem. (N.F.)* <u>22</u>, 27.

ZANKER, V., (1952), *Z. physik. Chem.* <u>199</u>, 225.

ZANKER, V., (1952), *Z. physik. Chem.* <u>200</u>, 250.

ZANKER, V., (1954), *Z. physik. Chem. (N.F.)* <u>2</u>, 52.

ZANKER, V., (1971), *XV. Symp. Ges. Histochemie.* Dusseldorf.

ZEIGER, K., and Schmidt, W., (1957), *Z. Zellforsch.* <u>45</u>, 578.

FLUOROMETRIC RECOGNITION OF CHROMOSOMES AND CHROMOSOME REGIONS

Torbjörn Caspersson

*Institute for Medical Cell Research & Genetics, Medical Nobel Institute,
Karolinska Institutet, Stockholm, Sweden*

INTRODUCTION

Recently considerable attention has been focused on the application of fluorescence procedures to the problem of chromosome identification in mammals, particularly in man. Although the methods are not much more than two years old, a great number of laboratories are using them for medical and/or general cell-biological purposes. This may be considered a very good illustration of what *quantitative* fluorescence work — the central theme of this meeting — can contribute to different fields in biology and medicine. This whole field rests on microfluorometric measurements, which also provided the necessary basis for the now well established system used for identification of the human karyotype. The gist of the method is to produce a banding pattern over the chromosome observable by fluorescence techniques — by binding certain fluorochromes to chromosomal constituents.

The chemistry of the fluorescence procedures used for chromosome identification is discussed by Dr. Ed. Modest in another paper in these proceedings; the present discussion is focused on consideration of fluorometric techniques developed for this field and statistical and other techniques used for the analysis of the fluorescence patterns obtained from the chromosomes. Here the aim has been not only to furnish statistical information about the constancy of the patterns in man, but also to analyze certain types of chromosome aberrations of clinical importance. These techniques have logically grown out of earlier efforts in Stockholm to perform quantitative cytochemical work on chromosomes and so it makes sense at this point briefly to review the groundwork leading up to the present study.

Among the primary requisites for any kind of cytochemical analytical work on metaphase chromosomes are the recognition and identification of the individual chromosome region to be studied. These steps are absolutely essential to this type of work, as several different methods have to be applied one after the other on the same chromosome region, and the results of the analyses in different metaphase plates have to be compared.

In most cases the identification presents great technical difficulties. Chromosome size and morphology differ very much in different animal and plant species. In most objects of interest, however, the size is rather close to the resolution limit of the optical microscope, and the chromosomes carry few morphological markers. Thus, very often, it is not even possible to identify whole chromosomes with conventional cytological techniques. This is the case for most of the mammals, and is especially true for man, in whom only 7 of the 24 chromosome types can be recognized by conventional staining techniques.

The rapid development of mammalian cytogenetics, including clinically applied human cytogenetics, in the last decade has emphasized the need to develop better chromosome recognition procedures. At present the larger part of the general biomedical, cell biological work is centered around man and a small number of higher mammals. It is logical, therefore, in seeking improved chromosome region identification techniques, to concentrate on these species, in spite of the obvious difficulties involved. Regrettably enough, all of these species have quite small chromosomes, but large numbers of them, and very few morphological markers.

Quantitative cytochemical chromosome work goes back to the late thirties when the first ultramicrospectrophotometers were developed. The main problem then was the relation between the protein and nucleic acid metabolisms (Caspersson 1950). During the following two decades, quantitative cytochemical work with optical procedures concerned mainly interphase cells. In the early sixties, however, our biophysics group, where Gösta Lomakka deserves special mention, carried through quite an ambitious program for the development of ultramicrospectrophotometric, ultramicrointerferometric, and micro-X-ray-absorbometric techniques to the very limit of resolution in order to get chemical information from chromosome parts and other nuclear structures. It

108

PLATE I

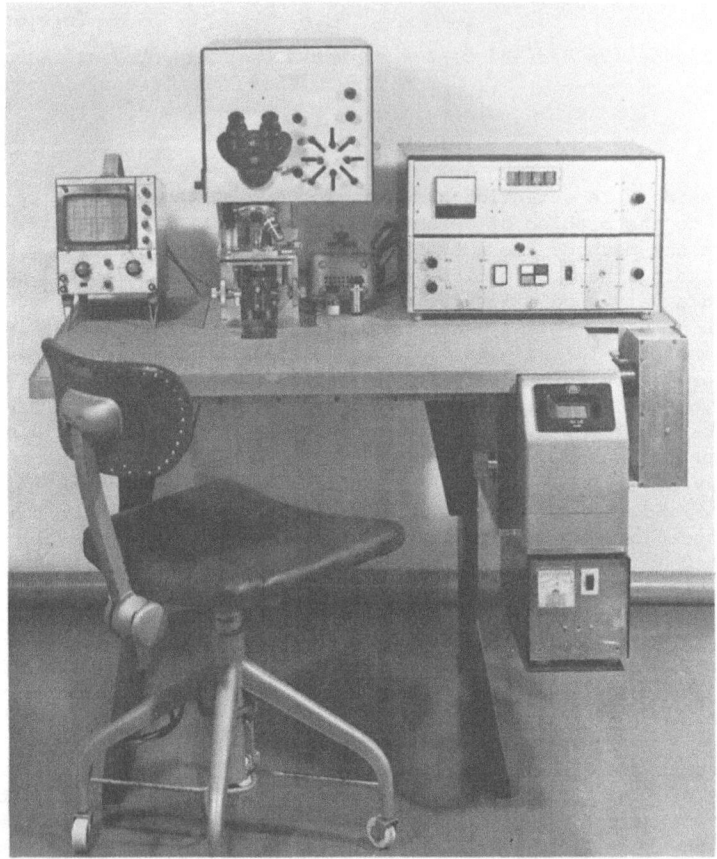

Figure 1. *Rapid scanning ultramicrospectrophotometer for ultraviolet and visible light, and with preselection of measuring field*

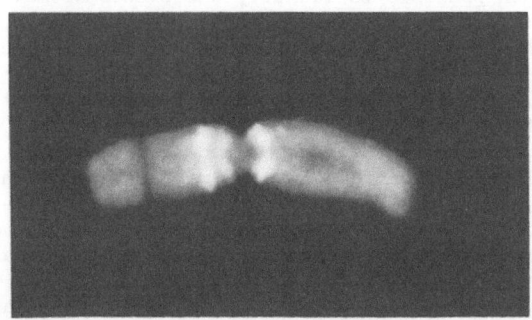

Figure 3. *Quinacrine mustard-stained M-chromosomes from* Vicia faba. *Fluorescence photomicrograph, 2500 x*

proved possible for all techniques to get very close to the limit of resolution of the light optical microscope (Caspersson and Lomakka 1970). Then, inevitably, chromosome identification got to be a crucial problem in cytochemical work.

Extensive work was performed over a period of about two years to try to use the determination of DNA in the individual chromosomes for identification purposes (Caspersson and Lomakka 1970; Carlson et al. 1963). With the most refined ultra-microspectrophotometric procedures (Figure 1. see Plate I) this type of chromosome classification could be done in some objects. For those in the center of interest, e. g., man and the mouse, the differences between many chromosomes were so small that the technique would have been useless in practical work.

One step forward was to determine the DNA-distribution pattern along the chromosome by scanning the chromosome cross-wise with a very small measuring aperture moved in a very tight scanning pattern (Carlson et al. 1963; Heneen and Caspersson 1973; Caspersson et al. 1970b). Chromosomes from several species with large chromosomes were tried, but showed little detail. The first promising indications came from rye, with seven chromosome pairs of a size comparable with those met in the human karyotype. The different rye chromosomes are of the same length and width, but have a few morphological markers which make it possible for an experienced cytologist to distinguish between them. In collaboration with the cytologist Waheeb Heneen, we measured the DNA-distribution patterns. As demonstrated by Figure 2, these measured patterns were so characteristic that they could serve for identification purposes.

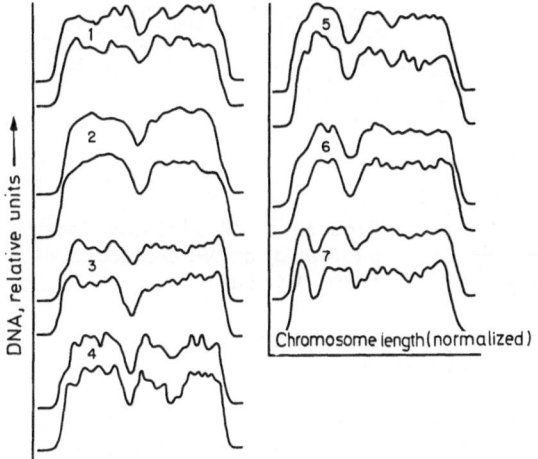

Figure 2. *DNA-distribution patterns for the seven chromosomes of rye, measured by high-resolution scanning ultramicro-spectrophotometry*

Similar measurements were then made on human chromosomes, also in extensive experiments. The system worked fine for the larger chromosomes in the karyotype, but was not very reliable for the smaller chromosomes. In addition, measurement was extremely difficult and the instrumentation so expensive that the technique could hardly be of any practical value.

The importance of these studies is that they showed definitely *that it would be possible to use the pattern of distribution of chemical substances, such as DNA along the chromosomes, for chromosome and chromosome region identification purposes,* but that better, and not least much more convenient methods had to be developed.

This led our thoughts to fluorescence techniques. These have several great advantages over absorbometric procedures. One is their extreme sensitivity. For instance Rigler (1969) showed in one of our ultramicrofluorometers in model experiments that by binding acridine orange to DNA, amounts of DNA down to below 10^{-15} g could be determined. It is also a well known fact that the discriminative power of a fluorescence microscope for particles adjacent to one another is almost twice that of conventional bright field microscopy in transmitted light (due to the difference between "Selbstleuchter" and "Nicht-Selbstleuchter" according to Abbe). Furthermore, the determination of the amounts of fluorescing substances in an irregular object is obviously much simpler than absorbometric quantitative work.

The way chosen was to look for fluorescing substances binding to the DNA. For several different reasons, we started with base-binding substances. One reason was simply that phosphate-binding and also carbohydrate-reacting dyes (Feulgen) were known to show very few details in metaphase chromosomes. This led to the choice of certain acridine compounds - in the first instance quinacrine mustard (QM) and quinacrine (Q), of which the former gave best results (see E. Modest, these proceedings).

These compounds bind to DNA, and thus the pattern of fluorescence is influenced by the DNA-distribution pattern. As stated previously, the measurement of such fluorescence patterns is very much easier than absorbometric DNA-pattern measurements. However, other factors also influence the fluorescence patterns. Figure 3 (see Plate I) shows strongly fluorescent regions in a very large metaphase chromosome from a plant, *Vicia faba*. The DNA-distribution pattern in Figure 4 clearly demonstrates that in the bands the fluorescence per unit DNA is considerably higher than in the rest of the chromosomes. This illustrates the fact that the Q- and QM-patterns are influenced by more factors than mere DNA distribution. Such factors are presence of proteins blocking the binding capacity of the DNA in some chromosome regions, and include enhancement or diminishing of the acridine fluorescence by adjacent DNA base pairs (AT and GC, respectively) (Rigler 1969; Pachmann and Rigler 1972; Weissblum and de Haseth 1972). Figure 5 (see Plate II) shows a high resolution ultramicrofluorometer specially developed for detailed study of such patterns (Caspersson et al. 1969b; Caspersson et al. 1969a).

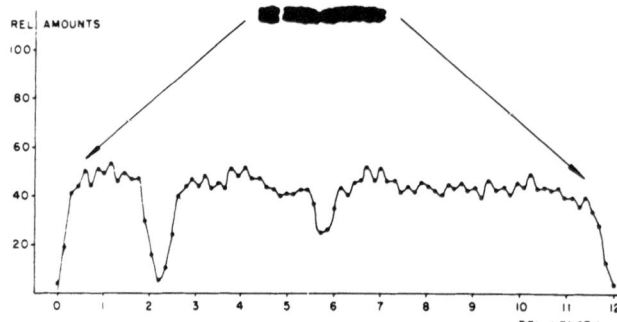

Figure 4. *DNA-distribution pattern of an M-chromosome from* Vicia

Several technical problems arise in ultramicrofluorometry that close to the limit of resolution of the microscope. Firstly, very high-class optics have to be used in order to catch the fine details of the patterns. Secondly, the exposure of the object to the exciting radiation must be as low as possible because of fluorescence fading, which runs relatively fast in these very minute structures. Thirdly, great precision is required both to define the place in the preparation where measurement has to be done, and to set the size and the shape of the area to be measured.

METHODS

In the instrument depicted in Figure 5, these conditions have been met in the following ways. A very high-quality fluorescence optical system is used, with epi-illumination and a 1.30 numerical aperture lens, working at full aperture. The Leitz and the Zeiss systems that have been tested have given the same good results. A phase-contrast system is used for all preparatory work, starting with the search for a suitable place in the object, up to and including the setting of the measuring diaphragms to give the desired location, size, and shape of the area to be measured.

In order to facilitate the diaphragm setting, the optical arrangement outlined in Figure 6 is used. The image from the microscope, by phase or fluorescence, is projected at high magnification by a special prism system over a movable stage. On this stage, diaphragms of suitable sizes and shapes can be placed in order to cut out the image of the region to be measured. The light from this region then goes by way of an optical system, including a simple monochromator arrangement, to the photo-multiplier. By moving the stage, scanning measurements can also be made - for instance determination of the fluorescence patterns of chromosomes. Crucial to the practical work is an arrangement whereby the measuring area can be directly and very conveniently

PLATE II

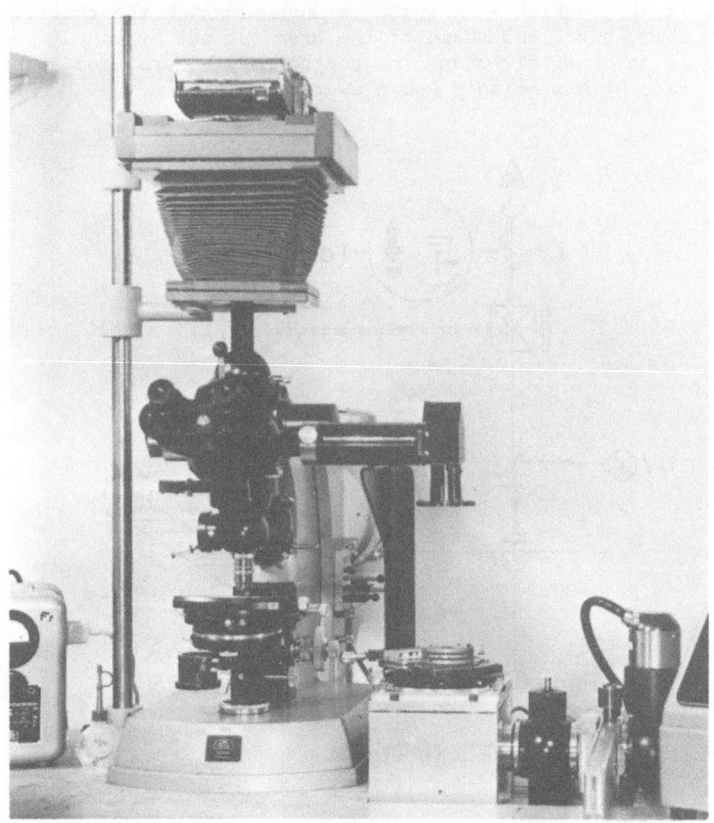

FIGURE 5. *High resolution ultramicrofluorometer, developed for chromosome measurements with arrangement for very precise preselection of measuring field and with low irradiation load on object*

observed by the eye. The diaphragm on the stage is illuminated from behind by a red lamp. This red light is then led back to the image the observer sees in the microscope. The visual image in the microscope is thus the conventional fluorescence image, in the case of Q, it is greenish - on which is superimposed the small, brightly red spot showing the place, size, and shape of the area cut out by the diaphragm. Thus, the diaphragm can be set entirely under the control of the eye, and scanning measurements can also be made with precisely known sweeps.

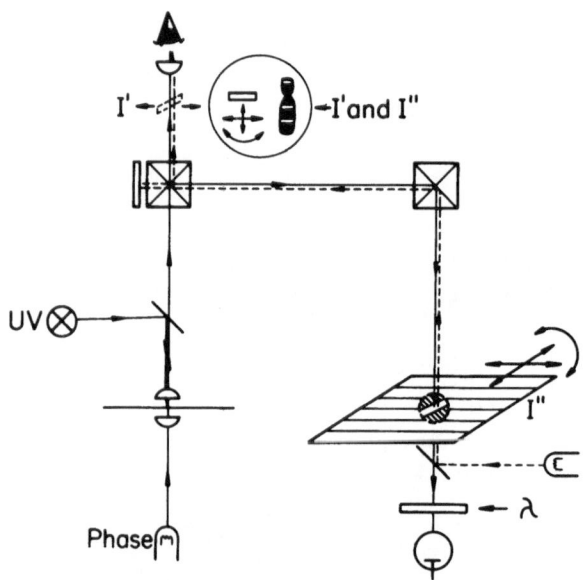

Figure 6. *Diagram for the instrument in Figure 5*

Measurements were performed with this instrument on a variety of objects, mainly plants. Intensely fluorescent regions such as those demonstrated earlier occur in many species and can sometimes be used for chromosome identification. In most species, however, they are so few that they are of little use in the general identification problem.

Man belongs to this type in which only very few intensely fluorescent spots occur regularly. In the studies on human blood material (Caspersson et al. 1970d; 1970c; 1971c), however, another very faint pattern of fluorescent banding was observed (Figure 7. see Plate III). It was so faint that in the beginning it could not be studied visually. Nonvisual measurements, however, indicated that the pattern covered all the chromosomes, and that it was different in different chromosome types. Obviously here was a way to achieve chromosome identification, but in order to study it, it would be necessary to measure a very great number of metaphases. This would have been a very difficult task in the fluorometer described.

In order to assemble rapidly a large number and range of fluorescence patterns, a less precise but very much faster fluorometer system (Caspersson et al. 1970d) was devised (Figures 8. see Plate IV, and Figures 9 and 10). It was based on microdensitometry of fluorescence photomicrographs of Q- or QM-stained metaphase plates. In it, both negatives and positives can be used; the latter, with the aid of a reflectometer arrangement. The advantage of positives is that one can easily cut them up and make a preliminary sorting of the chromosome pictures to be measured. In routine work, aiming at studies of large numbers of chromosomes, this saves much time.

A very large number of fluorescence patterns have been studied in this instrument. Based on material composed of about 30,000 curves from this instrument, it is now possible to identify all human chromosomes accurately. (Caspersson et al. 1970c; 1970d; 1971c; 1972; and Caspersson and Zech 1972). The current, internationally accepted numbering and definition of human chromosomes are based on these measurements.

PLATE III

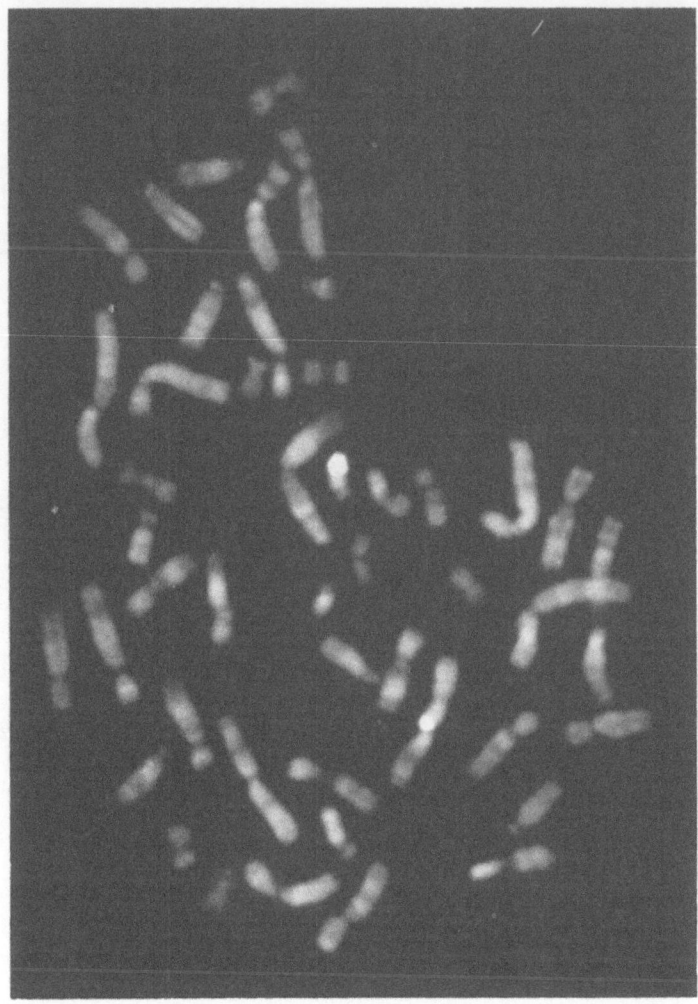

FIGURE 7. *Human metaphase plate after QM staining. 2400 x. Observe the fine fluorescence pattern over all of the chromosomes*

PLATE IV

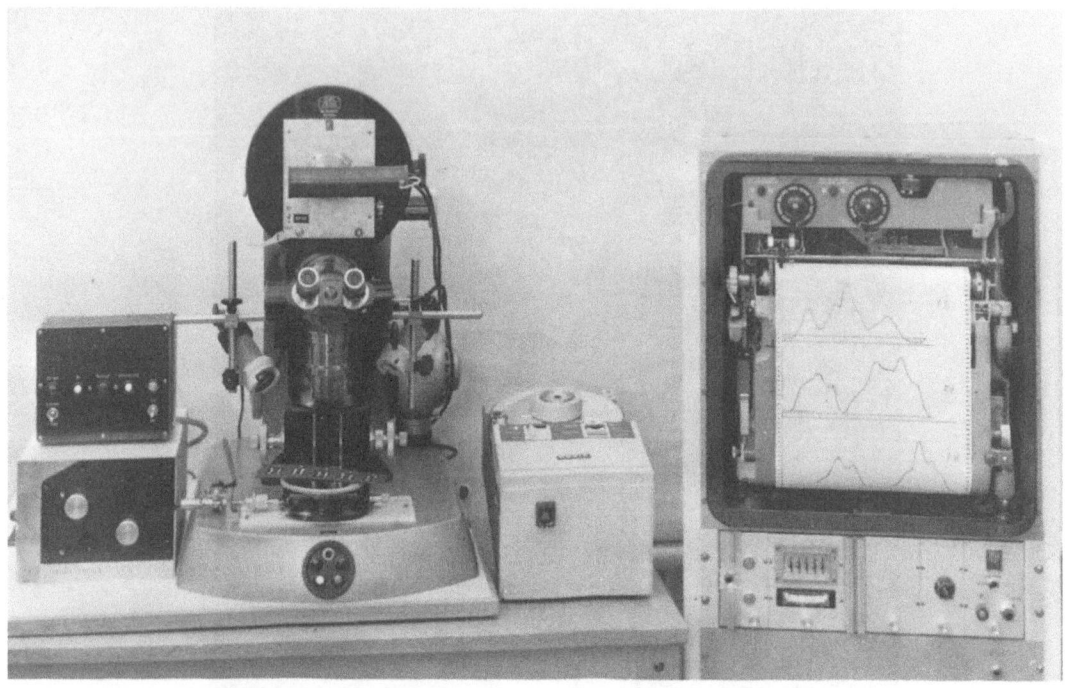

Figure 8. *Combined reflectometer and densitometer for rapid measurements of fluorescence patterns in prints or photographic negatives*

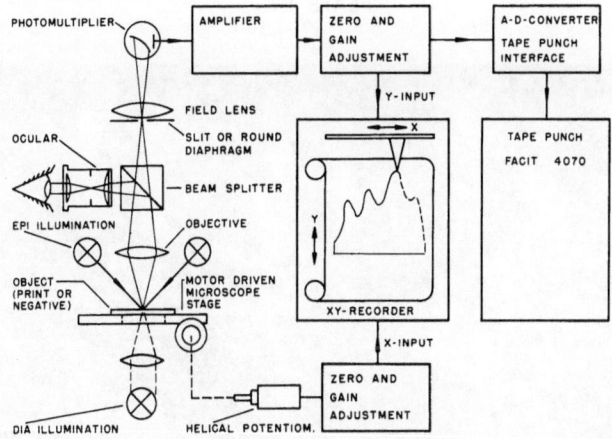

Figure 9. *Diagram of the optical and mechanical parts of the instrument in Figure 8*

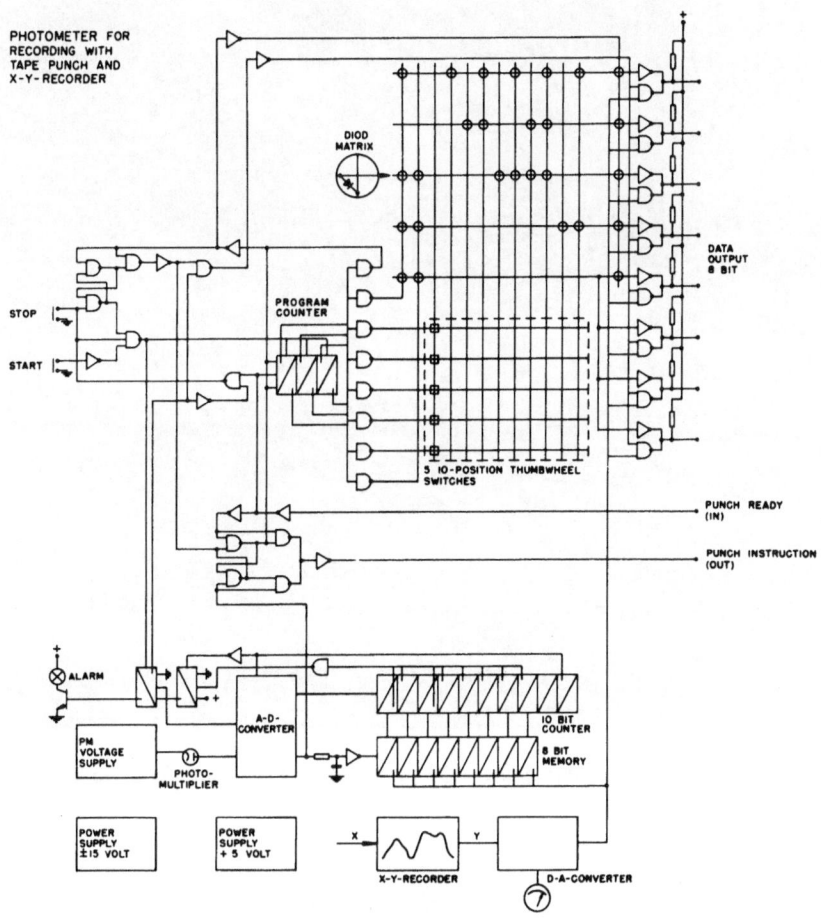

Figure 10. *Electrical circuits for the instrument in Figure 8*

PLATE V

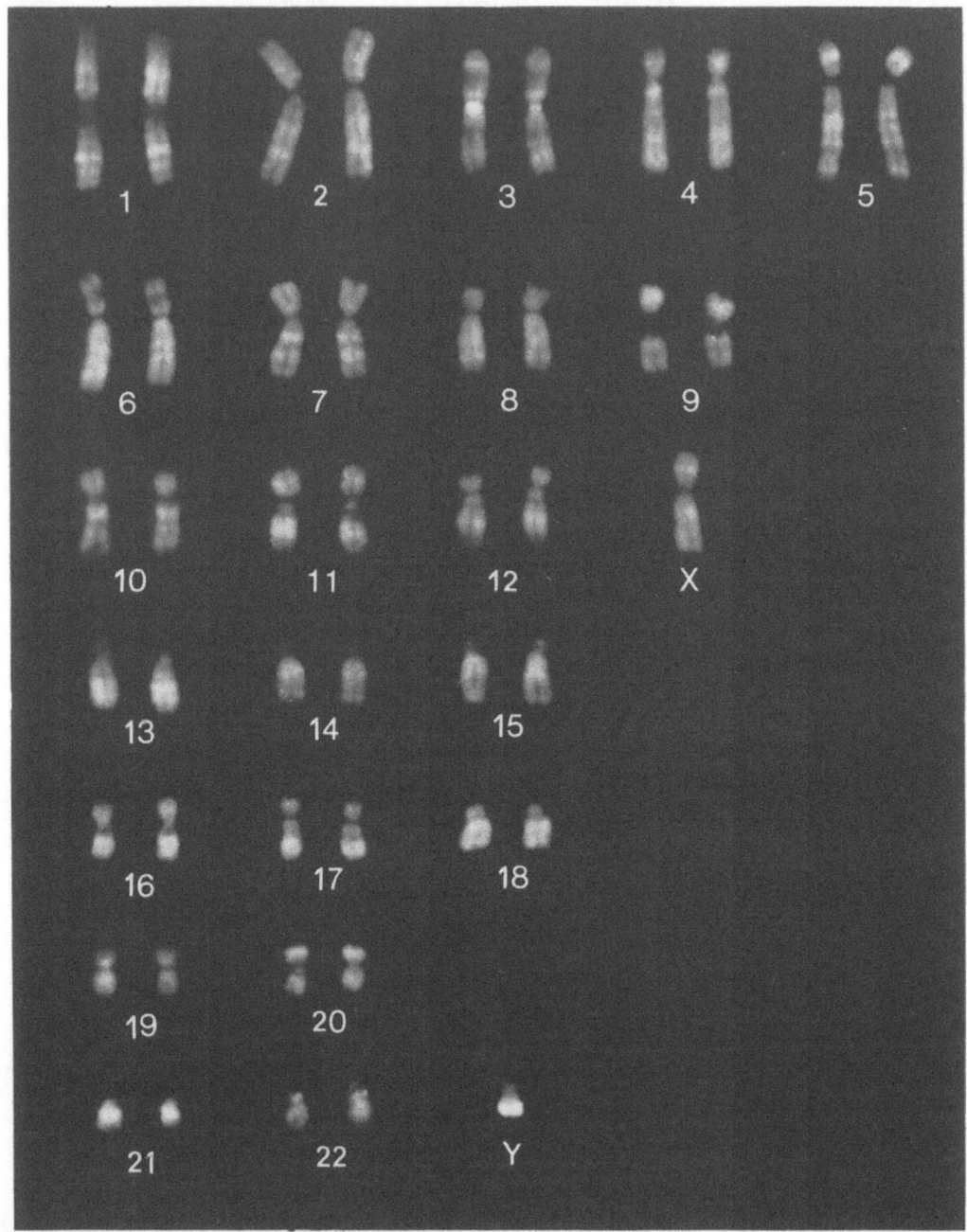

Figure 11. *The human karyotype. QM preparation. Fluorescence photography,*
2500 x

The whole identification system is built upon the fluorometric measurements. The present cytological technique also permits chromosome identification to be done visually on fluorescence photographs. Figure 11 (see Plate V) shows the human karyotype, derived from a series of prints from a negative, exposed and developed so that the intensity differences are considerably enhanced. Thus in present-day routine laboratory work, chromosome classification is done mainly visually. For statistical studies, of course, the pattern measurements are necessary. In the analysis of chromosome aberrations, fluorometry is a great advantage.

Figure 12 shows the typical patterns of the whole human karyotype, and is the accepted norm for chromosome numbering at present. In numbering the chromosomes we followed, as far as possible, the principle agreed upon earlier, that the shorter the chromosome the higher the number. In addition, the numbers were kept that had earlier been assigned to 13 of the 24 chromosome types by means of morphology and autoradiography.

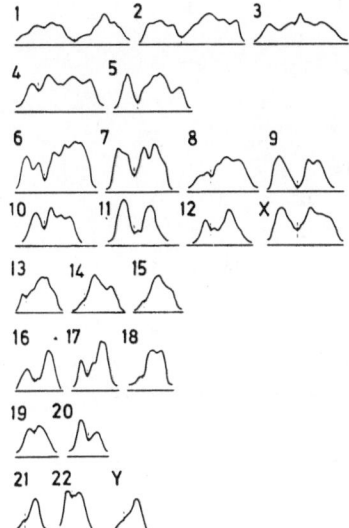

Figure 12. *The human karyotype. QM preparation. Typical fluorescence distribution patterns of chromosomes*

In order to study the statistical significance of the patterns, computerized procedures were used (Caspersson et al. 1972; 1971b; Möller et al. 1972). The first effort concerned the C-group, the 8 chromosomes of which have practically the same length and are indistinguishable from each other by earlier techniques. It was found that in Fourier analysis of the patterns, these chromosomes could be described by a surprisingly low number of coefficients; this simplifies the statistical work considerably. In material covering about ten sets of each of the chromosomes, the statistical significance of the patterns was shown to be very high. Using the same material, the computer could sort all the original patterns correctly. This led to the idea of using the patterns to sort chromosomes automatically.

In order to assemble a large amount of data, the fluorescence pattern recording device was equipped with a digital output and a tape punch (see Figure 10). It was thus possible to work quite rapidly, especially as the tape could be read directly by the computer. The confusion matrix in Figure 13 illustrates the present state of the technique. The accuracy of this computerized pattern identification is close to 90%, but can be improved considerably. These patterns are the background of work done in collaberation with Dr. Castleman at the Jet Propulsion Laboratory (JPL), Pasadena, to develop an entirely automated chromosome recognition system using the automatic microscope and film scanner, connected to a digital computer, earlier developed at the JPL (Caspersson et al. 1971a). What we are especially interested in is not the actual chromosome sorting - that can be done visually very quickly - but the possibility of sorting out aberrant chromosomes.

FROM / TO	1	2	3	4	5	6	7	8	9	10	11	12	13	14	15	16	17	18	19	20	21	22	Y	X
1	37					1																		
2		36																						
3			37	2		1	1																	
4			2	36	2																			
5				2	39																			
6				1		33	1						1											
7							38		1															
8						1		33	1	5		1		1										
9								2	34		3													2
10										37														
11										2	38													
12												38												
13				1		1							33	1			1				1			1
14													2	28	8		1							
15													5	8	31					1				
16																38	2		1					
17													1			1	38		1					
18																		39		1				
19													1						32	2	7	1		
20																			2	34		1		
21																					36	3	2	
22													2			1	1	2	3	4	4	31		
y													2								1		8	
x																								19
SUM	37	36	39	41	41	37	40	36	38	42	41	39	44	41	39	40	41	42	38	41	42	44	12	22
ERROR			2	5	2	4	2	3	4	5	3	1	11	13	8	2	3	3	6	7	6	13	4	3

Figure 13. *Confusion matrix of a series of 913 chromosomes, sorted by aid of a computer program*

Apart from the possibility of recognizing individual chromosomes, the pattern recognition technique also offers opportunities to identify individual chromosome regions. This is of special significance in a) mutagenesis work, b) the study of medically important aberrations, and c) cytochemical chromosome analysis with different kinds of methods. Fluorescence techniques are especially suitable for this purpose because quantitative recordings of pattern details can be made easily and analyzed mathematically by computer.

The selection of suitable bands was made in the following way (Möller et al. 1972). In order to get information on the location of the fluorescent bands which are most significant for the identification of *chromosome regions*, the average patterns for all chromosome types were first calculated by computer (about 1,000 chromosomes). Because of small variations in contraction within the different parts of the same chromosome, the average curves provide only a general outline of the banding pattern. In order to identify finer details, the individual pattern curves were subjected to mathematical contrast enhancement by computer. It was then possible, by means of digital filtering, to detect weak bands that had been partially hidden by adjacent higher bands; these could then be identified as clear maxima in their proper places. Different degrees of contrast enhancement were tested. Figure 14 depicts the whole human karyotype with greatly enhanced contrasts, clearly showing the precise location of even very faint bands.

All the contrast-enhanced individual patterns for each chromosome type were compared and the prominent bands were superimposed on the averaged pattern curves. In this way it was possible to get precise information on the location of the constantly observable bands; these locations could then serve as reference points for the identification of different chromosome regions.

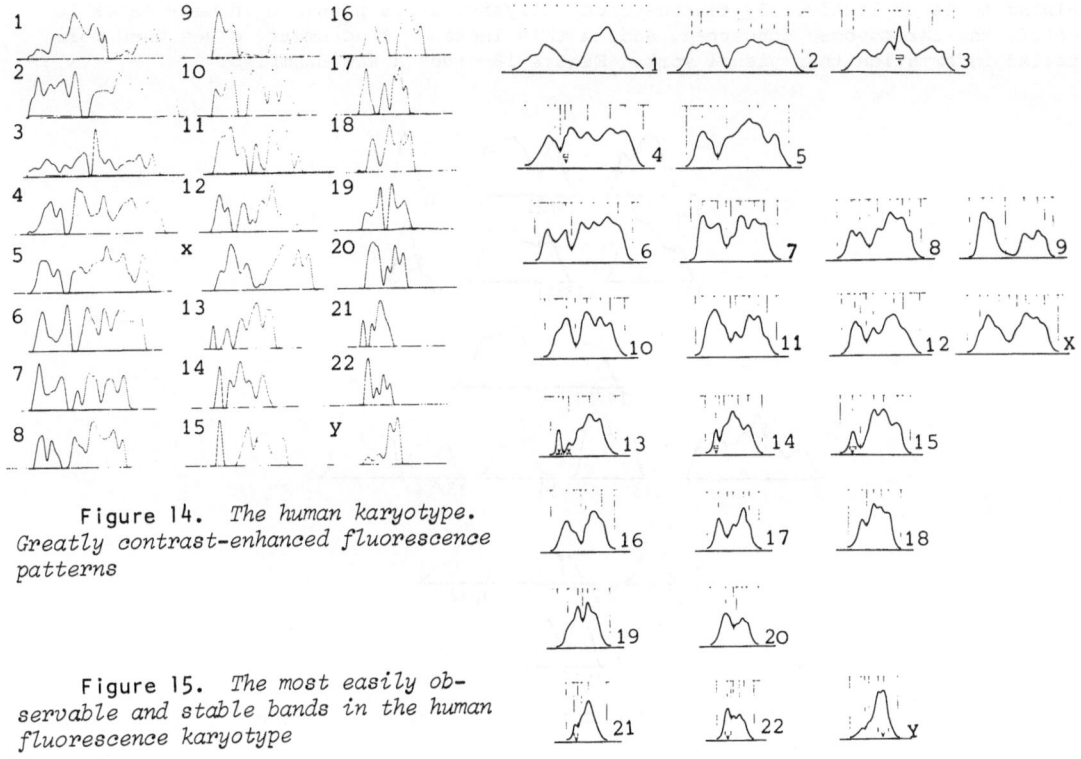

Figure 14. *The human karyotype.*
Greatly contrast-enhanced fluorescence
patterns

Figure 15. *The most easily ob-*
servable and stable bands in the human
fluorescence karyotype

APPLICATIONS

The outcome of this analytical work on the human karyotype is presented in Figure 15. The locations of the bands which were shown to be constantly reproducible by the analytical work described above are marked with vertical lines. The several short variant regions, subject to individual variations, are marked "v" in the Figure.

The biomedical applications of these procedures are beyond the scope of this report, but it might be appropriate to comment on them briefly here. When the fluorometric analysis had shown the patterns to be constant and typical for all the human chromosome types, and when the cytological techniques had improved so that the patterns could be made clearly visible in photographs, it was natural to rely mainly on visual analysis of photographs. Quite a large number of laboratories use these methods today; some of them also use fluorometry.

Generally, concerning the application of fluorescence techniques in human cytogenetics, it can be said that the finding that all normal human chromosomes exhibit a specific and reproducible fluorescence banding pattern makes it possible to identify chromosome abnormalities in much greater detail than ever before. Extra chromosomes can be specifically identified rather than merely assigned to a group of chromosomes. Minor chromosome anomalies are not overlooked so easily, and the nature of structural re-arrangements can be clarified to a much greater extent even without meiotic studies. Thus the more accurate cytogenetic diagnosis will allow a more detailed study of genotype-phenotype correlation and will lead to improved genetic counseling in chromosome abnormality cases.

The new techniques are perhaps of the greatest practical importance in cases with structural chromosome aberrations. There exist a number of different types of such aberrations which may involve any of the autosomes or sex chromosomes, and many of these have been analyzed by fluorescence techniques. The medical literature in this field is already quite large and is rapidly growing.

One area in which fluorometry comes into its own is that of prenatal diagnosis; here the possibility of analyzing translocations is very important. Conventional staining tells us little. By fluorescence analysis, it is possible in many cases to identify the chromosomes concerned, and in this instance fluorometry gives much more detailed information than visual work. Figure 16 gives a few examples.

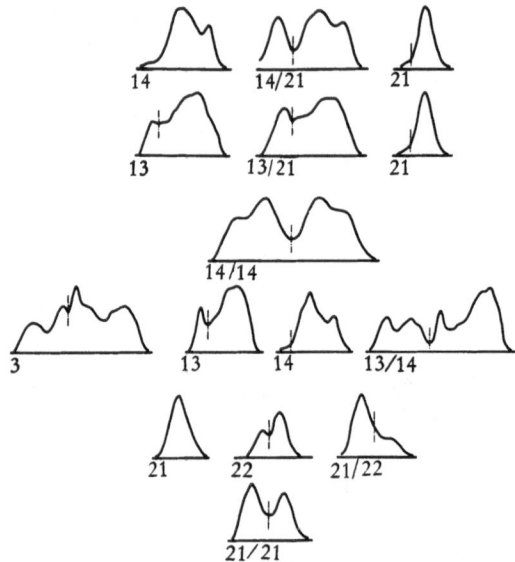

Figure 16. *QM-stained translocations involving the D and G chromosomes*

Several mammals other than man have been studied, and it has been possible to identify the complete karyotypes in the same way as for man. Of special interest, moreover, is the work on somatic cell hybrids; here biochemical and immunological studies in combination with chromosome identification are beginning to make it possible to relate the location of some genes to certain chromosomes.

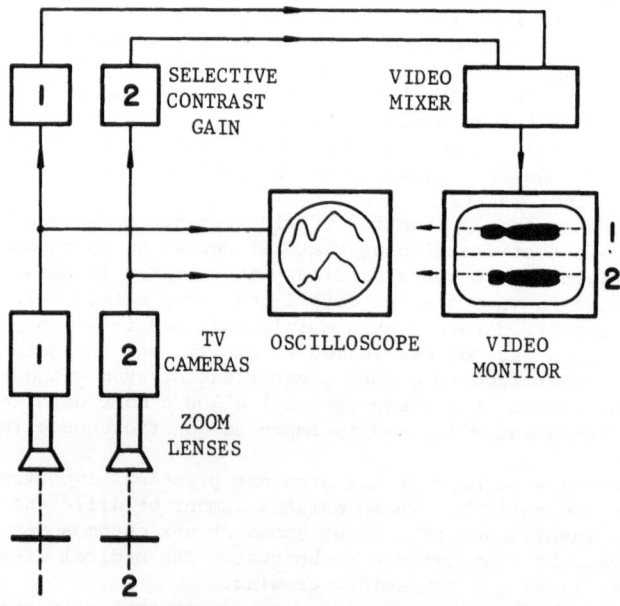

Figure 17. *Diagram for TV-based chromosome analysis*

PLATE VI

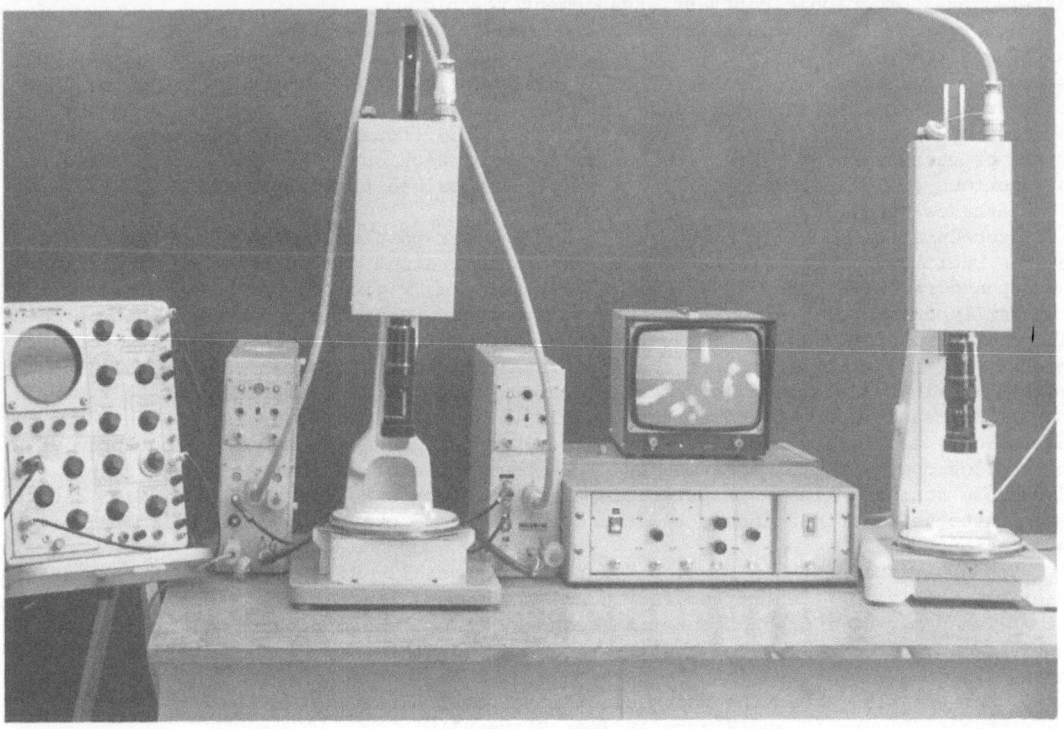

Figure 18. *TV-based arrangement for rapid analysis of normal and pathological karyotypes*

The fluorescence patterns of the mammalian chromosomes are very faint, a condition which creates no difficulty at all for strict fluorometric work. For visual chromosome sorting however, it is a complication, and it was quite some time before it was mastered by photographic contrast enhancement techniques.

For some of the most burning problems today concerning chromosomes, for instance the identification of small chromosome aberrations caused by environmental agents such as ionizing radiation or chemicals, it is necessary to study in great detail the chromosomes from a large number of metaphases. Detailed work with photographic contrast enhancement entails the preparation of a number of differently exposed, very hard prints of each negative - sometimes up to seven of each one. Because this process is so laborious and time consuming, it is simply impossible to analyze the great numbers of metaphases necessary.

Procedures based on TV techniques offer another possibility to enhance contrasts. We have built up a system in which a TV camera is trained on the negative of a fluorescence photograph (Caspersson et al. 1970a) and the TV picture can be observed in an arbitrarily chosen optical enlargement on a monitor. Contrast enhancement can be obtained electronically in any region of intensity, which means that, regardless of the intensity of the fluorescence of the individual chromosome one can rapidly examine its pattern in minute detail. The camera can also be fitted to a microscope. This means a great increase in speed and convenience in the analysis of metaphase plates for aberrations, as the practical work has already shown. The instrument is equipped with an oscilloscope, which simultaneously gives the course of intensity along a freely adjustable scanning line. In principle this corresponds to the registration of fluorescence patterns described above, and the curve produced provides additional detailed information.

The apparatus has been developed further, especially with a view to the important practical problem of identifying very small chromosome aberrations. This has been accomplished by the introduction of a second TV camera whose picture can be relayed to the same monitor as that of the first camera. In this way pictures of two chromosomes, e.g. members of one and the same pair, can be juxtaposed with optional enlargement, thus greatly increasing the possibilities of identifying small aberrations in the patterns (Figure 17 and Figure 18 (see Plate VI)).

The application of quantitative fluorescence techniques in the study of the chemical organization of the nuclear elements is a rapidly growing field in which many new developments may be expected in the near future. Chromosome identification is only one sector of this field. Analysis of the mode of binding of different types of DNA- and/or protein-binding fluorescent agents to different parts of the chromatin can also be expected to give a deeper insight in the interactions between the different components of the gene-carrying elements.

ACKNOWLEDGEMENTS

For the work carried out in Stockholm, I wish to acknowledge the continuous collaboration of several colleagues, especially Drs. Lore Zech and Gösta Lomakka.

REFERENCES

CARLSON, L., Caspersson, T., Foley, G. E., Kudynowski, J., Lomakka, G., Simonsson, E., and Sören, L., (1963), *Exp Cell Res.* **31**, 589.

CASPERSSON, T., (1950), In: *Cell Growth and Cell Function.* Norton, New York.

CASPERSSON, T., Castleman, K. R., Lomakka, G., Modest, E. J., Möller, A., Nathan, R., Wall, R. J. and Zech, L., (1971a), *Exp Cell Res.* **67**, 233.

CASPERSSON, T., Lindsten, J., Lomakka, G., Möller, A., and Zech, L., (1972), Int. Rev. Exp. Pathology. **11**, 1-72. Richter and Epstein (eds.), Acad. Press.

CASPERSSON, T., Lindsten, J., Lomakka, G., Wallman, H., and Zech, L., (1970a), *Exp Cell Res.* **63**, 477.

CASPERSSON, T., and Lomakka, G., (1970), In: *Introduction to Quantitative Cytochemistry II.* Wied, G. L., and Bahr, G. F., (Eds.) Academic Press Inc, New York, 27-56.

CASPERSSON, T., Lomakka, G., and Möller, A., (1971b), *Hereditas.* <u>67</u>, 103.

CASPERSSON, T., Lomakka, G., and Zech, L., (1971c). *Hereditas.* <u>67</u>, 89.

CASPERSSON, T., Simonsson, E., and Zech, L., (1970b), *Exp. Cell Res.* <u>63</u>, 243.

CASPERSSON, T., and Zech, L., (1972), In: *Perspectives in Cytogenetics. - The Next Decade.* Wright, S.W., Crandall. B. F., and Boyer. L., (Eds.) C.C. Thomas Springfield, 163.

CASPERSSON, T., Zech, L., and Johansson, C., (1970c), *Exp. Cell Res.* <u>62</u>, 490.

CASPERSSON, T., Zech. L., Johansson, C., and Modest, E. J., (1970d), *Chromosoma.* <u>30</u>, 215.

CASPERSSON, T., Zech. L., Modest, E. J., Foley, G. E., Wagh, U., and Simonsson, E., (1969a), *Exp Cell Res.* <u>58</u>, 141.

CASPERSSON, T., Zech, L., Modest, E. J., Foley, G. E., Wagh, U., and Simonsson, E., (1969b), *Exp. Cell Res.* <u>58</u>, 128.

HENEEN, W., and Caspersson, T., (1973), *Hereditas.* in press.

MÖLLER, A., Nilsson, H., Caspersson, T., and Lomakka, G., (1972), *Exp. Cell Res.* <u>70</u>, 475.

PACHMANN, U., and Rigler, R., (1972), *Exp. Cell Res.* <u>72</u>, 602.

RIGLER, R., (1969), *Exp Cell Res.* <u>58</u>, 460.

WEISSBLUM, B., and de Haseth, P. L., (1972), *Proc. Nat. Acad. Sci. USA.* <u>69</u>, 629.

CHEMICAL ASPECTS OF THE FLUORESCENCE ANALYSIS OF CHROMOSOMES

Edward J. Modest and Sisir K. Sengupta

The Children's Cancer Research Foundation and the Departments of Pathology and Biological Chemistry, Harvard Medical School, Boston, Massachusetts

INTRODUCTION

In the previous paper Dr. Caspersson has described a collaborative project be-tween our two institutions that has produced some very useful information on identi-fication and study of chromosomes by fluorescence analysis. About ten years ago, when he had developed the necessary microfluorometric instrumentation, Dr. Caspersson con-ceived that it might be possible to label different regions of metaphase chromosomes with specific DNA-binding agents that are fluorescent. This research program has achieved far-reaching success (Caspersson et al., 1968; 1969a; 1969b; 1970a; 1972) : our chromosome fluorescence banding method is now the primary international identification standard for human metaphase chromosomes (Paris Conference, 1971); is in use for the detailed analysis of chromosomes and chromosome structure; and has led to the development of nonfluorescent banding techniques for human chromosomes, including the use of Giemsa stain and trypsin (Sumner et al. 1971; Patil et al. 1971; Schnedl 1971; Drets and Shaw 1971; Lomholt and Mohr 1971; Dutrillaux and Lejeune 1971; Wang and Fedoroff 1972; Seabright 1971).

CHEMICAL ASPECTS

The first proposal was to design and synthesize fluorescent analogs of actinomycin D, known to be DNA-guanine specific (Goldberg et al. 1962). While the actinomycin chemistry was being worked out, we selected as the first agent quinacrine mustard (Figure 1) (Jones et al. 1957), on the basis that this compound would satisfy our initial requirements, namely: a fluorescent aminoacridine nucleus that can intercalate into bihelical DNA; basic nitrogen atoms that can form ionic bonds with DNA phosphate; and an alkylating group in the side chain capable of forming covalent bonds with DNA-guanine. This compound has worked quite well in metaphase chromosomes of various species from plants through human. The analogous antimalarial quinacrine (Figure 2) has the same fluores-cence properties but cannot form covalent bonds since it has no alkylating group in the side chain. In our first studies (Caspersson et al. 1968), quinacrine mustard produced three intensely fluorescent bands near the centromere of the large *Vicia faba* M chromosome (Figure 3. Plate I). Initially, quinacrine failed to give these bands in *Vicia*, probably because it was used at too low a staining concentration. In later work (Caspersson et al. 1969a), we found that quinacrine gives the same bands quali-tatively, although much less effectively (Caspersson et al. 1970b). Thus, the primary mode of binding for both compounds is probably by intercalation of the acridine nucleus into bihelical DNA; the selective alkylation of DNA-guanine by quinacrine mustard is secondary.

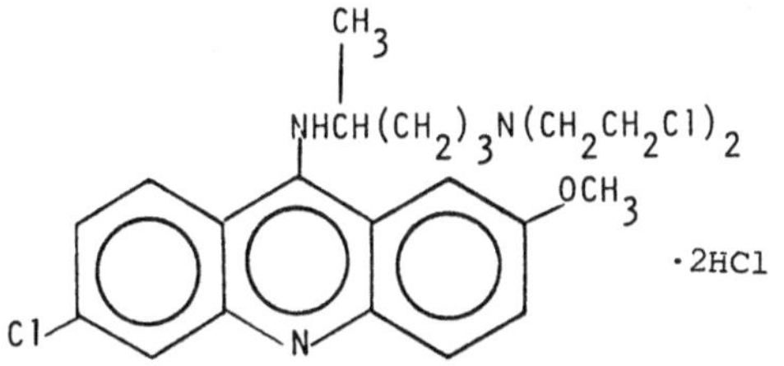

Figure 1. *Structure of quinacrine mustard dihydrochloride (QM, FLU 1)*

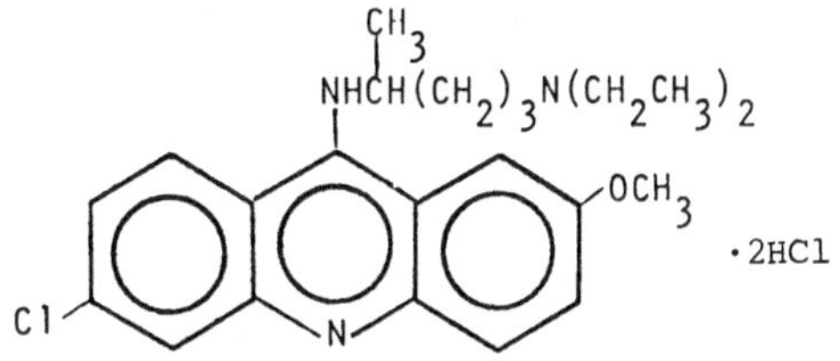

Figure 2. *Structure of quinacrine dihydrochloride (Q, FLU 6)*

PLATE I

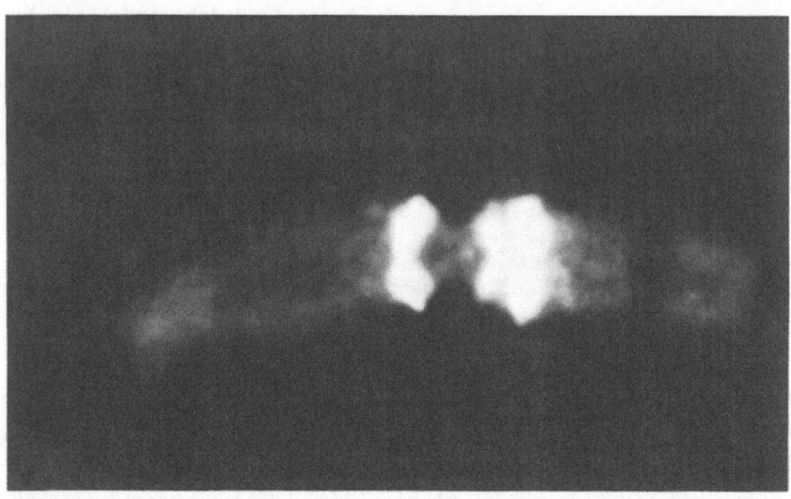

Figure 3. *Quinacrine mustard treated* Vicia faba *M chromosome, X 2000.*
(From Caspersson et al., 1968)

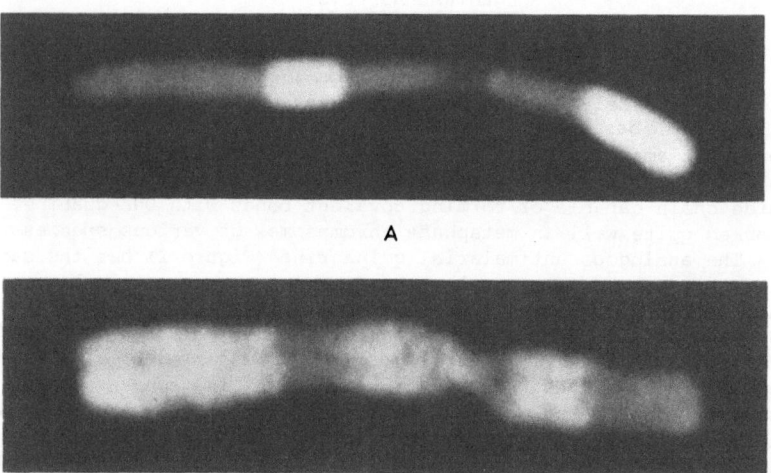

Figure 10. **Scilla sibirica** *metaphase chromosome treated (A) with ethidium bromide (EB), and (B) with quinacrine mustard (QM), X 2000. Note reversal of fluorescent banding pattern. (From Caspersson et al., 1969 b)*

PLATE II

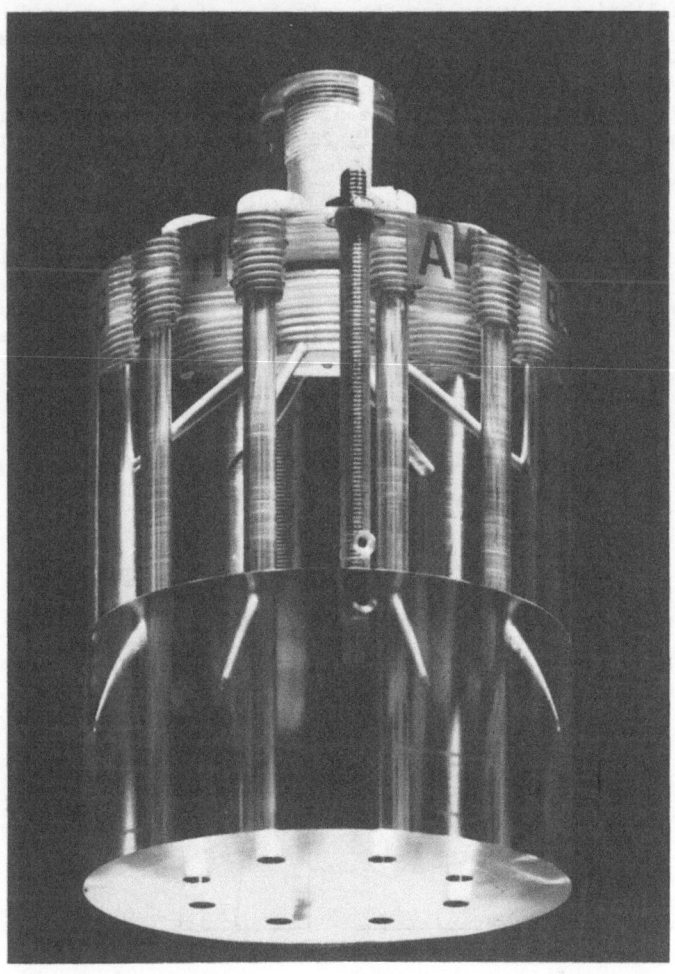

Figure 4. *Paulus ultrafiltration apparatus*

We have determined the binding properties of several fluorochromes to DNA and other macromolecules *in vitro* by ultrafiltration and by other techniques (Caspérsson et al. 1970b; Sengupta et al. 1971a; Modest and Sengupta (in preparation)). The ultrafiltration studies, equivalent to equilibrium dialysis for determination of apparent binding constants and apparent nucleotide:ligand ratios, were carried out with a Paulus ultrafiltration apparatus (Figure 4. Plate II), (Paulus 1969).

The ultrafiltration data for the binding of quinacrine mustard and quinacrine to calf thymus DNA are presented in Figure 5. Scatchard plot calculations indicate that quinacrine mustard binds approximately 25 times as strongly to calf thymus DNA as does quinacrine (K_{ap}^{QM} : K_{ap}^{Q} = 3.94 × 10^8 : 1.6 × 10^7 M^{-1}), but fewer quinacrine mustard molecules bind per unit length of DNA (ligand:nucleotide ratio for quinacrine: quinacrine mustard = 0.72:0.058). The biphasic nature of both curves in Figure 5 suggests more than one mode of binding for each compound. The similarity of the two curves in Figure 5 indicates strongly that the primary binding mechanism of both compounds is the same; i.e., intercalation. If the binding mode of quinacrine mustard were exclusively alkylation, equilibrium binding curves would not be measurable.

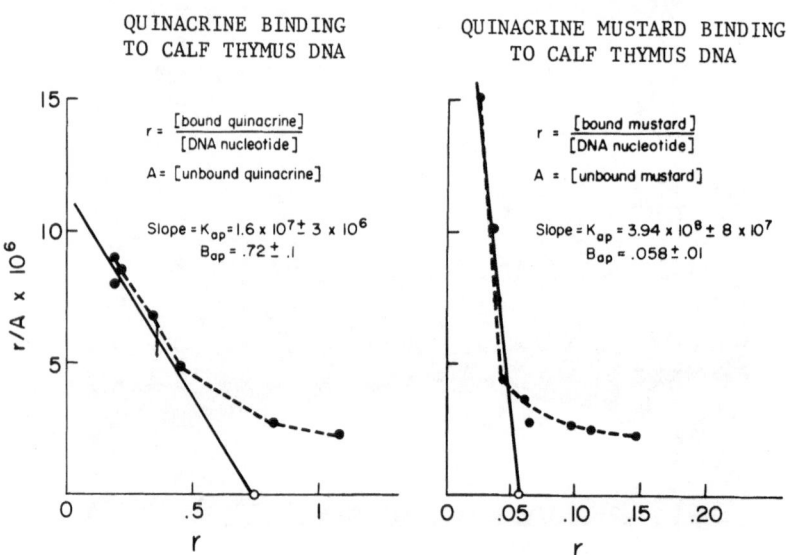

Figure 5. *Scatchard plots (by ultrafiltration) for binding of quinacrine mustard (QM) and quinacrine (Q) to calf thymus DNA in 0.01 M phosphate buffer at pH 7. K_{ap} = apparent binding constant in moles^{-1}. B_{ap} = apparent number of binding sites per nucleotide. (From Modest and Sengupta 1973)*

In addition to these studies, we have observed that quinacrine shows no base specificity when bound to calf thymus DNA or poly d(AT); this is in contrast to quinacrine mustard, which, under our conditions, binds only to guanine-containing polynucleotides in bihelical configuration. This observation of lack of DNA-base binding specificity for quinacrine is in accord with the work of O'Brien et al. (1966), who used spectral shift methodology. Furthermore, the binding of quinacrine mustard, as opposed to that of quinacrine, is relatively irreversible, in that ethanol or dilute acid under mild conditions will extract quinacrine but not quinacrine mustard from the acridine-DNA complex. Hydrochloric acid under forcing conditions is required to break the quinacrine mustard-DNA complex. Neither fluorochrome binds significantly to calf thymus histone.

Figure 6 shows the structure of actinomycin D. In order to preserve the DNA-binding capacity and DNA-guanine specificity of the molecule in the process of intro-

ducing a fluorophoric group, we deduced on the basis of published information (Goldberg et al. 1962; Müller and Crothers 1968), that substitution could be introduced at
position 7 (and perhaps at position 8) of the phenoxazinone chromophore without
endangering the selective binding properties of the molecule. We decided to introduce
an amino group at position 7 (Sengupta et al. 1971b), since such an electron-donating
group can enhance both the fluorescence and absorption properties of a conjugated
aromatic ring system (Udenfriend 1962).

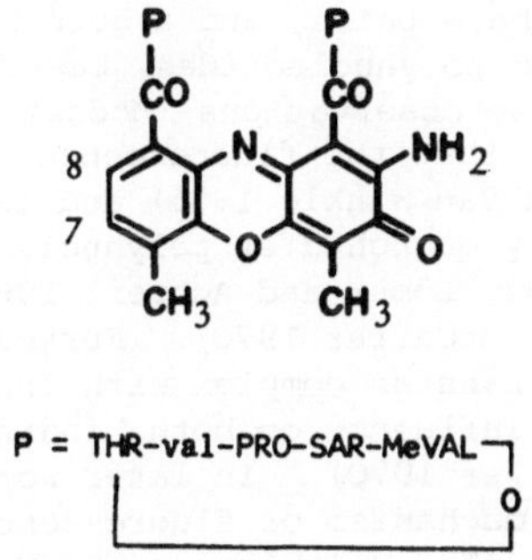

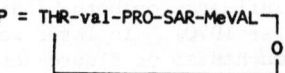

P = THR-val-PRO-SAR-MeVAL
O

BINDING CHARACTERISTICS:

1. Binds to double helical DNA

2. Binds specifically to guanine

3. Does not bind to RNA or protein

BINDING GROUPS:

1. 2-Amino group

2. 3-Oxo group

3. The peptide lactones (P)

Figure 6. *DNA binding properties and structure of actinomycin D*

After appropriate chemical studies with model compounds, 7-aminoactinomycin D was synthesized via direct nitration of actinomycin D, followed by reduction (Figure 7).
This compound binds to bihelical DNA with guanine specificity in the same way as does
actinomycin D, and exhibits strong fluorescence in the red (emission 585 nm at 465 nm
excitation in 0.1 M phosphate buffer, pH 6.8). Melting-temperature curves for 7-amino-
actinomycin D and actinomycin D bound to calf thymus DNA (Figure 8) show that both

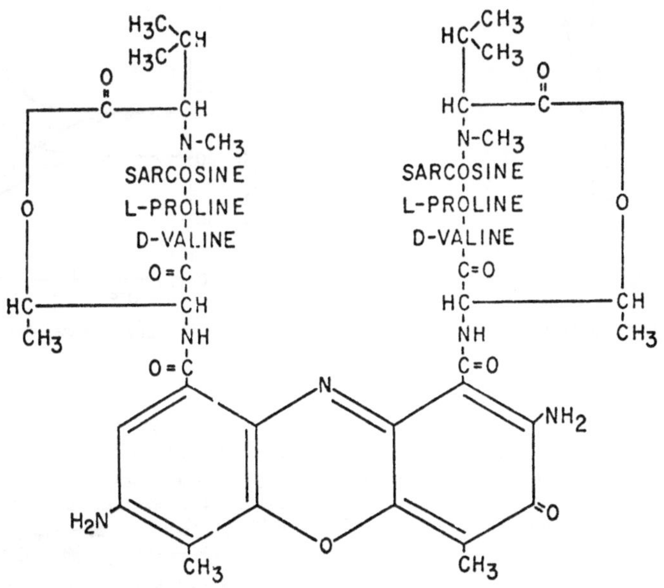

Figure 7. *Structure of 7-aminoactinomycin D (FLU 402)*

compounds produce the same increase in melting temperature (11.5°). Ultrafiltration experiments indicate that the two compounds have comparable binding constants and nucleotide:ligand ratios. Neither compound binds to calf thymus histone. 7-aminoactinomycin D produces fluorescent DNA seqments in *Drosophila* polytene chromosomes (Modest and Sengupta, in preparation).

Recent publications by Rigler (1969; Pachmann and Rigler 1972), and by Weisblum and De Haseth (1972) report an apparent enhancement (or at least an increased efficiency) of fluorescence of certain aminoacridines (including quinacrine, quinacrine mustard and proflavine) when bound to dAT base pairs, and a quenching of fluorescence for these fluorochromes by guanine-containing polynucleotides, based on *in vitro* measurements. We have confirmed and extended these observations (Modest and Sengupta, in preparation).

Earlier publications reported that the fluorescence of acriflavine when complexed with DNA is quenched (Heilweil and Van Winkle 1955) and later that proflavine or acriflavine fluorescence is selectively quenched at polynucleotide guanine sites but not at AT base pairs (Tubbs et al. 1964; Löber and Achtert 1969; Finkelstein and Weinstein 1967; Thomes et al. 1969; Chan and McCarter 1970). Further, it was postulated that DNA-guanine quenches by a charge transfer complex with the aminoacridine and that the adjacent base pairs have a strong influence on both binding and fluorescence efficiency (Tubbs et al. 1964; Chan and McCarter 1970). In later work, these studies are being extended to quinacrine and to the mechanism of fluorescence banding in the chromosome (Rigler 1969; Pachmann and Rigler 1972; Weisblum and De Haseth 1972). Rigler has used fluorescence polarization to confirm binding of proflavine or quinacrine to the DNA employed (Pachmann and Rigler 1972).

In Figure 9 we indicate graphically, following Weisblum's methodology, changes in relative fluorescence intensity for quinacrine at a fixed concentration in the presence of calf thymus DNA or poly d(AT) at increasing concentrations. In agreement with the published data, quenching of fluorescence is observed with calf thymus DNA and enhancement with poly d(AT). Similar but smaller effects are seen with quinacrine mustard (Figure 9); the reduced quenching for quinacrine mustard/calf thymus DNA may reflect alkylation of DNA-guanine with attendant reduction in the efficiency of alkylated guanine residues in charge transfer complex formation.

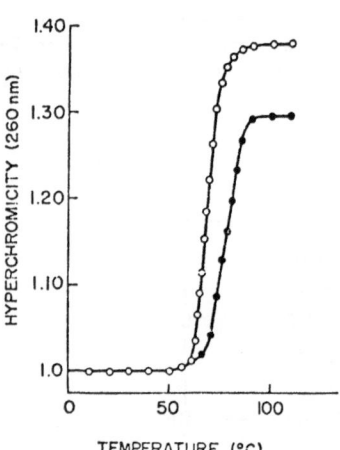

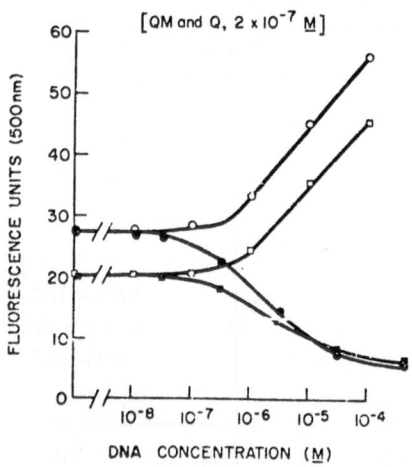

Figure 8. *Thermal denaturation profile of calf thymus DNA in the absence (○—○) and in the presence (●—●) of 7-aminoactinomycin D (FLU 402). Measurements were done in 0.01 M phosphate buffer at pH 7.0 (From Modest and Sengupta 1973)*

Figure 9. *Effect of polynucleotides on quinacrine mustard (QM) and quinacrine (Q) fluorescence in 0.1 M phosphate buffer at pH 6.8. Excitation wavelengths 415 nm for QM and 430 nm for Q. ■—■, QM-calf thymus DNA. □—□ , QM-poly d(AT). ●—●., Q-calf thymus DNA. ○—○ , Q-poly d(AT). (From Modest and Sengupta 1973)*

Acridine orange shows fluorescence enhancement both with d(AT) base pairs and with polynucleotide guanine (Modest and Sengupta (in preparation); Weisblum and De Haseth 1972), but does not give chromosome banding. Ethidium bromide, reported by LePecq and Paoletti (1967) to exhibit fluorescence enhancement in the presence of polynucleotide guanine, gives "reverse" banding in metaphase chromosomes in a limited number of species: *Scilla sibirica* (Caspersson et al. 1969b), *Vicia faba* (Vosa 1970), Chinese hamster (Modest and Yerganian (in preparation) b) and the Australian wallaby (Pearson et al. 1971). In these species, the ethidium bromide fluorescent bands are precisely the reverse, mirror image of the quinacrine mustard bands: those regions that are fluorescent with quinacrine mustard are less fluorescent with ethidium bromide and vice versa. Figure 10 (Plate I) shows the fluorescent banding of the large *Scilla* metaphase chromosome with quinacrine mustard and with ethidium bromide (Caspersson et al. 1969b).

Despite the fact that ethidium bromide gives reverse fluorescent banding, we raise the argument that since the fluorescence of neither acridine orange (Weisblum and De Haseth 1972) nor ethidium bromide (Modest and Sengupta, in preparation) (Figure 11) is quenched by polynucleotide guanine, the fact that banding, even though reverse banding, is seen with ethidium bromide argues against the simplistic postulate that fluorescence banding in general is entirely a function of the base-dependent enhancement/quenching phenomenon. In Figure 11 we show that the fluorescence of ethidium bromide is enhanced when bound either to calf thymus DNA or to poly d(AT), and that the degree of enhancement appears to be inversely proportional to the guanine content.

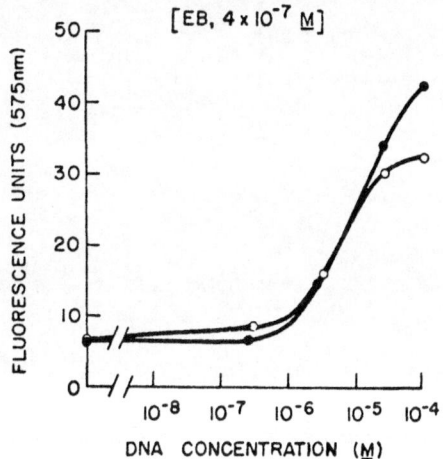

Figure 11. *Effect of polynucleotides on ethidium bromide (EB) fluorescence in 0.1 M phosphate buffer at pH 6.8. Excitation wavelength 465 nm.* ●——●, *poly d(AT).* ○——○, *calf thymus DNA*

An understanding of the nature of the fluorescent bands in metaphase chromosomes is of the utmost importance for many reasons, including the use of this methodology to provide information on the ultrastructure of the metaphase chromosome. There are clearly two kinds of different fluorescent banding in human metaphase chromosomes (Figure 12. Plate III): (1) faint, reproducible identification bands (present in all normal metaphase chromosomes of all individuals) (Caspersson et al. 1970a; Caspersson et al. 1972), and (2) intense, variable polymorphic regions (consistent in certain metaphase chromosomes of a given individual but variable from individual to individual) (Caspersson et al. 1970a; Caspersson et al. 1972; Evans et al. 1971). The identification banding correlates to some extent with DNA distribution along the chromosome, and is influenced by the distribution of DNA and protein along the chromosome. The second kind of banding visualizes a special kind of DNA, perhaps supercoiled or d(AT) rich or highly repetitive. In connection with the latter point, Southern has found recently that highly repetitive satellite DNA of the guinea pig contains a 6-base sequence - CCCTAA. It may well be that fluorescence enhancement in the chromosome depends on the

occurence of the sequence
```
 -A-A-C-
  : : :
  . . .
 -T-T-G-
```
in a refinement of the proposal of Chan and McCarter (1970).

PLATE III

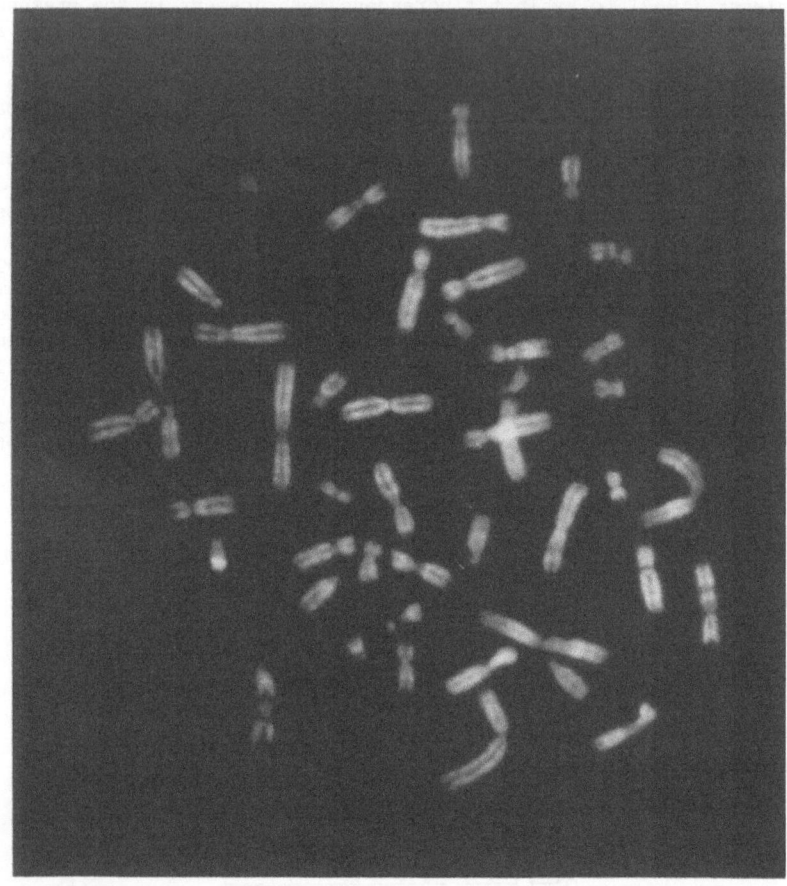

Figure 12.   *Human metaphase chromosomes stained with quinacrine mustard (QM)*

FURTHER STUDIES

It is important now to undertake a careful study of human metaphase chromosomes banded with a differentially staining fluorochrome and to correlate, insofar as possible, distribution by several parameters: absorption spectrum, fluorescence, and possible enhancement/quenching effects. In an approach to such work with human chromosomes, Dr. Caspersson and his colleagues are now studying *Vicia faba* M chromosomes stained with various fluorochromes, including proflavine. The bright fluorescent bands in the *Vicia faba* M chromosome (Figure 3) do not reflect DNA distribution along the chromosome (Caspersson et al. 1968; Caspersson et al. 1969a). Proflavine, which has more favorable visible absorption characteristics than quinacrine mustard, shows fluorescence banding in *Vicia* qualitatively similar, to that produced by quinacrine mustard, although less efficient. Preliminary results indicate that some kind of fluorescence enhancement/quenching phenomenon may be important in the chromosome itself, since proflavine distribution along the *Vicia* chromosome, as measured by absorption scanning, does not correlate with the intense fluorescent bands observed. Similar studies are in progress with human chromosomes (Caspersson et al. (in preparation)). Elsewhere an electron microscopic comparison of DNA-protein distribution in human metaphase chromosomes with banding is being done (Bahr 1972).

Further work may be expected to lead to more definitive explanation and analysis of the nature and mechanism of chromosome fluorescence banding in chromosomes, and may help explain chromosome ultrastructure and the genetic implications thereof.

ACKNOWLEDGEMENTS

Supported in part by Research Grant C6516 and Research Career Development Award K3-CA-22,151 from the National Cancer Institute, National Institutes of Health, U. S. Public Health Service, Bethesda, Maryland.

REFERENCES

BAHR, G. F., (1972), *Abstracts of Papers*, Tenth Annual Somatic Cell Genetics Conference, Aspen, Colorado, 18 January.

CASPERSSON, T., Farber, S., Foley, G. E., Kudynowski, J., Modest, E. J., Simonsson, E., Wagh, U., and Zech, L. (1968), *Exptl. Cell Res.*, 49, 219.

CASPERSSON, T., Lindsten, J., Lomakka, G., Möller, A., and Zech, L., (1972). *Int. Rev. Extpl. Pathol.*, (Richter, G.W., and Epstein, M.A., eds.) Academic Press, N.Y., 11,1.

CASPERSSON, T., Zech, L., Johansson, C., and Modest, E. J., (1970a), *Chromosoma*, 30, 215.

CASPERSSON, T., Zech, L., Modest, E. J., (1970b). *Science*, 170, 762.

CASPERSSON, T., Zech, L., Modest, E. J., Foley, G. E., Wagh, U., and Simonsson, E., (1969a), *Exptl. Cell Res.*, 58, 128.

CASPERSSON, T., Zech, L., Modest, E. J., Foley, G. E., Wagh, U., and Simonsson, E., (1969b), *Exptl. Cell Res.*, 58, 141.

CASPERSSON, T. et al., in preparation.

CHAN, L.M. and McCarter, J.A., (1970), *Biochim. Biophys. Acta*, 204, 252.

DRETS, M. E., and Shaw, M. W., (1971), *Proc. Nat. Acad. Sci. (U.S.)*, 68, 2073.

DUTRILLAUX, B., and Lejeune, J., (1971), *Compt. Rend. Acad. Sci. Paris*, 272, 2638.

EVANS, H. J., Buckton, K. E., and Sumner, A. T., (1971), *Chromosoma*, 35, 310.

FINKELSTEIN, T., and Weinstein, I. B., (1967), *J. Biol. Chem.*, 242, 3763.

GOLDBERG, I. H., Rabinowitz, M., and Reich, E., (1962), *Proc. Nat. Acad. Sci. (U.S.)*, <u>48</u>, 2094.

HEILWEIL, H. G., and Van Winkle, Q., (1955), *J. Phys. Chem.*, <u>59</u>, 939.

JONES, R. Jr., Price, C. C., and Sen, A. K., (1957), *J. Org. Chem.*, <u>22</u>, 783.

LePECQ, J. B., and Paoletti, C., (1967), *J. Mol. Biol.*, <u>27</u>, 87.

LOMHOLT, B., and Mohr, J., (1971), *Nature New Biology*, <u>234</u>, 109.

LÖBER, G., and Achtert, G., (1969), *Biopolymers*, <u>8</u>, 595.

MODEST, E. J. and Sengupta, S.K., (1973), *Chromosome Identification - Techniques and Applications in Biology and Medicine*, (Caspersson, T., and Zech, L., eds.), Nobel Symposium XXIII, Academic Press, New York, pp. 323-26.

MODEST, E. J., and Sengupta, S. K., in preparation.

MODEST, E. J., and Yerganian, G., in preparation.

MÜLLER, W., and Crothers, D. M., (1968), *J. Mol. Biol.*, <u>35</u>, 251.

O'BRIEN, R. L., Olenick, J. G., and Hahn, F. E., (1966), *Proc. Nat. Acad. Sci. (U.S.)*, <u>55</u>, 1511.

PACHMANN, U., and Rigler, R., (1972), *Exptl. Cell Res.*, <u>72</u>, 602.

PARIS CONFERENCE (1971), Conference on Standardization in Human Cytogenetics. Birth defects: Original Article Series, The National Foundation, New York, <u>8</u>, 7.

PATIL, S. R., Merrick, S., and Lubs, H. A., (1971), *Science*, <u>173</u>, 821.

PAULUS, H., (1969), *Anal. Biochem.*, <u>32</u>, 91.

PEARSON, P. L., Bobrow, M., Vosa, C. G., and Barlow, P. W., (1971), *Nature*, <u>231</u>, 326.

RIGLER, R., (1969), *Exptl. Cell. Res.*, <u>58</u>, 460.

SCHNEDL, W., (1971), *Chromosoma*, <u>34</u>, 448.

SEABRIGHT, M., (1971), *Lancet*, <u>2</u>, 971.

SENGUPTA, S. K., Ramsey, P. G., Modest, E. J., (1971a), *Proc. Am. Assoc. Cancer Res.*, <u>12</u>, 90.

SENGUPTA, S. K., Tinter, S. K., Ramsey, P. G., and Modest, E. J., (1971b), *Fed. Proc.*, <u>30</u>, 342.

SOUTHERN, E.M., (1970), *Nature*, <u>227</u>, 794.

SUMNER, A. T., Evans, H. J., and Buckland, R. A., (1971), *Nature New Biology*, <u>232</u>, 31.

THOMES, J. C., Weill, G., and Daune, M., (1969), *Biopolymers*, <u>8</u>, 647.

TUBBS, R. K., Ditmars, W. E. Jr., and Van Winkle, Q., (1964), *J. Mol. Biol.*, <u>9</u>, 545.

UDENFRIEND, S., (1962), *Fluorescence Assay In Biology and Medicine*, Academic Press, New York. p.21.

VOSA, C. G., (1970), *Chromosoma*, <u>30</u>, 366.

WANG, H. C., and Fedoroff, S., (1972), *Nature New Biology*, <u>235</u>, 52.

WEISBLUM, B. and De Haseth, P.L., (1972), *Proc. Nat. Acad. Sci. (U.S.)*, <u>69</u>, 629.

FLUORESCENCE CYTOCHEMICAL STUDIES OF CHROMOSOMES: QUANTITATIVE APPLICATIONS OF FLUORESCEIN MERCURIC ACETATE

Ronald R. Cowden and Sherrill K. Curtis

Department of Anatomy, Albany Medical College, New York

INTRODUCTION

In the last few years there has been a renewed interest in nonhistone proteins of chromosomes. In no small measure this has been due to the development of preparative biochemical procedures which have opened wider opportunities for the characterization of these proteins. Just as the nomenclature for nonhistone proteins in the biochemical literature reflects the tenuous state of the art, a similar degree of confusion exists in the cytological literature, particularly if some of the earlier cytochemical studies of chromosomal proteins are considered. In the absence of any assurance that designations applied to cytological preparations conform in detail with their biochemical counterparts, it would seem desirable for the present to accept certain conventions based on operative definitions. Within this context, it is proposed that proteins demonstrable within chromosomes after quantitative removal of histones be considered "nonhistone" proteins plus "residual" proteins, and that proteins demonstrable after both nucleic acids and histones have been removed be designated "residual" proteins. These proposals are made in the full knowledge that the 0.2N hydrochloric acid treatment normally used to remove histones from cytological preparations probably removes some nonhistone chromosomal proteins. It is also apparent that a substantial fraction of the nonhistone protein is so intimately bound to nucleic acids that removal of nucleic acids results in removal of all or most of these proteins as well.

From experience gained in classical light microscopy as well as in biological ultrastructure, we are aware that fixation can affect the preservation of certain substances, can substantially alter the structure of intracellular macromolecular complexes, or can even produce artifacts. In approaching the cytochemical examination of chromosomes, it is incumbent on the investigator to assess the effects of fixation on chromosome structure. For this reason, some electron micrographs illustrating some of these fixation effects will be presented along with the cytochemical information.

It would be reasonable to assume that while chromosomes in general, as carriers of genetic information, might display an enormous degree of evolutionary conservatism in their fundamental organization, there is substantial latitude for special variation. Wide variations in amounts of satellite DNA, the presence of multiple sex chromosomes, and cases of chromatin loss from cells other than the germ cell line have been repeatedly reported in the cytological literature. The polytene chromosome, that rather special case of endopolyploidy, is an extreme case of chromosomal specialization and should be contrasted to generalized chromosomes in any examination of chromosome structure.

The original intention of these investigations was to examine the intrachromosomal distributions of nonhistone and residual chromosomal proteins as earlier defined. In this laboratory over the past decade, cytochemical studies were undertaken with a variety of visible protein end-group methods. The results of these studies will be briefly summarized here. It became obvious in the course of the investigations of chromosomal proteins, that visible chromophore methods did not offer the sensitivity of fluorescence methods, and that particular end groups could only be approached cytochemically with fluorescence methods.

MATERIALS AND METHODS

The cytochemical work described in this paper was undertaken on squashes of *Drosophila* polytene chromosomes, smears of amphibian blood, and paraffin sections of grasshopper tests and sea cucumber ovaries. Preparations of *Drosophila virilis* chromosomes were obtained from Drs. Ellen and Robert Rasch. The *Drosophila* salivary glands had been fixed in absolute ethanol-acetic acid (henceforward referred to as "3:1") and

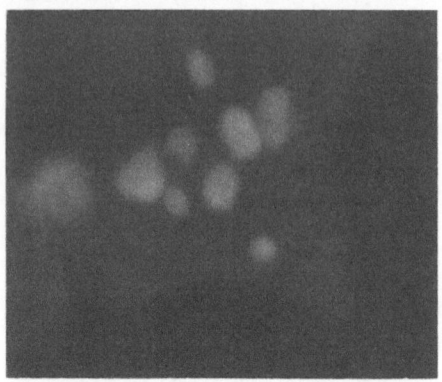

Figure 1. *Unextracted grasshopper meiotic chromosomes reduced with thioglycolate and fluorochromed with fluorescein mercuric acetate. BG-12 excitation filter, 530 nm barrier filter. Note: This same combination was used in all fluorescent photographs in this study.* 2,500X

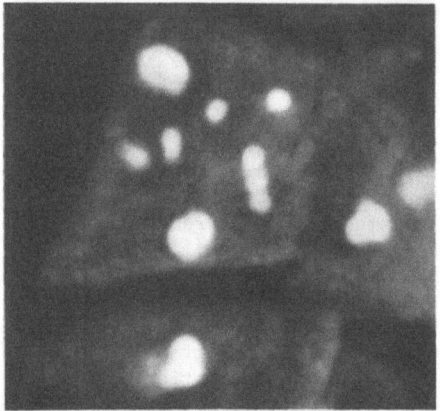

Figure 2. *Grasshopper meiotic chromosomes pretreated with 0.2N HCl. Note substantial increase in fluorescence.* 2,500X

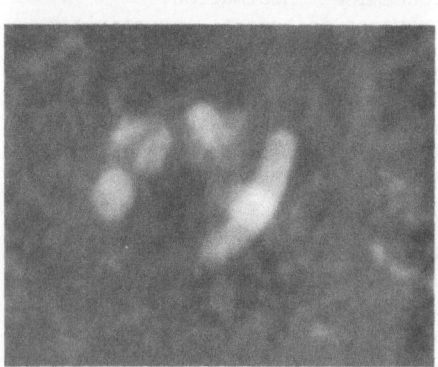

Figure 3. *Similar to Figures 1 and 2, but pretreated with RNase, DNase, and 0.2N HCl.* 2,500X

squashed under coverslips in 45-50% acetic acid. The coverslips had been removed by the dry-ice method, and the preparations had been air-dried for shipment. Smears were made of the peripheral blood of the giant salamander, *Amphiuma tridactylum*, allowed to air dry, and fixed in 3:1. Additional smears were made of the peripheral blood of the newt, *Diemyctylus viridescens*, in which an anemia had been provoked by injection with phenylhydrazine hydrochloride. On days nine to twelve after injection, abundant mitotic figures were found in the peripheral blood. The smears were air-dried and fixed in 3:1 or absolute methanol. Sea cucumbers *(Stichopus badionotus)* were collected in the vicinity of the Bermuda Biological Station, and their ovaries were fixed in 3:1, conventionally embedded in paraffin, and sectioned at 4μm. In addition, various species of grasshoppers were collected in Florida, Louisiana, and Colorado. No special attempt was made to identify the species since they were regarded simply as sources of large meiotic chromosomes. The testes were removed from the grasshoppers, fixed either in 3:1 or 10% calcium acetate formalin (Lillie 1965), embedded in paraffin, and sectioned at 3-5μm.

With the exception of sections of *Stichopus* ovaries, the smears, polytene chromosomes, and sections either were stained directly without any pretreatment, or were subjected to one of the following pretreatments:

a) 0.2N hydrochloric acid (HCl) for six hours at room temperature to remove histones;

b) a sequence of RNase (0.1 mg/ml, pH 6.0-6.5, for two hours), DNase (0.04 mg/ml in 0.003M magnesium sulfate at pH 6.0-6.5, for two two-hour changes and a third, overnight change), and 0.2N HCl for six hours;

c) a sequence of 5% trichloroacetic acid (TCA) for fifteen minutes at 90°C. and 0.2N HCl for six hours. *Stichopus* ovaries were examined without pretreatment.

The fluorescence methods used included the fluorescein mercuric acetate procedure of Cowden and Curtis (1970) for protein sulfhydryl (SH) groups or, after reduction with thioglycolate, for protein disulfide (SS) plus SH groups; brilliant sulfoflavine at pH 2.8 used according to Ruch (1970) to fluorochrome ionizable basic groups of proteins; and the salicylhydrazide-zinc method of Stoward and Burns (1967) for the fluorescent demonstration of C-terminal carboxyl groups. In all instances, the preparations were mounted in fluorescence-free mounting media, and photographs or measurements were conducted using fluorescence-free immersion oil.

The staining procedures used for conventional bright-field microscopy included the Ritter and Berman (1963) modification of the Morel-Sisley reaction for tyrosine; the p-dimethylaminobenzaldehyde-nitrite (DMAB) method of Adams (1957) for tryptophan; the acrolein-Schiff method of Van Duijn (1961) for proteins in general; and the mercury orange method of Bennett and Watts (1958) for protein-bound SH groups or after reduction with thioglycolate, for SS plus SH groups. Basic groups of proteins were demonstrated according to principles developed by Deitch (1955) using the acidic dye, Biebrich scarlet, at pH 2.8.

Microfluorometric measurements were made with an admittedly makeshift instrument: Zeiss photomicroscope II equipped with an incident Fl II fluorescence condenser and a stabilized XBO-150 xenon arc source. Since the microscope did not allow insertion of a standard condenser below the stage, any general orientation was accomplished by the use of uncondensed transmitted light from a 60-watt tungsten source. A Bausch and Lomb 620 nm filter was mounted over the transmitted light exit port. Fluorescence measurements were made by shifting to the incident mode. The measurements were recorded with a Reichert microspectrophotometer system on the highest sensitivity setting, to which a digital voltmeter was attached. All fluorometric values were expressed in arbitrary units. With the exception of experiments involving fluorescence fading and regeneration, care was taken to wait five seconds before recording data. Further, the slides were frequently moved in order to reduce the effects of fluorescence fading in neighboring cells, thus increasing the reproducibility of the measurements.

The grasshopper testes used for electron microscopy were freed of connective tissue, fixed for two hours either in 3:1 or 10% calcium acetate formalin, and separated into individual follicles. The follicles then were either embedded directly or subjected to various pretreatments before embedment. For each of the fixatives the following pretreatments were used:

a) two six-hour changes of 0.2N HCl at room temperature;

b) a sequence of two-hour changes of RNase (0.1 mg/ml at pH 6.0-6.5), four two-

138

PLATE II

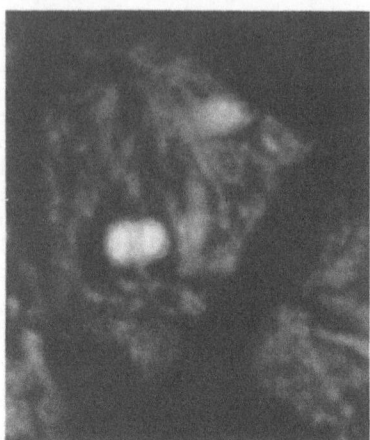

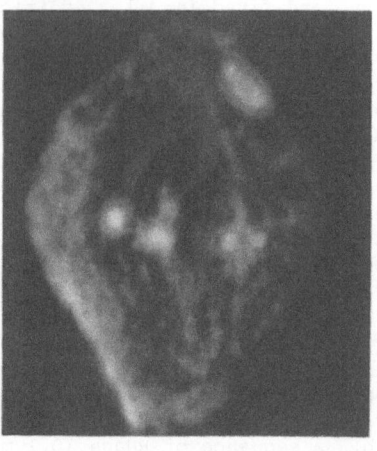

Figure 4. *Grasshopper meiotic chromosomes fluorochromed with pH 2.8 brilliant sulfoflavine after four hours pretreatment with DNase, RNase pretreatment, and extraction in 0.2N HCl. Note both chromosomal "coat" and intensely fluorescent central region.* 1,200X

Figure 5. *Meiotic chromosomes from the same preparation displaying further reduction of central chromosomal protein as a result of enzymatic and 0.2N HCl action.* 1,200X

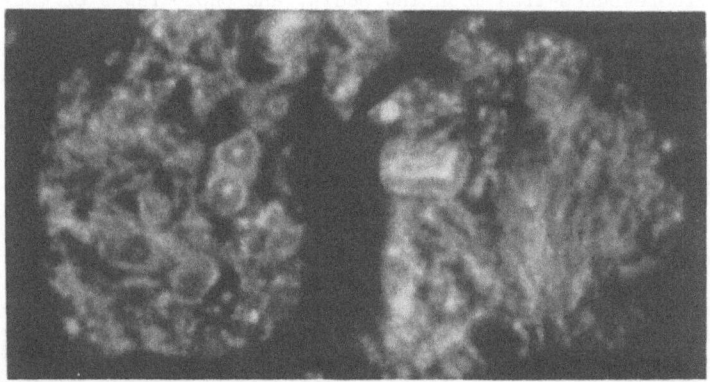

Figure 6. *Similar to Figures 4 and 5, but displaying further reduction in central chromosomal protein material.* 1,200X

hour changes of DNase (0.04 mg/ml in 0.003M magnesium sulfate at pH 6.0-6.5), and two six-hour changes of 0.2N HCl; and

c) a sequence of 5% TCA at 90°C. for one hour and two six-hour changes of 0.2N HCl. Some of the follicles in each group were conventionally embedded in paraffin, while the remaining follicles were post-fixed in cold 2% osmium tetroxide for one hour, dehydrated through absolute ethanol, cleared with propylene oxide, and embedded in epon (Luft 1961) Ultrathin sections of epon-embedded follicles were cut with glass knives on a Sorvall MT-1 ultramicrotome. The sections were mounted on uncoated grids, doubly stained with saturated uranyl acetate and lead citrate (Reynolds 1963), and examined with a Hitatchi HU-11C electron microscope. For purposes of orientation, adjacent sections were cut approximately 1 μm in thickness, mounted on microscope slides, and stained with the toluidine blue-pyronine procedure of Ito and Winchester (1963). Paraffin-embedded follicles were subjected to two procedures:

a) the alkaline fast green procedure of Alfert and Geschwind (1953) for histones, and

b) the azure B procedure of Flax and Himes (1952) for nucleic acids.
The histochemical tests were performed on paraffin-embedded material solely to determine the effectiveness of the extraction procedures. In all cases, these procedures were found to be adequate — i.e., 0.2N HCl treatment resulted in removal of histones demonstrable by the Alfert and Geschwind (1953) procedure; and treatment with either nucleases or hot TCA resulted in extraction of nucleic acids demonstrable by azure B at pH 4.0.

OBSERVATIONS

Qualitative

Since histones, with the possible exception of the "basic keratin" of mouse (and possibly other mammalian) spermatozoa are known to contain trivial levels of sulfur amino acids (Hendricks and Mayer 1965), it would appear that a method based on the demonstration of chromosomal SH and SS groups would offer a direct approach to the examination of chromosomal proteins other than histones. This possibility was further strengthened by two earlier observations: Sandritter and Krygier (1959) demonstrated that intranuclear concentrations of SH groups were almost exclusively localized in chromosomes, chromatin, and nucleoli of HeLa cells; Bachmann and Cowden (1965) then indicated that mercury orange demonstrable protein SH per nucleus in polyploid rat and frog hepatocytes was related to ploidy. The relationship between total nuclear SH and ploidy is significant because nuclear size overlapped DNA classes and was directly related to total nuclear protein content as measured by the low pH acid-dye method of Deitch (1955). Thus, nuclear protein SH and presumably SS groups do not display the same pattern of quantitative or topological distribution in nuclei as do proteins in general. The pH dependence of acid-dye binding to intranuclear proteins was considered in greater detail by Alvarez and Cowden (1966).

Conditions in polytene chromosomes will be considered later. The examination of grasshopper meiotic chromosomes in unpretreated preparations indicated that the chromosomes were nearly selectively fluorochromed (Figure 1. Plate I). A dramatic increase in fluorescence was obtained when preparations were pretreated with 0.2N HCl (Figure 2 Plate I). At least this suggests that some substances, probably histones, were removed. As a consequence of their removal, a substantially larger portion of SH and SS groups of chromosomal proteins became available for binding to fluorescein mercuric acetate. When both classes of nucleic acids were removed as well as histones, there was a substantial reduction in fluorescence. Careful examination of these preparations revealed that this fluorescence was predominantly localized around the periphery of the chromosomes (Figure 3. Plate I). Support for the probable peripheral localization of a chromosomal protein "coat" in 3:1-fixed chromosomes will be presented both in combination with discussions of the other techniques and from evidence obtained by electron microscopy.

Brilliant sulfoflavine used at pH 2.8, because of its enhanced sensitivity over visible acidic dyes, proved to be particularly useful in the demonstration of chromosomal proteins. It also presented special advantages in the assessment of the rigor with which controls must be evaluated in experiments which depend on selective removal of chromosomal constituents. In all preparations fixed in 3:1, there was a distinct

PLATE III

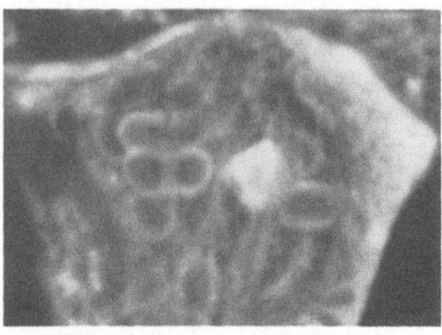

Figure 7. *Grasshopper meiotic chromosomes which received three changes of DNase pretreatment over 24 hours, RNase pretreatment, and extraction in 0.2N HCl fluorochromed with brilliant sulfoflavine at pH 2.8. While chromosomal "coats" are present, no central material is observed.* 1,250X

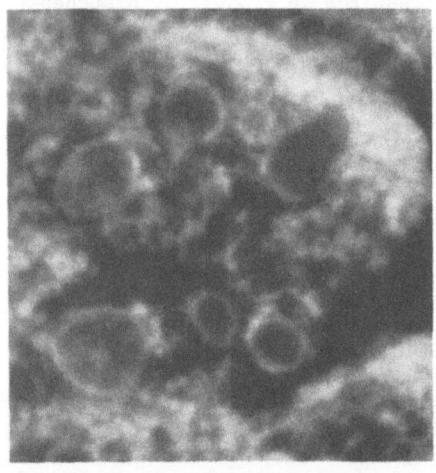

Figure 8. *Similar to Figure 7, but received hot 5% TCA-0.2N HCl pretreatment prior to fluorochroming with pH 2.8 brilliant sulfoflavine. Note presence of chromosomal "coat" and pronounced swelling of chromosomes.* 1,250X

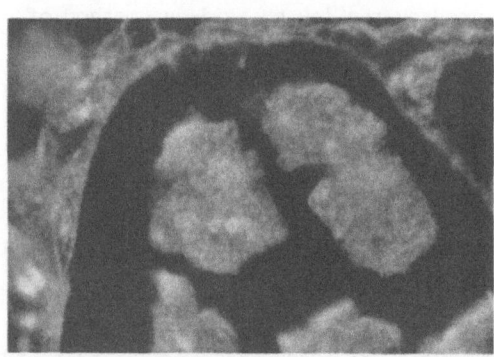

Figure 9. *Grasshopper spermatocyte in meiotic metaphase, pretreated with 0.2N HCl and fluorochromed with the C-terminal carboxyl method. Note only chromosome "coats" are fluorochromed.* 450X

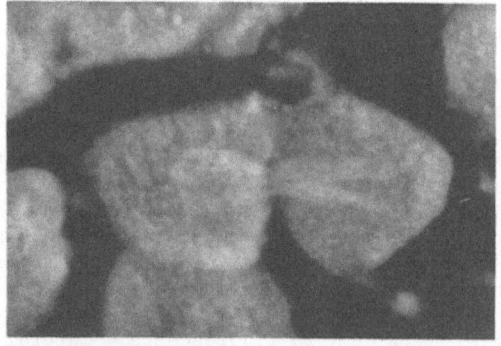

Figure 10. *Similar to Figure 9, but showing affinity of C-terminal carboxyl method for spindle elements.* 450X

chromosomal "coat", a conception of chromosome structure that was laid to rest intel-
lectually by Kaufmann et al. (1960). The existence of chromosomal "coats" would seem
even less likely in the light of electron micrographs of chromosome whole mounts prepared by the critical-point method (DuPraw 1965; 1970). Nevertheless, protein material
at the periphery of chromosomes consistently appeared in 3:1-fixed preparations. While
the protein "coat" was seen in all untreated and pretreated 3:1-fixed chromosomes, the
length and conditions of nuclease pretreatment clearly affected the retention of internal chromosomal proteins. Pretreatment for four hours followed by 0.2N HCl extraction
resulted in a spectrum of variations in the same preparations ranging from trivial
removal of central protein as demonstrated by brilliant sulfoflavine (Figure 4. Plate
II) to sequentially greater removal of internal chromosomal protein (Figures 5 and 6.
Plate II). This is particularly evident in Figure 6 where both cross and longitudinal
sections clearly display the chromosome "coat" and a substantially reduced central zone
of protein material. Treatment with three sequential DNase extractions followed by
0.2N HCl, however, resulted in complete removal of the central chromosomal material in
almost every instance, but persistence of chromosomal "coats" (Figure 7. Plate III).
Essentially the same pattern as that seen in Figure 7 was encountered in 3:1-fixed
chromosomes pretreated with hot 5% TCA then 0.2N HCl (Figure 8. Plate III), except
that some faint suggestion of residual central core material was seen in some chromosomes in this group. Substantial swelling of doubly extracted chromosomes was also
evident, but the magnitude of swelling was greater in preparations subjected to the
TCA-0.2N HCl sequence. In view of the more violent treatment experienced by these
chromosomes, such dimensional distortions might have been anticipated.

Because the chromosomal proteins other than histones are frequently referred to
as "acidic" chromosomal proteins, the use of cytochemical methods designed to demonstrate carboxyl side chains would appear to offer a rational way to demonstrate these
proteins cytochemically. In point of fact, Smetana and Busch (1966) pretreated both
normal liver cells and cells from Walker carcinoma 256 after formalin fixation with
hot 5% TCA and an extended Feulgen-type hydrolysis (1N HCl at 60°C.). The pretreatment
was done on the premise that it would remove nucleic acids (as well as some aldehyde
bridges which are known to interfere with histone extraction), and that the HCl treatment would remove histones leaving behind some "acidic" chromosomal proteins which they
demonstrated using pH 9.0 toluidine blue. It is worth noting that this approach, within the conventions of the current study, represents a demonstration of "residual"
chromosomal protein, since nucleic acid-associated protein would have been removed by
the nucleic acid extraction step. The demonstration of carboxyl groups still could
stand as a useful approach to dealing with these proteins cytochemically even though
histones contain substantial proportions of dicarboxylic monoamino acids.

Because of the possible importance of protein carboxyl groups in nonhistone chromosomal proteins, the fluorescent C-terminal carboxyl method of Stoward and Burns (1967)
was used on grasshopper testes. Frankly, the results were surprising. This method
failed to fluorochrome the inner portion of unpretreated and 0.2N HCl-pretreated grasshopper meiotic chromosomes, but the chromosomal "coats" were brightly fluorescent
(Figure 9). Plate III). In sections pretreated with nucleases and 0.2N HCl, the same
pattern was observed, but the chromosomal "coat" fluorescence was weaker. On the other
hand, substantial fluorescence of spindle elements was noted (Figure 10. Plate III),
and in some oocyte preparations, not included in the main body of the experiments described, nucleoli were brightly fluorescent. Thus, this method might prove to be a
particularly useful technique for studying both nucleoli and spindle fibers.

The rather extensive studies of extracted chromosomes using conventional (nonfluorescent) protein end-group methods can be summarized in almost embarrassingly
short form. With the exception of the DMAB-nitrite method for protein tryptophan,
chromosomes were uniformly and intensely stained by all of the visible chromophore
end-group techniques cited in this study. The DMAB-nitrite method resulted in general
cytoplasmic and nucleoplasmic staining, but no staining of chromosomes. While the
method is highly specific, it may well have sensitivity limitations. Further, the two
major steps of the reaction are carried out in concentrated HCl and thus may be quite
destructive. Hence, the absence of chromosomal staining by this method, while observed
repeatedly, does not constitute conclusive evidence for the absence of protein tryptophan in chromosomal proteins. With the other methods, intense and uniform staining
of chromosomes was obtained in preparations pretreated with 0.2N HCl (Figure 11.
Plate IV). In preparations pretreated with nucleases plus 0.2N HCl or hot 5% TCA plus

PLATE IV

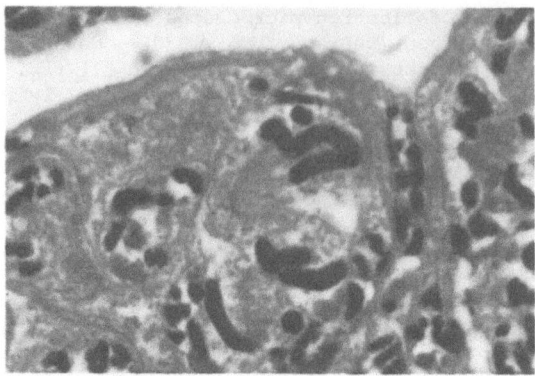

Figure 11. *Grasshopper chromosomes pretreated with 0.2N HCl and stained with Biebrich scarlet at pH 2.8. Chromosomes are intensely and uniformly stained. 500 nm interference filter. 900X*

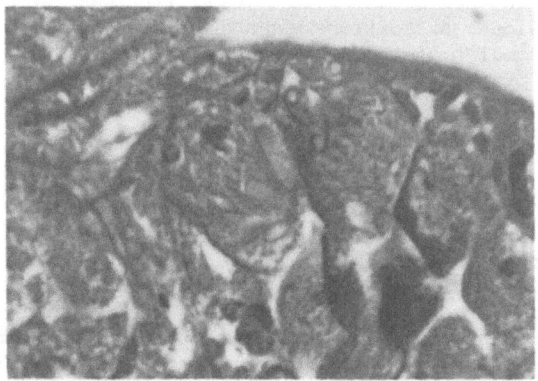

Figure 12. *Similar to Figure 11, but pretreated with hot 5% TCA and 1N HCl at 60°C. for 20 minutes before staining. Note chromosomal "coats" in cross and longitudinal sections, but loss of central chromosomal protein 500 nm interference filter. 900X*

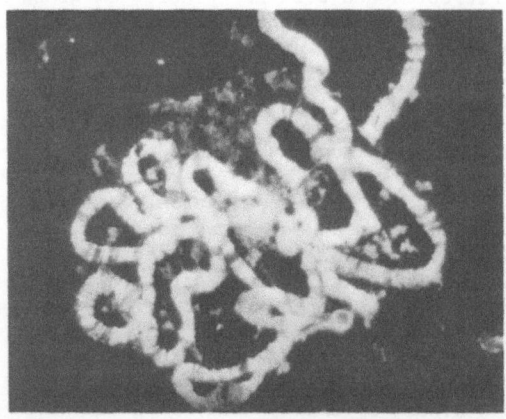 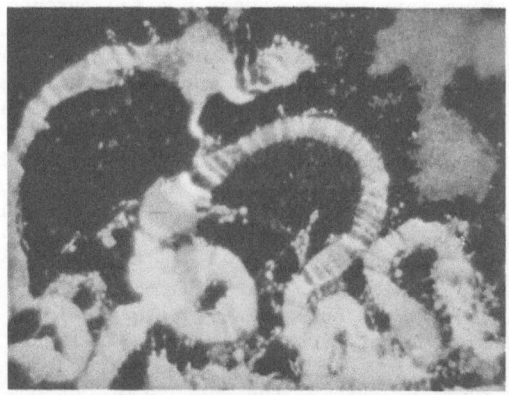

Figure 13. *Unpretreated dipteran polytene chromosomes fluorochromed with fluorescein mercuric acetate after thioglycolate reduction. 450X*

Figure 14. *Similar to Figure 13, but preparation was pretreated with 0.2N HCl before fluorochroming. 450X*

0.2N HCl, the chromosomal "coat" was evident, but the central chromosome as seen in cross, oblique, and longitudinal sections was either completely free of protein or displayed sustantially reduced levels of protein (Figure 12. Plate IV). These observations applied to preparations stained by the acrolein-Schiff method, by mercury orange, and by the modified Morel-Sisley reaction. These results are presented in confirmation of the data obtained through the use of fluorescent cytochemical method. Both grasshopper testicular material fixed in 10% calcium acetate formalin and mitotic newt erythroblasts also displayed essentially empty chromosomes after the hot 5% TCA-HCl sequence, but there was no suggestion of a chromosomal "coat" in these preparations.

Polytene chromosomes, particularly those from Dipteran tissues, offer an example of extreme chromosomal specialization. The work of Swift (1962) and Rudkin (1963) offers evidence that these chromosomes contain substantial levels of proteins other than histones. In the early phase of dissecting chromosomes by selective extraction methods, McDonnough et al. (1952), reported that the banding pattern of *Drosophila* polytene chromosomes was maintained intact after digestion with DNase and a Feulgen-type hydrolysis in 1N HCl. The chromosomes and their banding patterns were clearly demonstrable with mordant-type basic dyes. In the current series of studies, the most surprising effect was the total absence of fluorochroming in *Drosophila* chromosomes after *any* pretreatment when stained by the C-terminal carboxyl method of Stoward and Burns (1967). Polytene chromosomes were intensely fluorochromed by fluorescein mercuric acetate after reduction with thioglycolate (Figure 13. Plate IV), and displayed some diminution in fluorescence after pretreatment with 0.2N HCl (Figure 14. Plate IV). While substantial diminution was observed in preparations pretreated with both nucleases (24 hours and three changes of DNase) plus 0.2N HCl, a considerable amount of protein remained in these chromosomes, chiefly concentrated in the "banded" regions (Figure 15. Plate V). The same results were obtained in polytene chromosomes fluorochromed by brilliant sulfoflavine at pH 2.8. Figure 16 (Plate V) depicts the results of nucleic acid and histone extraction on *Drosophila* polytene chromosomes fluorochromed by brilliant sulfoflavine at pH 2.8. The contrast between banded and interband regions was highest in doubly extracted chromosomes. These observations suggest that polytene chromosomes contain substantially higher levels of "residual" proteins within the definition of this study, than do the more usual forms of meiotic or mitotic chromosomes.

To conclude the qualitative aspects of this presentation, it would be desirable to compare the effects of fixation in 3:1 with those of 10% calcium acetate formalin at the ultrastructural level. A more detailed consideration of chromosome ultrastructure will be covered elsewhere, but at this point, it might be well to note that in doubly extracted chromosomes fixed in 3:1 and subsequently handled as described in the Materials and Methods section, an unquestionable chromosomal "coat" was present (Figure 17. Plate VI). This "coat" had the general appearance of a flocculant precipitate and lacked a fibrillar or membranous structure. The "coat" was completely absent in 10% calcium acetate formalin-fixed chromosomes which had been sequentially treated in hot 5% TCA and 0.2N HCl (Figure 18 Plate VI). It would appear to be of further significance that chromosomes pretreated in this manner and fixed in 3:1 exhibited swelling and distortion of the internal fine fibrillar network, which presumably was associated with the fundamental filamentous structures of the chromosome (see Hyde 1965). Dimensional preservation was better in the 10% calcium acetate formalin-fixed chromosomes which had been extracted with hot 5% TCA and 0.2N HCl, and the regularity of the fine filamentous network was likewise better preserved. This could suggest that aldehyde fixation in conjunction with appropriate extractions would offer the optimal approach to dealing with residual proteins at the ultrastructural level. Further, the filamentous elements involved are very small, far beneath the resolving power of the light microscope. They could only be detected under conditions in which the concentration in aggregates would generate contrast.

Quantitative

Kaufman et al. (1971) have indicated that fluorescein isothiocyanate-antibody conjugates fade during excitation, but if left in darkness for 48 hours display regeneration of fluorescence and shifts in wavelength of emission. Regeneration of fluorescence in faded fluorescent antibody preparations left in the dark for a day or so has been known in practice for many years by workers in the field of cellular

PLATE V

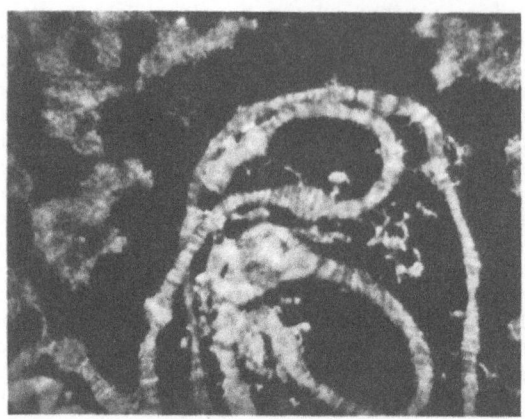

Figure 15. *Similar to Figures 13 and 14, but preparation was pretreated three times over 24 hours with DNase, with RNase, and with 0.2N HCl before fluorochroming with fluorescein mercuric acetate after thioglycolate.* 450X

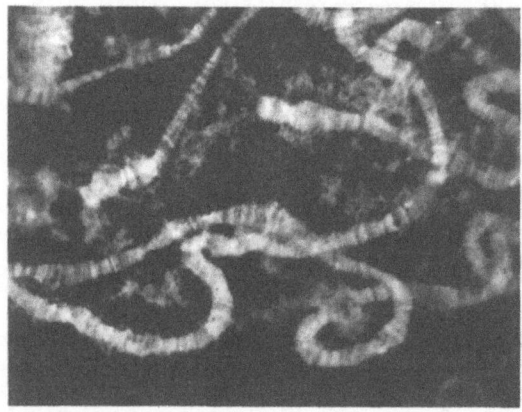

Figure 16. *Dipteran polytene chromosome squash which received the same pretreatment as that in Figure 15, but which was fluorochromed with brilliant sulfoflavine at pH 2.8.* 450X

PLATE VI

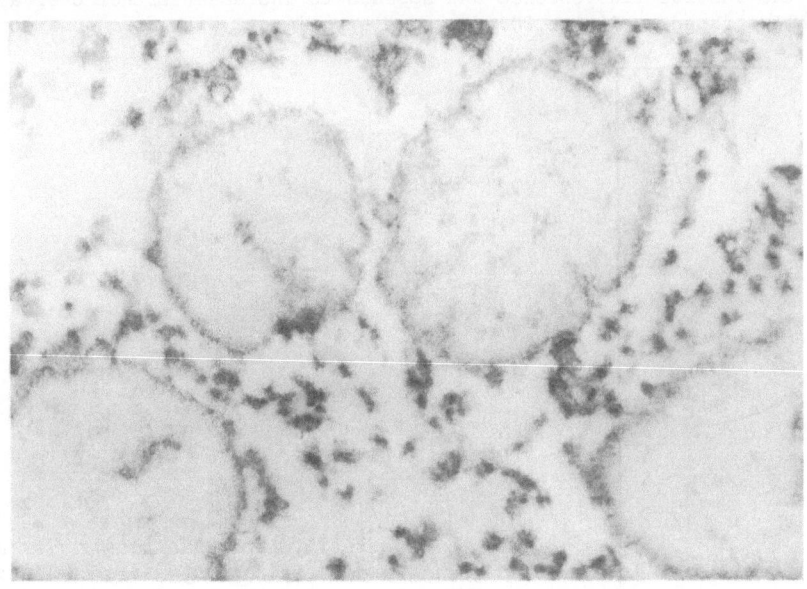

Figure 17. *Electron micrograph of grasshopper meiotic chromosomes from follicles fixed in 3:1 and doubly extracted. Note presence of flocculant, structureless chromosomal "coat" and distortion of fine internal filamentous network.* 22,313X

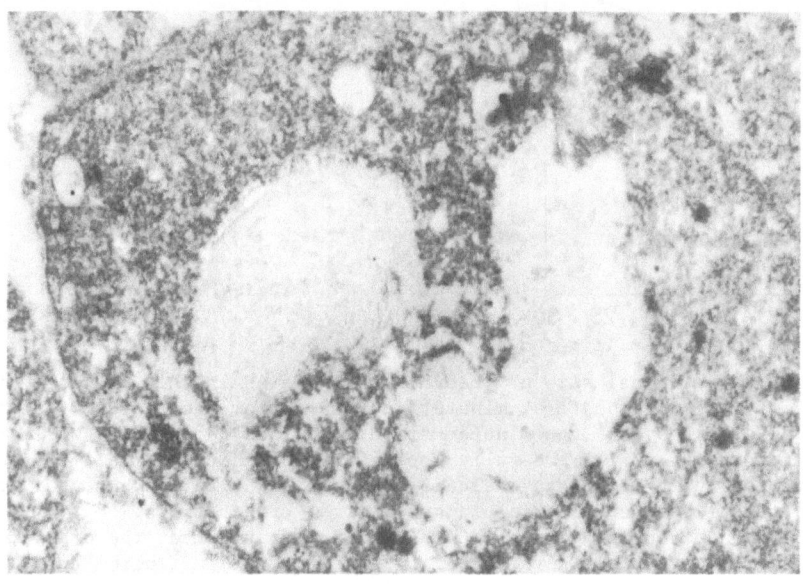

Figure 18. *Similar to Figure 17, but fixed in 10% calcium acetate formalin. Note absence of chromosomal "coat" and better preservation of a fine intrachromosomal filamentous network.* 12,666X

immunology. In the case of fluorescein mercuric acetate, fading does occur as predicted (Figure 19). After 20 hours in darkness, the fluorescence intensity becomes greater than the initial fluorescence and appears to increase further over a period of time, as illustrated in Figure 20. In some experiments with *Stichopus* ootids

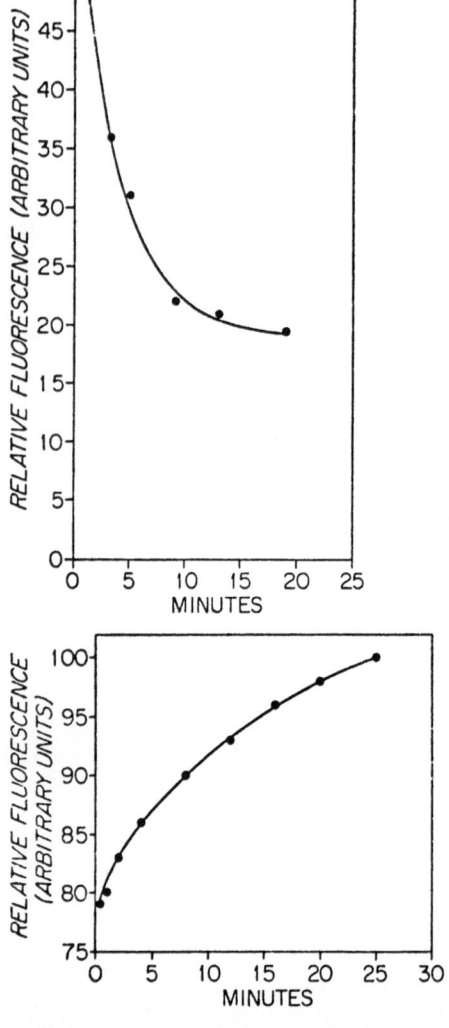

Figure 19. *Graph depicting fluorescence fading under excitation, using BG-12 excitation filter and a measurement interference wedge setting of 545 nm.   Spot size about 10 μm.* Amphiuma *erythrocyte nuclei reduced with thioglycolate and stained with fluorescein mercuric acetate*

Figure 20. *Graph depicting "regeneration" and continued increase in fluorescence of the same preparation after storage in darkness for 20 hours.   Conditions as in Figure 19*

(which represent the terminal size of preovulatory gametes in this species), fluorescence intensities fell for the first minute, increased from the first to the twentieth minute, remained stable over a ten-minute period, and then decreased. While the curves for *Amphiuma* erythrocyte nuclei are fully consistent for that material, they do not of necessity represent the complete evaluation of the quantitative behavior of fluorescein mercuric acetate-stained material in cytological preparations.

The fluorescein mercuric acetate method, and for that matter the other fluorescent mercurials, unquestionably have value in quantitative cytofluorometry. A histogram of measurements on 0.2N HCl and thioglycolate-pretreated *Amphiuma* erythrocyte nuclei fluorochromed with fluorescein mercuric acetate (Figure 21) corresponded well with a Feulgen DNA histogram presented by Cowden (1965) for basophil leucocytes from this species. This correspondence might have been more striking if a better instrumental system had been employed. Similarly, sections of *Stichopus* ootids were compared after fluorescein mercuric acetate fluorochroming with and without thioglycolate pretreatment. The data obtained indicate that the technique will separate protein SH groups from SH plus SS groups (Figure 22). The significance lies not in the spread of values (which arose chiefly from our inability to restrict the area of excitation properly), but rather in the fact that even with this system, there was no overlap between the values obtained in the two groups of measurements.

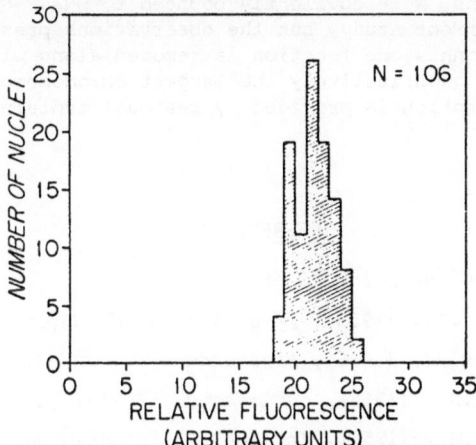

Figure 21. *Histogram of microfluorometric measurements of* Amphiuma *erythrocyte nuclei from smears fixed in 3:1, extracted with 0.2N HC , reduced with thioglycolate, and fluorochromed with fluorescein mercuric acetate*

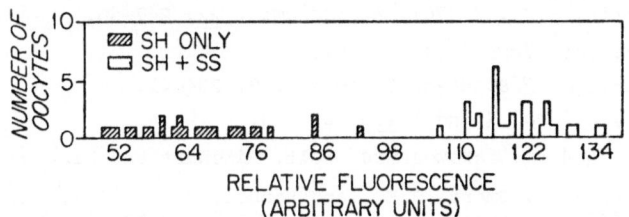

Figure 22. *Graph in histogram form of 10 μm spot measurements on ootids of the sea cucumber,* Stichopus badionotus, *made with and without thioglycolate reduction. In spite of the spread in values, note that the two groups do not overlap*

This method certainly requires more critical standardization with appropriate instrumental packages, but doubtless it will see considerable service in quantitative cytofluorometry.

CONCLUDING REMARKS

It would be difficult to attempt to discuss the information presented here without becoming entrapped in a general review of chromosome structure. Recent reviews have appeared by Hyde (1965) and DuPraw (1970), while Elgin et al. (1971), have reviewed molecular biological aspects of chromosomal proteins. The findings presented here have at least indicated that fluorescence techniques can be of significant value in the exploration of chromosomal proteins, and that quantitative applications of these methods may prove to be even more useful. In the light of these findings, polytene chromosomes must be considered highly specialized chromosomes. Autoradiographic studies, such as that of Cave (1968) on polytene chromosomes, must be evaluated in the full understanding that these chromosomes contain unusually high levels of "residual" protein. Results of such investigations cannot be extrapolated without reservation to chromosomes in general. The role of fixation as manifested by the 3:1-induced chromosome "coat" artifact points to the importance of examining the possible contribution of artifacts to quantitative investigations.

Perhaps the most fundamental bit of information acquired in the course of these experiments — results that must come as no surprise to those experienced in chromosome cytochemistry — was that a substantial quantity of nonhistone chromosomal protein is intimately bound to nucleic acids, and as a consequence is extracted along with the nucleic acids. Heilgartner (1968) proposed that "residual nuclear proteins," in this instance dilute NaOH-soluble proteins remaining after salt extraction of DNA, histones, and doubtless other proteins, were covalently bounded to DNA. Such suggestions extend beyond the scope of the current study, but the observations presented leave little doubt that a significant nonhistone fraction is removed along with the nucleic acids, and that this certainly is quantitatively the largest chromosomal nonhistone protein fraction. The obvious exception is provided by residual proteins in polytene chromosomes.

REFERENCES

ADAMS, C.W.M., (1957), *J. Clin. Path.*, 10 56-66.

ALFERT, M. and Geschwind, I.I., (1953), *Proc. Nat. Acad. Sci. U.S.A.*, 39, 991-99.

ALVAREZ, M.R. and Cowden, R.R., (1966), *Histochemie*, 7, 22-27.

BACHMANN, K. and Cowden, R.R., (1965), *Chromosoma*. 17, 181-93.

BENNETT, H.S. and Watts, R.M., (1958), In: *General Cytochemical Methods*, (Danielli, J.F. ed.,) Vol. I, 317-74. Academic Press, New York.

CAVE, D.M., (1968), *Chromosoma*, 25, 392-401.

COWDEN, R.R., (1965), *Zeit. Zellforsch.*, 67, 219-33.

COWDEN, R.R. and Curtis, S.K., (1970), *Histochemie*. 22, 247-55.

DEITCH, A.D., (1955), *Lab. Invest.*, 4, 324-51.

DUIJN, P. van., (1966), *J. Histochem. Cytochem.*, 9, 234-41.

DuPRAW, E.J., (1965), *Nature*, (Lond.) 206, 338-43.

DuPRAW, E.J., (1970), *DNA and Chromosomes*, Holt, Rinehart and Winston, New York.

ELGIN, C.R., Froehner, S.C., Smart, J.E., and Bonner, J., (1971), In: *Advances in Cell and Molecular Biology* (Dupraw, E. J. ed.), Vol. I, 2-57.

FLAX, M. H., and Himes, M. H., (1952), *Physiol. Zool.*, 25, 297-311.

HEILGARTNER, C.A., (1968), *Exp. Cell. Res.*, 49 520-32.

HENDRICKS, D.M., and Mayer, D. T., (1965), *Exp. Cell. Res.*, 40, 402-12.

HYDE, B.B., (1965), *Biophys. Chem.*, 15, 131-48.

ITO, S., and Winchester, R.J., (1963), *J. Cell Biol.*, 16, 541-77.

KAUFMAN, G.I., Nester, J.F., and Wassermann, D.E., (1971), *J. Histochem. Cytochem.*, 19, 469-76.

KAUFMANN, B.P., Gay, H., and McDonald, M.R., (1960), *Inter, Rev. Cytol.*, 9, 469-76.

LILLIE, R.D., (1965), *Histopathologic Technique and Practical Histochemistry*, McGraw-Hill, New York.

LUFT, J., (1961), *J. Biophys. Biochem. Cytol.*, 9, 409-14.

McDONNOUGH, E.S., Rowan, M., and Mohn, N., (1952), *J. Hered.*, 43, 3-7.

REYNOLDS, E.S., (1963), *J. Cell Biol.*, 17, 208-12.

RITTER, C., and Berman, J., (1963), *J. Histochem. Cytochem.*, 11, 590-602.

RUCH, F., (1970), In: *Introduction to Quantitative Cytochemistry*, (Wied G. L. and Bahr, G. F., eds.)., Vol. II, 431-50. Academic Press, New York.

RUDKIN, G.T., (1963), In: *The Nucleohistones*, (Bonner J.T., and P.O.P.Ts'o, eds.), 184-92. Holden Day, San Francisco.

SANDRITTER, W., and Krygier, A., (1959), *Z. Krebsforsch.*, 62, 596-610.

SMETANA, K., and Busch, H., (1966), *Cancer Res.*, 26, 331-37.

STOWARD, P.J., and Burns, J., (1967), *Histochemie*, <u>10</u>, 230-33.

SWIFT, H., (1962), In: *The Molecular Control of Cellular Activity*, (Allen, J.M. ed.), McGraw-Hill, New York.

Notes Added in Proof

Recently Zirkin (1973) published a cytochemical study of non-histone chromosomal proteins in frog hepatocyte nuclei and cultured bovine kidney cells. As in the case of Smetana and Busch (1966), he avoided ethanol-acetic acid (3:1) fixation and employed doubly extracted preparations; that is, hot TCA and HCl extraction prior to demonstration of "non-histone chromosomal proteins". It is his contention that non-histone proteins are lost in the transition from interphase to metaphase. This approach would be expected to demonstrate "residual chromosomal proteins" within the context of the definitions used in this study since it is noted that a very substantial fraction of non-histone protein is extracted along with DNA by either TCA or DNase. His findings agree very well with those of Smetana and Busch (1966), those presented in this study for doubly extracted preparations, and our ultrastructural study of extracted grasshopper meiotic chromosomes (Cowden and Curtis 1973). Comings (1972) has questioned the suggestion that there are major differences between the composition of non-histone proteins in "inactive" and "active" chromatin or interphase and metaphase chromosomes (Elgin and Bonner 1970). In appropriately fixed material extracted with HCl alone, a substantial amount of demonstrable protein is present in both interphase and metaphase chromosomes. This finding is consistent with the available biochemical findings.

COMINGS, D.E., (1972) In: *Advances in Human Genetics* (H. Harris and K. Hirschhorn, eds.), Vol. III, 237-431. Plenum Press, New York.

COWDEN, R.R. and Curtis, S.K., (1973), *Histochem. J.*, <u>5</u> - in press.

ELGIN, S.C.R. and Bonner, J., (1970), *Biochemistry*, <u>9</u>, 4440-7.

ZIRKIN, B., (1973), *Exptl. Cell Res.*, <u>78</u>, 394-98.

MICROFLUORIMETRIC COMPARISON OF CHROMATIN DURING CYTODIFFERENTIATION

Barton L. Gledhill* and Graham LeM. Campbell**

* *School of Veterinary Medicine, University of Pennsylvania,*
Present Address: Bio-Medical Division, Lawrence Livermore Laboratory
University of California, Livermore. California.
** *The Wistar Institute of Anatomy and Biology, Philadelphia, Pa.*

INTRODUCTION

Cytodifferentiation is a continuous process of nucleocytoplasmic interactions which regulate the changes leading to a terminally differentiated state. Classical literature abounds with elegant descriptions of the morphological changes and staining characteristics of cells as they undergo this process. However, our knowledge concerning the regulatory mechanisms associated with specific biological processes and functions is severely limited. One of the main thrusts toward elucidation of these mechanisms has focused attention on the functional and structural properties of nuclear chromatin. Conventional biochemical techniques have been used primarily for the extraction and characterization of nuclear components. In such efforts, many investigators have been concerned with determining the influence of nuclear protein fractions on template activity of DNA and chromatin. Biophysical techniques, such as X-ray diffraction and circular dichroism, also have been used to study chromatin. Both biochemical and biophysical studies, however, generally require the use of large populations of cells and usually do not permit the multiparametric investigation of single cells, which in our opinion, is critical for the thorough understanding of regulatory mechanisms.

Recent work by the Stockholm group (Bolund et al. 1970; Bolund 1971; Killander and Rigler 1969; Rigler 1966; 1969; Rigler and Killander 1969; Rigler et al. 1969; Ringertz et al. 1970) permits cytochemical investigations of the functional and *in situ* structural properties of chromatin to be carried out at the level of the single cell. One of the tools found particularly useful in characterizing changes in chromatin involves the use of the fluorescent dye, acridine orange (AO). The binding properties of this dye have been used to identify differences in the secondary structure of nucleic acids and to investigate potential changes in the sensitivity of chromatin to thermal denaturation. Such information is particularly valuable when obtained from cells that have the same origin but have marked differences in both their nuclear morphology and their synthetic capabilities.

Two systems involving cytodifferentiation that clearly demonstrate such differences are mammalian spermiogenesis and avian erythropoiesis. Some of the work described in this review (Darzynkiewicz et al. 1969; Gledhill et al. 1966; 1971; Ringertz et.al. 1970) was performed in the Stockholm laboratories and has centered on the changes in chromatin which occur during mammalian spermiogenesis. More recently we have been investigating the alterations that occur in chromatin during primitive erythropoiesis of the chick embryo (Campbell and Gledhill 1973). In both cases, the conspicuous aggregation of nuclear chromatin is associated with and, we postulate, also related to the total repression of genomic activity. However, these two systems are relatively dissimilar in their biological activities and functions. The characteristics of these systems complement each other, and when investigated in parallel, offer a unique opportunity for understanding the mechanisms involved in chromatin aggregation.

SPERMIOGENESIS

Morphological descriptions of the various stages in spermatogenesis are now available for several mammals (Courot et al. 1970; Leblond and Clermont 1952). Mammalian spermatogenesis can be divided into two phases for purposes of classification: *spermatocytogenesis* and *spermiogenesis*. First, the spermatogonium mitotically divides into daughter spermatogonia, most of which undergo further mitotic development into primary spermatocytes. The primary spermatocyte then enters meiosis, which, after the

production of two secondary spermatocytes, ends with the production of four haploid
spermatids. Next, during spermiogenesis, the spermatids undergo a remarkable meta-
morphosis and differentiate into spermatozoa. One of the most fascinating features of
spermatogenesis is the condensation of the nucleus. As the spherical nucleus is con-
densed to approximately 10% of its original volume, the typical loose, fibrillar,
nuclear chromatin of the young round spermatid aggregates into the characteristic
electron-dense, homogeneously opaque mass that is devoid of visable structural detail.

Cytochemical Changes

During the study of the nuclear changes which occur during spermiogenesis in the
bull (Gledhill et.al. 1966), a marked increase was found in both stainability of basic
proteins with alkaline bromphenol blue and the content of protein-bound arginine measured after staining with the Sakaguchi reaction (Figure 1). These alterations in

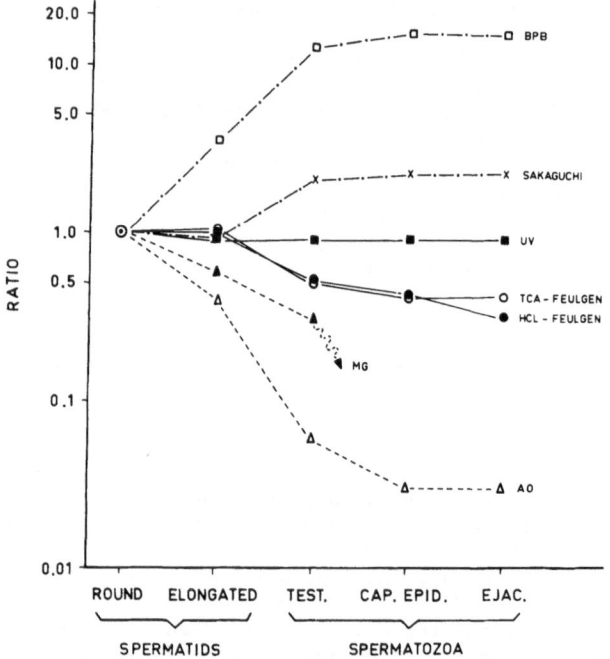

Figure 1. *A summary of the alterations in chromatin seen during spermiogenesis
in the bull. For each cytochemical parameter, the mean values for the different cell
types have been related to the corresponding mean values for round spermatids to
yield a ratio plotted on the logarithmic scale of the ordinate (From Gledhill et al. 1966)*

reactivity indicate that both qualitative and quantitative changes in the nuclear protein occur late in spermiogenesis. Concurrently there is a decrease in the number of
nuclear binding sites available for basic dyes such as AO and methyl green (See Figure
1, Figure 2 [on Plate I], and Table I). The reactivity of nuclear DNA to the classic
Feulgen reaction (employing either TCA or HCl hydrolysis) also is reduced markedly
(Figure 1) as spermatids mature, although there is no change in DNA content as measured by UV microspectrophotometry. The most obvious interpretation of these data is
that during spermatid differentiation, the negatively charged phosphate groups of the
nuclear DNA are masked by new proteins which contain more arginine and are more basic
than those proteins typically bound to DNA in somatic cells.

Thermal denaturation

Encouraged by the results of these investigations, an extension of the work was
undertaken (Ringertz et al. 1970). The sensitivity of DNA to heat denaturation in

PLATE I

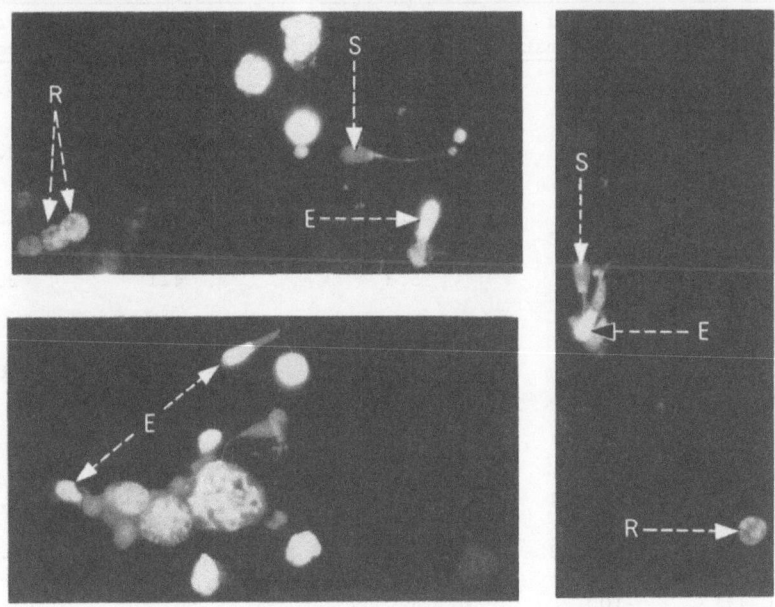

Figure 2. *Photomicrographs of typical areas of a testicle-scrape smear after staining with acridine orange. Examples of round (R) and elongated (E) spermatids and testicular spermatozoa (S) are identified (original magnification about x800; long wavelength UV) (From Gledhill et al. 1966)*

TABLE I

BINDING OF ACRIDINE ORANGE TO BULL SPERMATIDS AND SPERMATOZOA

Characteristic	Round Spermatids	Elongated Spermatids	Testicular Sperm	Ejaculated Sperm
F_{530} at 22°C [a]	70.0	35.3	11.7	5.3
F_{530} at 100°C [a]	38.9	20.4	14.6	10.4
F_{590} at 22°C [a]	9.8	5.3	2.1	1.0
F_{590} at 100°C [a]	10.9	5.3	4.1	2.9
ΔF_{530} 22° to 100°C (%)	-44	-42	+25	+96
ΔF_{590} 22° to 100°C (%)	+11	0	+95	+190
α at 22°C [b]	0.14	0.15	0.18	0.19
α at 100°C [b]	0.28	0.26	0.28	0.28
$\Delta \alpha$ 22° to 100°C (%)	100	73	56	47

[a] Arbitrary units.

[b] $\alpha = F_{590}/F_{530}$. Recalculated from Ringertz et al. 1970.

individual RNase-treated bull spermatids and spermatozoa was analyzed by measuring the increase in nuclear absorption at 265 nm following short-term exposure to various temperatures between 22 and 100°C. Changes were also measured in the fluorescence emission spectrum of similarly treated nuclei stained with AO.

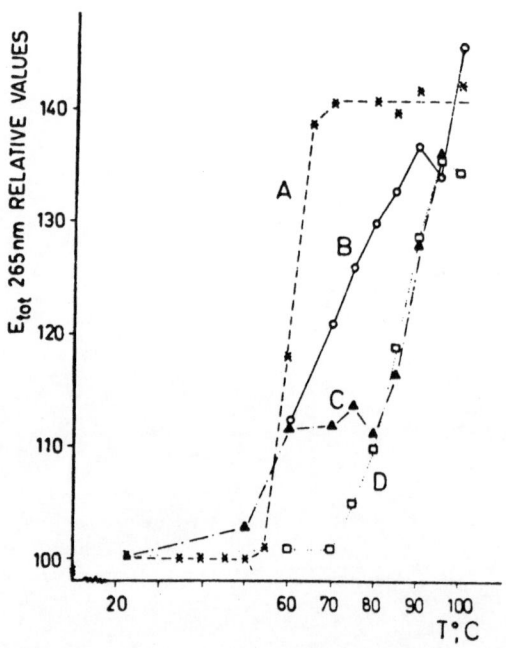

Figure 3. *Increase in relative extinction at 265 nm following exposure of DNA extracted from calf thymus (A), round spermatids (bull) (B), hen erythrocyte nuclei (C), and ejaculated bull spermatozoa (D) to increasing temperature (From Ringertz et al. 1970)*

Figure 3 illustrates the increase in relative extinction at 265 nm following exposure of DNA extracted from calf thymus, round spermatids from the bull, hen erythrocyte nuclei, and ejaculated bull spermatozoa. The increase in relative extinction at 265 nm is shown plotted against increasing temperature in order to construct a so-called "melting profile" (Rigler et al. 1969). DNA in intact cell nuclei melts at higher temperatures than does extracted calf thymus DNA. Furthermore, DNA present in chromatin undergoes a more gradual thermal denaturation than does purified DNA in solution. The denaturation process appears to begin at 55°C in calf thymus DNA and hen erythrocytes; whereas in ejaculated spermatozoa, denaturation begins at roughly 75°C. The shape of the curve for round spermatids suggests that the denaturation process begins at about 55°C, but because of too few experimental points, this curve is not plotted between 22 and 60°C.

As the spermatid differentiates into a spermatozoon there is a considerable decrease in the amount of AO bound (Figure 1: Table I). The same effect is seen when the cells are heated in the presence of formaldehyde. When formaldehyde is present, renaturation of thermally denatured chromatin during rapid cooling is prevented by reaction of the formaldehyde with the amino groups of nucleotide bases. This reaction blocks restoration of hydrogen bonding between DNA base pairs as the chromatin returns to room temperature (Rigler et al. 1969). When, after correction for background for each cell, the value for total emitted fluorescence at 590 nm (F_{590}) is divided by the value for total fluorescence at 530 nm (F_{530}), the quotient, called alpha (α) by Rigler (1966), is said to be a measure of the proportion of single-stranded to double-stranded nucleic acid present. Alpha, on theorectical grounds, is independent of the quantity of nucleic acid present. Analysis by AO-microfluorimetry of the degree of strandedness of chromatin is based on the fact that AO binds in a monomer form to double-stranded nucleic acid, giving rise to a green fluorescence (F_{530}), whereas single-stranded or denatured nucleic acids bind the dye in an aggregated form, which gives rise to a red fluorescence (F_{590}). The AO melting profiles for bull spermatids and spermatozoa are shown in Figure 4, and in a fashion similar to the UV melting profiles, illustrate the progressive stabilization of DNA that occurs during mammalian spermiogenesis.

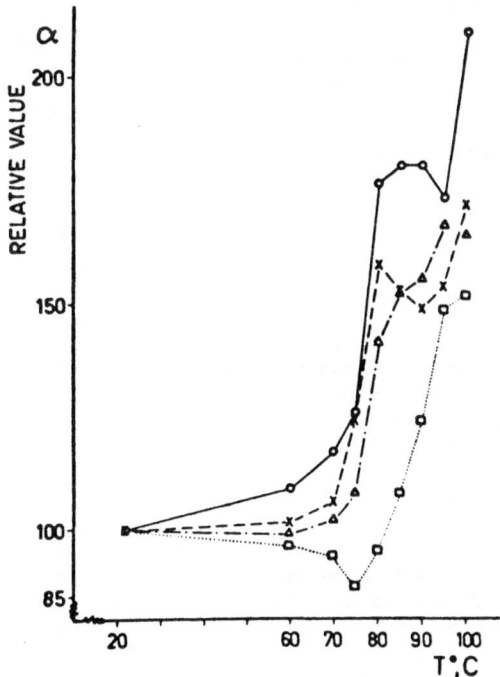

Figure 4. *Increase in relative value of* α *(F_{590}/F_{530}) following exposure of acridine orange stained, round (O) and elongated (X) spermatids, testicular (Δ) and ejaculated ($\square$) spermatozoa to increasing temperature. Although the* α *values at 22°C differ from each other, they have been equated and arbitrarily assigned a value of 100. The curves indicate a progressive stabilization of DNA to heat denaturation during spermiogenesis and epididymal maturation (From Ringertz et al. 1970)*

Repression of the Haploid Genome

The various activities of chromatin during mammalian spermatogenesis are summarized in Figure 5. Scheduled (replicative) synthesis of DNA is terminated abruptly in
the preleptotene primary spermatocyte. The ability of nuclear chromatin to perform unscheduled DNA synthesis, i.e., to incorporate tritiated thymidine after irradiation
with UV light, persists in the spermatogenic cells, however, and reaches its maximum
at approximately the same time that the rate of RNA synthesis is maximal (Gledhill
and Darzynkiewicz 1973). As the ability of chromatin to bind AO and actinomycin D
decreases, its sensitivity to thermal denaturation also decreases (Ringertz et al.
1970b). These changes appear to be related to the molecular rearrangement which occurs
during the aggregation of chromatin fibers and nuclear condensation, and seem closely
intertwined with the cell's temporary loss of genetic potentiality.

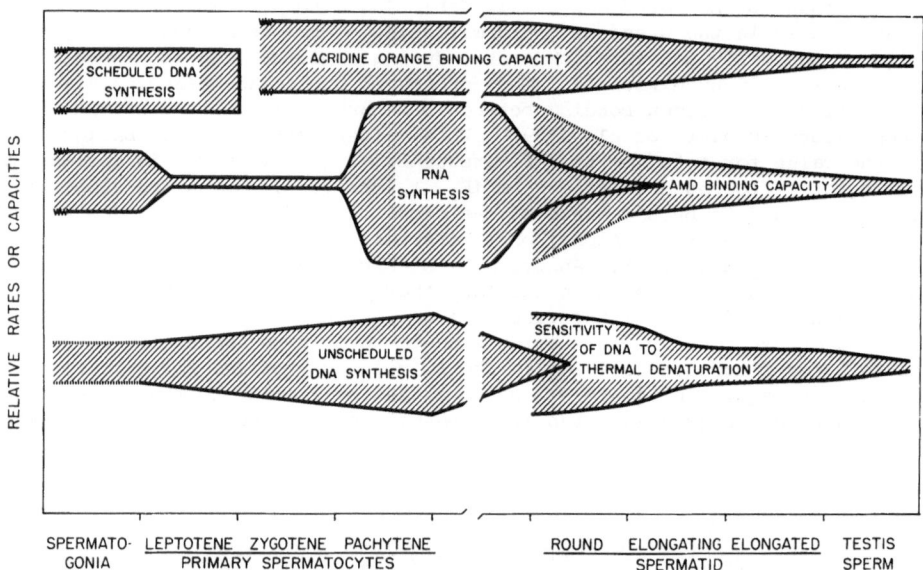

Figure 5. *A schematic depiction of some of the relative activities of chromatin
during mammalian spermatogenesis. Scheduled DNA synthesis (replicative) ends abruptly,
while the ability of the chromatin to perform unscheduled synthesis of DNA reaches
its maximum at the same time that synthesis of RNA is maximal. When the ability of
chromatin to bind actinomycin D and acridine orange decreases, the sensitivity of DNA
to thermal denaturation decreases; these changes are the result of the molecular
rearrangement occurring during chromatin aggregation of spermiogenesis*

Heterokaryocytes

Some of our more recent investigations have been concerned with the mechanisms
which initiate the reversal of the many concurrent processes that lead to condensation
and genetic inactivation of the spermatozoal nucleus. By introducing the fully con-
densed spermatozoal nucleus into a cytoplasmic milieu that can support nucleic acid
and protein synthesis, we are able to study the changes in the structure of chromatin
which signal nuclear reactivation (Gledhill et al. 1972). These experiments are analogous to the now classic experiments by Harris and his colleagues (for references, see
Harris 1970) and by the Stockholm group (Bolund et.al. 1969), in which hen erythroid
nuclei were reactivated in heterokaryocytes. We have fused spermatozoa from rabbit
ejaculates with SV_{40} transformed hamster somatic cells (F5-1 line). A substantial
degree of chromatin disaggregation occured when fusion was performed in the presence of
lysolecithin (Gledhill et al. 1972), but no morphologic alteration of spermatozoal
heads was seen when Sendai virus was used as the fusing agent. Lysolecithin, we
believe, modifies the investing membranes of the spermatozoal heads sufficiently to
allow disaggregation of chromatin once introduced into somatic cell cytoplasm (Figure
6). DNA synthesis was initiated in many of the disaggregated spermatozoal nuclei.

PLATE II

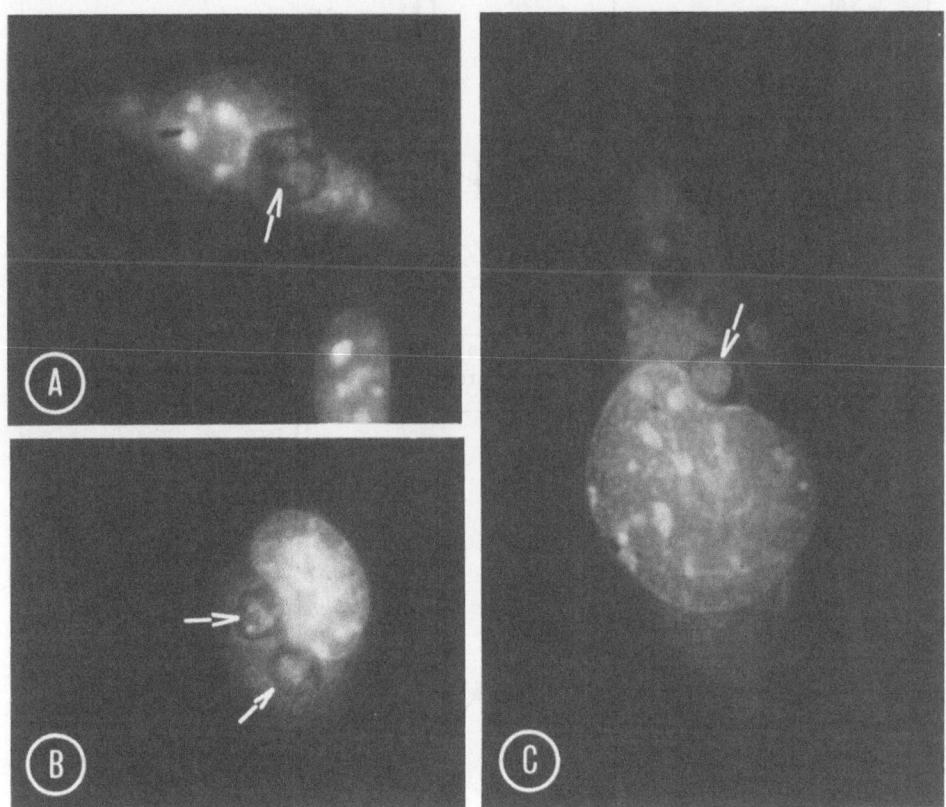

Figure 6. *Photomicrographs of typical rabbit spermatozoa:hamster F5-1 somatic cell heterokaryocytes stained at 22°C with acridine orange, 16 hours after fusion induced by lysolecithin. Partially disaggregated chromatin clumps (arrows) originating from cauda epididymidal (A) and ejaculated (B,C) spermatozoa are seen; deformation of the F5-1 nucleus by the spermatozoal chromatin is evident (original magnification x500: excitation light 365 nm).(From Gledhill et al. 1972)*

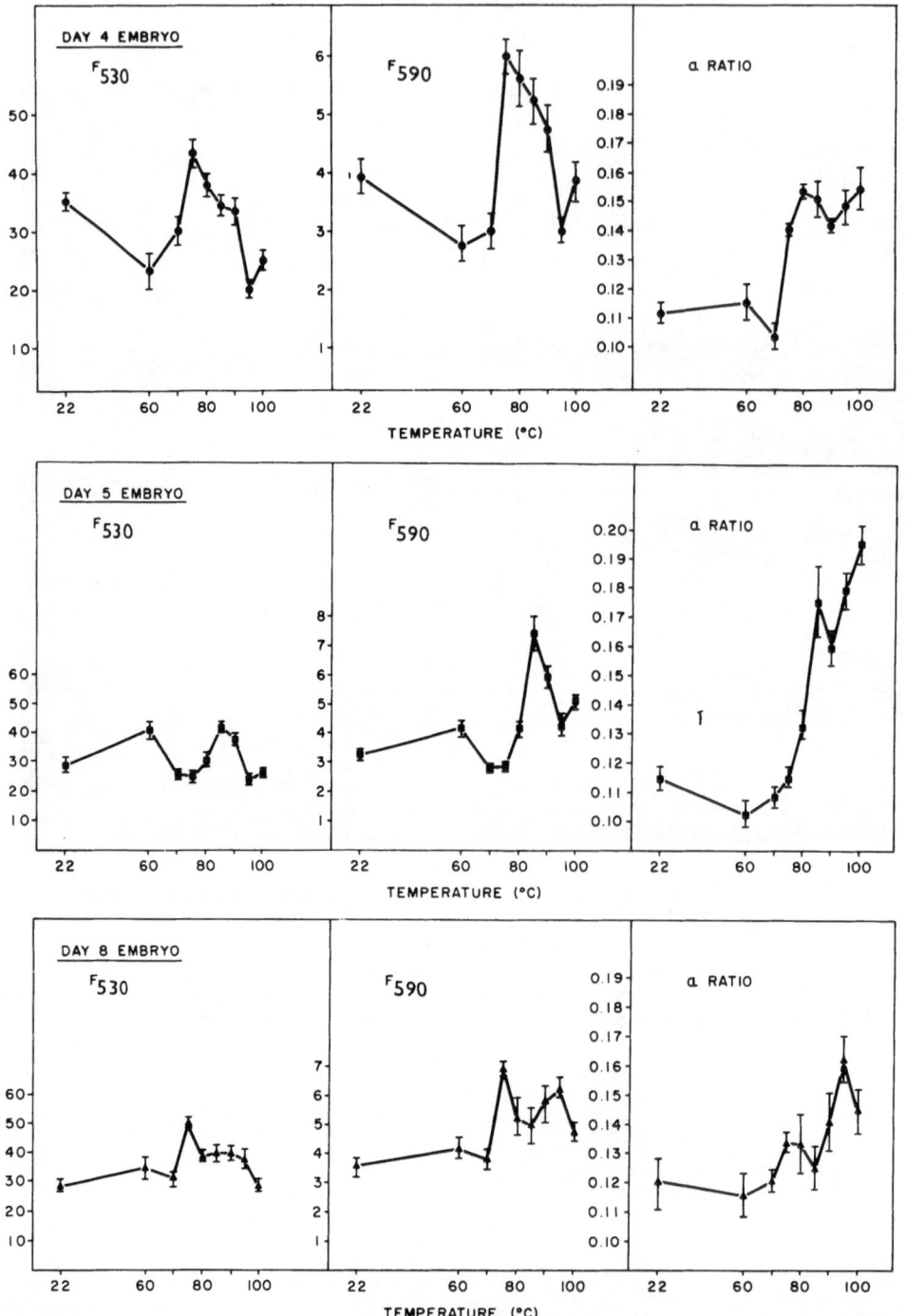

Figure 7. *Melting profiles from typical experiments on erythroblasts from day-4 chick embryos, day-5 chick embryos, and on erythrocytes from day-8 chick embryos. Relative values for total background-corrected fluorescence at 530 nm and 590 nm, and relative values for α (F₅₉₀/F₅₃₀) for acridine orange stained cells are given on the respective ordinates; temperature in degrees C is given on the abscissa. Each point represents the mean of 10 measurements; the 95% confidence limit is given for each point as a vertical line (From Campbell and Gledhill 1973)*

Microfluorimetry of AO-stained, disaggregated, spermatozoal chromatin indicated that it was considerably more "active" than the chromatin of unfused spermatozoa (Gledhill et al 1972). This work is still in its initial stages. We think that by using this technique in conjunction with other techniques, such as AO-microfluorimetry, we may learn more about the mechanisms which ultimately control the activity of the genome.

PRIMITIVE ERYTHROPOIESIS

Primitive erythroid cells comprise a discrete population that appears after 35 hours incubation of the chick embryo. This cohort of cells passes through a series of morphologically well-defined generations that is marked by six cell cycles and a nondividing postmitotic state (Weintraub et al. 1971). During passage, there are changes in nuclear morphology which include a decrease in size, an apparent increase in basophilia, and an increase in extent of chromatin condensation. Erythroid cells from day-4 embryos are in the fifth cell cycle and have an S period of 7 hours. Erythroblasts from day-5 embryos are in the sixth and terminal cell cycle, with an S period of approximately 11 hours. Erythrocytes from day-8 embryos are nondividing postmitotic cells, although some of them may be synthesizing hemoglobin (Campbell et al. 1971).

We have used AO-microfluorimetry to quantitate the changes in RNase-treated chromatin of primitive erythroid cells of the chick embryo. The sensitivity to thermal denaturation of the chromatin in these cells varies with their mitotic history. Chromatin of mature erythrocytes is less susceptible to this type of denaturation than is that of dividing erythroblasts, while the sensitivity between selected successive cell generations of dividing erythroblasts diminishes as the cells mature(Campbell and Gledhill 1973).

There generally was no increase in α value for day-4 erythroblasts until approximately 70°C, after which there was a constant increase until 80°C (Figure 7a). At this temperature, α reached a maximal increase (50% increase over the original value at 22°C), and thereafter remained relatively constant. The day-5 erythroid cells showed no increase in their α values until approximately 75°C. Thereafter a constant increase occurred until α was approximately doubled (Figure 7b). In general, the increase in α value began at 70-75°C and reached a maximum at 90-95°C for day-8 erythrocytes (Figure 7c). The average maximum increase in α for these erythrocytes was only approximately 30%, as compared with an average 50% increase for day-4 and day-5 erythroblasts (Figure 8).

INTERPRETATION AND SIGNIFICANCE

From the data reviewed here, it is clear that, in general, the changes in the melting profiles of chromatin in spermiogenic and erythroid cells are similar. As a consequence of maturation, there is a stabilization of the chromatin to thermal denaturation. The temperature at which round spermatids show half their maximal change in α is similar to the temperature at which the fifth-generation erythroblasts show half their maximal change. Likewise, the temperature at which the ejaculated spermatozoa and the nondividing erythrocytes show half their maximal change is roughly the same (compare Figures 4 and 8).

There is a further similarity between spermiogenic and erythroid cells in their AO-binding characteristics. We noted, as have others (Kernell et al. 1971), a decrease in AO-binding capacity, as reflected by reduced F_{530} and F_{590} values, associated with increased nuclear condensation. The spermiogenic cells, however, show this effect to a much greater extent; the F_{530} values for ejaculated spermatozoa are only 8% of the values for round spermatids. Similarly, F_{590} values for ejaculated sperm are barely 27% of the values for round spermatids. Mature erythrocytes have F_{530} and F_{590} values that are 61% and 51%, respectively, of those for the fifth-generation erythroblasts.

There is, on the other hand, a marked difference in the relative amount of change in α between spermiogenic cells and erythroid cells. At all stages of maturity the spermiogenic cells show larger relative changes in α as a consequence of thermal denaturation (range 100% to 47%, Table I) than do the maturing erythroid cells (41% to 32%, Table II). The complete significance of this difference between spermiogenic and erythroid cells is not clear, but we suspect that it reflects the loss of nuclear components as a result

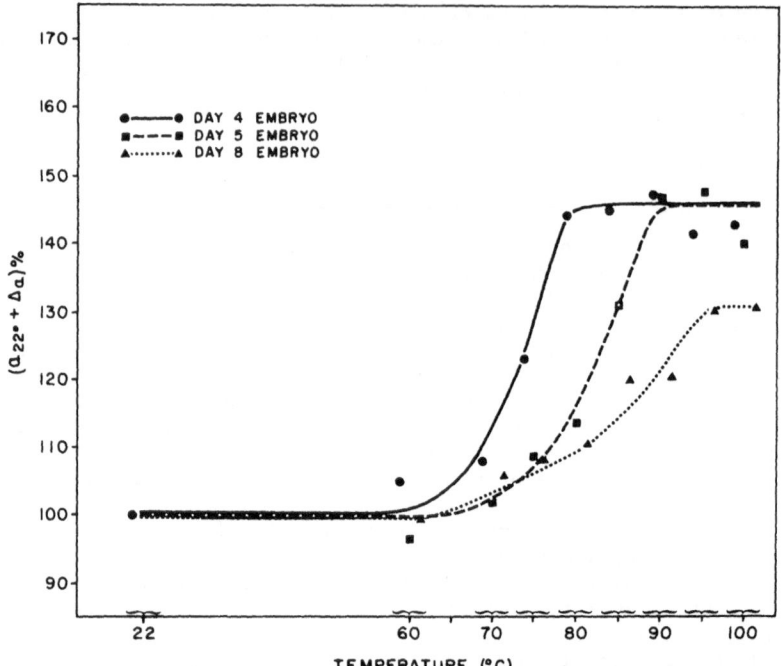

Figure 8. *A composite of melting profiles for acridine orange stained erythroid cells from day-4 (●), day-5 (■) and day-8 (▲) chick embryos. Each point on each melting profile represents the mean value for 80 measurements (10 cells × 8 experiments). For each curve, the mean α value (ordinate) for each temperature (abscissa) has been normalized so the α value at 22°C was arbitrarily set equal to 100 (From Campbell and Gledhill 1973)*

of heating the erythroid cells in a salt solution, and is related to differences in susceptibility to thermal denaturation.

TABLE II

MICROFLUORESCENT CHARACTERISTICS OF AO-STAINED AND THERMALLY DE-NATURED CHROMATIN IN PRIMITIVE ERYTHROID CELLS OF THE CHICK EMBRYO.

	Erythroblasts		Mature Erythrocytes (Mean ± S.E.M)
Characteristic	5th Generation (Mean ± S.E.M.)	6th Generation (Mean ± S.E.M)	
F_{530} at 22°C [a]	41.0 ± 1.0	38.9 ± 1.7	25.1 ± 0.8
F_{530} at 100°C [a]	22.1. ± 0.7	27.0 ± 1.1	33.2 ± 1.3
F_{590} at 22°C [a]	6.3 ± 0.4	4.6 ± 0.2	3.2 ± 0.1
F_{590} at 100°C [a]	3.5 ± 0.1	4.5 ± 0.2	5.4 ± 0.2
α at 100°C [b]	141	139	132
Δα (max−min)	47	50	33
Approx. temp. (°C) of 1/2 Δα	75	80−85	85−90

[a] Arbitrary Units: N=80 cells

[b] $\alpha = F_{590} / F_{530}$. These values have been normalized (α at 22°C = 100) for use in comparison with Figure 8. From Campbell and Gledhill, 1973.

Another consequence of thermal denaturation is that different stages of maturity in both the spermiogenic and erythroid progressions show varying responses in their ability to bind AO after heating to 100°C. Round and elongated spermatids show a decrease in F_{530} values, but show little change in F_{590} values. The erythroid cells from both the fifth and sixth generations also show a decrease in F_{530} values after treatment at 100°C, whereas only the erythroblasts from the fifth generation show a decrease in F_{590} values. In contradistinction, both the spermatozoa and the non-dividing erythrocytes show an increase in F_{530} and F_{590} values after thermal denaturation at 100°C.

An understanding of these differential effects on the various stages of cellular maturity is a key point to the understanding of thermal denaturation of chromatin. The decrease in F_{530} and F_{590} values can be interpreted as reflecting the loss of nuclear components due to the interaction of temperature and salt solutions. Salt solutions at a molarity of 0.15 M and higher are known to extract proteins from isolated nuclei and chromatin (Pardon and Richards, 1971). Although a 0.15 M NaCl + 0,015 M Na-citrate solution is used in the heat denaturation process (Rigler et al. 1969), it is not yet clear if proteins are extracted by the interactive effect of the salt and high temperature and if these proteins differ between cell types and stages of maturity. We postulate, however, that the changes in AO binding and thermal denaturation not only reflect differences in the nuclear protein composition as a consequence of or as a corollary to cellular differentiation, they also indicate the secondary structure of nucleic acids.

ACKNOWLEDGMENTS

Much of the data reported in this review was the result of collaborative efforts in conjunction with a number of colleagues to whom we are indebted.

The production of this paper has been supported in part by United States Public Health Service Research Grants HD-03577 from the National Institute of Child Health and Human Development, RR-05540 from the Division of Research Resources, and CA-10815 from the National Cancer Institute.

REFERENCES

BOLUND, L., Ringertz, N.R., and Harris, H., (1969), *J. Cell Sci.*, **4**, 71-87.

BOLUND, L., Darzynkiewicz, Z., and Ringertz, N.R., (1970), *Exptl. Cell Res.*, **62**, 76-89.

BOLUND, L., (1971), *Cytochemical Properties of Chromatin and Nuclei in Relation to Genome Activation and Repression.* Academic Dissertation, Karolinska Institutet, Stockholm.

CAMPBELL, G, LeM., Weintraub, H., Mayall, B.H., and Holtzer, H., (1971), *J. Cell Biol.*, **50**, 669-81.

CAMPBELL, G. LeM., and Gledhill, B.L., (1973), Chromosoma., in press.

COUROT, M., Hochereau-de Reviers, M.T., and Ortavant, R., (1970), In: *The Testis* (A.D. Johnson, W.R. Gomes, and N.L. VanDemark, eds.) Academic Press, New York, 339-432.

DARZYNKIEWICZ, Z., Gledhill, B.L., and Ringertz, N.R.,(1969), *Exptl. Cell Res.*, **58**, 435-38.

GLEDHILL, B.L., Gledhill, M.P., Rigler, R. Jr., and Ringertz, N.R. (1966), *Exptl. Cell Res.*, **41**, 652-65.

GLEDHILL, B.L., Darzynkiewicz, Z., and Ringertz, N.R., (1971), *J. Reprod. Fert.*, **26**, 25-38.

GLEDHILL, B.L., and Darzynkiewicz, Z., (1973), *T. Exptl. Zool.*, **183**, 375-82.

GLEDHILL, B.L., Sawicki, W., Croce, C.M., and Koprowski, H., (1972), *Exptl. Cell Res.*, 73, 33-40.

HARRIS, H., (1970), *Cell Fusion*. Harvard University Press, Cambridge.

KERNELL, A.M., Bolund, L., and Ringertz, N.R. (1971), *Exptl. Cell Res.*, 65, 1-6.

KILLANDER, D. and Rigler, R., Jr. (1969), *Exptl. Cell. Res.*, 54, 163-70.

LEBLOND, C.P., and Clermont, Y., (1952), *Ann. N.Y. Acad. Sci.*, 55, 548-73.

PARDON, J.F., and Richards, B.M., (1971), In: *Biological Macromolecules* (S.M. Timasheff and G.D. Fasman, eds.) in press.

RIGLER, R., Jr.(1966), *Acta. Physiol. Scand.*, 67, Suppl. 267.

RIGLER, R., Jr. (1969), *Ann. N.Y. Acad. Sci.*, 157, 211-24.

RIGLER, R., JR. and Killander, D., (1969), *Exptl. Cell. Res.*, 54, 171-80.

RIGLER, R., Jr., Killander, D., Bolund, L., and Ringertz, N.R. (1969), *Exptl. Cell. Res.*, 55, 215-24.

RINGERTZ, N.R. Bolund, L., and Darzynkiewicz, Z., (1970), *Exptl. Cell Res.*, 63, 233-38.

RINGERTZ, N.R., Gledhill, B. L., and Darzynkiewicz, Z. (1970 b), *Exptl. Cell. Res.*, 62, 204-18.

WEINTRAUB, H., Campbell, G. LeM., and Holtzer, H., (1971), *J. Cell Biol.*, 50, 652-68.

IMMUNOFLUOROMETRIC DETERMINATIONS OF SURFACE IMMUNOGLOBULINS OF INDIVIDUAL NORMAL AND MALIGNANT HUMAN LYMPHOID CELLS

D. Killander, U. Hellström, B. Johansson, E. Klein,
A. Levin and P. Perlmann.

*Radiumhemmet, Karolinska Hospital, Institute for Medical Cell Research
and Genetics and Department of Tumor Biology, Karolinska Institute
and the Immunology Division, Wenner-Gren Institute, Stockholm, Sweden*

INTRODUCTION

There is accumulating evidence for presence of immunoglobulins (Ig) on the surface of normal lymphocytes derived from peripheral blood. These membrane-bound Ig probably function as surface receptors which combine with antigen by an immunologically specific reaction (Mäkelä, et al. 1970).

Immunofluorescence is a widely used technique for studies on membrane-bound Ig. So far the technique has been used mostly qualitatively, i.e., cells have been judged visually as to whether they have fluorescent membrane staining or not. In the present work, quantitative immunofluorescence has been applied for determination of membrane-bound Ig. Thus, the fluorescence intensity of individual cells was measured in a microspectrofluorimeter after "staining" with fluorescein-conjugated antisera to Ig.

EXPERIMENTS AND RESULTS

First, model experiments were made to test that the fluorescence readings were meaningful, i.e., proportional to the amount of surface-bound Ig. Human erythrocytes were coated with varying amounts of I^{131}- labeled IgM and "stained" with FITC-conjugated antiserum to human IgM. A strong correlation was found between the quantity of protein antigen, as measured by its radioactivity, and the intensity of fluorescence — a finding which demonstrates that the fluorescence values are good estimates of the amount of membrane-bound IgM (Figure 1).

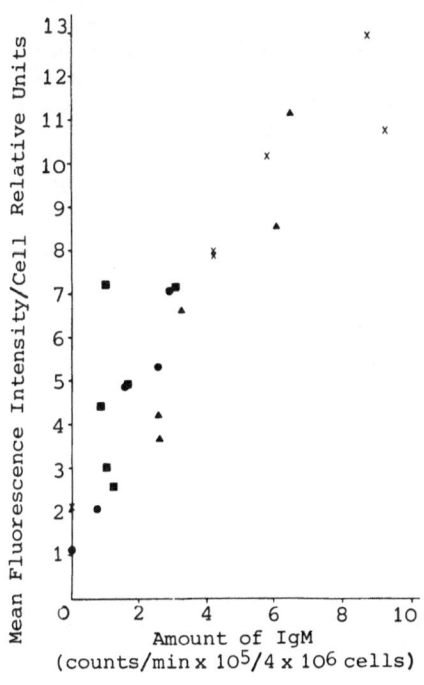

Figure 1. *Fluorescence intensities of formalinized human erythrocytes coated with varying amounts of I^{131}-labeled human IgM and reacted with fluorescein-conjugated anti human IgM in four experiments. Each fluorescence value, plotted against amount of IgM (expressed in cpm), is the mean of 20 cells measured in a microspectrofluorimeter described by Caspersson et al. 1965. Cells were excited individually with transmitted light. Background readings taken from adjacent cell free areas were subtracted from the cell values giving net fluorescence intensities. Prism monochromators were used for both excitation at 365 nm and emission at 520 nm. Further details are described elsewhere (Killander et al. 1970). For calibration of the instrument a standard crystal of uranyl glass (Schott GG17) of about the same size as the cells was used. (From Killander et al. 1970)*

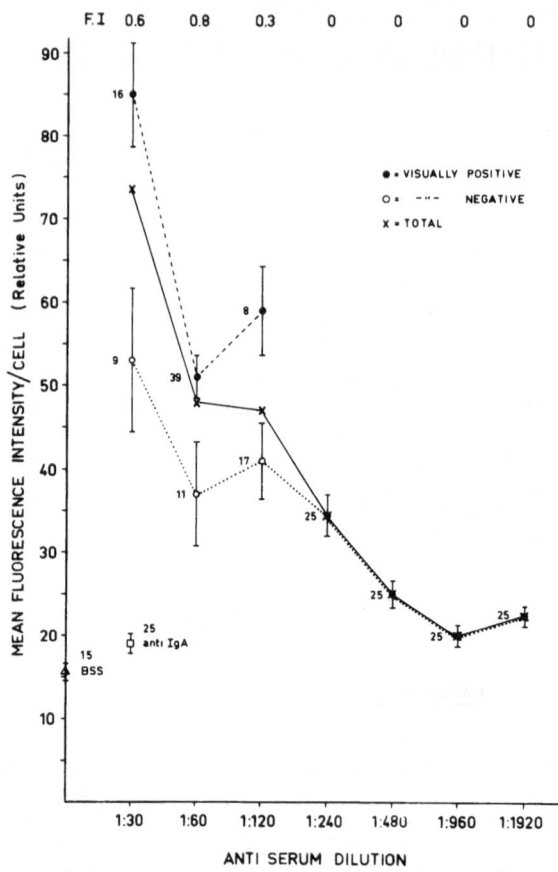

Figure 2. *Mean fluorescence (± standard error of the mean) intensities of Burkitt lymphoma cells* in vitro *(Daudi) reacted with different dilutions of fluorescein-conjugated anti human IgM. The measured cells were visually judged positive or negative (number of each is indicated). The fluorescence index (F.I.) indicating the fraction of cells judged positive is shown at the top. The cells were measured in a Zeiss photomicroscope II, equipped with an RCA IP28 photomultiplier. Individual cells were centered for measurement by use of phase contrast (Light source: 12V, 60W tungsten lamp). For measurement of fluorescence, cells were excited in incident light using an Osram HBO 100 W/2 superpressure mercury lamp, a light modulator, a 2.5 mm BG 38 heat filter, and a KP 500 or a 4 mm BG 12 filter. A II E epicondensor and an FL reflector (475) were used and cells were enclosed by a fixed aperture of appropriate size. A 100X Neofluar oil immersion phase objective (N.A. = 1.3) was used. Emitted light passed through a barrier filter eliminating light below 500 nm. The fluorescence intensity at 520 nm was measured by inserting an interference filter in front of the photomultiplier. The same standard system was used as described in text figure 1. The fluorescence values were taken from a Zeiss MPM reading instrument or from a computer system involving an Optilab Multilog 801 and a printer driver 832 (Bo Philip Instrumentation, Box 138, Vällingby, Sweden) by which background readings were subtracted automatically and net fluorescence values printed out. Fading of fluorescence was less than 10% during a 2 minute excitation time. (From Levin et al., 1971)*

Qualitative and quantitative immunofluorescence was compared using cultured Burkitt lymphoma cells (Daudi) with natural presence of IgM on their surfaces. The Daudi cells were reacted with varying dilutions of FITC-conjugated anti-human IgM. After fluorescence intensity of a cell was measured, a determination made by two independent observers as to whether the cell was "positive" (exhibited visual membrane

165

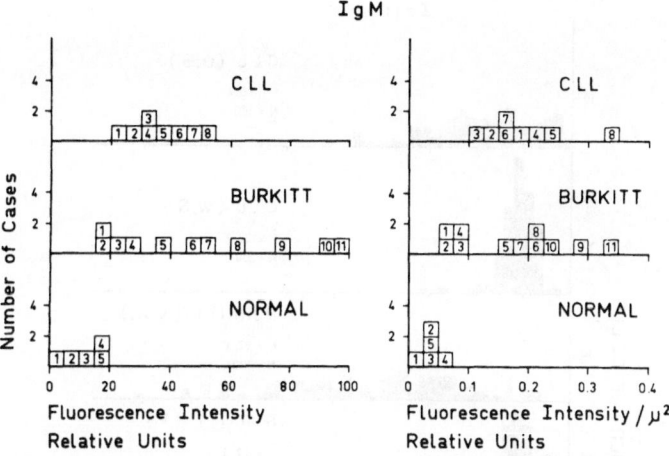

Figure 3. *Mean fluorescence intensities of total cells (left histograms) and expressed per μ² surface area (right histograms). Normal and malignant lymphoid cells were reacted with fluorescein-conjugated anti human IgM as previously described (Klein et al. 1968, 1971 a). Each square represents a mean of 100 measured cells from one patient. After fluorescence measurements (cf text Fig. 2) the diameters of the same cells were determined using a microscale in the microscopic eyepiece. Assuming that the cells were spheric, the total surface area could be calculated*

fluorescence) or "negative." With increasing dilutions of antiserum to IgM, the fraction of positive cells decreased as did the fluorescence intensities (Figure 2). Cells visually judged positive gave higher values compared with those classified negative. However, negative cells stained with low-antiserum dilution had fluorescence much higher than either the zero level represented by autofluorescence (BSS), or the intensity of cells reacted with FITC-conjugated antihuman IgA. These results indicate that it is possible to detect and measure low levels of specific immunofluorescence, not detected visually, by means of microspectrofluorometry.

Previous qualitative immunofluorescence studies have shown increased presence of cell membrane-bound Ig on Burkitt lymphoma cells (Klein et al. 1968; Klein et al. 1971) and on peripheral lymphoid cells from patients with chronic lymphocytic leukemia (CLL) (Klein et al. 1971a; Johansson and Klein 1970; Wilson and Nossal 1971). A number of cases were studied in more detail by quantitating the fluorescence intensities of individual cells after staining with standard antiserum dilutions. In agreement with previous results, Figure 3 shows that as judged from the higher mean fluorescence values there was more IgM attached to the surface of lymphoid cells in the majority of CLL and lymphoma patients than there was in the peripheral lymphocytes of healthy individuals. A few Burkitt patients (Nos. 8,9,10 and 11) had values exceeding the maximal CLL value (Figure 3, left). However, since the Burkitt cells were larger than the CLL lymphoid cells, the surface areas of the cells were determined and fluorescence intensities per unit surface area were calculated. Expressed in this way, the fluorescence values of the Burkitt patients 8-11 fell within the range of the CLL patients (Figure 3, right). Still the majority of lymphoma and leukemia patients exhibited higher values than the normal controls (Figure 3, right). The distributions of fluorescence values of individual cells in normal and malignant lymphoid cell populations are shown in Figure 4. The majority of cells in the normal cell populations had low values (Figure 4, bottom). However, a few normal cells exhibited high values well in the range of fluorescence intensities of the leukemia and lymphoma cell populations (Figure 4). Examples of two different types of distributions are seen in the malignant cell populations; one wide (CLL, O.S.,and Burkitt, N.K.) and one quite narrow (CLL, W.R., and Burkitt, P.). In addition to microfluorimetric determination of the cell surface-bound IgM, other parallel tests

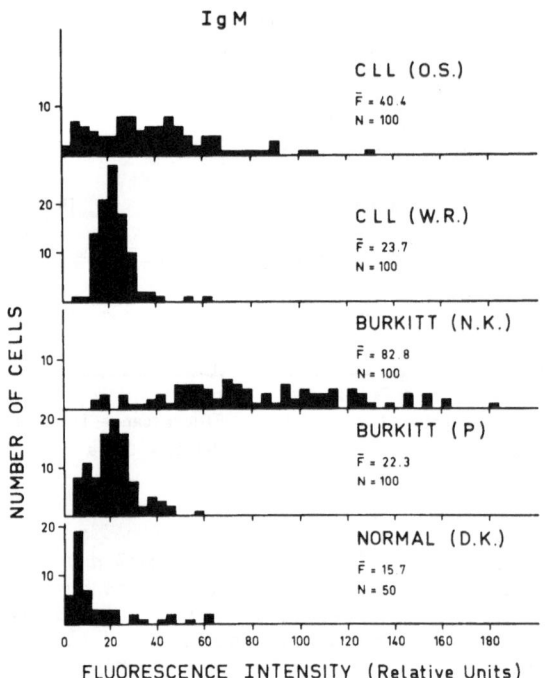

Figure 4. *Frequency distributions of fluorescence intensities of individual cells of normal and malignant lymphoid cell populations reacted with fluorescein-conjugated anti human IgM. Same measuring technique as described in legend for Figure 2*

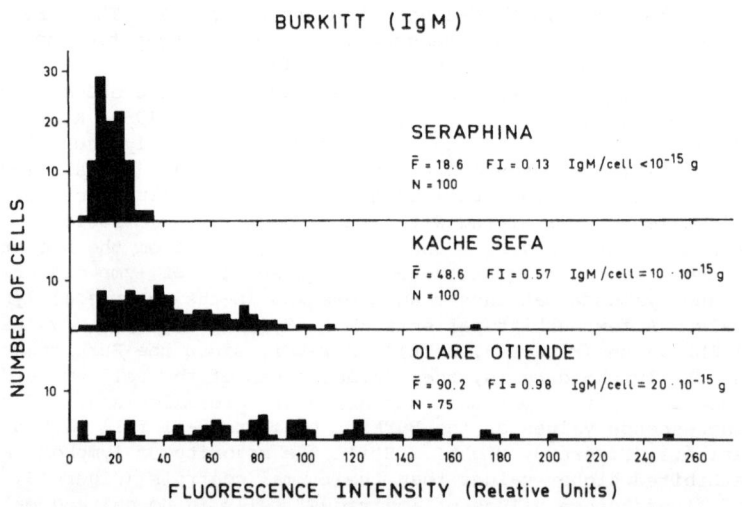

Figure 5. *Frequency distributions of fluorescence intensities (F) of individual cells of three Burkitt lymphoma patients reacted with fluorescein-conjugated anti human IgM. Same measuring technique as described in the legend for Figure 2. FI = fluorescence index. The amounts of IgM were determined by a micro-absorption test. (Note in proof: Seraphina has IgG on the cell membrane and is thus a negative control for IgM)*

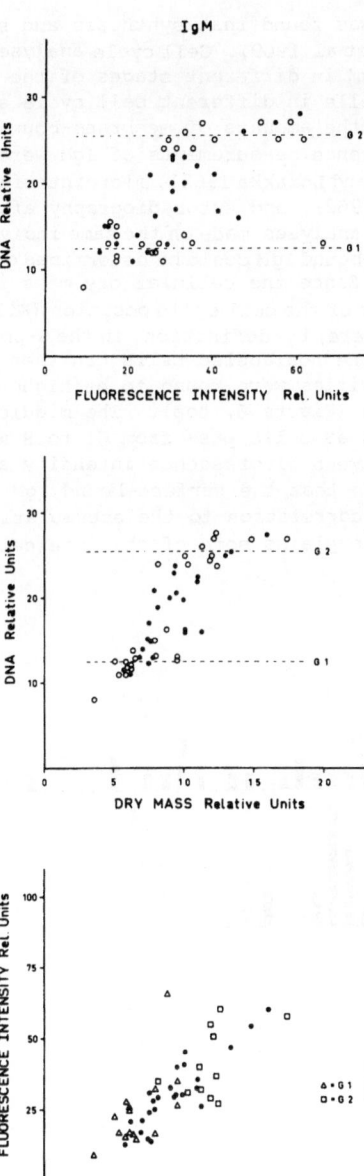

Figure 6. *Fluorescence intensities of Burkitt lymphoma cells* in vitro *(Daudi) reacted with fluorescein-conjugated anti human IgM, Feulgen DNA content and dry mass of the same cells. The cells were incubated for 20 minutes with* [3]*H-thymidine (1 μC/ml) before fluorescence staining. Labeled cells (solid symbols). Further details are described elsewhere (Killander et al. 1972)*

were made in some Burkitt cases. The fraction of cells with visually positive surface fluorescence (F.I.) was determined by different observers and the mean amount of IgM was determined by a microabsorption test (Klein et al. 1971a). The results obtained with these three different techniques correlated well (Figure 5), i.e., with increasing fluorescence intensity values higher FI and larger amounts of IgM were found.

In Ig-secreting cells it was found that synthesis and secretion of the Ig vary during the cell cycle (Takahashi et al.1969). Cell cycle analyses were made to determine the amount of membrane-bound IgM in different stages of the cell cycle and to investigate whether the position of cells in different cell cycle stages might contribute to the intercellular variation in the amounts of membrane-bound IgM found in lymphoid cell populations. The fluorescence measurements of IgM were combined with Feulgen DNA cytophotometry (Caspersson and Lomakka 1962), microinterferometric dry-mass determinations (Caspersson and Lomakka 1962), and autoradiography after pulse labeling with ^{3}H-thymidine - all cytochemical analyses made on the same individual cells. In this way the cycle position and the membrane-bound IgM could be determined of nontreated and asynchronously proliferating cells *in vitro*. Since the cellular dry mass increases throughout interphase the cell mass is one measure of the cell cycle position (Killander and Zetterberg 1965). Cells labeled with ^{3}H-thymidine are, by definition, in the S-phase of the cell cycle. Knowing the amount of Feulgen DNA the nonlabeled cells can then be identified as G1 or G2 cells. The fluorescence intensities were found to be highest in G2 cells, lowest in G1, and intermediate in S-cells (Figure 6, top). The middle graph (Figure 6) illustrates the increase in dry mass as cells pass from G1 to S and to G2. A strong correlation (r=0.79) was found between fluorescence intensity and dry mass (Figure 6, bottom). These results indicate that the surface-bound IgM increases continuously throughout interphase in close correlation to the accumulation of dry mass (=total cellular proteins). This also explains some of the intercellular variability in IgM

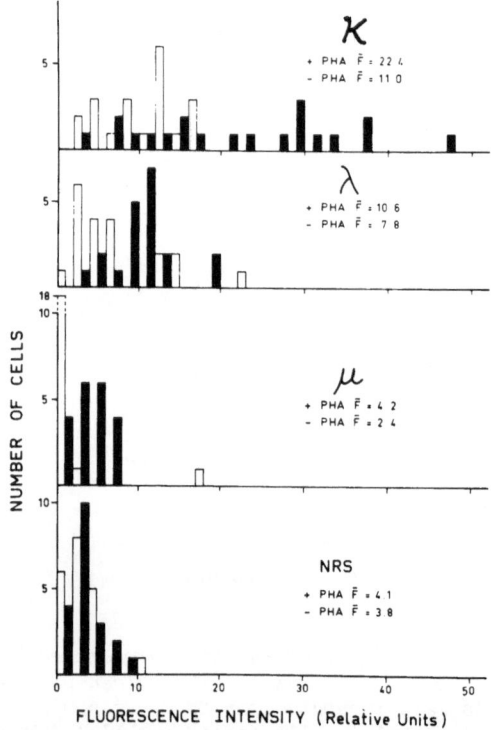

Figure 7. *Frequency distributions of fluorescence intensities of individual human normal nonstimulated (empty bars) and 24-hour PHA-stimulated (solid bars) lymphocytes stained by indirect technique for light Ig chains (κ and λ) and heavy chains (μ). The cells were first reacted with rabbit Ig specific for human κ, λ or μ chains and then with fluorescein-conjugated sheep- anti rabbit Ig. A control, i.e., cells stained with normal rabbit serum (NRS) is shown. Same measuring technique as described in the legend for Figure 2. Further details are described elsewhere (Hellström et.al.1971; Hellström et.al. to be published)*

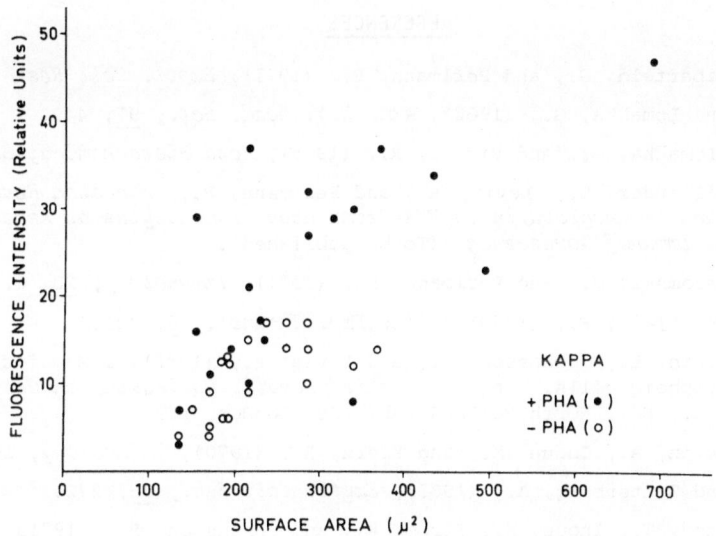

Figure 8. *Fluorescence intensities plotted against surface areas of non-stimulated and 48-hour PHA-stimulated normal human lymphocytes reacted for K chains (cf. legends for Figures 3 and 7)*

content found in proliferating lymphoid cell populations; high values contributed by cells in later stages of the cell cycle and low values to cells in early stages. An additional possible explanation might be that the lymphoid cell populations are mixtures of various types of cells with different amounts of membrane-bound Ig.

After stimulation of normal lymphocytes by phytohemagglutinin (PHA), for example, increased amounts of membrane-bound Ig have been found, as shown in qualitative immunofluorescence (Hellström et al. 1971) and electron microscopy (Biberfeld et al. 1971) studies. However, only the reactivity of light chains of membrane-bound Ig was increased. No changes in the heavy-chain reactivity were observed. For further studies on the PHA-induced increase in surface-bound Ig, we have applied the quantitative immunofluorescence method. Figure 7 shows that nonstimulated cells (white bars) stained with anti-light-chain sera (anti K and anti λ) exhibit much higher fluorescence intensities than cells stained for heavy chains (anti μ). Increased fluorescence intensities were found 24 hours after PHA treatment, particularly in the case of K chains, where more than half the number of cells exceeded the highest values of nonstimulated cells. However, a slight increase in absolute values was also noted in cells stained for heavy chains (μ). In order to obtain a measure of the density of Ig on the cell membrane, the surface area of the fluorescence-measured cells was determined. Figure 8 shows that a great many PHA-stimulated cells with the same surface areas as non-stimulated cells (150-350 μ^2), and which thus still have not undergone blast transformation, exhibit higher fluorescence values after staining for K chains than the non-stimulated cells. Thus, the density of membrane-bound Ig seems to increase prior to cell enlargement and blast transformation after PHA treatment.

ACKNOWLEDGEMENTS

This work was supported by the Swedish Cancer society, the Swedish Natural Science Research Council, the Karolinska Institute Funds, and conducted under contract No. NIH-NCI-E-69-2005 within the Special Virus Cancer Program of the National Cancer Institute, National Institutes of Health, USPHS. The development of biophysical instruments used in this work was supported by grants from the Swedish Natural Research Council to Professor T. Caspersson.

REFERENCES

BIBERFELD, P., Biberfeld, G., and Perlmann, P., (1971), *Exptl. Cell Res.*, 66, 177.

CASPERSSON, T. and Lomakka, G., (1962), *Ann. N.Y. Acad. Sci.*, 97, 449.

CASPERSSON, T., Lomakka, G., and Rigler, R., (1965), *Acta Histochemica, Suppl.* 6, 123.

HELLSTRÖM, U., Killander, D., Levin, A., and Perlmann, P., *Increased Reactivity of Membrane Bound Immunoglobulin in PHA Stimulated Lymphocytes as Assessed by Quantitative Immunofluorescence.* (To be published).

HELLSTRÖM, U., Zeromski, J., and Perlmann, P., (1971), *Immunology*, 20, 6.

JOHANSSON, B., and Klein, E., (1970), *Clin. Exp. Immunol.*, 6, 421.

KILLANDER, D., Klein, E., Johansson, B., and Levin, A., (1972), IgM moieties on malignant lymphoid cells. In: *Proc. third Lepetit Colloquium on Cell Interactions* (Silvestri, L., Ed). North-Holland Publ. Co. London, 119.

KILLANDER, D., Levin, A., Inoue, M., and Klein, E., (1970), *Immunology*, 19, 151.

KILLANDER, D., and Zetterberg, A., (1965), *Exptl. Cell Res.*, 38, 272.

KLEIN, E., Eskeland, T., Inoue, M., Strom, R., and Johansson, B., (1971a), *Ann. N.Y. Acad. Sci.*, 177, 306.

KLEIN, E., van Furth, R., Johansson, B., Ernberg, I., and Clifford, P., (1971), Proc. Cambridge Symp. *Oncogenesis and Herpes Type Viruses.*

KLEIN, E., Klein, G., Nadkarni, J. S., Nadkarni, J. J., Wigzell, H., and Clifford, P., (1968), *Cancer Res.*, 28, 1300.

LEVIN, A., Killander, D., Klein, E., Nordenskjöld, B., and Inoue, M. (1971), *Ann. N.Y. Acad. Sci.*, 177, 481-89.

MÄKELÄ, O., Sell, S., Greaves, M. F., Wigzell, H., Ada, G. L. and Paul, W. E., (1970), *Antigen Binding Lymphocyte receptors.* In Transplantation reviews 5, (Möller, G., Ed.) Munksgaard, Copenhagen.

TAKAHASHI, M., Yagi, Y., Moore, G. E., and Pressman, D. (1969) *J. Immunol.*, 103, 834.

WILSON, J. D., and Nossal, G. J. V., (1971), *Lancet*, 7728, 788.

CYTOFLUOROMETRY OF BIOGENIC MONOAMINES IN THE FALCK-HILLARP METHOD. STRUCTURAL IDENTIFICATION BY SPECTRAL ANALYSIS

Anders Björklund and Bengt Falck

Department of Histology, University of Lund, Lund, Sweden

INTRODUCTION

The basis for the detection of biogenic β-phenylethylamines and β-(3-indolyl) ethylamines in the Falck-Hillarp method is their transformation into intensely fluorescent isoquinoline and β-carboline molecules. This is achieved through combined cyclization-dehydrogenation reactions with gaseous formaldehyde under nearly dry reaction conditions. The chemical mechanisms underlying this fluorophore formation have been extensively studied and are now well understood (Corrodi and Hillarp 1963, 1964; Corrodi and Jonsson 1965 a, b; Jonsson 1966, 1967a, b; Björklund and Stenevi 1970; Björklund et al. 1973).

The first step in the reaction between the monoamines and formaldehyde is a so-called Pictet-Spengler cyclization reaction (I→II in Figure 1). This results in the formation of 1,2,3,4-tetrahydroisoquinoline and 1,2,3,4-tetrahydro-β-carboline molecules(II) which have in general either no or only very weak visible fluorescence (Hess and Udenfriend, 1959; Corrodi and Hillarp, 1963, 1964; Corrodi and Jonsson 1965 b). The fluorophore formation occurs in a second step where the tetrahydro derivatives are dehydrogenated to 3,4-dihydro compounds. In a recent study by Björklund, Falck, Lindvall, and Svensson (1973), it was shown that this dehydrogenation reaction can proceed in two alternative ways: either through an auto-oxidation to the 3,4-dihydro derivatives (III in Figure 1) (cf. Corrodi and Hillarp 1963; 1964; Corrodi and Jonsson 1965b), or through a second, acid-catalyzed reaction with formaldehyde to yield the 2-methyl-3,4-dihydro compounds (IV in Figure 1). The second reaction pathway, demonstrated by Björklund, Falck, Lindvall, and Svensson (1973), offers a full explanation of the previous observations that the fluorophore formation from the initially formed tetrahydroiso-quinolines and tetrahydro-β-carbolines requires formaldehyde and is catalyzed by proteins or amino acids (Falck et al. 1962; Corrodi and Hillarp 1963; 1964; Corrodi and Jonsson 1965b). This also suggests that the 2-methyl-3,4-dihydro molecules might be the predominating fluorescent end-products in the Falck-Hillarp histochemical reaction (cf. Björklund, Falck, Lindvall, and Svensson 1973).

The spectra of a fluorescent compound is a physical entity directly related to the molecular configuration. For this reason, the fluorescence spectra have a high analytical value and can be used for the identification of fluorescent compounds. As fluorescence is a very sensitive means of detection, the microspectrofluorometric analysis applied to fluorescence histochemistry provides possibilities for the identification of extremely low amounts of substances at their cellular storage sites. The spectral characterization of the formaldehyde-induced monoamine fluorophores has been carried out in a number of investigations during recent years (van Orden et al. 1965; Caspersson et al. 1966; Jonsson 1967a; Jonsson and Sandler 1969; Björklund et al. 1968a; b; 1970; 1971a; b; 1972a). As a result of these studies, the reliability and reproducibility of the microspectrofluorometric analysis of monoamine fluorophores is now so well established that this technique must be considered indispensable for the basic fluorophore characterization in the histochemistry of biogenic monoamines. The following is a short presentation of the structural requirements for fluorophore formation in the Falck-Hillarp method and of the relation of fluorescence emission and excitation spectra to the molecular structure of monoamine fluorophores. First, by way of introduction, some aspects are discussed of the application of the microspectrofluorometric technique to the analysis of biogenic monoamines.

CONSIDERATIONS ON INSTRUMENTATION AND TECHNIQUE

Instrumentation

The microspectrofluorometer is basically a fluorescence microscope with mono-chromators inserted in the pathways of the excitation and emission light. For the complete spectral characterization of the amine fluorophores, the instrument must permit the registration of both types of spectra and it should be possible to register the excitation spectrum between about 300 and 460 nm. Thus, a quartz glass optical system is necessary for the exciting light. Further, it should be possible to record the emission spectrum from about 400 nm throughout the visual part of the spectrum. Such instruments are now commercially available, *e.g.*, from Leitz and Zeiss, and micro-spectrofluorometers that have been used in amine histochemistry have been described in the literature (Caspersson et al. 1965; Thieme 1966; Ritzén 1967; Björklund et al. 1968a, 1972b; Pearse and Rost 1969; van Orden 1970; Rost 1971). These papers should be consulted for construction details.

Both the excitation and emission spectra are distorted by a number of instru-mental factors, and the registered spectra (i.e., the uncorrected measured values) usually deviate considerably from the true spectra (see Ritzén 1967). The *emission spectra* measured by the instrument are influenced by the barrier filters used, the spectral transmission characteristics of the optics, the transmission and band width of the analyzing monochromator as function of wavelength, and the spectral sensitivity of the photomultiplier. Similarly, the *excitation spectra* are influenced by the spectral transmission characteristics of the excitation monochromator and optics, and also by the varying intensity of the light source at different wavelengths. Alto-gether, these factors will make the recorded values distorted in a way that is charac-teristic for each individual instrument. For this reason, all published spectra should be corrected and expressed in standardized units, which will allow direct comparison between spectra obtained in different laboratories.

The correction of the emission spectra requires the preparation of an instrument calibration curve, which can be obtained, for instance, from measurements with a calibrated tungsten lamp, or from measurements of the fluorescence of a reference fluorescent solution with a known emission spectrum. For the correction of the ex-citation spectra, some kind of device for the continuous measurement of the spectral intensity distribution of the exciting light is necessary. This is because the charac-teristics of the exciting light will vary between individual lamps and throughout the lifetime of a particular lamp. The correction procedures employed in our laboratory have been introduced and described by Ritzén (1967) (see also Björklund et al. 1972b).

Recording of Excitation Spectra in the Short Wavelength Ranges

For excitation spectra down to about 300 nm, the optical pathway for the exciting light must be entirely made up of quartz glass optics (including the condenser) and the excitation monochromator must have high transmittance in this **short wavelength range**. For this reason, we have found grating monochromators superior to prism monochromators for the recording of excitation spectra of monoamine fluorophores. The specimens cannot be mounted on glass microscope slides in the normal way; instead, either quartz slides or non-fluorescent cover slips must be used. When cover slips are used, the specimens are mounted upside down, on the under surface of the slip, and liquid paraf-fin is used as immersion liquid between specimen and condenser front lens. Ritzén (1967) has pointed out that the quartz glass slides commonly used for ultraviolet microscopy (Suprasil 1®) emit a fluorescence at about 370-390 nm, which is excited near 310 nm. When the specimen is carried on such slides, their emission will excite the monoamine fluorophores in the specimen, giving rise to a visible fluorescence. Thus, a small peak will appear around 310 nm in the excitation spectra of the mono-amine fluorophores. As this peak is not always recorded from the blank, an artifact is produced in the spectrum. This artifact is not produced when cover slips are used.

Fluorescence Fading

Many of the monoamine fluorophores show photodecomposition under irradiation with ultraviolet or blue-violet light. If this occurs during the registration of the spec-tra it will naturally distort the shape of the spectrum. Consequently, the spectral

recordings should be made at such a speed that the fading during registration is negligible. This is usually no problem in case of the monoamine fluorophores, but is of
special significance for the histamine-OPT-fluorescence. This fluorescence fades
very rapidly, and we have had to employ recording on an oscilloscope screen to obtain
a sufficiently rapid recording (see Håkanson et al. 1971; and Björklund et al.
1972b). This type of fluorescence is not dealt with in the present paper. If a significant fading of the fluorescence occurs during the course of the spectral registration, the spectrum can possibly be corrected afterwards, provided that the rate of
photodecomposition of the substance is known.

Reabsorption of the Emitted Light

Reabsorption of the emitted light will occur in the specimen when absorption
(excitation) and emission spectra overlap, or when a second, nonfluorescent substance
with appreciable absorption in the emission range is present in the specimen. This
reabsorption will then lower the emission curve in the wavelength region where it
occurs, and in this way will distort the spectrum. However, as stated by Ritzén (1967),
reabsorption errors are usually small in microfluorometry, because the specimens are
generally very thin.

MOLECULAR REQUIREMENTS FOR FLUOROPHORE FORMATION IN THE FALCK-HILLARP METHOD

The Pictet-Spengler cyclization, which underlies the fluorophore formation in the
Falck-Hillarp method, involves the reaction of formaldehyde with a free amino group
(Figure 1). For this reason, tertiary amines or amides will not condense with formaldehyde, and thus only primary or secondary β-phenylethylamines and β-(3-indolyl)
ethylamines can be expected to yield fluorescence.

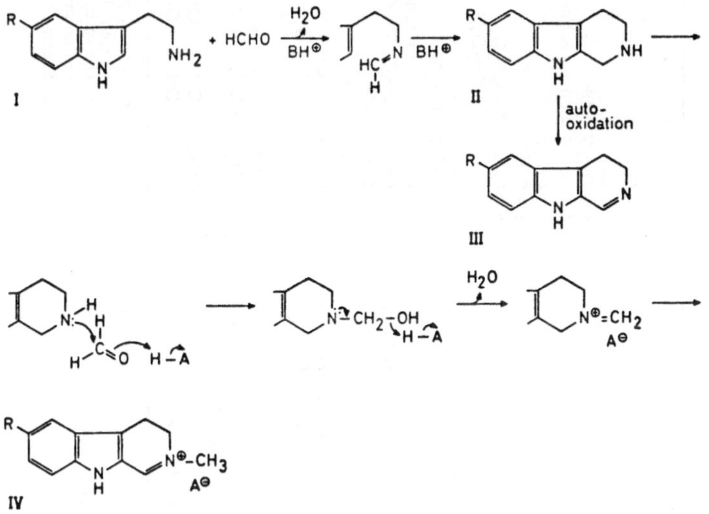

Figure 1. *Sequence of reactions between indolylethylamines and formaldehyde.
(Tryptamine: R = H; 5-hydroxytryptamine: R = OH). For explanations, see text. From
Björklund, Falck, Lindvall and Svensson (1973)*

There is much evidence that only compounds with high electron density at the
ring carbon atom where the closure takes place will cyclize with formaldehyde under
the mild histochemical conditions of the standard Falck-Hillarp method. For this
reason, probably only 3-hydroxylated β-phenyl-ethylamines and β(3-indoly)ethylamines
will give a good yield of fluorophores in this reaction (see Tables I and II; Falck
et al. 1962; Jonsson 1967b; Björklund and Stenevi 1970).

Among the phenylethylamines, 3,4-dihydroxylated compounds, such as dopamine (5)
(figures refer to compound numbers in Tables I and II), noradrenaline (13), and DOPA
(18), and 3-hydroxylated compounds, such as *m*-tyramine (3) and metaraminol, give the

TABLE I

Excitation and emission maximum of some phenylethylamine (PEA) derivatives after formaldehyde treatment according to the Falck–Hillarp technique, obtained from authentic substances enclosed in dried protein droplets. Data from Björklund, Nobin, and Stenevi (1971) and Björklund, Falck, and Stenevi (1971)

Compound	Structure of the corresponding 3,4-dihydroisoquinoline fluorophore[a]	Relative fluorescence yield	Exc. max. (nm)	Em. max. (nm)
1. PEA (phenylethylamine)	–	0	–	–
2. 4-hydroxy PEA (p-tyramine)	7 = OH	0	–	–
3. 3-hydroxy PEA (m-tyramine)	6 = OH	58	385	415 or 510[c]
4. 2-hydroxy PEA (o-tyramine)	5 = OH	0	–	–
5. 3,4-dihydroxy PEA (dopamine)	6 = OH, 7 = OH	109	320 and 410	475
6. 3,4,5-trihydroxy PEA	6 = OH, 7 = OH, 8 = OH	9	335 and 400–430[d]	500–530[d]
7. 4-methoxy PEA	7 = CH_3O	0	–	–
8. 3-methoxy PEA	6 = CH_3O	2	–	–
9. 3,4-dimethoxy PEA	6 = CH_3O, 7 = CH_3O	0	–	–
10. 3,4,5-trimethoxy PEA (mescaline)	6 = CH_3O, 7 = CH_3O, 8 = CH_3O	1	–	–
11. 3-methoxy-4-hydroxy PEA	6 = CH_3O, 7 = OH	0	310 and 370[e]	470[e]
12. 4-methoxy-3-hydroxy PEA	6 = OH, 7 = CH_3O	227	405	460
13. 3,4,β-trihydroxy PEA (noradrenaline)	4 = OH, 6 = OH, 7 = OH	100	320 and 410	475
14. 3-methoxy-4,β-dihydroxy PEA (normetanephrine)	4 = OH, 6 = CH_3O, 7 = OH	0	310 and 370[e]	470[e]
15. 4,β-dihydroxy PEA (octopamine)	4 = OH, 7 = OH	0	–	–
16. α-carboxy PEA (phenylalanine)	3 = COOH	0	–	–
17. α-carboxy-4-hydroxy PEA (p-tyrosine)	3 = COOH, 7 = OH	0	–	–
18. α-carboxy-3,4-dihydroxy PEA (DOPA)	3 = COOH, 6 = OH, 7 = OH	120[b]	320 and 410	475
19. α-carboxy-3,4,β-trihydroxy PEA (dihydroxyphenylserine, DOPS)	3 = COOH, 4 = OH, 6 = OH, 7 = OH	10[b]	380	470[b]
20. α-methyl-3,4-dihydroxy PEA (α-methyl-dopamine)	3 = CH_3, 6 = OH, 7 = OH	70[b]	320 and 410	475
21. α-methyl-3,4,β-trihydroxy PEA (α-methyl-noradrenaline)	3 = CH_3, 4 = OH, 6 = OH, 7 = OH	100[b]	320 and 410	475
22. N-methyl-3,4,β-trihydroxy PEA (adrenaline)	2 = CH_3, 4 = OH, 6 = OH, 7 = OH	47	320 and 410	475

a The positions of the substituents refer to figure below

b Data from Jonsson (1967)

c The two peaks are observed under different conditions. Data from Corrodi and Jonsson (1966) and Jonsson and Ritzén (1966)

d The spectra of 3,4,5 trihydroxy PEA show pronounced concentration-dependent variations. Data from Baumgarten et al. (1972)

e Spectra recorded at high amine concentrations (very weak fluorescence)

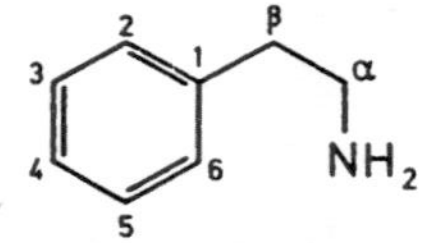

β-PHENYLETHYLAMINES 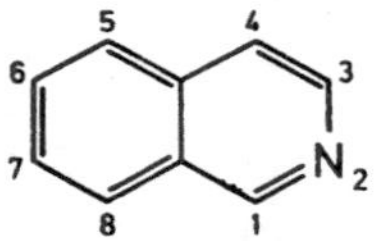   ISOQUINOLINES

TABLE II

Excitation and emission maxima of some indolylethylamine (tryptamine) derivatives after formaldehyde treatment according to the Falck-Hillarp technique, obtained from authentic substances enclosed in dried protein droplets. Data from Björklund, Nobin, and Stenevi (1971) and Björklund, Falck, and Stenevi (1971)

Compound	Structure of the corresponding 3,4-dihydro-β-carboline fluorophore [a]	Relative fluorescence yield	Exc.max. (nm)	Em.max. (nm)
23. tryptamine (β(3-indolyl)ethylamine)		9	370	495[b]
24. 5-hydroxytryptamine	6 = OH	33	(315)* 385 and 415[c]	520-530
25. 6-hydroxytryptamine	7 = OH	195	385(420)*	505
26. 5,6-dihydroxytryptamine	6 = OH, 7 = OH	14	310 and (380)* 405	(<430)* 500
27. 5-methoxytryptamine	6 = CH₃O	14	(330)* and 380(410)*	505[d]
28. α-carboxytryptamine (tryptophan)	3 = COOH	8	375	435 or 500[c]
29. α-carboxy-5-hydroxytryptamine (5-hydroxytryptophan)	3 = COOH, 6 = OH	3	310, 385 and 415	520 - 530
30. α-carboxy-5-methoxytryptamine (5-methoxytryptophan)	3 = COOH, 6 = CH₃O	3	(385)* 410	505[g]
31. α-methyl-5-hydroxytryptamine	3 = CH₃, 6 = OH	20[e]	410	520[e]
32. N-methyltryptamine	2 = CH₃	3	370	495[b]
33. N-methyl-5-hydroxytryptamine	2 = CH₃, 6 = OH	22	315 and (390)* 415	505
34. N-methyl-5-methoxytryptamine	2 = CH₃, 6 = CH₃O	3	(310) 380	<410 and 500[f]
35. N-acetyl-5-hydroxytryptamine	-	0	-	-
36. N-acetyl-5-methoxytryptamine	-	0	-	-
37. N,N-dimethyltryptamine	-	0	-	-
38. N,N-dimethyl-5-hydroxytryptamine (bufotenin)	-	0	-	-

* Figure within brackets denotes position of shoulder.

a The positions of the substituents refer to figure below.

b The emission maximum of the tryptamine and the N-methyltryptamine fluorophores exhibit a concentration-dependent variation from 450 - 520 nm (cf. Björklund et al. 1968 b).

c The different peaks are observed under different condition: (cf. Björklund et al. 1968 b).

d The emission maximum of the 5-methoxytryptamine fluorophore is higher (520 -530 nm) under more intense reaction conditions, and at high amine concentrations.

e Data from Jonsson and Sandler (1969).

f Björklund, unpublished data.

g Data from Björklund et al. (1970).

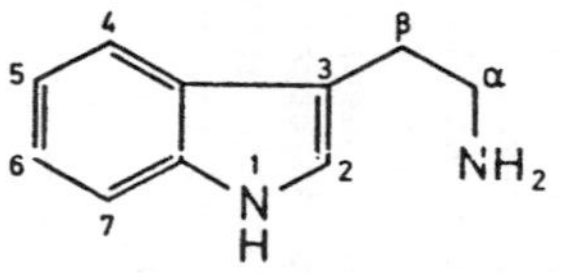

β (3-INDOLYL)-ETHYLAMINES 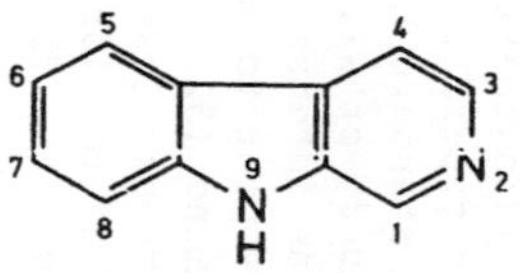   β-CARBOLINES

highest fluorescence yields. The 3-methoxylated phenylethylamines (*e.g.*, 3-methoxy-
4-hydroxyphenylethylamine (11) and normetanephrine (14) give only very low fluorescence
yields in the standard reaction, probably because a 3-methoxy group activates less in
the Pictet-Spengler reaction than a 3-hydroxy group does (Schöpf and Bayerle 1934;
Kovács and Fodor 1951; Whaley and Gowidachari 1951; Jonsson 1967b; Björklund and
Stenevi 1970). The fluorescence yields of 3-methoxylated phenylethylamines can be
greatly increased, however, through catalysis of the formaldehyde cyclization reaction
by hydrochloric acid (Björklund and Stenevi 1970; Björklund et al. 1971b). The
phenylethylamines without substitution in the 3 position (*e.g.*, phenylalanine (16),
p-tyrosine (17), *p*-tyramine (2), and octopamine (15) have a very low reactivity in the
Pictet-Spengler reaction, and thus give no, or hardly any detectable visible fluorescence in the formaldehyde gas treatment (cf. Björklund and Stenevi 1970).

The secondary phenylethylamine adrenaline (22) requires more intense reaction
conditions (higher reaction temperature and/or longer reaction time) than the corresponding primary amine, noradrenaline (Corrodi and Hillarp 1963; Falck et al. 1963;
Norberg et al. 1966). This has been ascribed to a higher requirement of energy in
the dehydrogenation step in the reaction with adrenaline.

In general, *primary indolylethylamines* have a low-to-moderate fluorescence yield
in the standard histochemical fomaldehyde reaction (Table 2; Jonsson 1967a; b;
Björklund et al. 1968b; 1971a;b). An exception is 6-hydroxytryptamine (25) which
gives an intense fluorescence (Johnsson and Sandler 1969; Björklund et al. 1971b).
With respect to *secondary indolylethylamines*, N-methylated compounds, such as N-methyltryptamine (32) and N-methyl-5-hydroxytryptamine (35), give fluorescence yields that
are somewhat less than those of their corresponding primary amines. Acid-catalysis
according to Björklund and Stenevi (1970) and Björklund et al. (1971b) generally
increases the fluorescence yields of tryptamine derivatives in the reaction with
formaldehyde.

The N-acetylated compounds (*e.g.*, N-acetyl-5-hydroxytryptamine (35) and N-acetyl-
5-methoxytryptamine, melatonin (36) have been found to give no or only very weak visible fluorescence after formaldehyde treatment (Björklund, unpublished results;
Björklund et al. 1968b); these compounds are amides and cannot be expected to enter
the flurophore-forming Pictet-Spengler reaction (cf. above).

Substituents on the side chain can influence not only the fluorescence yield but
also the spectral properties of the fluorophores, cf. below; however, they have much
less dramatic effects than substituents on the benzene ring of the monoamines. With
both phenylethylamines and indolylethylamines, the introduction on the α-carbon atom
(see Figures in Tables 1 and 2) of a carboxyl group (which gives the corresponding
amino acid) or of a methyl·group (which gives the nonbiogenic, but pharmacologically
active α-methylated amines) either will not influence, or will somewhat reduce the
fluorescence yield of the compound (Jonsson 1967a; Björklund et al. 1971a;b). This
means that the Falck-Hillarp method is useful also for the detection of certain amino
acids, such as DOPA and 5-hydroxytryptophan.

SPECTRAL PROPERTIES OF MONOAMINE FLUOROPHORES

A. Fluorophores derived from phenylethylamines

These fluorophores have most often a blue or violet fluorescence color, with
the emission peak maximum in the region 415-475 nm. A yellowish fluorescence with the
emission peak maximum at about 510 nm has been recorded from 3,4,5-trihydroxypheny-
lethylamine (6) (Ehinger and Falck 1969; Baumgarten et al. 1972) and 4-methoxy-
3,5-dihydroxyphenylethylamine (Björklund, unpublished), and also from 3-hydroxylated
phenylethylamines under certain conditions (*e.g.*, *m*-tyramine, 3) (Corrodi and Jonsson
1966; Jonsson and Ritzén 1966).

The Tautomerism of the 6-hydroxy Group

The dihydroisoquinoline fluorophores that have a 6-hydroxy group - which is the
3-hydroxy group of the original phenylethylamine - show a pH-dependent tautomerism
(Corrodi and Hillarp 1964; Jonsson 1966; Björklund et al. 1968a; 1972c). The
6,7-dihydroxy-3,4-dihydroisoquinolines formed from the catecholamines, dopamine and

noradrenaline (II and V in Figure 2), and the 6-hydroxy-3,4-dihydroisoquinolines formed from 3-hydroxylated phenylethylamines are thus in a pH-dependent equilibrium with their corresponding tautomeric quinoidal forms (III and VI in Figure 2). The quinoidal form predominates in the pH range between 6 and 10, i.e., at such pH values as occur in tissues, whereas the nonquinoidal form is predominant at lower pH values. This pH-dependent transition between two tautomeric states of the fluorophores is reflected in characteristic changes of their spectral properties. Thus, the quinoidal form of the catecholamine fluorophores has its excitation/emission peak maxima at 320 and 410/475 nm, and the non quinoidal form at 320 and 370/490-500 nm (Corrodi and Jonsson 1965a; Björklund et al. 1968a; 1972). Similarly, the maxima of the main excitation/emission peaks of the 6-hydroxy-3,4-dihydroisoquinolines formed from 3-hydroxylated phenylethylamines (lacking a 4-hydroxy group) are at 385/415 or 510 nm for the quinoidal form (at neutral or alkaline pH) and at 345-360/415 nm for the non-quinoidal form (at acid pH) (Corrodi and Jonsson 1966; Jonsson and Ritzén 1966).

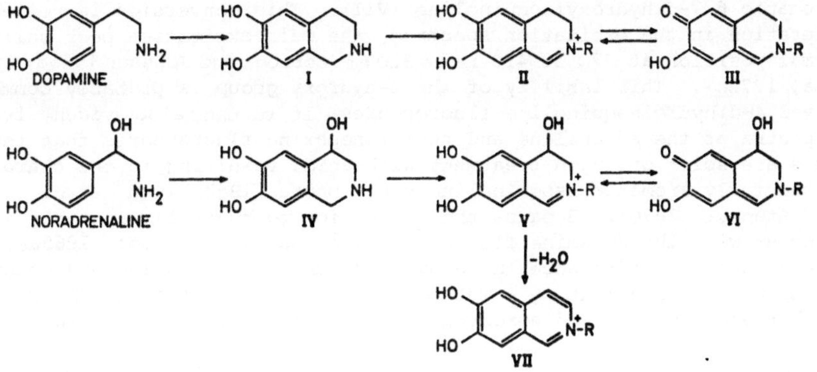

Figure 2. *The fluorophore formation from catecholamines by formaldehyde. Two alternative forms of the dehydrogenated isoquinolines are possible: either the 6,7-dihydroxy-3,4-dihydroderivatives (R = H), or the 6,7-dihydroxy-3,4-dihydro-2-methyl-derivatives (R = CH₃). The figures illustrate the pH-dependent tautomerism of the 6-hydroxy group (II, V and III, VI), and the formation of the fully aromatic 6,7-dihydroxyisoquinoline (VII) through the splitting off of the 4-hydroxy group*

The quinoidal form of the 6,7-dihydroxy-3,4-dihydroisoquinolines, formed from the catecholamines, has been shown to be the 6-quinone form (Jonsson 1966; III and VI in Figure 2), and it seems that the 7-hydroxy group cannot exhibit this form of tautomerism. Thus, the fluorophores having a methoxy group in the 6 position and a hydroxy group in 7 position, i.e., fluorphores formed from 3-methoxy-4-hydroxy phenylethylamines, do not exhibit this kind of pH-dependent tautomerism; these fluorophores will have similar spectral properties at both neutral and acid pH. This is exemplified by the fluorophore formed from 3-methoxy-4-hydroxyphenylethylamine (3-methoxytyramine) (11), which has its main excitation/emission peaks at 310 and 370/470 nm, at both acid and neutral pH. The spectra of the 6-methoxy-7-hydroxydihydroisoquinoline fluorophore, formed from this compound (11), are thus similar to those of the nonquinoidal form of the 6,7-dihydroxydihydroisoquinoline fluorophore formed from dopamine (5).

Influence of other Substituents on the Benzene Ring of the Isoquinoline Fluorophores

The most significant influence on the fluorophore formation is obviously exerted by the substituents in the 3-position of the phenylethylamines. The activating effect of 3-hydroxy or 3-methoxy groups in the cyclization reaction, and the tautomerism exhibited by a hydroxy group in this position have been mentioned above. However, from the spectral data in Table I (obtained after standard formaldehyde treatment) and the data presented by Björklund et al. (1971b) (acid-catalyzed formaldehyde treatment), it can be seen that substituents in other positions of the benzene ring also influence the spectral properties of the resulting fluorophores. There is, moreover, a clear tendency for the emission peak gradually to shift to higher wavelengths as more hydroxy or methoxy groups are introduced in the benzene ring. Thus one hydroxy or methoxy substituent (in 3-position) results in blue-violet fluorophores with the emission maxima at about 400-430 nm, two substituents result in blue fluorophores (em. max. 440-480 nm), and three substituents result in green or yellow fluorophores (em. max. about 490-510 nm). The yellow fluorophore obtained from 3-hydroxylated phenylethylamines, having only one substituent on the benzene ring, (Corrodi and Jonsson 1966; Jonsson and Ritzén 1966) is under certain conditions an exception to this general pattern.

The Labile 4-hydroxy Group

The phenylethylamines having a hydroxyl group in β-position, such as noradrenaline (13), adrenaline (22), and normetanephrine (14), will form 3,4-dihydroisoquinoline fluorophores with a 4-hydroxy group (see Figure in Table I). Corrodi and Jonsson (1965a) have found that this group on the fluorophore is labile and can be split off by treatment with acid. They demonstrated that such treatment will convert the noradrenaline fluorophore, 4,6,7-trihydroxy-3,4-dihydroisoquinoline (V in Figure 2), to the fully aromatic 6,7-dihydroxyisoquinoline (VII). This conversion is reflected in a marked alteration in the excitation spectrum, the main excitation peak shifting from its normal position at 370 or 410 nm to 320 nm (Corrodi and Jonsson 1965a; Björklund et al. 1968a; 1972c). This lability of the 4-hydroxy group is probably common for all 4-hydroxy-3,4-dihydroisoquinoline fluorophores; it is thus also evident from the excitation spectra of the adrenaline and normetanephrine fluorophores that their 4-hydroxy groups are split off upon treatment with acid, resulting in the characteristic spectrum of the fully aromatic isoquinoline fluorophores (Björklund et al. 1968a; Björklund and Stenevi 1970). Because the conversion to the fully aromatic isoquinoline cannot occur with the dopamine fluorophore (Corrodi and Jonsson 1965a), it has been possible to use the difference in behavior of the noradrenaline and dopamine fluorophores upon acid treatment in a method for the microspectrofluorometric differentiation between dopamine and noradrenaline in tissue sections (Björklund et al. 1968a; 1972a).

Influence of other Substituents on the Pyridine Ring of the Isoquinoline Fluorophores

The introduction of a methyl group in the pyridine ring does not seem to affect the spectral properties of the fluorophores. Thus, a methyl group on the α-carbon atom of the side chain, such as in α-methyl-dopamine (20), α-methylnoradrenaline (21), and α-methyl-dopa, will not alter the spectral properties of the fluorophore (Table I; see Jonsson 1967 a). Also, the adrenaline fluorophore (22), which has a 2-methyl group, has spectral properties very similar to those of the noradrenaline fluorophore (13), which lacks this group (Table I). The carboxyl group of amino acids should appear in the 3-position of the isoquinoline fluorophore. However, although this carboxyl group could be split off during the formaldehyde treatment, it is not known to what extent this might occur. It is reasonable to suppose that this is not the case with the carboxyl group of the 3,4-dihydroxyphenylserine (DOPS) fluorophore (19), as its spectrum is unlike that of the noradrenaline fluorophore (13), which lacks the carboxy group (lower excitation maximum; Table I). On the other hand, the spectra of the DOPA (18) and the dopamine (5) fluorophores are very similar (Table I), and so in this case, a decarboxylation is quite feasible.

B. Fluorophores Derived From Indolylethylamines

Spectrally, it is convenient to divide the indolylethylamine fluorophores into three main groups (cf. Björklund et al. 1971a):

1. Fluorophores without Substituents on the Indole Nucleus

Data for such compounds have been published for tryptamine (23), N-methyltryptamine (32), and tryptophan (28) (Björklund et al. 1968b). The fluorophores of these compounds have one main excitation peak at 370-375 nm, and an emission peak at 495-500 nm (Table II). Tryptophan has, in addition, a blue fluorophore (exc. max/em. max = 375/435 nm) which is usually predominant at low-to-moderate concentrations of the amino acid.

2. 5-hydroxytryptamine (24) and 5-hydroxytryptophan (29)

These fluorophores have a main double peak in the excitation spectrum; one peak occurs at about 380-390 nm, and the other at about 410-420 nm. The emission peak reaches its maximum at 520-530 nm, and exhibits a concentration-dependent shift to about 550 nm at higher amine concentrations (Table II, Björklund et al. 1968b).

3. Substances with other Substituents on the Indole Nucleus

At present, data are available for the fluorophores derived from 6-hydroxytryptamine (25), 5,6-dihydroxytryptamine (26), 5-methoxytryptamine (27), N-methyl-5-methoxytryptamine (34), and 5-methoxytryptophan (30). N-methyl-5-hydroxytryptamine (33) also belongs to this group. These fluorophores show the characteristic double peak in the excitation spectrum described for 5-hydroxytryptamine and 5-hydroxytryptophan above, but the emission peak reaches its maximum at 495-505 nm. Among the fluorophores of this group, the emission peak of the 5-methoxytryptamine fluorophore shows a shift toward 520-525 nm at higher concentrations.

INFLUENCE OF SUBSTITUENTS ON THE PYRIDINE RING OF THE β-CARBOLINE FLUOROPHORES

As can be seen from the spectral data in Table II, substituents on the side chain of the indolylethylamine derivatives (which will yield fluorophores with substituents on the pyridine ring; see figure in Table II) have variable effects on the spectral properties of the fluorophores.

The carboxyl group of amino acids will yield flurophores with a carboxyl group in 3-position; however, a decarboxylation during the formaldehyde treatment must also be taken into account in case of the β-carboline fluorophores. The 3-carboxyl group has no clear spectral influence in the case of the 5-hydroxytryptophan (29) and 5-methoxytryptophan (30) fluorophores, whereas clear differences are recorded between the tryptophan and tryptamine fluorophores (cf. Björklund et al. 1968b). An introduction of a methyl group on the nitrogen atom (2-position) or in the 3-position is achieved in the fluorophores of N-methylated and α-methylated indolylethylamines. With respect to the compounds of this type that are included in Table II (compounds 31-34), only the N-methyl-5-hydroxytryptamine fluorophore (33) has spectral properties that are clearly different from those of the nonmethylated analogue.

CONCLUDING REMARKS

Microspectrofluorometry is an indispensable tool for the basic characterization of biogenic amines in the Falck-Hillarp method. The number of known biogenic phenylethylamines and indolylethylamines is rapidly increasing at present, and there is now a wide variety of these amines which have been isolated and identified chemically from animal tissues (cf. Björklund et al. 1971a). Thus, there is a great demand for refined analytical histochemical techniques for the strict identification of intracellular fluorogenic compounds. Although microspectrofluorometry of biogenic amines is a technique still in its developmental stage, its present-day applications have high analytical value. Above all, microspectrofluorometric analysis now allows a reasonably secure identification of the most common monoamines, dopamine, noradrenaline, and 5-hydroxytryptamine, and in the past few years, it has been widely used in our laboratory, as well as in others, for this purpose.

It should be remembered, however, that microspectrofluorometry is a technique with obvious limitations: certain closely related compounds cannot as yet be distinguished from each other by means of spectrofluorometers. Such compounds are dopamine and DOPA or 5-hydroxytryptamine and 5-hydroxytryptophan. Also, similar to most other histochemical or chemical analytical techniques, the spectral analysis cannot offer a final and unequivocal structural identification, but only a characterization of the molecule. This means that, in every instance where a strict identification is desired, the microspectrofluorometry should be used in conjunction with other, independent analytical methods. High precision and extreme sensitivity are the most attractive features of cytofluorometry of biogenic monoamines; when used with adequate caution and insight, this method offers unique possibilities for the morphological, biochemical, pharmacological, and functional studies of individual cells and cell systems.

ACKNOWLEDGEMENTS

The research reported in the present review has been supported by grants from the United States Public Health Service (NS 06701-06) and from the Magnus Bergvall and Åke Wiberg Foundations; it was carried out within a research organization sponsored by the Swedish Medical Research Council (B72-14X-712-07B and B72-14X-56-08B).

REFERENCES

BAUMGARTEN, H.G., Björklund, A., Holstein A. F., and Nobin, A., (1972), *Z. Zellforsch.*, **126**, 483.

BJÖRKLUND, A., Ehinger, B., and Falck, B., (1968a), *J. Histochem. Cytochem.*, **16**, 262.

BJÖRKLUND, A., Ehinger, B., and Falck, B., (1972a), *J. Histochem. Cytochem.*, **20**, 56.

BJÖRKLUND, A., Falck, B., and Håkanson, R., (1968b), *Acta Physiol. Scand. Suppl.*, **318**,

BJÖRKLUND, A., Falck,B., and Håkanson, R., (1970), *Anal. Biochem.*, **35**, 264.

BJÖRKLUND, A., Falck, B., and Owman,C., (1972b), In: *Methods in Investigative and Diagnostic Endocrinology.*, (Rall,J.E., and Kopin, I.J, Eds.) Vol. 1; North-Holland, Amsterdam.

BJÖRKLUND, A., Falck, B., and Stenevi, U., (1971a), *Progr. Brain Res.*, **34**, 63.

BJÖRKLUND, A., Falck,B., Lindvall,O., and Svensson, L.Å., (1973), *J. Histochem. Cytochem.*, **21**,17.

BJÖRKLUND, A., Nobin, A., and Stenevi, U., (1971b), *J. Histochem. Cytochem.* **19**, 286.

BJÖRKLUND, A., and Stenevi, U., (1970), *J. Histochem. Cytochem.*, **18**, 794.

CASPERSSON, T., Hillarp, N.-Å., and Ritzén, M., (1966),*Exp. Cell. Res.*, **42**, 415.

CASPERSSON, T., Lomakka, G., and Rigler, R. Jr., (1965), *Acta Histochem. Suppl* **6**, 123.

CORRODI, H., and Hillarp N.-Å., (1963), *Helv. Chim. Acta.*, **46**, 2425.

CORRODI, H., and Hillarp, N.-Å., (1964), *Helv. Chim. Acta.*, **47**, 911.

CORRODI, H., and Jonsson, G., (1965a), *J. Histochem Cytochem.*, **13**, 484.

CORRODI, H., and Jonsson, G., (1965b), *Acta Histochem.*, (Jena) **22**, 247.

CORRODI, H., and Jonsson, G., (1966), *Helv. Chim. Acta.*, **49**, 708.

EHINGER, B., and Falck, B., (1969), *Histochemie*, **18**, 1.

FALCK, B., Häggendal, J., and Owman, C., (1963), *Quart. J. Exp. Physiol.*, **48**, 253.

FALCK, B., Hillarp, N.-Å., Thieme, G., and Torp, A., (1962), *J. Histochem. Cytochem.*, **10**, 348.

HAKANSON, R., Juhlin, L., Owman, C., and Sporrong, B., (1970), *J. Histochem. Cytochem.*, **18**, 93.

HESS, S.M., and Udenfriend, S., (1959), *J. Pharmacol. Exp. Ther.*, **127**, 175.

JONSSON, G., (1966), *Acta Chem. Scand.*, **20**, 2755.

JONSSON, G., (1967a), *Histochemie*, **8**, 288.

JONSSON, G., (1967b), *The Formaldehyde Fluorescence Method for the Histochemical Demonstration of Biogenic Monoamines.* M.D. thesis, Stockholm.

JONSSON, G., and Ritzén, M., (1966), *Acta Physiol. Scand.*, **67**, 505

JONSSON, G., and Sandler, M., (1969), *Histochemie.*, **17**, 207.

KOVACS, O., and Fodor, G., (1951), *Ber. Dtsch. Chem. Ges.*, <u>84</u>, 795.

NORBERG, K.-A., Ritzén, M., and Ungerstedt, U., (1966), *Acta Physiol. Scand.*, <u>67</u>, 260.

PEARSE, A. G. E., and Rost, F. W. D., (1969), *J. Microsc.*, <u>89</u>, 321.

RITZÉN, M., (1967), *Cytochemical Identification and Quantitation of Biogenic Monoamines - A Microspectrofluorimetric and Autoradiographic Study*. M.D. thesis, Stockholm.

ROST, F.W.D., and Pearse, A.G.E., (1971), *J. Microsc.*, <u>94</u>, 93.

SCHÖPF, C., and Bayerle, H., (1934), *Ann. Chem.*, <u>513</u>, 190.

THIEME, G., (1966), *Acta Physiol. Scand.*, <u>67</u>, 514.

VAN ORDEN, L., (1970), *Biochem. Pharmacol.*, <u>19</u>, 1105.

VAN ORDEN, L., Vugman, I., and Giarman, N. J., (1965), *Science.*, <u>148</u>, 642.

WHALEY, W. M., and Govindachari, T. R., (1951), In: *Organic Reactions*. (Adams, R., ed), John Wiley and Sons, Inc. New York. V. 6, 151-206.

MICROFLUORIMETRIC QUANTITATION OF BIOGENIC MONOAMINES

E. Martin Ritzén

Pediatric Endocrinology Unit, Department of Pediatrics,
Karolinska Sjukhuset and the Institute for Medical Cell
Research and Genetics, Medical Nobel Institute,
Karolinska Institutet, Stockholm

INTRODUCTION

Quantitation of fluorescence intensity at the microscope level can now be accomplished by using one of several highly sensitive and commercially available microfluorimeters on the market. However, exact quantitative measurement of fluorescence *intensity* does not necessarily prove to be a correct quantitation of the fluorescent substance. The results to be discussed below on the microfluorimetric quantitation of biogenic monoamines by the Falck-Hillarp formaldehyde method (Falck et al. 1962; Corrodi and Hillarp 1963; 1964) indicate that careful methodological studies are necessary before a new fluorescence method is used for quantitative purposes.

METHODS

The fluorescence measurements, including the recording of emission and excitation spectra in cells or parts of cells, were performed by means of a recording fluorescence microspectrograph previously described (Caspersson et al. 1965: Ritzén 1967b).

Standardization

A system for standardizing the measurements proved to be necessary for comparing results obtained at different times or by use of different instruments. The latter could only be achieved by measuring the fluorescence intensity of a fluorescent object with the size range of 10-50 µm. Such an object was prepared by grinding a piece of uranium glass (Schott GG 17 filter) into a fine powder. This powder was suspended in ethanol and allowed to sediment stepwise to yield fractions with decreasing particle size. Particles having a suitable fluorescence intensity were mounted under coverslip with Entellan®, after marking out one or two particles with non-fluorescent india ink.

The uranium glass particles proved to have a very stable fluorescence. During continuous six hours' irradiation in the microfluorimeter, less than 2% decrease in fluorescence was noted. Also, no appreciable decrease in fluorescence intensity was measured after three month's storage in the dark.

Protein Microdroplets

The monoamines found in cells are stored at high concentrations in special cytoplasmic storage granules. They are easily diffusible from these granules, and are not thought to be firmly bound to any macromolecular complexes. It was previously shown (Corrodi and Hillarp 1963) that the monoamine-formaldehyde reaction proceeds faster in a dried state in the presence of proteins than in solution. Thus, a suitable model for the study of monoamines should consist of a dry system, where the monoamine could be included at different concentrations into a protein matrix in the presence or absence of other catalysts for the monoamine-formaldehyde reaction. Protein microdroplets proved to be suitable for this purpose; monoamines (catecholamines or 5-HT) were dissolved in 5-10% solutions of bovine serum albumin (Armour Fraction V), and sprayed in a nitrogen atmosphere to form microdroplets on microscope slides. The small droplets dried instantaneously and were treated with formaldehyde vapours at 80°C. Dry mass of single droplets was measured by scanning microinterferometry (Caspersson

PLATE I

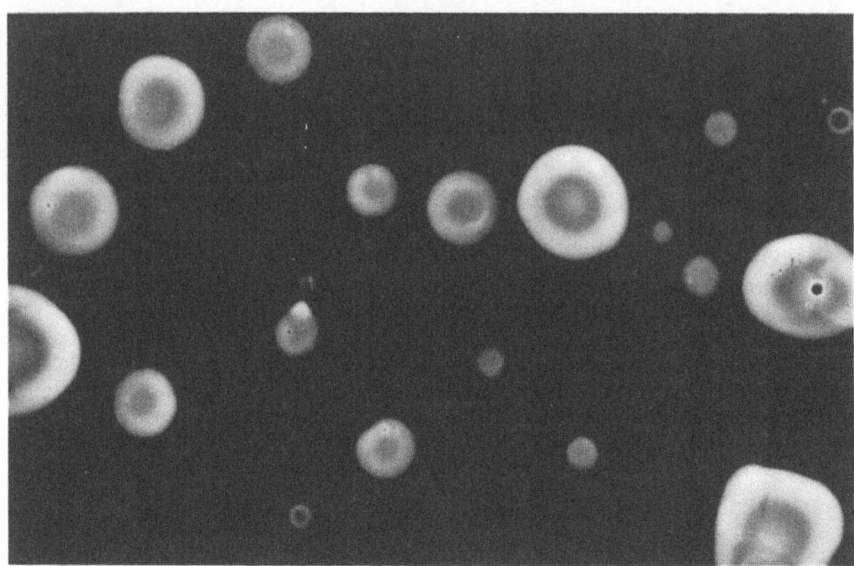

Figure 1, *Fluorescence photomicrograph of model microdroplets. The monoamine to be studied was dissolved in a 5% albumin solution which was then sprayed on microscope slides, allowed to dry, and treated with formaldehyde vapours. The dry mass of selected droplets was determined by scanning and integrating microinterferometry, and then the fluorescence measured could be related to the known content of monoamine in the droplet. Magnification 500 ×*

and Lomakka 1961) and from the known relation protein/monoamine, the content of monoamines could be determined. Following dry mass determination, the specimen was transferred to the microfluorimeter and the fluorescence of the whole droplet was recorded. When droplets of different sizes were compared, an excellent correlation between fluorescence intensity and dry mass was found (Ritzen 1966). (Figure 1, see Plate 1).

Formaldehyde Reaction

The conditions of reaction between the monoamine and formaldehyde could be kept constant from time to time only with great difficulty. Therefore, a standard preparation of monoamine-protein microdroplets was included into each batch of formaldehyde treatment, and the fluorescence of the unknown specimen could then be related to that produced by a certain amount of monoamine in the microdroplet. By varying the amount of water in the paraformaldehyde, the fluorescence intensity from a unit mass catecholamine-protein droplet could vary from 25 to 100% (Figure 2).

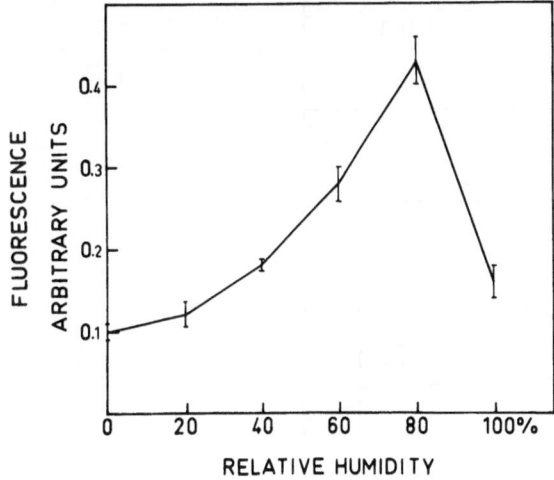

Figure 2. *Maximum reaction yield as function of the water content of paraformaldehyde. The paraformaldehyde used had been equilibrated in an atmosphere of varying humidity (Hamberger et al. 1965) (abscissa) before heated at 80° C with protein microdroplets containing 2.2×10^{-2} M noradrenaline (From Ritzén 1966)*

Photodecomposition

The rate of photodecomposition of the monoamine-formaldehyde complex turned out to be fast enough to limit the usefulness of high energy illumination systems. During a period of 5 minutes, the fluorescence intensity of the 5-HT-formaldehyde complex decreased to 40% of the original value, when illuminated in the microfluorimeter used in the present study. This necessitated rapid measurements and correction of spectra for that amount of fluorescence fading which occurred during spectral registration.

Fluorescence Quenching

As previously observed (Corrodi and Hillarp 1963) the fluorescence of the monoamine-formaldehyde complexes decreased dramatically when water was added to the system (Ritzén 1967a). There was an instantaneous drop of the 5-HT-formaldehyde fluorescence intensity to about 10% of the initial value, while the catecholamine fluorescence intensity decreased to about 15-20%.

This quenching effect was shared by another highly polar solvent, formamide. Organic solvents (ethanol, butanol, ethyl ether, xylene) caused no quenching of the monoamine-formaldehyde fluorescence. Quenching was observed even at very low concentrations of fluorophore. Direct comparisons of fluorescence quantum yield of the authentic catecholamine-formaldehyde product (6.7-dihydroxy-3.4-dihydroisoquinoline) in the solid state and in aqueous solution showed 8-9 times higher fluorescence in the solid state.

Concentration Quenching

When catecholamine or 5-HT was studied at increasing concentrations in the protein microdroplets, a linear relationship between fluorescence intensity and monoamine concentration was found up to a critical level. When the concentration of catecholamines exceeded 4.5 x 10^{-2} M or 5-hydroxytryptamine (5-HT) 9 x 10^{-2} M no further increase in fluorescence was measured (Figure 3,4). Such a deviation in linearity may be explained in several different ways. Incomplete reaction with formaldehyde was excluded, since the pure, authentic catecholamine-formaldehyde product showed the same concentration dependent quenching.

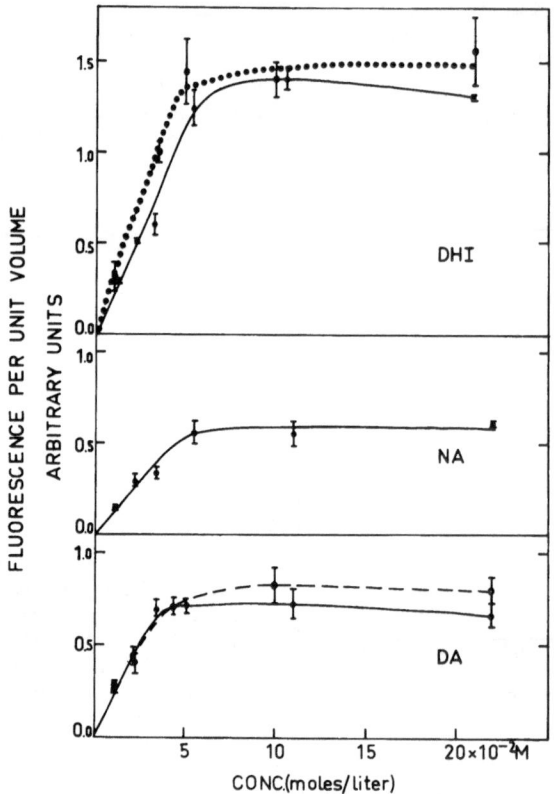

Figure 3. *Relation between fluorescence intensity and concentration of noradrenaline (NA), dopamine (DA) and authentic 6,7-dihydroxy-3,4-dihydroisoquinoline (the fluorescent reaction product in the dopamine-formaldehyde reaction, DHI) in dried microdroplets. NA and DA were treated with formaldehyde gas, DHI was measured directly without further treatment. Protein (——) and sucrose (·····) matrix. In one experiment (---), ATP was added to the solution at a molar ratio ATP/DA of 1/4. (From Ritzén 1966)*

In thin objects with negligible reabsorption of emitted light, the fluorescence F may be expressed as a function of the fluorescence quantum yield η, the extinction coefficient ε, the concentration of fluorophore c and the object volume v:

$$F = \eta(1-10^{-\varepsilon cv})\text{const.}$$

Thus, a change in either of η or ε would cause a deviation from the expected fluorescence intensity. The extinction coefficient was found to be the same throughout the range examined, even when the fluorescence was heavily quenched (Ritzén 1966). Other factors were kept constant, and thus a real change in fluorescence quantum yield was assumed, that is, the absorbed energy is disposed of by other routes than by emitting fluorescence. Since energy losses due to molecular collisions can be disregarded in the solid system used, and no evidence of non-fluorescent dimer or polymer formation was found, it was assumed that energy migration between closely situated molecules, eventually ending in absorption by non-fluorescent molecules, was the cause of decreasing fluorescence quantum yield. The critical concentration corresponds to an intermolecular distance of catecholamines of 33Å and for 5-HT of 26Å. Energy migration over such distances have frequently been observed (Förster 1951).

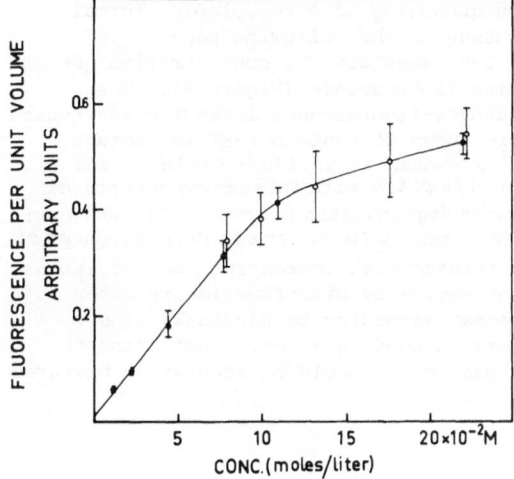

Figure 4. *Same as figure 3, but showing the relation between fluorescence intensity and concentration of 5-HT in microdroplets. (From Ritzén 1967a)*

It is interesting to compare this concentration quenching with other microfluorimetric techniques, where the low molecular weight fluorophore is bound to the macromolecules to be measured (e.g. acridine orange or fluorescent Schiff binding to DNA) (Figure 5).

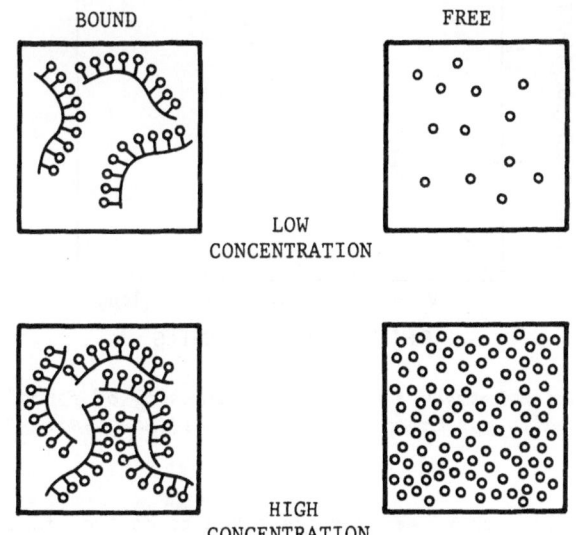

Figure 5. *Schematic representation of intermolecular distances between fluorescent molecules when bound to macromolecules at varying concentrations (left) and free in a diffusible state (right). In the former case, the close distance between the fluorophore molecules (00) is maintained even at low concentrations of the macromolecules, giving a constant fluorescence quenching*

In the latter case the distance between neighboring fluorescent molecules is determined by the proximity of the binding sites on the macromolecule. For a given species of macromolecules the distance between the fluorophore molecules is the same, whether there is a high concentration of macromolecules or not (Rigler 1966). Thus, the degree of concentration quenching (and the fluorescence quantum yield) is constant. In a case like the monoamine-formaldehyde fluorescence, where there is some degree of diffusion of the fluorophore within the storage site, increased concentration will mean decreased intermolecular distances. Variable concentration quenching of fluorescence should, therefore, be anticipated in all techniques involving measurement of low molecular weight, diffusible fluorophores.

The practical implications of concentration quenching of catecholamine-formaldehyde fluorescence will be discussed by Dr. Jonsson in the following paper.

The 5-HT-formaldehyde product was somewhat less sensitive to concentration quenching of the fluorescence than was the catecholamine fluorescence (Figure 4). The validity of the linear relationship 5-HT-formaldehyde-fluorescence intensity was tested in a study of rat peritoneal mast cells, that are known to contain 5-HT in storage granules up to a concentration of about 5×10^{-2} M (Moran et $al.$ 1962; Carlsson and Ritzén 1969). Rat peritoneal mast cells were treated in $vitro$ with increasing amounts of compound 48/80. This substance is well known to cause degranulation of mast cells and subsequent release of 5-HT as well as histamine from the granules (Moran et $al.$ 1962; Bloom et $al.$ 1970). The mean content of 5-HT in the samples treated with increasing concentrations of compound 48/80 was determined in two different ways - by microfluorimetry after formaldehyde treatment and by biochemical 5-HT assay according to Bogdanski et $al.$ (1956). The results obtained by both methods were in good agreement, indicating that the amount of 5-HT in individual rat peritoneal mast cells could be accurately measured by the microfluorimetric technique (Figure 6).

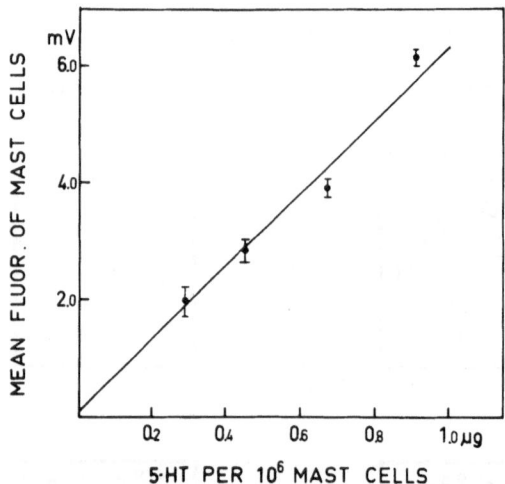

Figure 6. *Correlation between fluorescence intensity of formaldehyde treated mast cells and biochemically estimated content of 5-HT (abscissa). (From Ritzén 1967a)*

In conclusion, provided that measures are taken to control factors influencing the fluorescence yield, the microfluorimetric measurements of formaldehyde induced fluorescence provides an accurate method for quantitation of biogenic monoamines at the cellular level.

ACKNOWLEDGEMENTS

The editors of the journal Experimental Cell Research have kindly permitted reproduction of figures 3, 4 and 6 that previously have been included in original articles of that journal (see references).

REFERENCES

BLOOM, G.D., Diamant, B., Hägermark, O., and Ritzén, M., (1970), *Exp. Cell Res.* <u>62</u>, 61-75.

CARLSSON, S.A., and Ritzen, M., (1969), *Acta Physiol. Scand.* <u>77</u>, 449-464.

CASPERSSON, T., and Lomakka, G., (1962), *Ann. N.Y. Acad. Sci.* <u>97</u>, 449.

CASPERSSON, T., Lomakka, G., and Rigler, R., (1965), *Acta Histochem. Jena Suppl.* <u>6</u>, 123.

CORRODI, H., and Hillarp, N.Å., (1963), *Helv. Chim. Acta,* <u>46</u>, 2425.

_____ (1964), *ibid* <u>47</u>, 911.

FALCK, B., Hillarp, N.Å., Thieme, G., and Torp, A., (1962), *J. Histochem. Cytochem.,* <u>10</u>, 348.

FÖRSTER, Th., (1951), *Fluoreszenz Organischer Verbindungen,* Vandenhoeck-Ruprecht, Göttingen.

HAMBERGER, B., Malmfors, T., and Sachs, Ch., (1965), *J. Histochem. Cytochem.,* <u>13</u>, 147.

MORAN, N.C., Uvnäs, B., and Westerholm, B., (1962), *Acta Physiol. Scand.,* <u>56</u>, 26.

RIGLER, Jr. R., (1966), *Acta Physiol. Scand.,* <u>67</u>, Suppl. 267.

RITZÉN, M., (1966), *Exp. Cell Res.,* <u>44</u>, 505.

RITZÉN, M., (1967a), *Exp. Cell Res.,* <u>45</u>, 178.

RITZÉN, M., (1967b), "Cytochemical Identification and Quantitation of Biogenic Monoamines". *Thesis,* Karolinska Institutet, Stockholm.

QUANTITATION OF BIOGENIC MONOAMINES DEMONSTRATED WITH THE FORMALDEHYDE FLUORESCENCE METHOD

Gösta Jonsson

Department of Histology, Karolinska Institutet, Stockholm, Sweden

INTRODUCTION

The biogenic monoamines* have attracted a great deal of interest during the past few decades, especially in view of their role as transmitter substances in both the central and peripheral nervous systems. These bioactive compounds are considered to be involved in the regulation of a number of important functions: i.e., blood pressure, heart rate and contractile force, body temperature, mood, and extrapyramidal motor functions.

The introduction of the formaldehyde fluorescence technique by Falck and Hillarp for the histochemical demonstration of biogenic monoamines (Falck et al. 1962; Falck 1962; Carlsson et al. 1962) opened up quite new possibilities in this research field not only for morphological and distribution studies on monoamine-containing tissue structures, but also for studies on more dynamic processes (transmitter mechanisms and how they are affected, e.g., by psychoactive drugs). In the latter types of investigations, the quantitative aspects of the method are of utmost importance for the interpretation of the results obtained.

The present paper deals with the possibilities of using the Falck-Hillarp technique for quantitation of biogenic monoamines, with special reference to quantitation of noradrenaline (NA) in the postganglionic sympathetic adrenergic nerve terminals. Some principal features of the morphology of the adrenergic neuron and certain transmitter mechanisms will also be discussed briefly, in view of their importance in any consideration of the methodological problems in monoamine quantitation using the Falck-Hillarp technique. For technical details of the technique, chemical background, and specificity, the reader is referred to Falck and Owman (1965), Fuxe et al. (1970), Corrodi and Jonsson (1967, 1971).

THE ADRENERGIC NEURON

The present knowledge about the sympathetic adrenergic neuron is based on the extensive biochemical, pharmacological, ultrastructural, and histochemical investigations performed during the last decade. The description below deals exclusively with the sympathetic adrenergic neuron, but many of the basic properties are similar for most of the monoaminergic neurons.

A schematic drawing of the adrenergic neuron is shown in Figure 1. The cell-body has several dendrites and generally a fairly long axon (Norberg and Hamberger, 1964). Using the Falck-Hillarp technique, the cell bodies can be seen to display a fairly great variation in NA fluorescence intensity. The average NA concentration in the cell bodies of rat superior cervical ganglion has been calculated to be about 10-100 µg/g wet weight of the tissue (Norberg and Hamberger 1964; Dahlström et al. 1966). In the fluorescence microscope the main axon and nonterminal axons are seen to display a very weak fluorescence. Although it has not been possible to make any exact determination of the NA concentration in these structures, it has been estimated to be about 100-500 µg/g (Norberg and Hamberger 1964). In the effector organ, there is a considerable arborization of the axon into a large number of nerve terminals showing characteristic axonal enlargements, the so-called varicosities (Figure 2. See Plate I). These structures, which exhibit a very intense fluorescence, are considered to be specialized for uptake, storage, synthesis, and release of transmitter (Figure 3) (see, i.a., Malmfors 1965; Iversen 1967). The average NA concentration is also very high in the varicosities, and is

---

* *In this paper the term "biogenic monoamines" is used to denote the catecholamines, dopamine, noradrenaline, adrenaline, and the indolealkylamine 5-hydroxytryptamine (serotonin).*

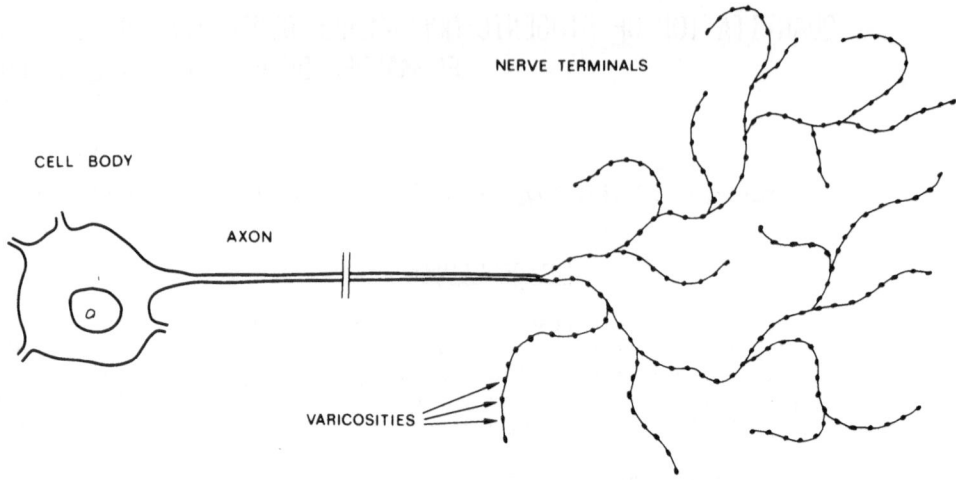

CELL BODY

AXON

NERVE TERMINALS

VARICOSITIES

Figure 1. *A schematic representation of the adrenergic neuron*

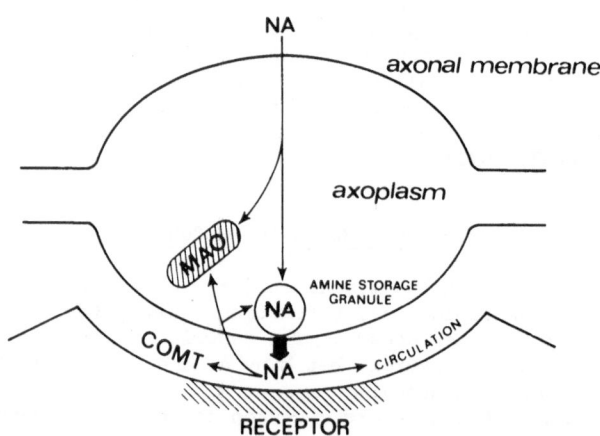

Figure 3. *Schematic drawing of an adrenergic nerve terminal (varicosity).*
NA = *noradrenaline;* MAO = *monoamine oxidase.* COMT = *catechol-O-methyl transferase*

calculated to be approximately 5,000 μg/g wet weight (Dahlström et al. 1966; Jonsson and Sachs 1971).

Within the adrenergic neuron, NA is considered to be stored in special submicroscopic granules, the so-called *amine storage granules.* Although at present there is no conclusive evidence, several investigations have suggested that most of the NA present intraneuronally is particle bound, in all probability to the amine storage granules (see Euler 1966; Carlsson 1966; Potter 1967). Studies at the ultrastructural level have demonstrated that the vast majority of the granules are localized in boutons which in all probability correspond to the varicosities (Hökfelt 1968; 1969). In the rat irides these axonal enlargements have been estimated to contain on the

PLATE I

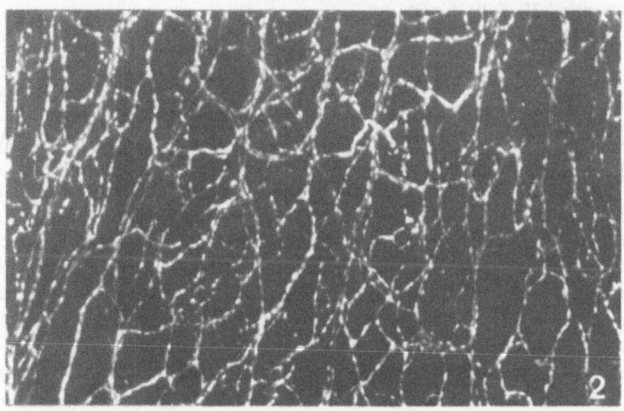

Figure 2. *Fluorescence photomicrograph of a rat iris whole-mount treated according to the technique of Falck and Hillarp. The fluorescent adrenergic nerves form a dense network with strongly fluorescent varicosities. 160x*

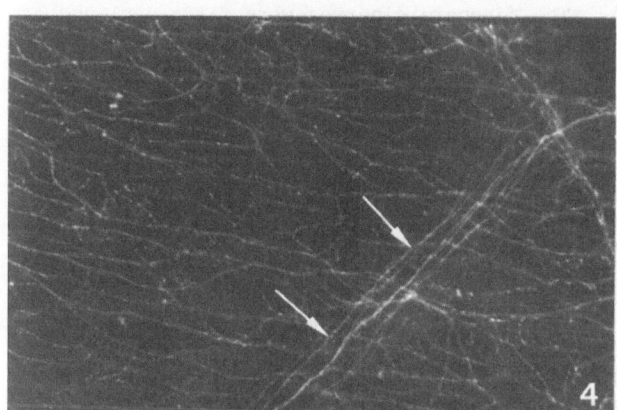

Figure 4. *Rat iris pretreated with reserpine + nialamide after* in vitro *incubation in $10^{-6}M$ ^{3}H-NA for 30 min. The adrenergic nerves show a smooth appearance with nondistinct varicosities. Note the intensely fluorescent nonterminal axons (→). 160x*

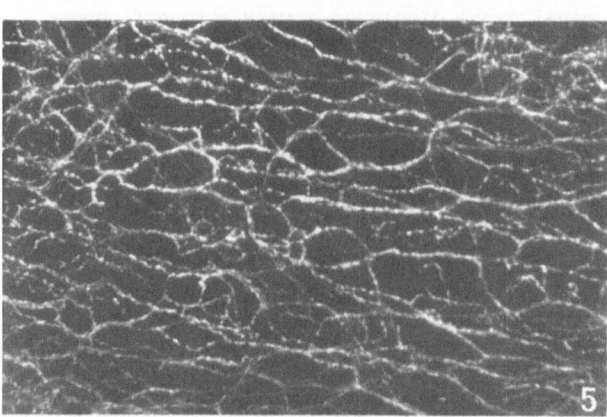

Figure 5. *Rat iris pretreated with α-methyl-p-tyrosine methylester (H44/68) after incubation in $10^{-6}M$ ^{3}H-NA for 30 min. The adrenergic nerves exhibit a normal appearance, cf. Figure 2. 160x*

average 300-600 granules. The varicosity has been found to contain about 5×10^{-3} pg NA (Dahlström et al. 1966). Assuming that all the NA present in the varicosities is bound to the amine storage granules, the average NA concentration in these granules would be in the order of 100,000 μg/g.

The NA present in the granules is considered to be bound to ATP and Mg^{++} in a protein complex (see Lagercranz 1971) from which it can be released by nerve impulses.

The whole adrenergic neuron has the property of taking up and accumulating both circulating NA and NA released by nerve impulses (Iversen 1967). It is now generally accepted that the adrenergic neuron possesses at least two uptake-accumulation mechanisms, one located at the level of the axonal membrane (the "membrane pump"), and one at the level of the amine storage granules (see Figure 3). The "membrane pump" uptake mechanism is extremely efficient (Jonsson et al. 1969); it is inhibited by certain drugs such as cocain and tricyclic antidepressants (e.g., desipramine), but is not affected by the monoamine depletor, reserpine (see, *i.a.*, Malmfors 1965; Carlsson 1965). After NA has been taken up intraneuronally, the amine is incorporated in the amine storage granules by a $ATP-Mg^{++}$-dependent uptake mechanism which is efficiently blocked by reserpine (Carlsson 1965). The main part of the NA taken up is thus trapped by the granules, while a small part is catabolized by MAO localized to the mitochondria.

NA FLUORESCENCE-CONCENTRATION RELATIONSHIP IN ADRENERGIC NERVE TERMINALS

Using a combined isotope and microfluorimetric approach, studies have been made of the fluorescence-concentration relationship for NA present in adrenergic nerve terminals in rat irides (Jonsson 1969). The adrenergic nerves were first depleted of their endogenous NA stores, by pretreatment with either reserpine (10 mg/kg *i.p.*, 16 hrs) or the tyrosine hydroxylase inhibitor, α-methyl-p-tyrosine methylester (H44/68, 500 mg/kg *i.p.*, 16 hrs). Treatment with these compounds produces an almost complete depletion of the endogenous NA, although by different mechanisms: in the case of reserpine, by interference with NA storage; in the case of H44/68, by interference with the NA synthesis (Jonsson and Sachs 1969). The irides from the pretreated animals were dissected out and thereafter incubated *in vitro* at +37°C in a Krebs-Ringer bicarbonate buffer (pH 7.4) containing various concentrations of ^{3}H-NA ($10^{-7}M$-$10^{-5}M$, 30-60 min) for partial replenishment of the nerves with ^{3}H-NA. In order to retain ^{3}H-NA taken up intraneuronally, the animals treated with reserpine also had to be pretreated with the potent MAO inhibitor, nialamide (100 mg/kg *i.p.*, 2 hrs), thereby eliminating rapid catabolism of the amine. The incubated irides were prepared as whole mounts, and were dried and exposed to formaldehyde gas for the histochemical demonstration of NA. The incubation in ^{3}H-NA produced a restitution of the adrenergic nerve plexus, with the gradual increase of fluorescence intensity in the nerves paralleling the increase of ^{3}H-NA concentration in the medium. The fluorescence morphology of the restituted nerves differed, however, for the two pretreatments. The nerves from animals pretreated with reserpine + nialamide showed a smoother appearance with nondistinct varicosities, and the nonterminal axons displayed a very strong fluorescence intensity (Figure 4. Plate I). In the case of H44/68, the nerves showed a normal fluorescence morphology, *viz.* distinct varicosities and weakly fluorescent nonterminal axons (Figure 5. Plate I, cf. Figure 2). These differences are related to differences in the *subcellular localization of NA*. After reserpine + nialamide, ^{3}H-NA taken up and accumulated within the nerves is mainly *extragranularly localized*, whereas after H44/68, the transmitter is stored in the amine storage granules *(granularly localized)* as under ordinary conditions (Jonsson et al. 1969; Jonsson and Sachs 1969; 1970; Sachs 1970).

The formaldehyde-induced fluorescence of NA present in the replenished nerves was measured in a microspectrofluorimeter (Jonsson 1969). The irides were then scraped off of the slides, dissolved, and the radioactivity taken up and accumulated determined by liquid scintillation spectrometry. Using specific chemical-analytical procedures, it was shown that in all cases more than 90% of total radioactivity constituted unchanged ^{3}H-NA. Furthermore, results obtained from denervation experiments have demonstrated that the radioactivity taken up and accumulated in irides during the experimental conditions in this study is confined to the adrenergic nerves (Jonsson et al. 1969). The results thus point to a direct correlation between microfluori-

metric-measured NA fluorescence intensity and the NA concentration in the same preparation. The results obtained are presented in Figure 6 and 7.

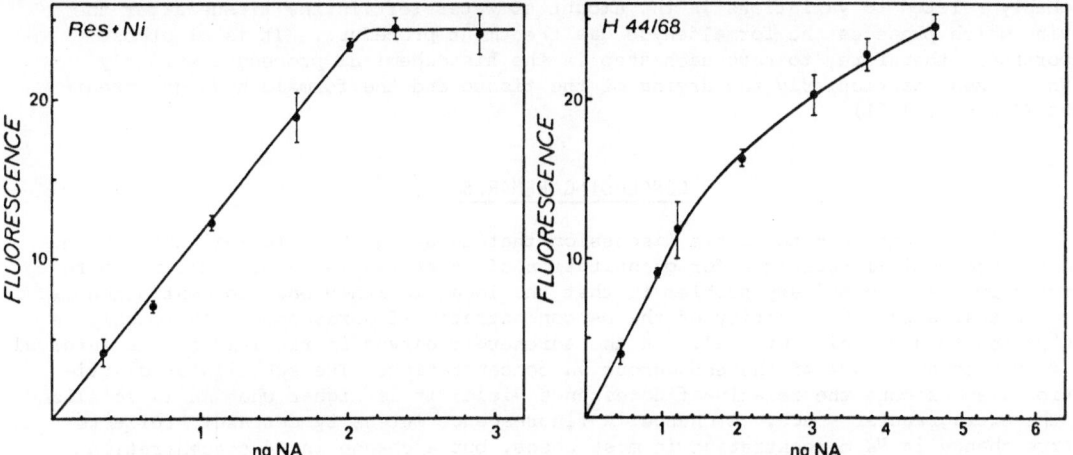

Figure 6. *Relation between fluores-*
cence intensity (arbitrary units) determined
microfluorimetrically and ^{3}H-NA *taken up*
and accumulated in vitro *in adrenergic*
nerves of rat iris. The rats had been
pretreated with reserpine + nialamide

Figure 7. *Relation between fluores-*
cence intensity (arbitrary units) determined
microfluorimetrically and ^{3}H-NA *taken up*
in vitro *in adrenergic nerves of rat iris.*
The rats had been pretreated with H44/68.
Figures 6 and 7 from Jonsson 1969

Figure 6 shows the *NA* fluorescence-concentration relationship in adrenergic nerves
of irides from rats pretreated with reserpine + nialamide. The [3]*H-NA* present in the
nerves in this case is mainly localized in the extragranular space. The fluorescence
is proportional to *NA* concentration up to about 2 ng/iris; this corresponds to about
40% of the endogenous *NA* level (Jonsson 1969). Above 2 ng/iris, there is a fairly
sharp threshold, and a marked deviation from linearity occurs. In the case of H44/68
where the [3]*H-NA* taken up and accumulated is stored in the amine storage granules, the
same phenomenon was observed, although the deviation from linearity was more gradual,
starting at about *NA* concentration of about 1.5 μg/iris (Figure 7). These results
show, furthermore, that the relative fluorescence yield per mole *NA* is higher after
reserpine + nialamide, and so indicate that the Falck-Hillarp technique is more sensitive for extragranularly stored *NA*. This view is also consistent with results
reported by van Orden et al. (1967; 1970). The deviation from linearity is in all
probability related to a concentration-dependent quenching of the fluorescence that
starts earlier, when the transmitter is granularly stored. This supposition is also
supported by the results obtained by Ritzén (1966), who, using microdroplet models and
microfluorimetric techniques, found that the NA fluorescence-concentration relation
is linear up to a value corresponding to about 8.000 μg/g wet weight of the tissue.
Above this value, he observed a concentration-dependent quenching of the fluorescence.

No exact information is available regarding the subcellular localization of the
NA fluorophor after processing for fluorescence histochemistry. Considering the
quenching value, i.e., the critical amine concentration when deviation from linearity
starts, obtained by Ritzén, (8.000 μg/g) and the fact that in the H44/68 case the
fluorescence is proporational to *NA* concentration up to about 30% of the endogenous
NA concentration (corresponding to a *NA* concentration in the granules of approximately
30,000 μg/g), it is reasonable to assume that even in the case when the amine is stored
mainly in the granules *in vivo*, the *NA* fluorophor diffuses out of the granules during
the histochemical procedure (Jonsson, 1967; 1971).
In this context it should also be mentioned that it is not possible to extrapolate the quenching value from one neuron system (e.g., rat iris) to another. This is
because different tissues require different humidities of the formaldehyde gas for
optimal fluorescence; These variances might in turn lead to various degrees of diffusion
of the fluorophor from the amine storage granules, and thus to changes in the quenching

value. It has also been found difficult to obtain exactly the same *NA* fluorescence yield from one experiment to another — a finding which indicates that the quenching value can vary to a certain extent even in the same neuron system. This variation is probably related to variations in the amount of water left in the tissue after the drying which precedes the formaldehyde gas treatment procedure. It is of utmost importance, therefore, to have each step in the histochemical procedure strictly standardized, particularly the drying of the tissue and the formaldehyde gas treatment (Jonsson, 1971).

CONCLUDING REMARKS

It is evident from the above discussion that in using the Falck-Hillarp fluorescence histochemical technique for quantitation of *NA* stored in sympathetic adrenergic nerve terminals the primary problem is that the local intraneuronal concentration is so high that a strict linearity of the NA concentration-fluorescence relationship is restricted to a certain interval. In the adrenergic nerves in rat irides this interval is from 0 to about 40% of the endogenous *NA* concentration. The subcellular distribution also affects the relative fluorescence yield; it is higher when *NA* is localized in the extragranular space. A change in fluorescence intensity certainly reflects a true change in *NA* concentration in most cases, but a change in *NA* concentration can escape detection when the remaining *NA* concentration is in the nonlinear part of the concentration-fluorescence relationship.

ACKNOWLEDGEMENTS

Part of the work reviewed in the present paper has been supported by research grants from the Swedish Medical Research Council (B72-14X-2295-05B), Magnus Bergvalls Stiftelse, Carl-Bertel Nathorsts Stiftelse and Karolinska Institutet (forskningsfonder).

REFERENCES

CARLSSON, A., (1965), In: *Handbuch der exp. Pharmakol.* (Eichler, O., and Farah, A., eds.), Springer-Verlag, Heidelberg. 529-92.

CARLSSON, A., (1966), *Pharmacol. Rev.* 18, 541-49.

CARLSSON, A., Falck, B., and Hillarp, N.-Å., (1962), *Acta Physiol. Scand.* 56, Suppl. 196.

DAHLSTRÖM, A., Häggendal, J., and Hökfelt, T., (1966), *Acta. Physiol. Scand.* 67, 289-94.

EULER von, U.S., (1966), *Acta Physiol Scand.* 67, 430-40.

FALCK, B., (1962), *Acta Physiol. Scand.* 56, Suppl. 197.

FALCK, B., Hillarp, N.-Å., Thieme, G., and Torp, A., (1962), *J. Histochem. Cytochem.* 10, 348-54.

FALCK, B., and Owman, C., (1965), *Acta Univ. Lund.*, Sectio II, No. 7.

HÖKFELT, T., (1968), *Z. Zellforsch.* 91, 1-74.

HÖKFELT, T., (1969), *Acta Physiol. Scand.* 76, 427-40.

IVERSEN, L. L., (1967), *The uptake and Storage of Noradrenaline in Sympathetic Nerves.* Univ. Press, Cambridge.

JONSSON, G., (1966), *Acta Chem. Scand.* 20, 2755-62.

JONSSON, G., (1967), *The Formaldehyde Fluorescence Method for the Histochemical Demonstration of Biogenic Monoamines.* A Methodological Study. M.D. Thesis, Stockholm.

JONSSON, G., (1969), *J. Histochem. Cytochem.* 17, 714-23.

JONSSON, G., (1971), *Progr. Histochem. Cytochem.* 2, 229-344.

JONSSON, G., Hamberger, B., Malmfors, T., and Sachs, C., (1969), *Europ. J. Pharmacol.* 8, 58-72.

JONSSON, G., and Sachs, C., (1969), *Acta Physiol. Scand.* 77, 344-57.

JONSSON,G., and Sachs, C., (1970), *Acta Physiol. Scand.* 80, 307-22.

LAGERCRANZ, H., (1971), *Acta Physiol. Scand.* Suppl. 366.

MALMFORS, T., (1965), *Acta Physiol. Scand.* 64, Suppl. 248.

NORBERG, K.-A., and Hamberger, B., (1964), *Acta Physiol. Scand.* 63, Suppl. 238.

POTTER, L. T., (1967), *Circ. Res.* 21.

RITZÉN, M., (1966), *Exp. Cell Res.* 44, 505-29.

SACHS, C., (1970), *Acta Physiol. Scand,* Suppl. 341.

VAN ORDEN, L. S., III, Bensch, K. G., Langer, S. Z., and Trendelenburg, U., (1967), *J. Pharmacol. Exp. Ther.* 157, 274-83.

VAN ORDEN, L. S., III, Schaefer, J. M., Burke, J. P., and Lodoen, F. V., (1970), *J. Pharmacol. Exp. Ther.* 174, 357-68.

THE IDENTIFICATION OF ARYLETHYLAMINES BY MICROSPECTROFLUOROMETRY OF ACID- AND ALDEHYDE-INDUCED FLUORESCENCE

F. W. D. Rost and A. G. E. Pearse

Royal Postgraduate Medical School, London, England

INTRODUCTION

For many years, we have been interested in methods for the histochemical demonstration of catecholamines, tryptamines, and other arylethylamines, particularly in relation to the storage of these substances and uptake of their precursors by cells of the endocrine polypeptide (APUD) series.

Histochemical methods for the demonstration of arylethylamines are usually based on condensation reactions with formaldehyde, leading to fluorescent products. Similar techniques have recently been developed using vapors of other aldehydes or a carboxylic acid instead of formaldehyde. The present paper is concerned chiefly with reviewing these techniques as carried out in our own laboratory.

The first observation of formaldehyde-induced fluorescence (FIF) has been attributed to Erös (1932) who observed fluorescence in enterochromaffin cells of the gut; however, it is more probable that he was observing the autofluorescence of lipofuscin. FIF was probably first observed by Baroni (1933), in malignant melanomas, although he did not report the fixative used. Eränkö (1951a; 1951b) reported fluorescence in "unstained" formalin-fixed sections of adrenal medulla, which correlated with the presence of noradrenaline demonstrated by the iodate method of Hillarp and Hökfelt (1954). The fluorescence was soon shown to be due to a reaction between noradrenaline and formaldehyde (Eränkö 1955). The FIF of enterochromaffin cells was reported by Shepherd et al. (1953), and was shown to be due to the presence of 5-hydroxytryptamine by Barter and Pearse (1953; 1955).

The chemical processes involved in FIF were further elucidated by Vialli and Traverso (1960) and a number of others, notably Corrodi, Hillarp, and Jonsson in their work from 1963 to 1967. The chemistry of the reactions has recently been reviewed by Jonsson and Sandler (1969). Briefly, it may be stated that the amine substance condenses with a molecule of formaldehyde to form a further ring, so that phenylethylamines become isoquinolines and indolylethylamines become β-carbolines. This is the Pictet-Spengler condensation reaction (Jonsson 1971; Pictet and Spengler 1911).

DEVELOPMENT OF HISTOCHEMICAL METHODS

At first, aqueous formalin was used for the demonstration of FIF. This method is of low sensitivity, because water-soluble amines, and some of their intermediate products, tend to be dissolved in the "fixative." Nonetheless, the original technique of Eränkö (1952) is adequate for some purposes (El-Badawi and Schenk 1967; Rost and Polak 1969), and there are improved variants in which the removal of amine substances is minimized by the use of Krebs-Ringer solution (Eränkö 1966) or phosphate buffer (Laties et al. 1967) containing formalin. The technique for the demonstration of FIF was greatly improved by the use of freeze-dried tissue exposed to formaldehyde vapor (Barter and Pearse 1955; Pearse 1956; Eränkö 1963; Falck and Torp 1961; Lagunoff et al. 1961). This method not only avoids loss of amines in the aqueous "fixative", but also yields better fixation of the tissues. The technique of freeze-drying followed by formaldehyde vapor has been standard practice for some years (Falck and Owman 1965; Eränkö 1967).

Rost and Ewen (1971), explored the possibilitites of employing substances other than formaldehyde as fixative vapors for the induction of fluorescence. A vapor fixative was chosen for two main reasons. Firstly, Tock and Pearse (1965) had already found that vapor fixation of freeze-dried material could minimize the loss of water-soluble materials; and secondly, it was thought likely that the vapor would be the monomeric form of the reagent, whereas solutions contain polymers, hydrated forms, and other derivatives. Rost and Ewen (1971) considered that the vapor fixative to be used for the demonstration of arylethylamines should have the following properties:

a) Ready condensation with the amine group.

b) No hindrance to cyclization, and preferably encouraging cyclization.

c) No fluorescence of its own.

d) Vaporization at a reasonable temperature, say 50°-80°C.

e) The vapor should not decompose in air to nonreactive products.

f)   Ideally, the fluorophores formed from the various arylethylamines should have different emission or excitation spectra, or both, or have distinct characteristics to enable the individual substances to be identified by microspectrofluorimetry or otherwise.

g) Either useful action as a general tissue fixative, or no fixative action.

Consideration was first given to the use of other aldehydes, in the Pictet-Spengler reaction. Acetaldehyde vapor could be obtained (Rost and Ewen 1971) by heating metaldehyde ("Meta") in a manner similar to that in which formaldehyde is obtained from paraformaldehyde. Glutaraldehyde vapor had been obtained by Tock and Pearse (1965) by extracting glutaraldehyde from aqueous solution into liquid paraffin, which was then heated. Rost and Ewen (1971) however, found it possible to generate the vapor by heating a strong (50%) aqueous solution of glutaraldehyde, the vapor being dried by phosphorous pentoxide. Vapors of acetaldehyde and glutaraldehyde were found to give good fluorescence with many amine substances both in models and in tissues (Rost and Ewen 1971).

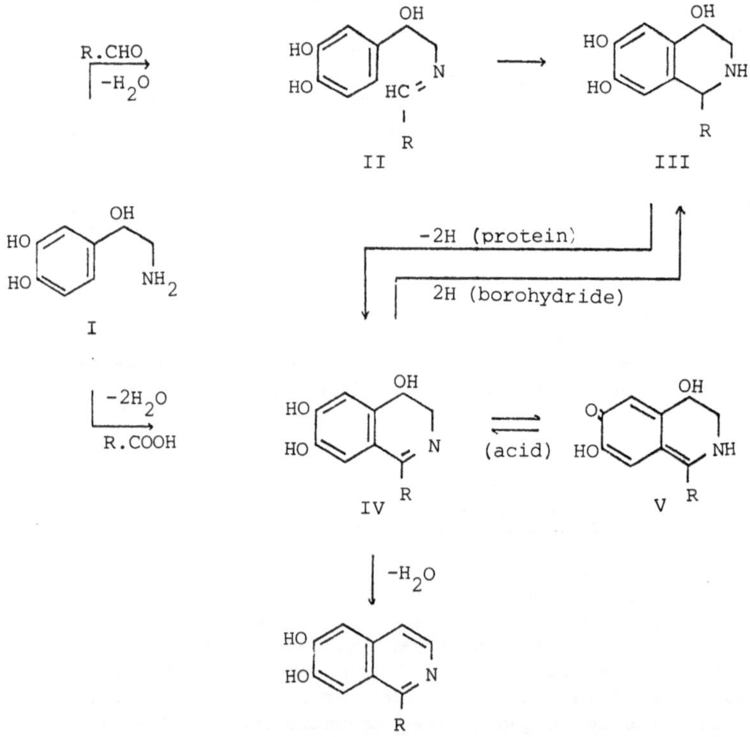

Figure 1. *Proposed general reaction of an aldehyde* (R·CHO) *or carboxylic acid* (R·COOH) *with a phenylethylamine, exemplified by noradrenaline (I), leading to the formation of a strongly fluorescent quinonoid (V). A tyramine or methoxy compound could react similarly. In the case of noradrenaline only, an irreversible reaction can occur from the isoquinoline (IV) to a fully resonant form (VI)*

In models, histamine was found to give a fluorescence after exposure to vapors of either acetaldehyde or glutaraldehyde (Eränkö 1967) but this method did not give sufficient sensitivity for use in tissues (Cross et al. 1971).

As an alternative to the Pictet-Spengler reactions employing an aldehyde, isoquinolines can be synthesized from phenylethylamines by condensation with a carboxylic acid and dehydration (Bischler and Napieralski 1893; Pictet and Gams 1909). Consideration was therefore given to the possibility that fluorophores could be induced from arylethylamines by exposing freeze-dried material to vapors of a carboxylic acid or acid anhydride. Vapors of formic, acetic, and other acids were investigated. Formic acid vapor was found to destroy the tissue. Acetic acid vapor at 60° to 80°C for 4 hours was found to give satisfactory fluorescence with arylethylamines in models and tissues. The probable reactions are illustrated in Figure 1.

Spectra

Of the microspectrofluorometers which have been used in connection with studies of induced fluorescence of arylethylamines, as far as we are aware the only corrected instruments are Caspersson's modified UMSP (Caspersson et al. 1965; Ritzén 1967), the Leitz microspectrographs in Lund (Björklund et al. 1968a and Londen (as described in a previous paper), and (corrected for emission spectra only) the microspectrofluorometer of Van Orden (1970). Of these instruments the London example (FATIMA) can be used with excitation by epiillumination at wavelengths below about 360 nm. On this basis, the only reports of excitation and emission spectra which can be accepted are those made with these instruments (Caspersson et al. 1966; Jonsson and Ritzén 1966; Björklund et al. 1968a, b; 1970; 1971; Van Orden 1970; Ewen and Rost 1972). The corrected excitation spectra can be compared with absorption spectra measured by Corrodi and co-workers (1964a; 1965a; 1965b).

Excitation and emission maxima measured with FATIMA (Ewen and Rost 1972) from models using formaldehyde, acetaldehyde, glutaraldehyde and acetic acid are listed in Table I.

TABLE I

Excitation *(Ex)* and emission *(Em)* peaks of fluorophores derived from various arylethylamines by treatment in models with vapors of formaldehyde, acetaldehyde, glutaraldehyde, and acetic acid. Each value represents the average of a number of readings, rounded off to the nearest 10nm. Only the longest-wavelength excitation peak is shown; all these substances except normetanephrine have an additional peak at about 320-330nm.

SUBSTANCE	FORM-ALDEHYDE		ACET-ALDEHYDE		GLUTAR-ALDEHYDE		ACETIC ACID	
	Ex	Em	Ex	Em	Ex	Em	Ex	Em
Noradrenaline	410	480	400	530	410	530	400	530
Adrenaline	410	480	410	520	410	530	390	530
Dopamine	410	500	400	530	400	540	400	530
Dopa	410	500	400	540	400	530	400	540
3-O-methyl-noradrenaline (Normetanephrine)	*	*	410	540	420	540	420	540
Tryptamine	*	*	400	530	400	530	400	540
5-Hydroxytryptamine	410	540	400	530	400	530	400	540
5-Hydroxytryptophan	410	540	400	540	410	580	410	560
5-Methoxytryptamine	400	560	410	560	400	560	400	560

* Little or no reaction

Emission maxima with formaldehyde were similar to those found by other workers. With glutaraldehyde, acetaldehyde, and acetic acid, the emission maxima for fluorophores induced from catecholamines and tryptamines were at longer wavelengths than those observed with formaldehyde, and there was less differentiation between catecholamines and tryptamines — all maxima were in the range 520-580nm.

Excitation maxima were observed at about 400-410nm in all catecholamines and tryptamines studied. An additional excitation peak at about 320-330nm was observed in nearly all cases. With acetic acid, there were marked differences in the ratios of the heights of the peaks. The observed ratios of the heights of the peak at 400-410nm to that at 320-330nm were: dopamine, 1:0.8; dopa, 1:1; noradrenaline, 1:2; adrenaline, 1:2; normetanephrine, 1:0.5; 5-HT, 1:1; 5-HTP, 1:1.

METHODOLOGY

Tissues are snap-frozen in melting difluorodichloromethane (Arcton 22, I.C.I. Ltd.); freeze-dried in a thermoelectric freeze-drier (Pearse 1963;1968) at -40°C overnight; exposed to vapor at 60-80°C for 4 hours; and vacuum-embedded in paraffin wax containing a high proportion of plasticizer (Ralwax 1, Raymond Lamb). Our freeze-drying technique has been illustrated in detail elsewhere (Campo-Aasen and Pearse 1966/67). Possibly because of the particular ambient humidity of our laboratory, we have not found it necessary to equilibrate the paraformaldehyde with water vapor before heating it to generate formaldehyde vapor. Acetic acid vapor is obtained by heating acetic acid, Analytical Reagent quality, which has been dried by freezing (at 4°C) and pouring off the remaining liquid; the vapor is further dried by phosphorous pentoxide placed in the vapor chamber. Tissue sections are cut and mounted on glass slides (0.8-1.0mm thick) under glass coverslips (Grade 1 1/2; 0.160-0.190mm thickness) with DPX. The DPX is sometimes diluted a little with xylene; it has been found unnecessary to dewax the section before mounting, as the wax dissolves in the solvent of the mountant. If excitation spectra are to be measured in the ultraviolet region at wavelengths shorter than 360nm, a quartz coverslip is used instead of glass; a quartz slide is not necessary because epiillumination is always employed.

Microspectrofluorometry is carried out with the microspectrofluorometer described in a previous paper given at this conference. Spectra are measured as a series of individual readings equally spaced in the quantum energy (frequency) spectrum; the specimen is usually not illuminated between readings. Owing to the high sensitivity of the instrument, it is only necessary to illuminate with weak light; therefore fading is less of a problem than with some other instruments. We have also experimented with the technique of measuring spectra repetitively from the same region, the exact time of each measurement being noted and extrapolations made to estimate the intensities of the fluorescence at each wavelength at the moment of commencement of irradiation. Since installing the automatic data logging system, however, we have not found this tedious method necessary.

The apparent fluorescence from a control area is always measured and subtracted from the test readings. Results are corrected for the reflectivity of the beamsplitter in the reference channel, the spectral bandwidth of the emission monochromator and the spectral sensitivity of the emission measuring photomultiplier. Graphs are plotted, either by hand or by a digital plotter attached to the computer, with a horizontal scale linear in photon quantum energy.

ACKNOWLEDGEMENTS

This work has been supported by grants from the Medical Research Council (for equipment); the Muscular Dystrophy Group of Great Britain; the Wellcome Trust; and the Smith, Kline, and French Foundation.

REFERENCES

BARONI, B., (1933), *Arch. Ital. Derm.*, 9, 543-86.

BARTER, R., and Pearse, A. G. E., (1953), *Nature*, (Lond.), 172, 810.

BARTER, R., and Pearse, A. G. E., (1955), *J. Path. Bact.*, 65, 25-31.

BJÖRKLUND, A., Ehinger, B., and Falck, B., (1968), *J. Histochem. Cytochem.*, 16, 263-70.

BJÖRKLUND, A., Falck, B., and Håkanson, R., (1968), *Acta Physiol. Scand.* Suppl. 318.

BJÖRKLUND, A., Nobin, A., and Stenevi, U., (1971), *J. Histochem. Cytochem.*, 19, 286-98.

BJÖRKLUND, A., and Stenevi, U., (1970), *J. Histochem. Cytochem.*, 18, 794-802.

BISCHLER, A., and Napieralski, B., (1893), *Ber. Dt. Chem. Ges.*, 26, 1903-08.

CAMPO-AASEN, I., and Pearse, A. G. E., (1966-67), *Medicina Cutanea.* 5, 485-92.

CASPERSSON, T., Hillarp, N.-A., and Ritzén, M., (1966), *Exptl. Cell. Res.*, 42, 415-28.

CASPERSSON, T., Lomakka, G., and Rigler, R., (1965), *Acta Histochem. Suppl.*, 6, 123-26.

CORRODI, H., and Hillarp, N.-A., (1963), *Helv. Chim. Acta*, 46, 2425-30.

CORRODI, H., and Hillarp, N.-A., (1964a), *Helv. Chim. Acta*, 47, 911-18.

CORRODI, H., and Hillarp, N.-A., (1964b), *J. Histochem. Cytochem.*, 12, 582-86.

CORRODI, H., and Jonsson, G., (1965a), *J. Histochem. Cytochem.*, 13, 484-487.

CORRODI, H., and Jonsson, G., (1965b), *Acta Histochem.*, 22, 247-58.

CORRODI, H., and Jonsson, G., (1966), *Helv. Chim. Acta.*, 49, 798-806.

CORRODI, H., and Jonsson, G., (1967), *J. Histochem. Cytochem.*, 15, 65.

CROSS, S. A. M., Ewen, S. W. B., and Rost, F. W. D., (1971), *Histochem. J.*, 3, 471-76.

EL-BADAWI, A., and Schenk, E. A., (1967), *Lab. Invest.*, 16, 672.

ERÄNKÖ, O., (1951a), *Nature* (Lond.) 168, 250-51.

ERÄNKÖ, O., (1951b), *Acta Physiol. Scand.*, 25, Suppl. 89, 22-23.

ERÄNKÖ, O., (1952), *Acta Anat.* 16, Suppl., 17, 1-60.

ERÄNKÖ, O., (1955), *Nature*, (Lond.) 175, 88-89.

ERÄNKÖ, O., (1963), Personal communication acknowledged in Falck and Torp, below.

ERÄNKÖ, O., (1967), *J. R. Microsc. Soc.*, 87, 259-76.

ERÄNKÖ, O., and Räisänen, L., (1966), *J. Histochem. Cytochem.*, 14, 690-91.

ERÖS, G., (1932), *Zbl. Allg. Path. Anat.*, 54, 385-91.

EWEN, S. W. B., and ROST, F. W. D., (1972), *Histochem. J.*, 4, 59-69.

FALCK, B., and Owman, C., (1965), *Acta Univ. Lund.*, Sectio II, No. 7.

FALCK, B., and Torp, A., (1961), *Med. Exptl*, (Basle). 5, 428-32.

HILLARP, N.-A., and Hökflet, B., (1954), *Acta Physiol. Scand.*, 30, 55-68.

JONSSON, G., (1971), *Progr. Histochem. Cytochem.*, 2, 299-334.

JONSSON, G., and Ritzén, M., (1966), *Acta Physiol. Scand.*, 67, 505-13.

JONSSON, G., and Sandler, M., (1969), *Histochemie.*, 17, 207-212

LAGUNOFF., Phillips, D. M., and Benditt, E. P., (1961), *J. Histochem. Cytochem.*, 9, 534-41.

LATIES, A. M., Lund, R., and Jacobowitz, D., (1967), *J. Histochem. Cytochem.*, 15, 535-41.

PEARSE, A. G. E., (1956), In: *Lectures on the Scientific Basis of Medicine 1954-55.* 4, 358-86.

PEARSE, A. E. E., (1963), *J. Sci. Instrim.*, 40, 176-77.

PEARSE, A. E. E., (1968), *Histochemistry, Theoretical and Applied.*, Vol. I. 3rd Edition. Churchill, (London).

PICTET, A., and Gams, A., (1909), *Ber. Dt. Chem. Ges.*, 42, 2943-52.

PICTET, A., and Spengler, T., (1911), *Ber. Dt. Chem. Ges.*, 44, 2030-36.

PRENNA, G., (1969), Personal communication acknowledged in Rost and Ewen (1971).

RITZÉN, M., (1967), *Cytochemical Identification and Quantitation of Biogenic Amines.*, M.D. Thesis, Stockholm.

ROST, F. W. D., and Polak, J. M., (1969), *Virchows Arch. A. Path. Anat.*, 347, 321-26

ROST, F. W. D., and Ewen, S. W. B., (1971), *Histochem. J.*, 3, 207-12.

SHEPHERD, D. M., West, G. B., and Erspamer, V., (1953), *Nature*, (Lond.) 172, 357.

TOCK, E. P. C., and Pearse, A. G. E., (1965), *J. R. Microsc. Soc.*, <u>84</u>, 519-37.

VAN ORDEN, L. S., (1970), *Biochem. Pharmac.*, <u>19</u>, 1105-17.

VIALLI, M., and Traverso, G., (1960), *Istituto Lombardo (Rend. Sc.)* <u>B94</u>, 167-84.

Part IV
Investigation of Enzyme Reactions and Transport Mechanisms

QUANTITATIVE ASPECTS OF RAPID MICROFLUOROMETRY FOR THE STUDY OF ENZYME REACTIONS AND TRANSPORT MECHANISMS IN SINGLE LIVING CELLS

Elli Kohen[*], Cahide Kohen[**], Bo Thorell[***] and Gilbert Wagener[****]

[*] *Papanicolaou Cancer Research Institute, Miami, Florida 33163*
[**] *Clinical Faculty, Department of Pathology, University of Miami, School of Medicine, Miami, Florida.*
[***] *Department of Pathology, Karolinska Institute, Stockholm, Sweden.*
[****] *Biomedical Instrumentation Laboratory, Veterans Administration Hospital, Miami, Florida*

INTRODUCTION

Fluorescent molecular probes are ideally suited not only for kinetic studies of enzyme reactions in localized cell structures (i.e.,mitochondria, nucleus, cytoplasm) (Chance 1970) but also for observations on transport mechanisms across intracellular membranes (Kohen, Siebert et al. 1971). Among intracellular fluorochromes, reduced pyridine nucleotides with their blue fluorescence (Sund 1968) provide a direct probe into the microenvironment of intracellular compartments (Kohen 1964) through the analysis of NAD-(NADP) reduction reoxidation transients (Kohen et al. 1970a; 1971a;1971b; 1972) resulting from rapid microelectrophoretic additions (Kohen et al. 1970a) of metabolites (i.e.,glucose-6-phosphate) to the cell cytoplasm or nucleus. The maximization of the signal-to-noise ratio in microfluorometric determinations can best be achieved through the use of an optimum choice of the fluorescence excitation source, optical filtering system and fluorescence detection system (Kohen et al. 1970a; 1971a; 1971b; 1972; Chance and Legallais 1959; Kaufman et al. 1971; Ploem 1971; Zatzick 1970; Jones et al. 1971) in addition to the properties of the fluorochrome *per se* (i.e. for reduced pyridine nucleotides high quantum yield when bound to proteins (Estabrook 1962),universal distribution in animal and plant cells, specific functions in various cell compartments (Kohen 1964) or metabolic pathways, association with close to 100 dehydrogenases (Sund 1968) etc.).

In recent years rapid microfluorometric studies (Kohen et al. 1970a; 1971a; 1971b; 1972) with a time resolution of around 10 milliseconds have utilized an optimized photon counting technique (Zatzick 1970; Jones et al. 1971) whose practical performance can be brought to the limits of theoretical predictions. While changes in the redox potential of NADH and NADPH can be easily detected in 3 to 5 μm regions of various tissue culture cells by such techniques, further technical improvements are required for the exploration of other intracellular fluorochromes; e.g., flavoproteins (Chance and Graham 1971), thiamin (Udenfriend 1969) or porphyrins (Runge 1966). Another possibility concerns the introduction of exogenous fluorescent or luminescent probes for the detection of changes in charge or water structure of membranes (Chance 1970) as well as for the detection of various ions (Shimomura et al. 1963) and their intracellular distribution. For all these compounds, quantitative fluorescence studies have to take into account changes in the quantum yields of the fluorescence which may be influenced by conformation and the microenvironment surrounding the fluorescent molecules (Radda 1973). Thus, quantitative microfluorometry offers an expanding field for the study of intracellular reactions, with special reference to the intact living cell, at a level of spatial and temporal resolution so far unattained by other techniques.

MATERIALS AND METHODS

Cell Types, Microelectrophoresis

Because it requires cell immobilization, the microelectrophoretic technique for the intracellular addition of metabolites necessitates the use of glass-grown tissue culture cells (Kohen 1964). Within this limitation, a large variety of mammalian cell cultures have thus far been used; e.g.,ascites (EL2 clone), human liver and conjunctiva, human astrocytoma, L cells and Chinese hamster fibroblasts.

The microelectrophoretic additions of metabolites into the living cell can be regulated in two ways (Kohen et al. 1970a; 1972): either through an adjustable voltage (0-22.5 volts) or through an interval timer allowing the delivery of metabolites within as little as 25 milliseconds (effective mixing time at 15-20 μm distant cellular sites, from 50 to 80 milliseconds). The timer also triggers a marker, which indicates the exact time and duration of microelectrophoretic additions on the same chart paper as that on which fluorescence changes are recorded.

Fluorescence Standards

Non-NADH Standards. To correct for changes in the excitation intensity, the fluorescence standard can be a material emitting in approximately the same wavelength as reduced pyridine nucleotides, such as MY^4PO (1-methyl-4-(5-phenyl-2-oxazolyl)-pyridinium p-toluene-sulfonate) synthesized by Ott, Hays and Kerr (1956).

NADH or NADPH Standards. Capillaries filled with NADH or NADPH can also be used as fluorescence standards. Such standards, however, will not permit exact quantitation, since it is not easy to extrapolate from their fluorescence emission to the emission of a given concentration of pyridine nucleotides within a localized cell region. The quantum yields of the free pyridine nucleotides in solution will be quite different (and generally lower) than their quantum yield within the living cell (Estabrook 1962). The latter conditions can be roughly simulated by the use of standards made of protein-bound (*e.g.*, albumin-bound) pyridine nucleotides.

Microfluorometers

To monitor NAD (NADP)-linked dehydrogenases, fluorescence excitation at 366 nm is provided by a mercury arc and Corning CS 7-54 + Schott UG 1 filters (Kohen et al. 1970a; 1971a; 1971b; 1972; Chance and Legallais 1959) (Figure 1). A Leitz Ultropak system (55x water-immersion

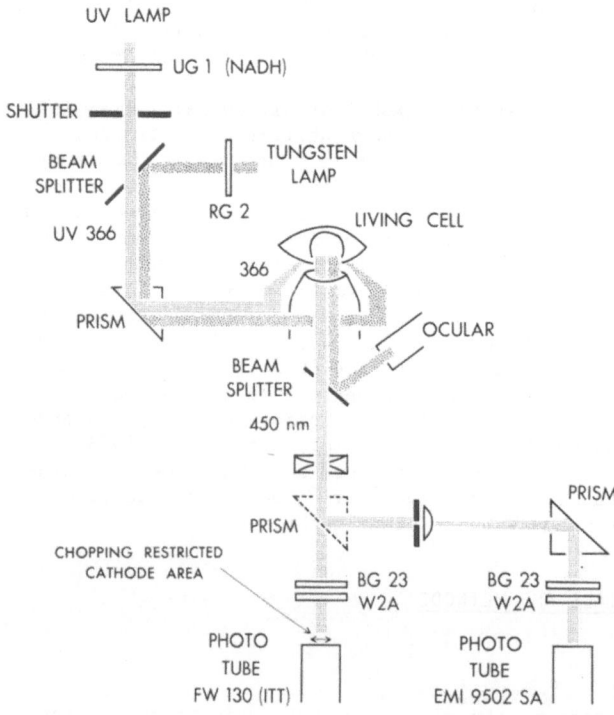

Figure 1. *Schematic diagram of equipment used for single-and two-channel microfluorometry. In both systems a beam splitter allows simultaneous illumination with red light for visual observations in the microscope and ultraviolet light for fluorescence excitation. A dichromatic filter reflects the red light towards the ocular, while the blue fluorescence emission is transmitted towards the phototube. An EM1 9502 SA phototube is used for single-channel work. The cell area under observation is limited by the size of the orifice (0.85 mm/165 x magnification) in a copper plate placed at the image plane of the microscope.*

For two-channel work the prism reflecting the fluorescence emission towards the camera aperture of the microscope is removed and the image of the cell structures under observation is directly focused on the chopping-restricted sensitive cathode area of an ITT FW 130 phototube. The diameter of the cell region under observation is defined by the size of the restricted cathode area (0.75 mm circle/70 x magnification) objective with ring condenser), for incident-light dark field illumination,mounted on an inverted Unitron metallurgical microscope leaves the top of the stage free (Kohen et al. 1969) as required for the microelectrophoretic introduction of specific metabolites (glucose-6-phosphate, malate, adenine nucleotides) in controlled amounts to the cell cytoplasm or nucleus. A beam-splitter system (Filtraflex-DC, Balzers) allows cell visualization with red light, while the blue fluorescence emission (450 nm) is transmitted towards the photocathode protected from the 405 nm mercury band and residual red light leakage by Wratten 2E and Zeiss BG 23 filters, respectively (Kohen et al. 1970). For single-channel microfluorometry, the cell region (d = 5 μm) to be observed is delimited by a 0.85 mm diaphragm in a copper plate placed at the camera aperture of the Unitron (Kohen et al. 1970a; 1971a). Photodetection by means of an "optimized photon counting technique" (Zatzick 1970; Jones et al. 1971) is based on (a) an EMl 9502 SA photomultiplier operated at a plateau voltage at which the photon count is zero order independent of gain; (b) a 200-MHz amplifier discriminator which can differentiate spurious bursts from genuine photoelectron bursts and (c) a multifunctional digital synchronous computer, followed by a digital-to-analog converter (Zatzick 1970). This technique permits the recording of intracellular fluorescence changes within a few milliseconds.

Thus, kinetic studies can be made on enzyme reactions in single living cells. NAD or NADP reduction-reoxidation transients (Kohen et al. 1970a; 1971a; 1971b; 1972) resulting from the microelectrophoretic addition of metabolites can be used to observe bottlenecks along enzymatic sequences, mitochondrial-extramitochondrial interactions and changes in the redox state of pyridine nucleotides following metabolic preloading. Small displacements in the NAD-NADH (or NADP-NADPH) equilibrium can be detected against a high cell-fluorescence background by operating the digital computer on a "background-subtract" mode.

The main transient parameters (Kohen et al. 1970a; 1971a; 1971b; 1972) (Figure 2) that can be analyzed in the fluorescence pulse correspond to: the lag which precedes any change in fluorescence, the rise half time $t_{1/2\ on}$, the rise time T, the time of rise and half decay $t_{1/2\ off}$, the reoxidation time RT (or the time it takes from maximal rise until the return to the original level), the amplitude Δ F. The rate of substrate utilization can be roughly evaluated as the quotient of the added substrate over $t_{1/2\ off}$. A time study on the onset of metabolic response after properly timed substrate applications at different sites can be used to evaluate the transfer time of metabolites across the nuclear or mitochondrial membranes (from around 35 msec for the former, up to several seconds for the latter).

Two-Channel Rapid Microfluorometry based on an electronic chopper, photon counter equipment has been developed for the study of two cellular sites or parameters simultaneously (Figure 1). For this purpose, an ITT FW 130 16-stage multiplier phototube is used with an end-window-type photocathode of restricted area (0.7mm circle) exhibiting a calibrated S20 spectral response. With the magnification used, the restricted sensitive area represents a 5 μm diameter region in the microscope field. Electronic chopping at an adjustable frequency (e.g.,0.1-1 KHz) is provided by a function generator which imposes a translational (nonsinusoidal) displacement of the restricted sensitive area along x and y coordinates. The restricted sensitive area alternates between two extreme positions, A and B, and spends very little time in between during a single cycle. Before starting the translational vibration, the resting position can be optimized along x and y coordinates by an external manual control. The fluorescence signals are gated in such a way that when a multifunctional photon counter is operated on the chopping mode, photon counts coming from one of the translational positions will be integrated in one channel, while those coming from the other position will be integrated in the other channel.

Microspectrofluorometry

The main problems in setting the limitations in microspectrofluorimetric studies of NADH or NADPH fluorescence emission are related to the difficulty of obtaining satisfactory emission intensities from small areas of tissue without irreversible damage being caused by the high excitation intensities.

A multichannel approach (Karasek 1972), operating with N separate detectors to observe N spectral elements, can reduce by a factor of N the time needed to achieve a

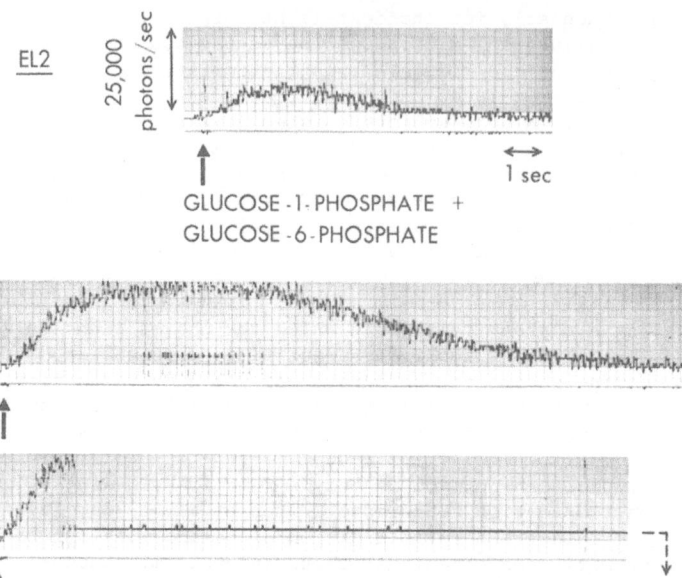

Figure 2. *Typical microfluorometric recordings obtained from the extramitochondrial space of an EL2 ascites cell with increasing durations of the microelectrophoretic current. (Metabolite added = glucose-6-phosphate + glucose-1-phosphate). Prior to microelectrophoretic additions of metabolites, the baseline was zeroed by entering the inital fluorescence of the cell in the background channel of a photon counter operated on a "background substract" mode. The time and duration of microelectrophoretic additions are indicated by the marker trace. Cell FEB 16-72/GIP G6P E10. Each arrow points to the start of a microelectrophoretic addition. The microelectrophoretic current was on for 100, 250, and 500 msec respectively. Increasing durations of the current correspond to increasing amounts of total substrate added (and consequently increasing durations of the transients as noticed). Two frames are required to show the last transient in full. When the increased fluorescence emission goes beyond 25000 photons/sec, the photon counter shifts automatically to a 10 times less sensitive scale (see last 500-msec addition). The fluorescence trace returns subsequently to the "25000 photons/sec" scale, in the decay (reoxidation) phase of the transient*

given signal-to-noise measurement. For the light levels encountered in microfluorimetric measurements, these conditions can be best realized by the incorporation within the microfluorimetric system of an "optical multichannel analyzer" composed of a silicon vidicon coupled to an appropriate image intensifier (Karasek 1972), a spectrometer, a computer console with two memories and a D/A converter for strip chart recording. The signal-to-noise ratio for the silicon vidicon system is provided by the formula:

$$\frac{S}{N} = \frac{P \times Q(\lambda) \times t}{N_c \sqrt{n}}$$

S is the signal, N the noise, P the photons/channel/sec, $Q(\lambda)$ the quantum efficiency, t the integration time, n the number of scans, N_c = noise/channel element/frame scan time.

RESULTS

The Sensitivity of Quantitative Fluorescence Studies on Pyridine Nucleotides in Cell Compartments

For quantitative studies of changes in the fluorescence emission of reduced pyridine nucleotides, the concentration of these nucleotides in localized cell regions must fall within the region of proportionality between fluorescence and concentration (Bowen and Wokes 1953). This sets an upper limit to the acceptable concentration of the fluorochrome, since fluorescence quenching will increase at higher concentrations. In the case of a living cell, two factors have to be taken into consideration: (a) the concentration of reduced pyridine nucleotides and (b) the thickness of the living cell (which in other fluorometric systems would have corresponded to the thickness of the microcuvette). The maximum concentration of reduced pyridine nucleotides in cell compartments must be such that no more than 5% of the exciting radiation will be absorbed. From Lambert and Beer's law (Bauman 1962):

$$log \frac{I_0}{I} = A = abc$$

(A = absorbance, a = molar absorption coefficient, b = optical path length, c = molar concentration), it is possible to calculate the absorbance at a specific NADH concentration in a cell region of given thickness. For NADH, the molar absorption coefficient (Kohen et al. 1970b) at 340 nm is 6.2×10^3. Independent assays from cell extracts (Borst and Colpa-Boonstra 1962) microspectrophotometric observations (Ritter and Thorell 1970) as well as calibration with a capillary standard of NADH set an upper limit of around 5 millimolar for the NADH concentration in cell structures. Actual concentrations, however, especially in the case of estimates made by fluorescence standard calibrations, may be much lower in view of a 10-12 times enhancement of the quantum yield, by binding to proteins or dehydrogenase enzymes (Estabrook 1962). Under these conditions, $log_{10} \frac{I_0}{I}$ would be roughly equal to 0.01-0.05 (i.e. $\sim$ 2.5-10% absorption of incident light). While such values are at (or above) the upper limit for which proportionality may be expected between fluorescence and concentration, a correction factor must be introduced since the actual excitation wavelength (Velick 1961) is the 366 nm emission line of a mercury arc rather than 340 nm (generally smaller absorptivity at 366 nm). Furthermore, the possibility that the actual NADH concentrations are smaller than estimated has been indicated above.

Therefore, it can be safely assumed that fluorescence changes in a localized cell compartment will indeed reflect genuine and proportional changes in the concentration of reduced pyridine nucleotides. This is particularly important when the homeostasis of the living cell is disturbed by the microelectrophoretic imposition of a metabolite pulse or a metabolic transient. Under these conditions it may be expected that a transient change (Kohen et al. 1970a; 1971a; 1971b; 1972) will occur in the concentration of NADH or NADPH or both, depending upon the metabolic state that leads to the appearance of a fluorescence pulse. While there may be some uncertainty as to the absolute concentrations (in view of considerable variations in quantum yield depending upon the conformation and microenvironment (Radda 1973) of the fluorescent molecules the relative changes should be very exact indeed, if we remain within the region of linearity between fluorescence and concentration.

Actual Signal-to-Noise Ratios and the Limits of Temporal Resolution in Rapid Microfluorometry

Total NADH in a cell region of 20 μm^2 (Leitz Ultropak optics) can be rather easily calculated from the above; it falls roughly in the range of 1 to 5×10^{-16} mole. Using the optimized photon counting technique mentioned above, this corresponds to an emission of 50,000 to 100,000 photoelectrons per second from the cathode of an EMI 9502 SA photomultiplier. The statistical fluctuations of photoelectrons on the cathode of the phototube (the "shot noise") can be estimated according to the square-root formula (Kohen et al 1970a; Velick 1961)

$$N = \sqrt{2 \times quanta/sec} \times \Delta f \ [sec^{-1}]$$

where N equals shot noise and Δf is the frequency bandwidth of the measuring system in Hz or $[sec^{-1}]$. If DC measurements are used for fluorescence detection instead of photon counting, the equivalent formula for noise will be:

$$N = \sqrt{2e \ i \ \Delta f}$$

where e is the elementary charge in coulombs, i is the primary photocurrent in amperes, and Δf is again the frequency bandwidth of the measuring system. Under these conditions, intracellular events occuring in a time interval of 10 milliseconds (see Table I for changes with various metabolites) and corresponding to photocathode emissions of 500 to 600 photoelectrons can be detected with a signal-to-noise ratio of 20 to 1. However, in progressing toward the limits of temporal resolution, a critical balance must be established between the fluorescence excitation intensities required and the

TABLE I

INTEGRATED AREAS[*] OF FLUORESCENCE TRANSIENTS

Substrate (pretreatment)	Cell Type				
	EL 2	EL 2 Giant	L	Chang Liver	Chang Conjunctiva
Glucose-1-phosphate	581700		804500		58500
Glucose-1-phosphate (Glycerol, 1-hour)	1056100				
Glucose-6-phosphate	424800	267900	155900	131500	93500
Glucose-6-phosphate (Ethanol, 24 hours)				69500	
6-phosphogluconate	282800	440200	95800		

* Areas in [(photon/sec^2) sec] sec = photons = radiant energy/Planck constant.

tolerance (Chance 1962) of the living cell to such intensities. Therefore, for spectral scan or scanning of finer intracellular compartments ($e.g.$, 1 μm^2) the temporal resolution must be restricted to maintain an acceptable signal-to-noise ratio.

Radiant Power of Reduced Pyridine Nucleotides in Localized Cell Structures

The maximum radiant power (Kaufman et al. 1971) detected at the photocathode from a 20 μm^2 cellular region emitting in the frequency range of 660 MHz (λ = 450 nm) can be calculated from the formula: radiant power = frequency of photon x Planck constant x number of photons per second, before corrections for the quantum efficiency $Q(\lambda)$. For 100,000 photons per second and a quantum efficiency of approximately 16% at the cathode of the EMI 9502 SA, the radiant power will be approximately 2.5 x 10^{-13} watt.

With DC-current measurements the equivalent formula for the fluorescence radiant power ($EMI\ Photomultiplier\ Tubes\ Catalog\ 1967$) is:

$$FP = \frac{i}{e} \cdot h\nu \cdot \frac{1}{Q(\lambda)}$$

where FP equals fluorescence radiant power, i (amperes) is the primary photo current, e is the elementary charge (coulombs), h is the Planck constant, ν is the frequency of the fluorescence emission, and $Q(\lambda)$ is the quantum efficiency of the cathode. For a maximum primary photocurrent of 5 x 10^{-15} amperes, the fluorescence radiant power will correspond to 10^{-13} watts. On the basis of empirical evidence the signal-to-noise ratio is found to be somewhat more favorable in the case of photon counting. A measure of the total number of photons (radiant energy/Planck constant) involved in the integrated area of the fluorescence transient is given in Table I for various experimental conditions.

The Quantitation of Reactants Introduced by Microelectrophoresis

Quantitative fluorescence studies in cell structures also depend on the quantitation of reactants introduced by microelectrophoresis. Theoretical estimates of the amounts (Kohen et al. 1970a) introduced can be made from the Avogadro and Faraday numbers.

Generally the microelectrophoretic current intensity used for an optimal response to metabolite does not exceed 10^{-8} amperes, which corresponds to approximately 10^{-13} equivalent per second.

When a metabolite such as the anionic portion of the disodium salt of glucose-6-phosphate is introduced microelectrophoretically, it is necessary to know the transference number (Edsall and Wyman 1958) of the anion (mobility of anion/mobility of anion + cation). Since the ionic mobility is proportional to the ionic conductance, the transference number can be calculated from conductance determinations. From comparative observations on the conductance of sodium chloride and disodium glucose-6-phosphate solutions, the transference number of glucose-6-phosphate - was roughly estimated to be around 0.15.

It was stated above that 10^{-8} amperes corresponds to 10^{-13} equivalent. Therefore 10^{-8} amperes should correspond to $.15 \times 10^{-13}$ equivalent glucose-6-phosphate or, rounding off, roughly 10^{-14} mole glucose-6-phosphate. It should be indicated that there is still some uncertainty as to the exact correspondence between microelectrophoretic current (Curtis 1964) and mole substrate released into the cell cytoplasm or nucleus. A second factor of uncertainty is the cell volume, as the surface dimensions of the cells investigated can be measured more accurately than can the thickness. However, when microelectrophoretic currents of increasing intensity or duration are applied subsequently to the same cell, the relative changes in concentration of the metabolite introduced can be determined quite accurately. For a cell volume of approximately 10 pl, 10^{-14} mole released in the cytoplasm should correspond to approximately a 1 mM initial concentration.

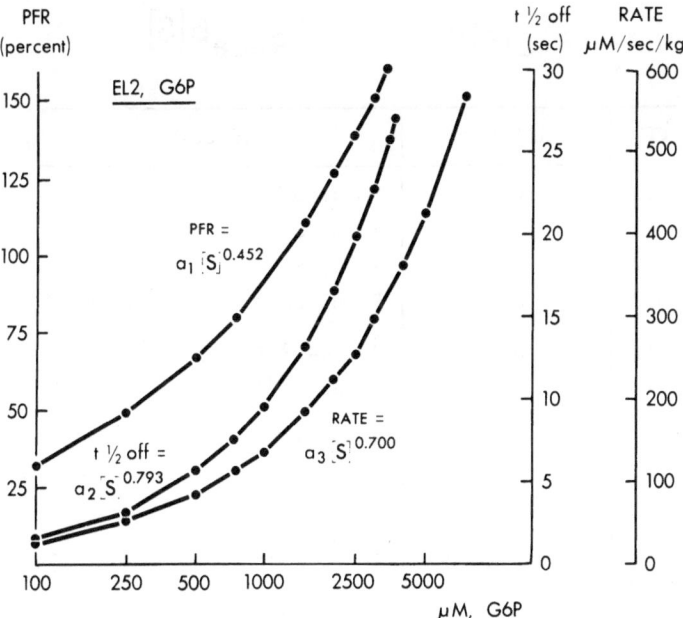

Figure 3. *The plots of three basic parameters (in ordinate) of NAD⁺ reduction-reoxidation transients in EL2 cells (substrate = glucose-6-phosphate), against substrate concentration (in abscissa; logarithmic scale). The parameters shown are the rate of metabolite (glucose-6-phosphate) conversion, the transient amplitude (PFR) and halftime ($t_{1/2off}$). The three curves correspond to power functions and they were obtained by curve fitting in a Hewlett-Packard computing calculator, and by entering consecutive parameter values obtained following repeated metabolite additions in increasing amounts. The relationship between the concentration and the parameter measured is indicated for each curve*

The Relationship of Transient Parameters to Concentration

The actual plots of various transient parameters against substrate concentration have been subjected to curve fitting tests in a Hewlett-Packard computing calculator. With various intermediates of glycolytic metabolism - *e.g.* glucose-6-phosphate, fructose-6-phosphate - the best fit is observed with the power curve (Figure 3, Table II), in which case the parameter will change with concentration according to the formula:

$$P = a \ [S]^b$$

where P represents the parameter, [S] the substrate concentration, and a and b are constants. However, with some substrates, *e.g.* 6-phosphogluconate (in EL2 ascites cells), an exponential relationship is observed according to the formula:

$$P = a \ e^{[S]b}$$

where e is the base for logarithm, [S] is the substrate concentration, and a and b are constants.

TABLE II

THE RELATIONSHIP BETWEEN THE RATE OF METABOLITE UTILIZATION (R) AND SUBSTRATE CONCENTRATION [S]

POWER CURVE $R = a[S]^b$	EXPONENTIAL $R = ce^{b[S]}$

CELL	SUBSTRATE	RATE CURVE	
EL2	G6P	$R = a_1 [S]^{0.70}$	
EL2	F6P	$R = a_2 [S]^{0.73}$	
EL2	FDP	$R = a_3 [S]^{1.80}$	
EL2	GAP	$R = a_4 [S]^{2.95}$	$R = c_1 e^{0.00017[S]}$
EL2	6-PG	$R = c_2 e^{0.00086[S]}$	
L	6-PG	$R = a_5 [S]^{1.15}$	
EL2	G1P	$R = a_6 [S]^{1.42}$	$R = c_3 e^{0.00061[S]}$
L	G1P	$R = c_4 e^{0.0011[S]}$	
EL2	G1P	$R = a_7 [S]^{1.42}$	
EL2	G1P (100) + G1, 6P (1)	$R = a_8 [S]^{1.66}$	
EL2	G1P	$R = a_9 [S]^{1.28}$	
EL2	G1P (4) + UTP (1)	$R = a_{10} [S]^{0.88}$	
EL2	M6P	$R = a_{11} [S]^{1.027}$	

a,b and c = constants, e = base for natural logarithms, G6P = glucose-6-phosphate, F6P = fructose-6-phosphate, FDP = fructose-1,6-diphosphate, GAP = glyceraldehyde 3-phosphate, G1P = glucose-1-phosphate, G1,6P = glucose-1,6-phosphate, UTP = uridine tri-phosphate, M6P = mannose-6-phosphate.

TABLE III

KINETIC PARAMETERS OF NADP$^+$ REDUCTION-TRANSIENTS IN VARIOUS CELL TYPES (Substrate= 6-phosphogluconate)

Cell Type	[S] optimal* (mM)	LAG (msec)	$t_{\frac{1}{2}on}$ (msec)	Rise Time (msec)	ΔF/Rise Time photons/sec^2	ΔF (Per-cent)	$t_{\frac{1}{2}off}$ (sec)	Reoxidation Time (sec)	Rate of Substrate Utilization (µM/sec/Kg)
EL2	20.	390 ± 140	1500 ± 280	4500 ± 990	1650	60	97.1 ± 22.5	115.1 ± 23.1	610
EL2 (Radia-tion Giant)	2.8	1900 ± 1100	3470 ± 1150	15500 ± 6590	400	50	101.9 ± 12.7	128.2 ± 16.5	40
L Cells	3.8	790 ± 360	1670 ± 300	5700 ± 1520	800	35	48.4 ± 18.7	35.8 ± 2.3	130
Chang Liver Cells	NO DETECTABLE RESPONSE								

$t_{\frac{1}{2}on}$ = transient rise half time, $t_{\frac{1}{2}off}$ = transient rise and decay half time, [S] = substrate concentration,

Standard errors according to the formula S.E. = $\pm \sqrt{\Sigma\, d^2/n(n-1)}$

* At substrate concentrations larger than optimal, there was either stabilization of kinetic parameters at a plateau level or signs of inhibition (transients with no return to baseline or minimal ΔF changes over baseline).

TABLE IV

EFFECT OF ANAEROBIOSIS ON THE 6-PHOSPHOGLUCONATE RESPONSE IN GIANT EL2 CELLS (X-RAY PRODUCED 1000r)

Experimental Conditions	[S] optimal (mM)	LAG (msec)	$t_{\frac{1}{2}on}$ (msec)	Rise Time (msec)	ΔF/Rise Time (photons/sec^2)	ΔF (per-cent)	$t_{\frac{1}{2}off}$ (msec)	Reoxidation Time (sec)	Rate of Substrate utilization (µM/sec/Kg)
AEROBIC	2.8	1900 ± 1100	3500 ± 1150	15500 ± 6600	400	50	101.0 ± 12.7	128.2 ± 16.5	27
ANAEROBIC	3.5	840 ± 270	13300 ± 6450	41200 ± 18000	490	160	247.5 ± 78.7	No return to baseline	21

Lags and Delays Along Metabolic Pathways or at the Level of Intracellular Membranes

An important aspect of rapid microfluorometric techniques concerns the determination of delays in early and overall temporal parameters that occur following the additions of metabolites. Typical examples are shown in Tables III and IV. It is noteworthy that longer lag periods (Kohen et al. 1971a; 1971b; 1972) are found for glycolytic intermediates on the far side of an isomerase, in relation to the terminal dehydrogenase coupled to the fluorescence reaction (NAD^+ or $NADP^+$ reduction). The present time resolution of the system makes it possible to determine such delays quite accurately. Furthermore, the delays involved in metabolite transfers across intracellular membranes (Kohen, Siebert, et al. 1971) (*e.g.* nuclear, mitochondrial) fall within the temporal resolution of the system (around 35 milliseconds at the nuclear membrane, up to a second or more at the mitochondrial membrane).

Fluorescence Emission Spectra of Intracellular Fluorochromes

The study of NADH (or NADPH) fluorescence spectra and spectral shifts can yield information concerning the actual status (*e.g.* free vs. bound) of reduced pyridine nucleotides in localized compartments of the living cell. In the formation of the dehydrogenase-NADH complex, generally a shift occurs in excitation and fluorescence maxima toward the ultraviolet (Velick 1969).

In preliminary trials with Ehrlich ascites cells or yeast cells, the first prototype of an optical multichannel analyzer (Karasek 1972) without any image intensifier, was used together with a low magnification (low numerical aperture) objective and a 6-millimeter diameter fiber to carry fluorescent light from the image plane at the camera aperture of the microscope directly onto the vidicon silicon surface. Bypassing the spectrometer, the fluorescence was spread over 200 channel elements. When corrections are made to account for actual emissions from a 5 μm cell region, the maximal counts recorded from individual channels correspond to approximately 1,500 electrons/sec at a quantum efficiency close to one. The noise for the vidicon system is equivalent to 15,000 electrons/sec. Thus, with the low-sensitivity vidicon, the signal-to-noise ratio does not exceed 0.1. However, these pilot experiments were made only to provide an indication as to whether a much more sensitive vidicon could be used for microfluorimetric determinations (in which case the signal would increase by three orders of magnitude, but the quantum efficiency would drop from about 100% to 10%). Under these conditions, the signal-to-noise ratio would be at its worst around 2.5 to 1 (taking into account the image intensifier noise) for an integration time of 1 sec, and a 10:1 ratio could be achieved in less than 10 sec.

DISCUSSION

The quantitative aspects of rapid microfluorimetry establish the main limitations on what is possible at this time. The amounts of reduced pyridine nucleotides (Borst and Colpa-Boonstra 1962, Ritter and Thorell 1970) and their changes in concentration during metabolic events (Kohen et al. 1970a; 1971a; 1971b; 1972) are such that with the present temporal resolution, considerable information can be derived as to the rate of utilization of various metabolites and the changes in the redox potential of intracellular pyridine nucleotides (Kohen et al. 1969; 1971a; 1971b) resulting from such additions. Thus it can be seen that in an enzymatic sequence such as the glycolytic sequence (Embden Meyerhof pathway) (Meyerhof and Wilson 1949) the metabolites on the far side of a known bottleneck in relation to the terminal dehydrogenase are utilized more slowly than the metabolites past the bottleneck (Hess and Brand 1965).

It is possible to get a first glimpse at the possible intracellular status of glycolytic enzymes (Hess 1971) by comparing the concentration dependence of fluorescence pulse parameters obtained with metabolites. In addition to rates of substrate utilization and parameters concerning the accumulation of reduced pyridine nucleotides or their removal by subsequent reoxidation, temporal parameters of two types can be established. It is possible to some extent to measure (a) the lags (Kohen et al. 1970a; 1971b; 1972) along various enzymatic pathways, and (b) the delays which occur at the level of intracellular membranes (Kohen and Siebert, 1971) (*e.g.* mitochondrial or nuclear membranes). Additional details will be given in the next paper.

The present resolution of the microfluorimetric system permits an evaluation of various interactions between intracellular organelles, as well as metabolite-modulated interdependent enzymatic pathways, in correlation with different physiological states of the cell or pathological changes. In principle, at long range the method should increase in versatility through several developments such as adaptation of various fluorescence probes (for the study of membranes, charge or ion distribution) (Chance 1970, Price and Radda 1972) and determination of fluorescence spectra and polarization, preferably by multichannel analysis.

ACKNOWLEDGEMENTS

The authors wish to express their indebtedness to Professor B. Chance, (Johnson Foundation, University of Pennsylvania) for encouragement, fruitful exchange, and his generous loan of instrumental parts incorporated in the present rapid microfluorometer, as well as related microelectrophoretic equipment. The authors are also thankful to Research Engineers Lennart Åkerman (Department of Pathology, Karolinska Institute, Stockholm) and Lars Nordberg (Department of Medical Engineering, Karolinska Institute, Stockholm), to Dr. Michael Zatzick (SSR Instruments Company, Santa Monica, California) and Dr. Robert Leif (Papanicolaou Cancer Research Institute, Miami, Florida) for useful advice and consultations. The drawings were prepared by Mr. G. K. Rietberg.

This work was supported by Grants P-518 and BC-15A (Louise Helene Neussel Memorial Grant for Cancer Research) from the American Cancer Society and Swedish Medical Research Council (B70-12X-630-06).

REFERENCES

BAUMAN, R.P., (1962) *Absorption Spectroscopy*, John Wiley & Sons, New York, 14.

BORST, P. and Colpa-Boonstra, J.P., (1962), *Biochim. Biophys. Acta*, 56, 216.

BOWEN, E.J., and Wokes, F., (1953), *Fluorescence of Solutions*, Longmans, Green, London.

CHANCE, B., and Legallais, V. (1959), *Rev. Sci. Instr.*, 30, 732.

CHANCE, B., (1962), *Ann. N.Y. Acad. Sci.*, 97, 431.

CHANCE, B., (1970), *Proc. Natl. Acad. Sci. U.S.*, 67, 560.

CHANCE, B., and Graham, N., (1971), *Rev. Sci. Instr.*, 42, 941.

CHANCE, B., Lee, C.P., and Blasie, J.K., eds., (1971), *Probes of Structure and Function of Macromolecules and Membranes*, Vol. 1, Academic Press, New York.

CURTIS, D.R., (1964), In: *Phys. Tech. in Biol. Res.*, (Nastuk, W.L., ed.), Vol. 5, Academic Press, New York, London, 144.

EMI Photomultiplier Tubes Catalog (Brochure ref. 3OM/6-67 (PMT) Issue).

EDSALL, J.T., and Wyman, J., (1958), *Biophysical Chemistry*, Vol. 1, Academic Press, New York.

ESTABROOK, R.W., (1962), *Analytical Biochem.*, 4, 231.

HESS, B., (1971), In: *Proceedings First European Biophysics Congress*, (Broda, E., Locker, A., and Springer-Lederer, H., eds.), Vol. 4, Verlag der Wiener Medizinischen Akademie, Vienna, 447.

HESS, B., and Brand, K., (1965), In: *Control of Energy Metabolism*, (Chance, B., Estabrook, R., and Williamson, J.R., eds.), Academic Press, New York, 111.

HEWLETT Packard Calculator, Stat-Pac, Vol. 1, Hewlett Packard Company, Loveland, Colorado.

JONES, R., Oliver, C.J., and Pike, E.R., (1971), *Applied Optics*, 10, 1673.

KARASEK, F.W., (1972), *Research/Development*, 23, 47, 48, 50.

KAUFMAN, G.I., Nester, J.F., and Wasserman, D.E., (1971), *J. Histochem. Cytochem.*, 19, 469.

KOHEN, E., (1964), *Exptl. Cell. Res.*, <u>35</u>, 303.

KOHEN, E., Kohen, C., and Thorell, B., (1969), *Biomedical Engin.*, <u>4</u>, 554.

KOHEN, E., Kohen, C., and Thorell, B., (1970a), *Mikrochimica Acta*, 1970, 1190.

KOHEN, E., Kohen, C., and Thorell, B., (1970b), In: *Sixth International Symposium on Microtechniques*, Vol. C, D, Verlag der Wiener Medizinischen Akademie, Vienna, 57.

KOHEN, E., Kohen, C., and Thorell, B., (1971a), *Biochim. Biophys. Acta*, <u>234</u>, 531.

KOHEN, E., Kohen, C., and Thorell, B., (1971b), In: *Proceedings First European Biophysics Congress*, (Broda, E., Locker, A., and Springer-Lederer, H., eds.), Vol. 4, Verlag der Wiener Medizinischen Akademie, Vienna, 465.

KOHEN, E., Kohen, C., and Thorell, B., (1972), *Mikrochimica Acta*, 1972, 103.

KOHEN, E., Siebert, G., and Kohen, C., (1971), *Hoppe Seyler's Z. Physiol. Chemie*, <u>352</u>, 927.

MEYERHOF, O., and Wilson, J.R., (1949), *Arch. Biochem.*, <u>21</u>, 22.

OTT, D.G., Hayes, F.N., and Kerr, V.N., (1956), *J. Am. Chem. Soc.*, <u>78</u>, 1941.

PLOEM, J.S., (1971), *Ann. N.Y. Acad. Sci.*, <u>177</u>, 414.

PRICE, N.C., and Radda, G.K., (1972), In: *Wenner-Gren Symposium on Structure and Function of Oxidation-Reduction Enzymes*, Wenner-Gren Center, Stockholm, August 23 - 27, 1970, (Akeson, A. and Ehrenberg, A., eds.), Pergamon Press Ltd., Oxford, 161.

RITTER, C. and Thorell, B., (1970), *J. Histochem. Cytochem.*, <u>18</u>, 49.

RUNGE, W.J., (1966), *Science*, <u>151</u>, 1499.

SHIMOMURA, O., Johnson, S.H., and Seige, Y., (1963), *Science*, <u>140</u>, 1339

SUND, H., (1968), In: *Biological Oxidations*, (Singer, T.P., ed.), Interscience Publishers, New York, 603.

UDENFRIEND, S., (1969), *Fluorescence Assay in Biology and Medicine*, Vol. 2, Academic Press, New York, 292.

VELICK, S.F., (1961), *Light and Life*, (McElroy, W.D., and Glass, B., eds.), John Hopkins University Press, Baltimore, 108.

ZATZICK, M.R., (1970), *Research/Development*, <u>21</u>, 16.

RAPID MICROFLUOROMETRY FOR BIOCHEMISTRY OF THE LIVING CELL IN CORRELATION WITH CYTOMORPHOLOGY AND TRANSPORT PHENOMENA

Elli Kohen,[*]   Bo Thorell,[**]   Cahide Kohen,[***]   and Moritz Michaelis[****]

[*] *Papanicolaou Cancer Research Institute, Miami, Florida 33/36.*

[**] *Department of Pathology, Karolinska Institute, Stockholm, Sweden.*

[***] *Clinical Faculty, Department of Pathology, University of Miami, School of Medicine, Miami, Florida.*

[****] *Ophtalmology Research Laboratory, University of Maryland, School of Medicine, Baltimore, Maryland*

INTRODUCTION

The living cell represents the ultimate biochemical system, the complexity of which is based on a structural microarchitecture forming compartments for a great variety of reacting substances (Chance 1963). For the purpose of description, more or less complicated "cell models" can be constructed (Zeigler and Weinberg 1970). Such models, however, say little about the dynamic state of the living cell, in which microcompartments containing pools of various chemicals form and disappear continuously. To probe this multireaction system, one obviously needs methods which can qualitatively and quantitatively analyze amounts and states of substances within very small and localized portions of the functioning cell.

A detailed and dynamic picture of intracellular carbohydrate metabolism can be obtained by monitoring the blue fluorescence of reduced pyridine nucleotides (Chance and Thorell 1959; Chance and Baltscheffsky 1958), in correlation with the activities of NAD- or NADP-linked dehydrogenases (Sund 1968), associated with various pathways. The fluorescence pulse, which corresponds to coenzyme reduction-reoxidation transients resulting from the intracellular addition of substrates or intermediates, provides a quantitative basis for the evaluation of the conversion rates (Kohen et al. 1970a; 1971a; 1971b; 1972). By the use of different substrates as well as cofactors and inhibitors, various pathways can be mapped under different functional and environmental conditions.

MATERIALS AND METHODS

Various studies have been carried out by Kohen et al. (1970a; 1970c; 1971a; 1971b; 1972) describing: the Ultropak darkfield system and the beam-splitter system for cell manipulations simultaneously with fluorescence determinations; the optimized photon counting technique for microfluorimetric observations at a time resolution of 10 milliseconds; and the microelectrophoretic technique for the intracytoplasmic or nuclear additions of metabolites (with microelectrophoretic ejection times as short as 25 milliseconds and effective mixing times at distant cellular sites in the range of 80 milliseconds). The quantitative aspects, the signal-to-noise ratio limitations involved, and the accuracy of determinations, as well as the various temporal or intensity parameters of the fluorescence emission have been discussed separately in this conference (Kohen et al. 1973).

For these experiments a variety of cell types at different functional states were used: *i.e.*, EL2 ascites cancer cells, L cells, Chang liver cells and Chang human conjunctiva cells.

Microfluorimetric observations can be carried out at three levels: single-site observations on a 5 µm cell region (single-channel microfluorometry (Kohen et al. 1970a; 1970c; 1971a; 1971b; 1972), simultaneously on two cell sites (2-channel microfluorometry) (Kohen et al. 1973), or with the determination of fluorescence spectra (microspectrofluorometry) (Kohen et al. 1973). By the use of substrate amounts at physiological levels, displacements are obtained of the NAD-NADH or NADP-NADPH equilibrium (metabolic transients) lasting for several seconds (Kohen et al. 1970a; 1970c; 1971a; 1971b; 1972). Moreover, the endocellular pattern of carbohydrate metabolism can be readily and characteristically changed by drug adaptation (Kohen et al.1971c),

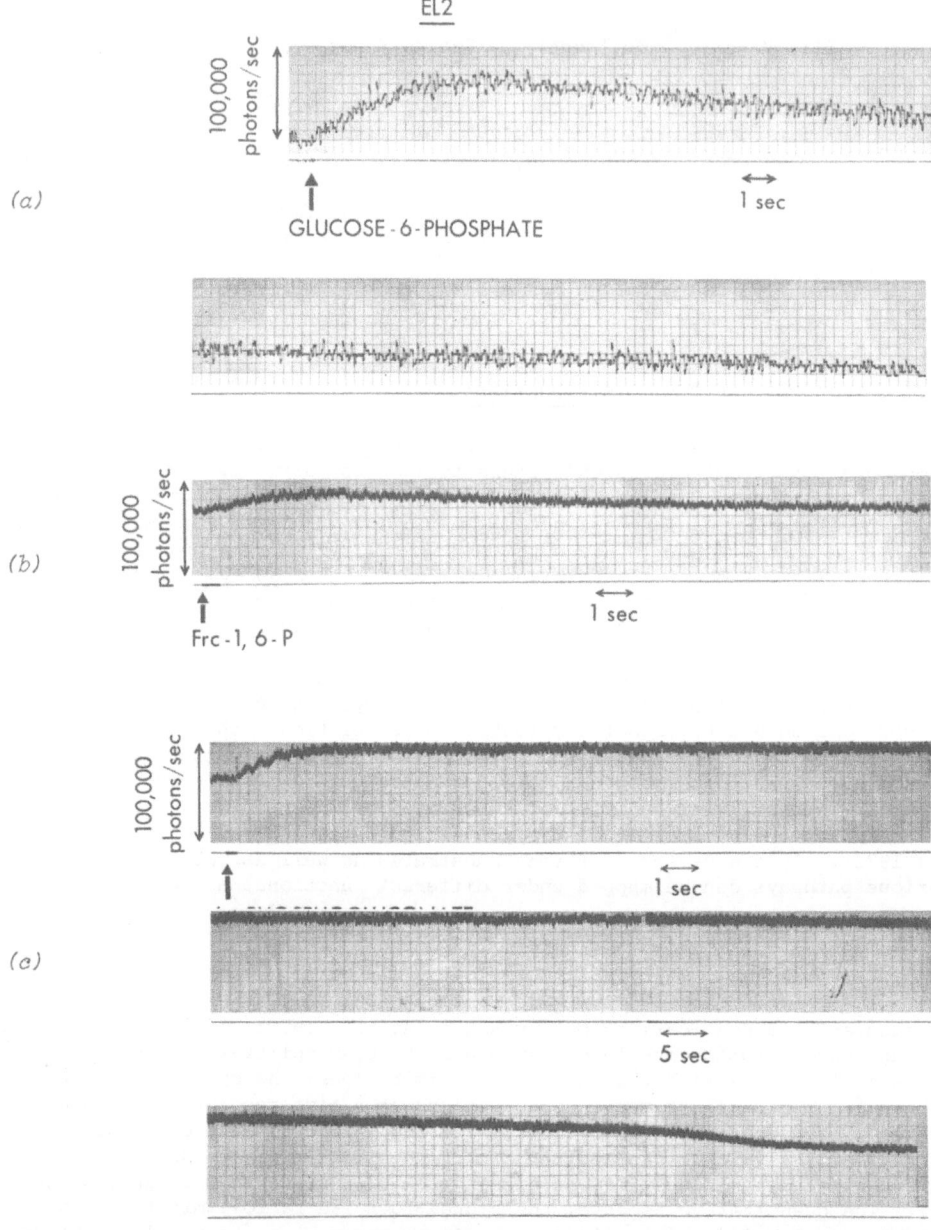

Figure 1. *Characteristic microfluorometric recordings obtained from the extra-mitochondrial space of EL2 cells, following the microelectrophoretic addition of three different metabolites: (a) a "multipotential" metabolite, glucose-6-phosphate which can be catabolized along various pathways (e.g., Embden Meyerhof glycolytic chain, hexose monophosphate shunt, etc.); (b) an intermediate of the Embden Meyerhof chain, fructose-1,6-diphosphate (frc-1,6-P); (c) a substrate of the hexose monophosphate shunt, 6-phosphogluconate. The time and duration of the microelectrophoretic additions are indicated by the marker trace. The time scale proceeds from left to right. Photon counts are corrected for background. It is noteworthy that the longest reoxidation times and transient duration are observed with 6-phosphogluconate, the lowest transient amplitudes and duration with fructose-1,6-diphosphate, and an intermediate transient duration with glucose-6-phosphate. Cells FEB 16-72/GPC2, APR 6-71)FDPE', and APR 1-71/6PGD'*

hormones (Kohen et al. 1970b) or simply by temperature (Kohen et al. 1971b).

In the interpretation of NAD or NADP reduction-reoxidation transients, the meaning of each temporal or intensity parameter has already been defined in the paper on quantitation in microfluorometry (Kohen et al. 1973). Each point in the fluorescence transient corresponds to the algebraic sum of all the reactions controlling the NAD or NADP reduction and reoxidation (Kohen et al. 1970a; 1970c; 1971a; 1971b; 1972). It seems that in the initial stage, reduction is predominant and thus a rapid increase in fluorescence is obtained, but following the rise halftime $t_{1/2}$ on, a certain slowing occurs in the rate of increase of fluorescence, as the reoxidative component starts having a larger contribution. Generally in most fluorescence pulses, the rise time is significantly shorter than the reoxidation time (Kohen et al. 1970a; 1970c; 1971a; 1971b; 1972) (Figure 1).

The rate of reduction can be estimated from the rise slope amplitude ΔF/rise time T, the rate of reoxidation from ΔF/reoxidation time = ΔF/RT. From the time of rise and half decay $t_{1/2}$ off, the rate of substrate utilization can be calculated as the quotient of added substrates over $t_{1/2}$ off.

RESULTS

Change in the Redox Potential of Intracellular Pyridine Nucleotides

Any change introduced by the addition of metabolites to the cell cytoplasm or nucleus corresponds to a disturbance in the homeostasis of the living cell. In EL2 ascites cells, under standardized conditions, the rate of NAD reduction with glucose-6-phosphate exhibits a remarkable constancy (Kohen, Siebert et al. 1972) and it is seven to eight times faster than reoxidation. An easy way to alter the transient displacements of the NAD-NADH (or NADP-NADPH) equilibrium is through the simultaneous introduction of other substrates or inhibitors capable of affecting the removal of reducing equivalents accumulated in the early phase of the transient. Lactate (Kohen et al. 1970a) can considerably slow down reoxidation by poising the equilibrium at the LDH reaction towards the side of reduction. One of the most responsive systems to lactate is found in the Chang human conjunctiva cells. Under these conditions, the area integrated under the fluorescence curve increases considerably, and thus the overall accumulation of reduced pyridine nucleotides can be 10 times as great as with glucose-6-phosphate alone.

An alternative way of acting on transient parameters is through "metabolic preloading", for instance, preincubating the cells for a short time with glycerol, xylitol or ethanol (Kohen, Siebert et al. 1972; Krebs 1968). The NAD-NADH balance shifts towards the reduction side (Kohen, Siebert et al. 1972; Krebs 1968; Krebs and Veech 1969) and there is both a considerable prolongation of all the temporal parameters of the fluorescence transient, and an increase in the maximum intensity. Longer incubations with ethanol (Krebs 1967) (24 hours instead of 1 hour) lead to "adaptive" changes (Orrenius 1965; Kenney 1970) (Table I).

Studies along Enzymatic Sequences or Side Chains: Lags, Bottlenecks, Rates at Various Enzymatic Steps

In the glycolytic sequence (Meyerhof 1949) (Table II) a considerable lag precedes the initation of the transient fluorescence changes following the addition of glucose-1-phosphate (from 700 milliseconds up to several seconds) (Kohen et al. 1970c; 1971a; 1971d; Kohen, Siebert et al. 1972). The lag is much shorter for the intermediates immediately next in the sequence (from 100 to 200 milliseconds; i.e., for all three of glucose-6-phosphate, fructose-6-phosphate, and fructose-1,6-diphosphate). The shortest lag is found with glyceraldehyde phosphate. There is no evidence of a significant difference in lag between fructose-6-phosphate and fructose-1,6-diphosphate. The full rise time can be one second (or more) longer for fructose-6-phosphate than it is for fructose-1-diphosphate, as the latter is past a known bottleneck at the phosphofructokinase.

From rates of substrate utilization of glycolytic metabolites, there is rather strong evidence for a bottleneck at the phosphofructokinase (Lardy 1965), this is most likely a result of allosteric control by activators and inhibitors. In the glycolytic chain, the maximal rates of intermediates prior to phosphofructokinase are all in the range of 100-300 µmole/sec/kg (Table II), while past the same enzyme the rates of sub-

TABLE I

KINETICS OF NAD$^+$ (NADP$^+$) REDUCTION-REOXIDATION TRANSIENTS IN
ETHANOL-GROWN (24 HOURS) CHANG LIVER TISSUE CULTURE CELLS AND CONTROLS (SUBSTRATE=Glucose-6-phosphate)

Treatment	[S] optimal* (mM)	LAG (msec)	$t_{1/2}$ on (msec)	Initial Fluorescence (Photons /sec)	$\frac{1}{2}$ ΔF/ $t_{1/2}$ on (Photons /sec^2)	Rise Time (msec)	ΔF/Rise Time (Photons /sec^2)	ΔF (per-cent)	$t_{1/2}$ off (sec)	Reoxidation T (sec)	Rate of Substrate utilization(μM/ sec/Kg)
Ethanol-grown (6mM)	2.9	No lag	390 ± 80	2600	8700	1150 ± 180	5950	260	9.6 ± 1.7	19.4 ± 2.8	400
Control	8.7	No lag	550 ± 50	4850	6500	2500 ± 500	2850	145	22.1 ± 3.1	34.4 ± 4.3	350

* Optimal concentrations were those at which a maximal ΔF was obtained and kinetic parameters (P) followed a power formula (P = a[S]^b). At larger substrate concentrations the parameters did not fit any regular formula and there was stabilization at a plateau level or signs of inhibition (transients with no return to baseline or minimal ΔF changes over baseline).

TABLE II

KINETIC PARAMETERS OF NAD+ (NADP+) REDUCTION-REOXIDATION TRANSIENTS IN EL2 CELLS

Metabolites	[S] optimal** (mM)	LAG (msec)	$t_{1/2}$ on (msec)	Rise Time (sec)	Δ F/Rise Time (photons/sec²)	$t_{1/2}$ off (sec)	Reoxidation Time (sec)	Rate of utilization (µM/sec/Kg)	n
Glc-1-P	4.0	700±130	2650±400	13900±4200	740	58.7±23.7	99.2±37.8	180	14
Glc-6-P	3.2	250±50	1500±250	5800±500	3400	24.1±2.9	37.6±6.0	315	15
Frc-6-P	1.5	110±30	850±200	5650±750	3100	23.0±3.1	31.8±3.6	110	5
Frc-1,6-P*	16.0	160±70	1000±200	2200±260	2700	12.1±1.5	22.8±4.2	2600	10
Glyceraldehyde* Phosphate	30.0	10±10	500±110	1150±340	4100	3.8±1.3	7.5±2.9	16130	7

$t_{1/2}$ on = transient rise half time, $t_{1/2}$ off = transient rise and decay half time, F = fluorescence pulse amplitude, n = number of determinations, standard errors according to the formula: S.E. = $\pm\sqrt{\Sigma} \, d^2/n(n-1)$.

* For all metabolites except those indicated by a *, the kinetic parameters were normalized according to a power formula of the type parameter = a [S]b calculated separately for each individual cell (whenever multiple additions could be repeated on a same cell). Whenever repeated additions were not possible, values were averaged from various cells. Anaerobiosis (plus incubation with 6mM ethionine (Farber 1963) to lower intracellular ATP) was required to detect transients with fructose-1,6-diphosphate, and glyceraldehyde phosphate.

** Optimal concentrations as in Table I.

Glc-1-P = glucose-1-phosphate, Glc-6-P = glucose-6-phosphate, Frc-6-P = fructose-6-phosphate, Frc-1,6-P = fructose-1,6-diphosphate.

strate utilization of fructose-1,6-diphosphate and 3-glyceraldehyde phosphate are in the range of 2,500-15,000 μmole/sec/kg. The sharp discrepancies between the transient times, the rates of substrate utilization, and the metabolite-concentration dependence at various steps of the glycolytic chain, are not in favor of a cooperative activity of glycolytic enzymes within a multienzyme complex; such a cooperative activity might have occurred in hypothetical superstructures, e.g.,"glycolytic bodies" (de Duve 1970; Hess 1971).

Microfluorimetric Indentification of Various Intracellular Pathways

When uridine-diphospho-glucose (UDPG) (Kalckar 1953) is added to EL2 or mouse L cells (anaerobic), it is either channeled towards glycogen storage, metabolized along the UDPG dehydrogenase pathway (Strominger et al. 1957) or converted to glucose-1-phosphate. In the first instance no NAD-NADH change is involved. In the case of glucose-1-phosphate formation, a lag similar to the above-observed glucose-1-phosphate lag should be seen. Finally in the case of catabolism along the UDPG dehydrogenase pathway, the lag would be unpredictable. In actuality, under the environmental and functional conditions present, channeling towards glycogen seems favored, since in the majority of EL2 or L cells, no NAD-NADH change follows the addition of UDPG. However, in a few of the L cells, a transient (NAD$^+$ reduction-reoxidation) is observed, preceded by a lag similar to that described for glucose-1-phosphate (Table III).

TABLE III

KINETIC PARAMETERS OF NAD (NADP) TRANSIENTS WITH GLUCOSE-1-PHOSPHATE[*] AND URIDINE-DIPHOSPHOGLUCOSE[*] IN L CELLS

Substrate	[S] optimal (mM)	LAG (msec)	$t_{\frac{1}{2}on}$ (msec)	Rise Time (msec)	ΔF /Rise Time (Photons /sec^2)	ΔF (per-cent)	$t_{\frac{1}{2}off}$ (sec)	Reoxidation T (sec)	Rate of utilization (μM/ sec/Kg)
Glucose-1-phosphate	5.6	1250 ±175	1950 ±290	6650 ±1300	1700	90	49.2 ±12.9	43.9 ±12.8	210
Uridine phospho-glucose	22.4	970 ±215	1350 ±180	3800 ±750	6350	195	25.0 ±4.2	55.1 ±17.2	1310

* Both in presence of uridine diphosphate (glucose-1-phosphate or uridine-diphosphoglucose/uridine diphosphate = 3/1).

Change of Fluorescence Pulse Parameters With Metabolite Concentrations

Information concerning direct in-situ activities and, hopefully, configuration of intracellular enzymes can be derived to some extent from the mathematical relationships between metabolite concentrations and various parameters of the fluorescence pulse. The parameters recorded from individual cells subjected to repeated additions of metabolites in gradually increasing amounts are submitted to various curve fitting tests in a computing calculator (Kohen et al. 1973). Generally the closest fits are with power curves of the type:

$$P = a \ [S]^b$$

where P is the parameter considered and $[S]$ is the substrate concentration, and a and b are constants. As far as rates of substrate utilization are concerned, a characteristic exponent is found for each intermediate of the glycolytic chain. However, there is always the possibility that when the added metabolite is several enzymatic steps away from the dehydrogenase involved in the reduction of glycolytic NAD$^+$, the true relationships involved at the fluorescence-linked step will be obscured by the interposed enzymatic steps. Thus the smallest exponents are found for intermediates of the glycolytic chain farthest away from the terminal dehydrogenase step (Velick 1958).

For both glucose-6-phosphate and fructose-6-phosphate the exponent (Kohen et al. 1973) is around 0.7, which means that the rate of substrate utilization for these metabolites increases more slowly than the metabolite concentration. Past the bottleneck at the phosphofructokinase (Lardy 1965), the exponent increases considerably (Kohen et al. 1973); i.e., 1.80 for fructose-1,6-phosphate and around 3.00 for glyceraldehyde phosphate, suggesting allostery.

The Possible Compartmentalization of Phosphorylated Carbohydrate Esters

Different and functionally distinct pools of phosphorylated carbohydrate esters (Kalant and Beitner 1971), *e.g.*, glucose-6- and -1-phosphate, are suggested by the much faster change in the rate of glucose-1-phosphate utilization with concentration. These observations could mean that glucose-1-phosphate may proceed along a different pathway than glucose-6-phosphate, and that the latter is not an obligatory intermediate in glucose oxidation (Kalant and Beitner 1971). The utilization of glucose-1-phosphate is further intensified when glucose-1,6-phosphate, the coenzyme for phosphoglucomutase is added (LeLoir 1951).

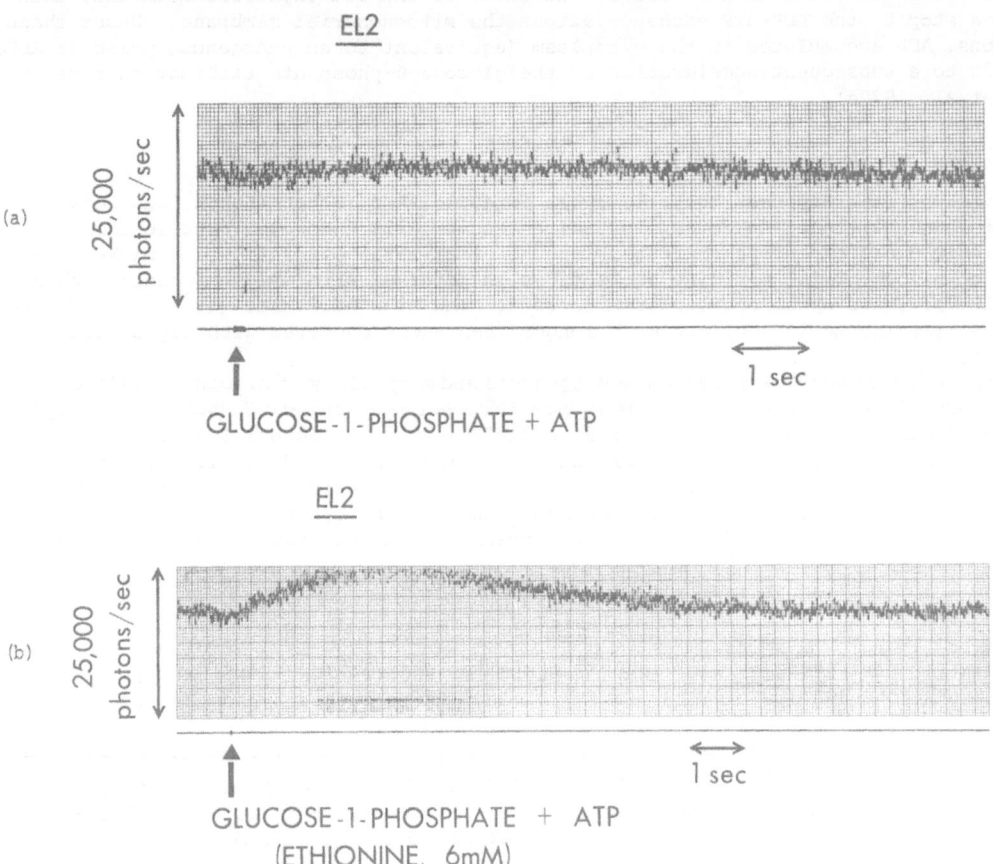

Figure 2. *Microfluorometric recordings from the extramitochondrial region of EL2 cells, following microelectrophoretic addition of glucose-1-phosphate + ATP(10:1). Conditions are as in Figure 1. The total lack of response to glucose-1-phosphate + ATP (Figure 2A), and the restoration of the transient following ethionine (Figure 2B) are noteworthy. Cells JAN 6-72/GIPATP and JAN 6-72/GIPATPAA2. When the fluorescence emission goes beyond 25,000 photons/sec (see Figure 2B) the photon counter shifts automatically to a 10-times-less-sensitive scale. The fluorescence trace returns to the "25,000 photons/sec" scale, in the decay phase (reoxidation) of the transient*

The Effect of Adenine Nucleotides and ATP Traps

ATP acts as an inhibitor in EL2 ascites cells (Kohen et al. 1970a; 1970b,; Lardy 1965). The simultaneous addition of ATP suppressed totally the normally observed huge response to glucose-1-phosphate. Subsequently a dramatic effect was observed after treating the cells with a compound that can act as an ATP trap, i.e., ethionine (Faber 1963) (through conversion to S-adenosyl ethionine). This led to an instantaneous restoration of the usual glucose-1-phosphate response (Figure 2). The interpretation is subject to caution,however, as ethionine can also lead to glycogen depletion secondary to ATP depletion (Lupu and Farber 1954).

In experiments with glucose-6-phosphate, the pretreatment with ethionine leads to an acceleration of all transient parameters (Kohen, Siebert et al. 1972) and increased rate of substrate utilization, but the effect is less quantitative than with glucose-1-phosphate, possibly due to compartmentation of glycolytic phosphate esters in (different) pools (Kalant and Beitner 1971).

A possibly more sensitive way to actuate on the adenine nucleotides is through inhibitors which can affect the intracellular compartmentalization of such compounds. The most useful so far has been atractylate (Klingenberg and Pfaff 1966; Kohen et al. 1970e), which inhibits adenylate translocase at the level of the mitochondrial membrane, thus putting a stop to the ADP-ATP exchange across the mitochondrial membrane. Under these conditions, ADP accumulates in the cytoplasm (equivalent to an endogenous pulse of ADP) and leads to a subsequent acceleration of the glucose-6-phosphate utilization rate (Kohen et al. 1970e).

Mitochondrial-Extramitochondrial Interactions. Intracellular Transport

Mitochondrial responses to Krebs cycle intermediates (Kohen, Kohen, and Jenkins 1966) are generally hard to detect in ascites and L cells. However occasionally in larger individuals of ascites cultures (around 5% of total population), a marked reduction of mitochondrial pyridine nucleotides is observed with malate (Chance, Kohen et al. 1967). Lags up to one sec are recorded. The rise half time (around 5 sec) and full rise time (up to 15 sec) are much longer than those observed with glycolytic intermediates.

Attempts were made to influence the biogenesis and morphology of mitochondria by exposure to unusually high concentrations of inorganic copper (Keyhani and Chance 1971)(300μg/1) or copper chelating agents, e.g., cuprizone ($10^{-4} - 10^{-5}$ M) (Suzuki 1969). In both instances ascites cells and L cells exhibited considerable cloudy swelling, with many granules taking up the Janus green stain. No detectable pyridine nucleotide responses were observed from these structures upon addition of Krebs cycle intermediates, but there were rather profound changes in the extramitochondrial space (Table IV). In both

TABLE IV

KINETIC PARAMETERS OF NAD$^+$ (NADP$^+$) REDUCTION-REOXIDATION TRANSIENTS IN CUPRIZONE-TREATED L CELLS (1 WEEK). SUBSTRATE=glucose-6-phosphate.

Treatment	[S] optimal (mM)	LAG (msec)	$t_{1/2}$ on (msec)	Rise Time (sec)	Δ F/Rise Time (Photons /sec^2)	Δ F (percent)	$t_{1/2}$ off (sec)	Reoxidation Time (sec)	Rate of utilization (μM/ sec/Kg)
Untreated	19	120 ±30	800 ±80	2550 ±180	5000	100	12.3 ±1.3	21.9 ±1.1	1460
Cuprizone 10^{-5}M	3.8	570 ±170	900 ±200	3160 ±665	2400	60	13.0 ±1.4	63.8 ±11.5	250

copper (Figure 3) and cuprizone-treated cells, the rates of utilization of glycolytic intermediates were decreased considerably (from a maximum of about 1,500 μmole/sec/kg in controls to about 250 μmole/sec/kg in the treated cells).

Observations on intracellular transport (Kohen, Siebert et al. 1972) suggest a weak barrier for the metabolites used at the level of the nuclear membrane, with delays in the range of 35-40 msec; while the mitochondrial membrane represents a much greater obstacle to metabolite penetration, with delays at least one order of magnitude longer. Dyes such as fluorescein (Kohen et al. 1971d) are transported across the cell, with a half-maximal buildup at distant cellular sites, within less than 80 msec.

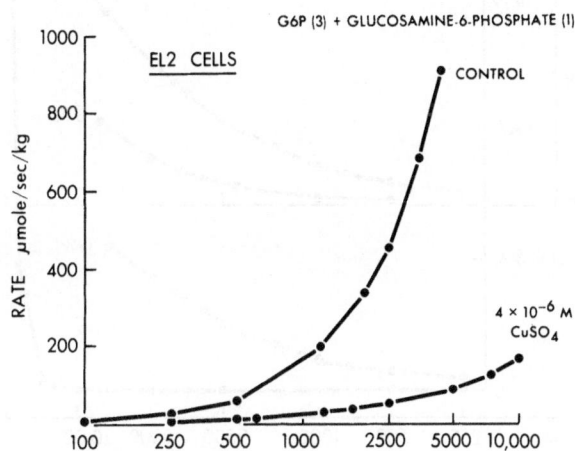

Figure 3. *The rate of glucose-6-phosphate conversion (ordinate) against the concentration of the added glucose-6-phosphate (abscissa) in the extramitochondrial space of EL2 controls and cells maintained for 48 hours in media with high levels of inorganic copper (300 μg liter). Glucose-6-phosphate was added together with glucosamine-6-phosphate in the ratio of 3 to 1. Glucosamine-6-phosphate was added as a competitive inhibitor of glucose-6-phosphate dehydrogenase, to channel the catabolism of glucose-6-phosphate along the Embden Meyerhof chain. It was observed that solutions of glucosamine-6-phosphate were weakly fluorescent, and to obtain significant fluorescence intensities from the cells, intracellular concentrations of glucosamine-6-phosphate over 10 mM had to be attained; in these experiments, however, the maximal concentrations of the metabolites were only 3 mM (see abscissa, glucose-6-phosphate/glucosamine-6-phosphate = 3/1*

Comparative Metabolic Studies in Different Cell Types

The general principles described above hold for all cell types studied, but characteristic differences in the various transient parameters are found, depending upon the cell type.

Under aerobic conditions, the redox state of extramitochondrial pyridine nucleotides can be easily altered in ascites cells by addition of glycolytic intermediates. Similar changes in L cells, and in Chang human liver and conjunctiva cells require anaerobiosis (Kohen et al. 1970c, 1971a, 1971d, 1972). As might have been expected, a specific metabolic profile emerges for each cell type (different substrate thresholds or transient kinetics due to different enzyme activities, configurations, or allosteric controls). The highest rate of 6-phosphogluconate utilization are observed in EL2 cells (Figure 4), and the highest utilization of glucose-6-phosphate in anaerobic L and conjunctiva cells (most likely due to release of mitochondrial control). These estimates of rates are based on the assumption that most of the metabolite is catabolized during the transient and not pooled within intracellular compartments.

Correlation Between Cell Cycles and Intermediate Metabolism

A glycogen cycle (Johannisson and Hagenfeldt 1971) has been described in correlation with

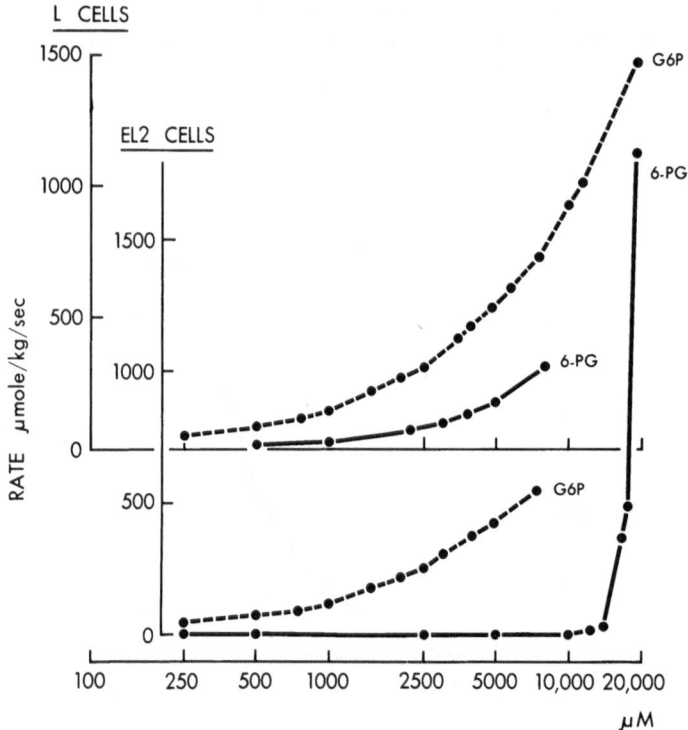

Figure 4. *Rates of glucose-6-phosphate and 6-phosphogluconate utilization in EL2 and L cells. Abscissa and ordinate are like those in Figure 3. The exponential increase in the rate of 6-phosphogluconate utilization beyond a substrate threshold (e.g., 10mM) is characteristic for EL2 cells*

the cell division cycle. In six-hour cultures of mouse L cells (following subdivisions of crowded cultures), the cells are stuffed with glycogen granules (periodic acid Schiff positive). Twenty-four hours after subdivision only few glycogen granules are found in 95-97% of these same cells. However, minimal differences are observed between the rates of glucose-1-phosphate utilization in 6- and 24-hour cultures of L cells. In older cultures (24-72 hours) there is a gradual increase of granule (glycogen)-rich larger cells (600-800 μm^2 cell area vs. 300-400 μm^2 for the average L cell), up to 5% of the total population. These larger cells exhibit a strong response to glucose-1-phosphate, in terms of NAD^+ ($NADP^+$) reduction and rate of utilization, contrary to the majority of smaller cells, which presumably are in the first half of the division cycle. The latter show no NAD-reduction following glucose-1-phosphate addition, indicating channeling towards glycogen storage. In cultures of both L cells and Chang liver cells, it is often observed that the larger cells are the only ones in fact which respond to glucose-1-phosphate.

Observations on Irradiated Ascites Cells

In X-ray-produced giant EL2 cells (Kohen, Kohen, and Jenkins 1966), radiation damage can appear in two degrees: (a) with lesser damage (*e.g.*, 1,000r) there is an overall active glycolysis, but some difficulty in the reoxidation of NADH, leading to a greater integrated area of the fluorescence pulse (NAD reduction-reoxidation transient); and (b) with further damage (*e.g.*, after 2,000r) there is an inhibited glycolytic activity (with glucose-6-phosphate as substrate) and again a slower removal of reducing equivalents (Figure 5, Table V). Similar but even more pronounced results are observed with 6-phosphogluconate. This is in agreement with earlier observations using X-ray and radiomimetics (Kohen et al. 1968a; 1968b), where glycolysis appeared activated or inhibited depending upon the severity of the damage (possibly as a result of alterations in mitochondrial and extramitochondrial lipoprotein membrane structures (Kohen et al. 1968b; Manteifel and Meisel 1966).

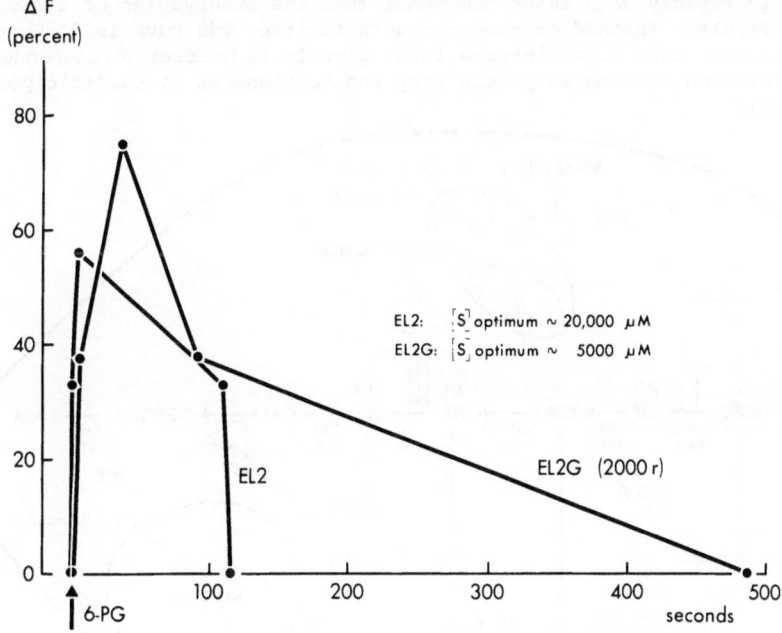

Figure 5. *NADP+ reduction-reoxidation transients with 6-phosphogluconate as substrate, in EL2 controls and EL2 cells two weeks after X-irradiation (2,000r). Each curve averaged from actual transients in at least six cells. A considerable prolongation of the reoxidation time is observed in the irradiated cells*

TABLE V

KINETIC PARAMETERS OF NAD+ (NADP+) REDUCTION-REOXIDATION TRANSIENTS IN IRRADIATED CELLS (SUBSTRATE = Glucose-6-phosphate)

Cell type	[S] optimal (mM)	LAG (msec)	$t\frac{1}{2}$ on (msec)	Rise Time (sec)	Δ F/Rise Time (photons /sec^2)	Δ F (percent)	$t\frac{1}{2}$ off (sec)	Reoxidation Time (sec)	Rate of substrate utilization (μM/ sec/Kg)
EL 2	3.3	200 ±50	1100 ±90	3800 ±200	5500	168	19.4 ±1.9	31.6 ±2.8	320
EL 2 G (2000r giant)	1.6	540 ±120	1060 ±125	4120 ±400	2000	66	42.4 ±10.0	60.9 ±11.7	60

DISCUSSION

Using appropriate substrates, as long as regular homeostatic mechanisms are maintained, the NAD-NADH equilibrium is poised toward oxidation and the NADP-NADPH equilibrium toward reduction (Krebs and Veech 1967). The steady-state redox potential of pyridine nucleotides, which is only temporarily disrupted by the introduction of metabolic transients, can be more permanently altered by metabolic preloading with ethanol, glycerol, etc.(Kohen, Siebert et al. 1972).

The bottleneck at the phosphofructokinase (Lardy 1965) is identified by the much greater rates of utilization of the glycolytic intermediates which are past the kinase. An important point of cell homeostasis is revealed by such intermediates, since the cell can cope so rapidly with these compounds that the observation of transients become quite difficult. Instead of cooperative activities (de Duve 1973; Hess 1971), it is more often characteristic of intracellular glycolysis to observe independent rates of substrate utilization at each enzymatic step and bottlenecks at specific points in the chain (Figure 6).

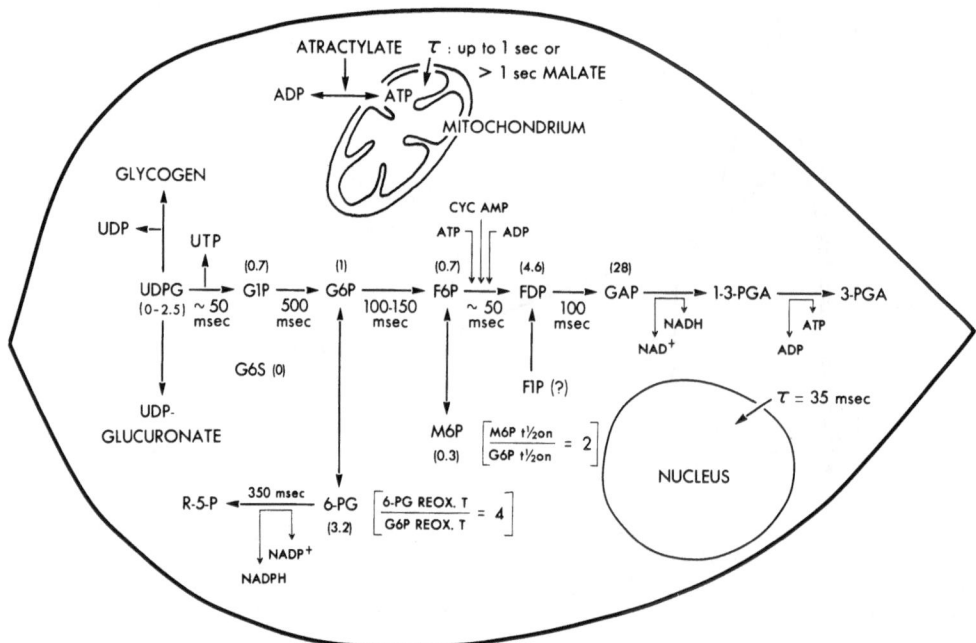

Figure 6. *A simplified diagram of metabolic compartments and carbohydrate pathways in EL2 cells and L cells, in correlation with microfluorometric observations. Lags along the Embden Meyerhof pathway, hexose monophosphate shunt, or side pathways are shown in msec, sofar as they could be determined. With the rate of glucose-6-phosphate conversion (added total glucose-6-phosphate/$t_{1/2}$ off) arbitrarily equated to one, the relative conversion rates for other metabolites are indicated by numbers in parentheses. G1P = glucose-1-phosphate, G6P = glucose-6-phosphate, F6P = fructose-6-phosphate, FDP = fructose-1,6-diphosphate, GAP = glyceraldehyde-3-phosphate, 1,3PGA = 1,3 phosphoglyceric acid, 3PGA = 3 phosphoglyceric acid, 6-PG = 6-phosphogluconate, M6P = mannose-6-phosphate, F1P = fructose-1-phosphate, UDPG = uridine-diphosphoglucose, UDP = uridine diphosphate, UTP = uridine triphosphate, G6S = glucose-6-sulfate, R5P = ribulose-5-phosphate, τ = metabolite transfer time at the level of intracellular membranes (e.g., mitochondrial or nuclear), REOX.T. = reoxidation time, $t_{1/2}$ on = rise half-time of the transient*

From lags and rates of utilization there is some evidence that Glc-1-P is altogether metabolized in a different pool than Glc-6-P. The rate of substrate utilization is related to concentration by a much larger exponent in the former case, as if these two phosphorylated sugar esters are dealt with differently within the cell. This may be the first direct evidence to the existence of independent metabolic pools (Kalant and Beitner 1971) within the extra mitochondrial space, a possibility which should further enhance the complexities of intracellular metabolic control. Other factors for consideration in the utilization of Glc-1-P are the glycogen metabolism and the correlation between the glycogen cycle of the cell and its mitotic cycle (Johannisson and Hagenfeldt 1971).

The role of mitochondrial-extramitochondrial interactions is apparent from the profound Pasteur effect observed in a variety of tissue culture cells (Kohen et al. 1970c; 1971a; 1971b). Even in malignant cells, such as ascites cells in culture or radiation-produced ascites giants which exhibit strong aerobic glycolysis, the rate of

NAD reduction and the accumulation of reducing equivalents are enhanced at anaerobiosis, while the ultimate removal of these equivalents is considerably slowed down.

For every cell type or even for variants of the same cell type, there is a characteristic pattern of mitochondrial-extramitochondrial interactions (Kohen et al. 1970e). Generally, extramitochondrial glycolytic activity remains dependent upon a certain degree of mitochondrial activity for the removal of reducing equivalents. Cases in which severe mitochondrial alterations are observed (such as "cloudy swelling": with copper or cuprizone treatment, X-rays) are accompanied by a concomittant inhibition in the rates of utilization of glycolytic intermediates in the extramitochondrial space.

The extramitochondrial pyridine nucleotides of practically all cells tried are very responsive to glycolytic intermediates. However, whenever a mitochondrial response is obtained, with Krebs-cycle intermediates, the relatively slow buildup of NADH accumulation (maximal ΔF at 10-20 seconds, instead of 2-5 sec for extramitochondrial NAD) seems to point to a membrane barrier.

These experiments illustrate the influence of organelle interactions (Kohen et al. 1970e) or metabolite modulation (Atkinson 1965) upon rates of substrate utilization along a specific pathway. With different states of cell functions,— *e.g.*, growth, differentiation or malignancy — pathways might be altered or modified. Such varying functional states can be defined by microfluorometry in qualitative and quantitative terms.

ACKNOWLEDGEMENTS

The authors wish to express their indebtedness to Professor Britton Chance (Johnson Research Foundation, University of Pennsylvania) for encouragement and fruitful exchange. They are also thankful to Research Engineers, Lennart Åkerman and Lars Nordberg (Karolinska Institute, Stockholm), Professor Zbynek Brada (Papanicolaou Cancer Research Institute) and Dr. Michael Zatzick (SSR Instruments Company, Santa Monica, California) for useful advice and consultations. The collaboration and advice of Dr. Tod S. Johnson, who helped in establishing the conditions, dose levels, and dose rates for the irradiation of EL2 cells, as well as the kind permission of Dr. Marvin Feldman for the use of the linear accelerator at the Radiation Therapy Section of the Veterans Administration, Miami, are thankfully acknowledged. The drawings were prepared by Mr. G. K. Rietberg.

This work was supported by grants P-518 and BC-15A (Louise Helene Neussel Memorial Grant for Cancer Research) from the American Cancer Society, and the Swedish Medical Research Council (B70-12X-630-06).

REFERENCES

ATKINSON, D.E., (1965), *Science,* 150, 851.

CHANCE, B., (1963), *Ann. N.Y. Acad. Sci.,* 108, 322.

CHANCE, B. and Baltscheffsky, H., (1958), *J. Biol. Chem.,* 233, 736.

CHANCE, B., Kohen, E., Kohen, C. and Legallais, V., (1967), In: *Advances in Enzyme Regulation,* (G. Weber, ed.), Vol. 5, Pergamon Press, New York, 3.

CHANCE, B. and Thorell, B., (1959), *J. Biol. Chem.,* 234, 3044.

de Duve, C., (1973), In: *Wenner-Gren Symposium on Structure and Function of Oxidation-Reduction Enzymes,* Wenner-Gren Center, Stockholm, August 23-27, 1970 (A. Åkeson and A. Ehrenberg, eds.), Pergamon Press Ltd., Oxford.

FARBER, E., (1963), *Adv. Cancer Res.,* 7, 383.

HESS, B., (1971), In: *Photo Synthesis, Bioenergetics, Regulation, Origin of Life,* Proceedings First European Biophysics Congress, (E. Broda, A. Locker, and H. Springer-Lederer, eds.), Vol. 4, Verlag der Wiener Medizinischen Akademie, Vienna, 447.

JOHANNISSON, E. and Hagenfeldt, K. (1971), *Karolinska Symposia on Research Methods in Reproductive Endocrinology, Third Symposium:* In: *In-Vitro Metabolism in Reproductive Cell Biology, Geneva,* January 25-27, (A. Diczfalusy, ed.), Forum, Copenhagen, 81.

KALANT, N. and Beitner, R., (1971), *J. Biol. Chem.,* 246, 504.

KALCKAR, H.M., (1953), *Biochim. Biophys. Acta,* 12, 250.

KENNEY, F.T., (1970), In: *Mammalian Protein Metabolism,* (H.N. Munro, ed.), Academic Press, New York, 131.

KEYHANI, E. and Chance, B., (1971), *FEBS Letters,* 17, 127.

KLINGENBERG, M. and Pfaff, E.,(1966), In: *Methods in Enzymology,* (J.M. Tager, S. Papa, E. Quagliariello, and E.C. Slater, eds.), Vol. 10 Elsevier, Amsterdam, 180.

KOHEN, E., Kohen, C. and Jenkins, W., (1966), *Exptl. Cell Res.,* 44, 175.

KOHEN, E., Kohen, C. and Thorell, B., (1968a), *Acta Pharm. Toxicol.,* 26, 556.

KOHEN, E., Kohen, C. and Thorell, B., (1968b), *Exptl. Cell Res.,* 49, 169.

KOHEN, E., Kohen, C. and Thorell, B., (1970a), *Biochim. Biophys Acta,* 198, 1.

KOHEN, E., Kohen, C. and Thorell, B., (1970b), *Exptl. Cell Res.* 59, 307.

KOHEN, E., Kohen, C. and Thorell, B., (1970c), *Mikrochimica Acta,* 1190.

KOHEN, E., Kohen, C. and Thorell, B., (1970d), *Symposium of Homologies in Enzymes and Metabolic Pathways and Symposium of Metabolic Alterations in Cancer,* (W.J. Whelan and J. Schultz, eds.), North Holland Press, Amsterdam, 481.

KOHEN, E., Kohen, C. and Thorell, B.,(1970e), In: *Wenner-Green Symposium on Structure and Function of Oxidation-Reduction Enzymes,* Wenner-Gren Center, Stockholm, August 23-27, (Å.Åkeson and A. Ehrenberg eds.) Pergamon Press, Ltd., Oxford, 557.

KOHEN, E., Kohen, C. and Thorell, B., (1971a), *Biochim. Biophys. Acta,* 234, 531.

KOHEN, E., Kohen, C. and Thorell, B., (1971b), *Exptl. Cell Res.* 64, 339.

KOHEN, E., Kohen, C. and Thorell, B., (1971c), *Hoppe-Seyler's Z. Physiol. Chemie.,* 352, 635.

KOHEN, E., Kohen, C. and Thorell, B., (1971d) In: *Photosynthesis, Bioenergetics Regulation, Origin of Life.* Proceedings First European Biophysics Congress, (E. Broda, A. Locker and H. Springer-Lederer, eds.) Vol. 4, Verlag der Wiener Medizinischen Akademie, Vienna, 465.

KOHEN, E., Kohen, C. and Thorell, B.,(1972), *Mikrochimica Acta,* 103.

KOHEN, E., Kohen, C., Thorell, B. and Wagener, G., (1973), This Volume.

KOHEN, E., Siebert, G. and Kohen, C., (1972), *Hoppe Seylers's Z. Physiol. Chemie,* 352, 927.

KREBS, H.A., (1968) In: *Advances in Enzyme Regulation,* (G. Weber ed.), Vol. 6, Pergamon Press, Oxford, 467.

KREBS, H.A. and Veech, R.L., (1969), In: *Advances in Enzyme Regulation,* (G. Weber ed.) Vol. 7, Pergamon Press, New York, 397.

LARDY, H., (1965), In: *Control of Energy Metabolism,* (B. Chance, R. Estabrook and J.R. Williamson, eds.), Academic Press, New York, 69.

LELOIR, L.F., (1951,In: *Phosphorus Metabolism,* Vol. 1, Johns Hopkins Press, Baltimore, 67.

LUPU, C.J. and Farber, E., (1954), *Proc. Soc. Exp. Biol.,* 86, 701.

MANTEIFEL, V.M. and Meisel, M.N., (1966), *Fed. Proc., Translation Suppl.,* 25, T 981-88.

MEYERHOF, O. and Wilson, J.R., (1949), *Arch. Biochem.,* 21, 22.

ORRENIUS, S., (1965), In: *Studies on the Drug-hydroxylating Enzyme of Rat Liver Microsomes,* Almquist and Wiksells Boktryckeri AB, Uppsala.

STROMINGER, J.C., Maxwell, E.S., Axelrod, J.,and Kalckar, H.M., (1957), *J. Biol. Chem.* 224, 79.

SUND, H., (1968), In: *Biological Oxidations*, (T. P. Singer, ed.) Interscience Publishers, New York, 603.

SUZUKI, K. (1969), *Science,* 163, 81.

VELICK, S.F., (1958), *J. Biol. Chem.,* 233, 1455.

ZEIGLER, B.P., and Weinberg, R., (1970), *J. Theor. Biol.,* 29, 35.

NEWER FLUOROMETRIC METHODS FOR THE ANALYSIS OF BIOLOGICALLY IMPORTANT COMPOUNDS

G. G. Guilbault

Department of Chemistry, Louisiana State University in New Orleans, New Orleans, Louisiana

GENERAL

In the past, manometric methods, pH procedures, and spectrophotometry have been used for determining enzyme activity. A detailed discussion of the use of enzymes for analysis can be found in review articles by Guilbault (1966; 1968; 1970). Spectrophotometry has been generally preferred because of its simplicity, its rapidity, and the capability of measuring lower enzyme and substrate concentrations. Spectrophotometry embraces the use of colorimetric methods where colored products are produced as a result of enzyme activity, and fluorescent methods where fluorescent compounds are produced as a result of enzyme activity. Fluorescent procedures are several orders of magnitude more sensitive than colorimetric methods and thus have replaced the colorimetric methods in numerous instances.

Previous fluorometric methods, although they have represented improvements over other prior art methods of determining enzyme activity, have not eliminated all of the problems associated with enzymic analyses. Fluorometric analysis depends on the production of a fluorescent compound as a result of enzyme activity between a substrate and enzyme. The rate of production of the fluorescent compound is related to both the enzyme concentration and the substrate concentration. This rate can be quantitatively measured by exciting the fluorescent compound as it is produced and by recording the quantity of fluorescence emitted per unit of time with a fluorometer.

The prior methods for fluorometrically measuring enzyme reaction rates, however, have been wet-chemical methods and rely on reacting a substrate solution with an enzyme solution. Wet-chemical methods involve time-consuming and wasteful preparation of costly substrate solutions and enzyme solutions. For example, when determining the presence and concentration of an enzyme, a substantial amount of substrate must be accurately weighed out and dissolved in a large amount of buffer solution — usually about 100 ml — to prepare a stock solution. The enzyme reaction is then usually carried out by measuring, for example, 3ml of stock substrate solution into an optical cuvette, adding a measured amount of the enzyme solution to the substrate solution, and recording the change in fluorescence emanating from the resultant solution per minute with a fluorometer. This standard wet-chemical method is costly and wasteful because trained laboratory technicians' time and relatively large quantities of expensive substrate are required.

SOLID-SURFACE FLUORESCENCE MONITORING SYSTEM

Solid-surface fluorometric methods, using a reagentless system, have been developed for the assay of clinically important enzymes, substrates, activators, and inhibitors. The method comprises forming a solid reactant film of one of the reactants on an inert silicone matrix pad, contacting the film of the first reactant with a solution of the second reactant (substance to be measured) to produce a fluorescent material, and monitoring the change of fluorescence with time to determine the concentration of the second reactant.

The reactant film is formed by dissolving the reagents in a solvent, depositing the reactant solution on the silicone pad so that the solution spreads evenly over the pad, and evaporating the solvent from solution. The reagent may be applied to the pad either from an acetone solution, if a substrate, or in a polymeric film such as polyacrylamide, if an enzyme. Either substrate or enzyme and/or coenzyme can be deposited on the pad in film form, depending upon whether the substance to be assayed is an enzyme or a substrate.

After the pad is formed (Figure 1), it is attached to a slide made of a rigid material, such as glass or metal, by applying an adhesive to the slide and overlying the pad on the adhesive. The pads are placed approximately 1 or 2 cm from the bottom of the slide (Figure 2).

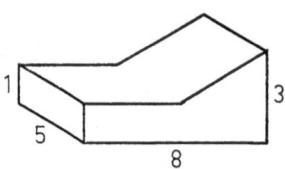

Figure 1. *Pad Support, Numbers in mm*

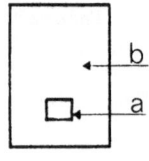

Figure 2. *Slide support, a) pad, b) slide (glass or metal)*

The fluorescent cell consists of a rectangular box made of ordinary sheet metal, wood, or plastic material (Figure 3), with portholes cut in to allow radiation to enter and leave the cell cavity. The top of the cell is provided with guides at opposite sides so that the slide containing the reagents can be reproducibly placed within the cell. In order to obtain as low a background as possible, the cell is painted with a black optical paint having a dull finish.

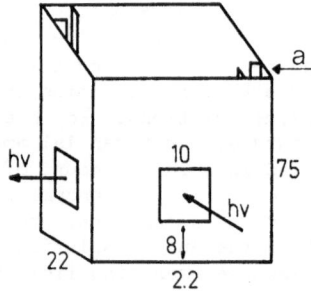

Figure 3. *Fluorescence Attachment Used to Hold the Substrate Pads. Numbers in mm, a) guides for slides*

Before the enzyme solution is added, the glass slide is put into the fluorometer and a background reading is taken. A blank rate is taken using all components of the system, except the unknown. The slide is then removed from the fluorometer and 20 μl of a solution of the substance to be assayed is applied to the slide. The recorder (100 mV input) is started immediately; the slide is placed back into the fluorometer; and the rate is recorded. A calibration plot of the change in fluorescence units per minute versus concentration is prepared and is used for subsequent analyses of enzyme or substrate.

After a run is completed, the pad can be peeled from the slide and the slide can be cleaned and stored for subsequent use.

The physical properties of the silicones and the stability of the reactant films of enzyme and substrate in the presence of silicones further make these materials ideally suited for use. The silicone materials can retain a reactant film on their surfaces for an indefinite time and permit the direct measurement of fluorescence from its surface when an appropriate second-reagent solution is dropped onto the first reactant film. Background interference due to light scattering and nonspecific fluorescence are minimal as compared with other materials.

The number of enzyme systems that could be monitored by this type of solid-surface device are as numerous as the number of enzyme systems known. A discussion follows of some of the more clinically important systems to be studied, for which methods have been developed.

ASSAY OF CLINICALLY IMPORTANT ENZYMES

Cholinesterase

Low levels of cholinesterase are found in individuals with anemia, malnutrition, and pesticide poisioning. High levels indicate a nephrotic syndrome.

Preliminary results have indicated that cholinesterase in solution can be monitored using a film of N-methyl indoxyl acetate on the surface of a silicone rubber pad:

$$N\text{-Methyl indoxyl acetate} \xrightarrow{ChE} N\text{-Methyl Indoxyl}$$

$$\text{(Fluorescent --}$$
$$\lambda_{ex} = 495 \text{ nm,}$$
$$\lambda_{em} = 525 \text{ nm)}$$

The rate of production of N-methyl indoxyl is followed and is proportional to the cholinesterase concentration.

Attempts have been made to develop methods for the direct assay of ChE in blood and body tissue. The reproducibility and accuracy obtainable using several configurations of pad and various experimental conditions have been compared and the optimum system developed. A linear region of 10^{-4} to 1 unit has been obtained (Guilbault and Zimmerman 1970).

Preliminary studies have indicated that pads made from either method were stable for at least 60 days (Table 1) if kept in a cold dark place.

TABLE I

PAD STABILITY DATA[a]

Day[b]	Rate, $\Delta F/min$[c]
0	3.10
2	3.04
4	3.08
10	3.20
30	3.00
60	3.00

a) A 500 µg/ml solution of cholinesterase was freshly prepared for each day. Twenty microliters of a 0.1 $\underline{M}$ substrate solution were applied to the pad and allowed to evaporate to dryness.
b) Days after preparation of pad.
c) Rates represent an average of at least three runs.

Alkaline Phosphatase

High levels of alkaline phosphatase are observed in rickets, Paget's disease, obstructive pandicea, and metastatic carcinoma.

The fluorogenic substrate, naphthol-As-Bi-phosphate, was used for the assay of alkaline phosphatase in serum. Guilbault et al. (1971) found that this substrate is the best for the direct assay of alkaline phosphatase in serum. A solution method was used and an accuracy and precision of about 2% was obtained. The deep green fluorescent naphthol As-Bi is formed and measured. Attempts have been made to monitor this reaction on a solid surface. A drop of serum to be assayed has been added directly to the naphthol AS-BI-phosphate on the surface and the rate monitored. No solutions are required. From 10^{-6} to 10 units are assayable.

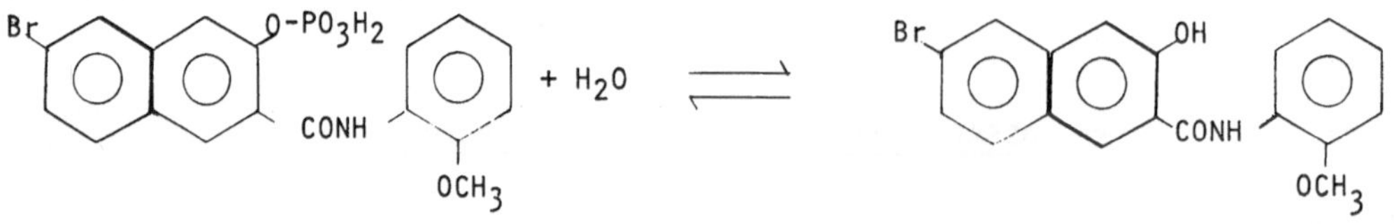

$$\text{Br} - \text{(naphthalene ring)} - O-PO_3H_2 \ \ \text{CONH} - \text{(phenyl)} - OCH_3 + H_2O \ \rightleftharpoons \ \text{Br} - \text{(naphthalene ring)} - OH \ \ \text{CONH} - \text{(phenyl)} - OCH_3$$

Naphthol AS-BI-Phosphate
(Nonfluorescent)

Naphthol AS-BI
(Fluorescent)

Further attempts have been made to extend the usefulness of this method for a wide variety of biological samples.

TABLE II

DETERMINATION OF SERUM ALKALINE PHOSPHATASE[a]

Units alkaline phosphatase taken	Units alkaline phosphatase found[b]
20	21.5
22	19
24	24.5
26	23
28	25
30	31
31	31
33	37
37	34
39	40
41	39
43	42
48	49
53	54[c]
61	60[c]
64	64[c]

a) Units expressed as μmole phenolphthalein liberated min^{-1} l^{-1} serum.
b) Obtained by multiplying the rate by 31.5. Average from three or more different samples - each analysis in duplicate.
c) One sample — average of two determinations.

The color of the silicone rubber,however, affects both the background and reaction rates. With any of the filter systems, the background and reaction rates increase in the order black<grey<clear<white. Each possible combination of pad color and filter was examined, and it was found that the most accurate results could be obtained if a combination of 7-54 filter and grey silicone rubber pads were used.

Table II shows the results obtained from a series of serum samples. Over 80 serum samples were analyzed with a relative error of ± 5%.

Figure 4 shows the correlation between the serum alkaline phosphatase values obtained from the solid-surface method and the alkaline phosphatase values obtained from a solution method employing phenolphthalein phosphate. The graph shows good agreement between the two methods over a wide range of serum values. If the serum values are expressed as μmole phenolphthalein liberated min^{-1} l^{-1} serum, then normal serum has 27-105 units.

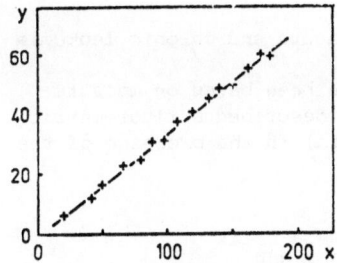

Figure 4. *Correlation between results obtained with naphthol AS-BI method and phenolphthalein method.* x *Naphthol AS-BI units,* y *Phenolphthalein Units*

Table III shows the alkaline phosphatase values obtained from highly jaundiced serum compared with the phenolphthalein-phosphate method. All the results from these samples were low and implied a limitation on the method. It is, however, possible to predict which samples will give low results — generally, those which are very dark yellow or orange-yellow in color. The light-yellow serum gave good results as did haemolyzed serum. The low values obtained from jaundiced serum is the main limitation of the method, but since it is possible to recognize these solutions, the analysis may be carried out by an alternative method.

TABLE III

ANALYTICAL RESULTS OBTAINED FROM JAUNDICED SERUM SAMPLES[*]

Units alkaline phosphatase taken	Units alkaline phosphatase found[**]
27	18
34	21
59	30
86	21
135	44
185	115
200	138

[*]   Units expressed as μmole phenolphthalein liberated min^{-1} 1^{-1} serum.
[**] Average of three results.

Lipase

A low-plasma lipase indicates vitamin A deficiency and some malignancies. Lipase assay is most important in diagnosis of pancreatitis.

Two fluorogenic substrates have been evaluated for the assay of lipase — N-methyl indoxyl myristate, and 4-methyl umbelliferone myristate, a substrate recently prepared in our labs.

The production of the green fluorescent naphthol AS-BI or the blue 4-methyl umbelliferone myristate can be followed on the surface.

4-Methyl Umbelliferone Myristate ———————⟶

4-Methyl Umbelliferone λ_{ex} = 340 nm

λ_{em} = 450 nm

A direct assay of lipase (10^{-4} to 10 units) in serum and pancreas can be effected.

Lactate Dehydrogenase

Levels of this enzyme are elevated in individuals with acute and chronic leukemia in relapse, myocardial infarctions, and carcinomatosis.

Fluorometric analysis of dehydrogenases in the past have been based on measurement of NADH fluorescence. Guilbault and Kramer (1964) have described a fluorometric method of coupling the NADH to an electron acceptor (resazurin) in the presence of the cofactor phenazine methosulfate (PMS).

$$\text{Lactate + NAD} \xrightarrow{\text{LDH}} \text{NADH}$$

$$\text{NADH + Resazurin} \xrightarrow{\text{PMS}} \text{Resorufin + NAD}$$
$$\text{(Non-Fluorescent)} \qquad \text{(Fluorescent)}$$

A semi-solid-state fluorescence method for the analysis of lactic dehydrogenase (LDH) has been developed (Guilbault and Zimmerman 1972). The commercially available enzyme in aqueous solution, and the enzyme in blood serum were studied. The serum analysis is compared with a method described by Bergmeyer, et al,(1963). As little as 160 units of LDH per milliliter of blood serum can be analyzed with an accuracy of better than 3%. The reaction system used was:

$$\text{Li Lactate + NAD} \xrightleftharpoons{\text{LDH}} \text{NADH + Pyruvic Acid}$$

The rate of production of the fluorescent product NADH was monitored.

TABLE IV

DATA AND RESULTS OF THE ASSAY OF SERUM LDH USING THE PROPOSED METHOD

Rate ΔF/min.	Concentration taken	u/ml found	Relative Error %
6.6	160	168	+4.9
7.3	190	189	-0.3
7.2	200	186	-6.5
10.1	280	276	-1.3
13.2	360	372	+3.3
13.9	380	394	+3.4
15.3	440	437	-0.6
16.0	460	459	-0.2
17.2	500	496	-0.8
18.8	560	546	-2.5
21.4	620	626	+1.0
		Average Relative Error	2.3 %

Some of the results obtained in the assay of LDH in serum using the solid-surface method are shown in Table IV. An average relative error of 2.3% was obtained.

ASSAY OF CLINICALLY IMPORTANT SUBSTRATES

Glucose

High levels are used to indicate diabetes.

Guilbault et al.(1968), have described a fluorometric system for glucose using glucose oxidase, peroxidase, and p-hydroxyphenylacetic acid. The rate of production of fluorescence is equivalent to the amount of glucose present.

$$\text{Glucose} \xrightarrow{\text{Oxidase}} H_2O_2$$

Nonfluorescent   Fluorescent

To develop a reagentless surface method for assay of glucose, the enzymes, glucose oxidase and peroxidase, as well as the substrate, p-hydroxyphenylacetic acid, were placed on the surface. The addition of glucose then triggers the reaction; fluorescence is produced and its rate of production is proportional to the glucose present in blood (Guilbault and Tang, n.p.). A range of $10^{-6} - 10^{-1}M$ glucose can be assayed with a deviation of only 2.5%.

ADVANTAGES OF SOLID-SURFACE METHOD

The advantages of this radically new system for the clinical laboratory are:
• The time required for analysis by present methods is 30-60 minutes to prepare reagents plus 10-20 minutes per assay. When a sample comes in, no matter what time of day or night, the technician must go through the regular preparation and analysis routine. The solid-surface method would have 10 bottles of pads, one for each of 10 common tests. The technician would need only to go to the bottle marked "urea test", pick out a pad, put it in the instrument on the strip holder, add a 10 µl sample of blood or urine, and read out the results directly in mg % urea. Total time, about 2 minutes.
• The accuracy of current methods (mostly spectrophotometric) is limited to an error of about 2-3% over a narrow range (0.15 to 0.85 Abs units). Because fluorescence methods are based on the production of a signal over zero signal, they are independent of scale reading, and have an accuracy of 1% over a 3-4 fold linear range of concentrations.
• Temperature must be controlled in all enzyme assays based on a kinetic approach. This is expensive and bothersome. The solid-surface fluorescence method proposed has no temperature dependence since the silicone rubber pad used to support the sample does not conduct heat. Thus, provided the sample of blood is at the same temperature as that used to prepare the calibration plot (generally room temperature, 25° C), the temperature of the environment does not affect the results. For example, in the assay of alkaline phosphatase, we obtained the same rate at 10°, 15°, 20°, 25°, 30°, 35°, and 40° C external temperature, provided we added the blood from a sample kept at 25° C. This represents a significant improvement.
• The reagents used in present clinical procedures are unstable, and new solutions must be prepared daily. o-Toluidine and peroxidase in glucose assay, or NAD in LDH assay are examples of this. Yet when the reagents are placed in *solid form* on a surface, they can be kept for months with *no deterioration*. In our ChE, alkaline phosphatase, and LDH assays we found we could use our pads for 90 days, with the same calibration plot useful everyday.
• Many clinical methods are subject to interferences because of (a) protein absorption in the near UV-visible, or (b) absorption of impurities in the reagents. By using reagents that fluoresce in the visible, and by using them in a solid form, these two interferences are eliminated.

REFERENCES

BERGMEYER, H., Bernt F. and Hass, B. (1963), In: *Methods of Enzymatic Analysis*, (H. Bergmeyer, ed.) Academic Press, New York: 736.

GUILBAULT, G.G., (1966), *Anal. Chem.*, <u>38</u>, 527R.

GUILBAULT, G.G., (1968) *Anal. Chem.*, <u>40</u>, 459R.

GUILBAULT, G.G., (1970) *Anal. Chem.*, <u>42</u>, 334R.

GUILBAULT, G.G., Brignoc, P., and Zimmer, M., (1968), *Anal. Chem.*, <u>40</u>, 1256.

GUILBAULT, G.G., and Kramer, D.N.,(1964), *Anal. Chem.*, <u>36</u>, 2497.

GUILBAULT, G.G., and Tang, D., unpublished results.

GUILBAULT, G.G., and Vaughn, A., (1971), *Anal. Chim. Acta*, <u>55</u>, 107.

GUILBAULT, G.G., and Zimmerman, R., (1970), *Anal. Letters*, <u>3</u>, 133.

GUILBAULT, G.G., and Zimmerman, R., (1972), *Anal. Chim. Acta*, <u>58</u>, 75.

MICROFLUOROMETRIC INVESTIGATIONS ON THE INTRACELLULAR TURNOVER OF FLUOROGENIC SUBSTRATES

M. Sernetz

Battelle-Institut E.V., Frankfurt am Main

INTRODUCTION

Rotman and Papermaster (1966) reported first results of investigations on "fluoro-chromasia", the phenomenon whereby living cells develop an appreciable intracellular fluorescence during incubation in a buffer containing the fluorogenic substrate, fluo-rescein-diacetate (FDA). The fluorescence is due to florescein (F), which by enzymat-ic hydrolysis is liberated from the nonfluorescent FDA through cellular esterases.

This paper contributes the results of further investigations on the kinetic analy-sis of the turnover of fluorogenic substrates in individual cells for the description of specific properties of different types of cells.

METHODS

Questions of instrumentation and methodology have been described elsewhere (Sernetz and Thaer 1970; 1972; 1973). Capillary cuvettes were used for calibration of the microscope fluorometer and standardization of measured fluorescence intensities; they were also used for the correlation of spectra obtained from cells with those obtained from measurements on solutions. Their application has been described in an earlier paper (Sernetz and Thaer 1973).

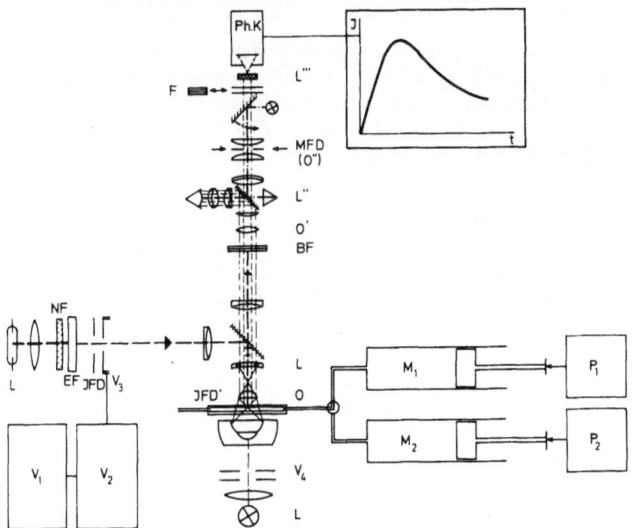

Figure 1. *Schematic representation of the microscope fluorometer equipped for measurement of intracellular turnover of fluorogenic substrates in single cells (for detailed description, see also Rotman 1966; Sernetz 1970;1972. L=light-source; IFD= field diaphragm; O=object plane or their conjugated planes; MFD=measuring field dia-phragm; Ph. K=Photocathode and recorder; EF, BF, NF=excitation filter, barrier filter and neutral filters; V_1 V_2 V_3=automatic shutter (Wild); L, V_4, O, L'=phase contrast equipment for simultaneous observation in transmitted light; O, M_1P_1, M_2P_2=flow chamber with automatic syringes for alternative supply with different media*

The individual intact cells to be investigated in a flow chamber were focused in the object plane of a recording microfluorometer or microscope-spectrofluorometer (Figure 1), equipped for incident-light excitation (Sernetz and Thaer 1970) and having an automatic shutter (Wild) to prevent cells from damage by continuous exciting radiation.

The microscope fluorometric measurements were carried out on monolayer cultures of L and HeLa cells. The cells were grown on cover slides, which served as an upper window of the flow chamber. In the case of cell suspensions such as lung macrophages and peritoneal mast cells, a flow chamber was filled with the suspension and put upside down for a while until the sedimented cells attached to the inner surface of the upper window. It was possible to exchange the medium repeatedly and quickly either by a syringe system as indicated in Figure 1 or by simply pipetting. Thus the experiments have been performed either at constant substrate concentrations or at an initial substrate concentration injected once at zero time.

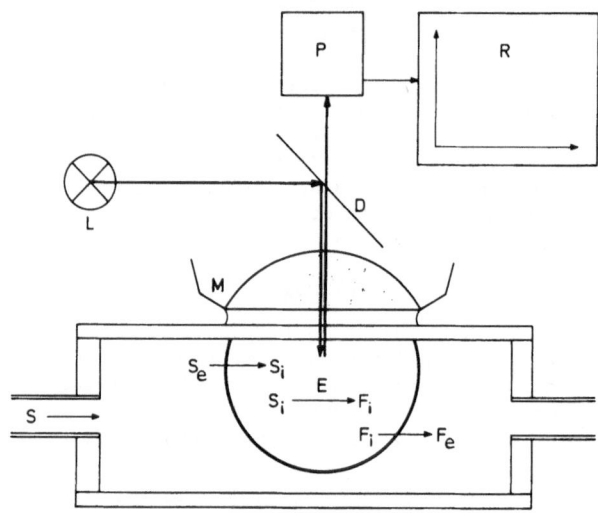

Figure 2. *Simplified model for the description of intracellular turnover of fluorogenic substrates, measured by microscope fluorometry in incident light in a flow chamber. S=fluorogenic substrate (fluorescein-diacetate); F=fluorescent product (fluorescein); E=cellular enzyme (esterases)(Index $_i$ for intra-cellular, Index $_e$ for extra-cellular);L , D, M, P, R= microscope fluorometer (as in Figure 1)*

Figure 2 shows a simplified model assumed for the overall process of intracellular turnover of fluorogenic substrates.

In the first step, the substrate S_e (FDA) permeates into the cell. In a second reaction, the intracellular substrate,S_i is hydrolyzed to yield the intracellular and fluorescent product, F_i, which in the third step of the process may again be released from the cell into the medium.

RESULTS and DISCUSSION

Figures 3a-c show photomicrographs of a single L cell taken immediately after incubation with 10^{-5} M FDA at successive intervals of about 30 seconds. Figure 4 shows both the fluorescence intensity measured on a single L cell during uptake of the fluorogenic substrate FDA and the elimination of the fluorescent product, fluorescein, as a function of time.

Figure 5 shows the reproducibility of the turnover rates during repeated exposure of the same cell to substrate-containing and substrate-free medium. The initial slopes are constant and the cells appear unaffected. It may also be taken from Figure 5 that the elimination can be measured as an isolated process after washing.

It is obvious that a cell cannot accumulate the product infinitely, especially if treated with fluorogenic substrate in high concentrations. As may be seen from Figure 6, a maximum is reached at which the cells seem to be saturated: repeated exposure to FDA gives no further rise and the slopes decrease.

Correlating the fluorescence intensities measured on single cells both with absolute amounts of fluorescein by means of the capillary method (Sernetz 1973) and with the estimated average cell volume, it was found that the intracellular concentrations

PLATE I

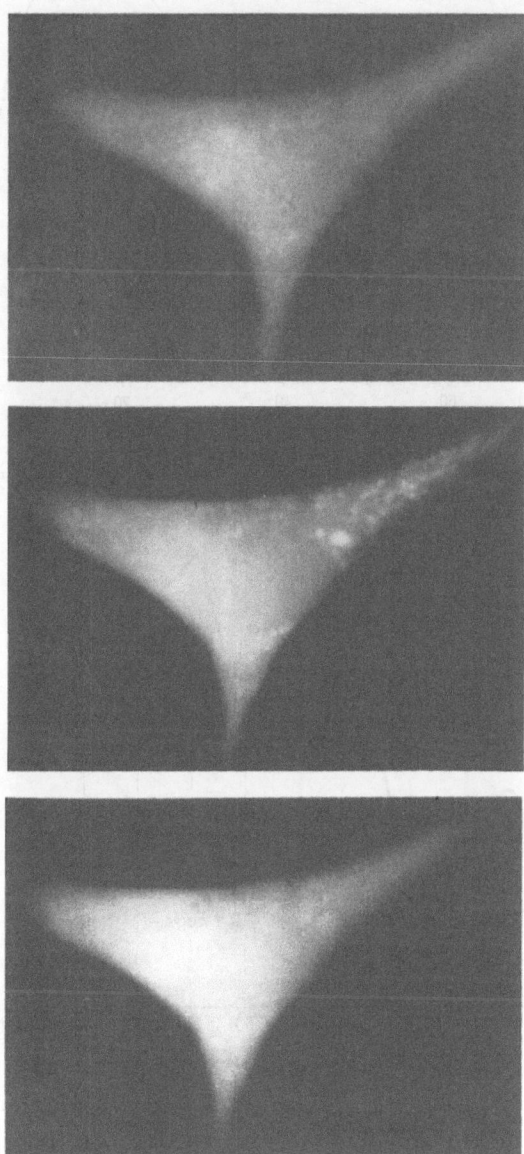

Figure 3, a-c. *Photomicrographs of a single fibroblast (L cell) taken at intervals of about 30 seconds after incubation with FDA; its increasing fluorescence corresponds to the initial phase of turnover indicated by the curve of Figure 4. Objective FL oil 105/1.32 (Leitz) × 2,000*

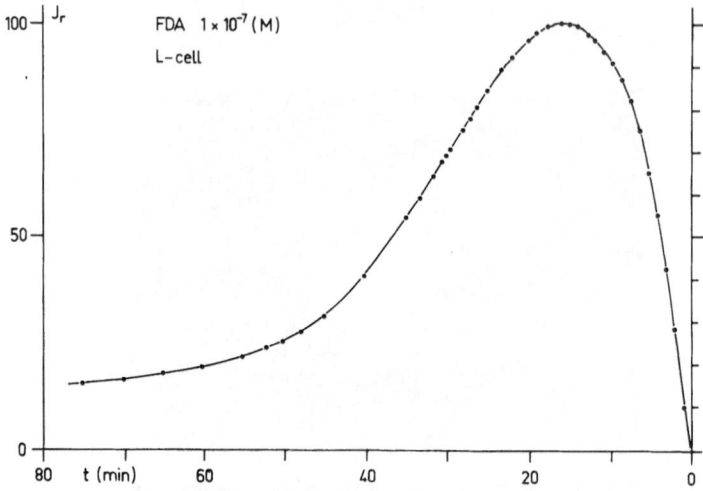

Figure 4. *Fluorescence ('fluorochromasia') in a single L cell incubated with*
10^{-7} M FDA, as a function of time

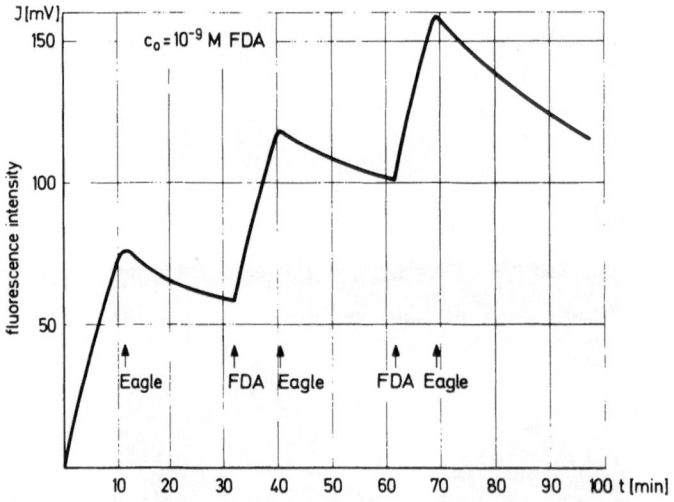

Figure 5. *Repeated incubation of a single L cell with 10^{-9} M FDA and alternating*
washing with substrate-free medium for demonstration of constant slopes

of F_i at maximum fluorochromasia may reach 10^{-4} molar. At these high intracellular
concentrations, morphological alterations of the cells have also been observed. With
respect to this finding, however, it should be kept in mind that phenolphthalein is
very often used for cell cultivation, in concentrations as high as 10^{-5} to 10^{-4} molar,
both as ph indicator and as constituent of the growth medium.

Figure 7 shows that a single cell reacts with exactly the same turnover rate to
repeated exposure to 10^{-7} M FDA, even after a period of four hours in the flow chamber.
The elimination characteristics are comparable as well. In this graph the fluorescence
intensity axis has been calibrated in moles of fluorescein, with capillary cuvettes
used for standardization. This calibration already provides for the reduced quantum
yield of intracellular fluorescein as discussed by Sernetz and Thaer (1972) and in the
following paragraphs. The intracellular mass turnover rates have also been calculated
in this manner in moles per time unit and cell.

As may be seen from Figures 5 and 7, the decrease in intracellular fluorescence
can be measured as a separated process by merely washing the cells with a substrate-

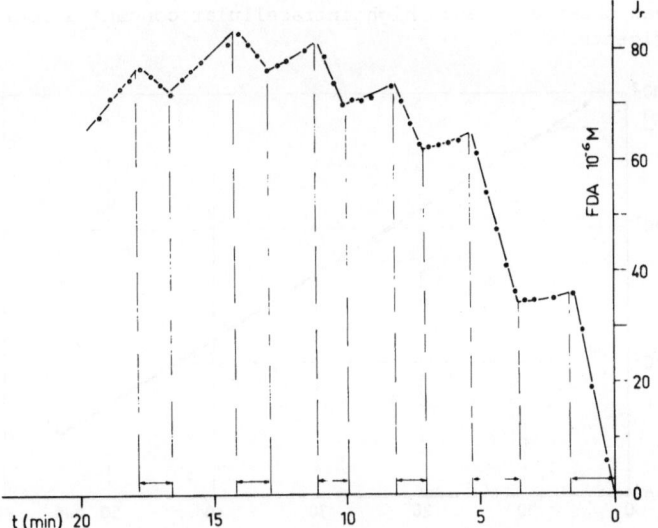

Figure 6. *Repeated incubation of a single cell with 10^{-6} M FDA and alternating washing with substrate-free medium for demonstrating the decrease in the initial slopes and saturation in the maximum of fluorochromasia*

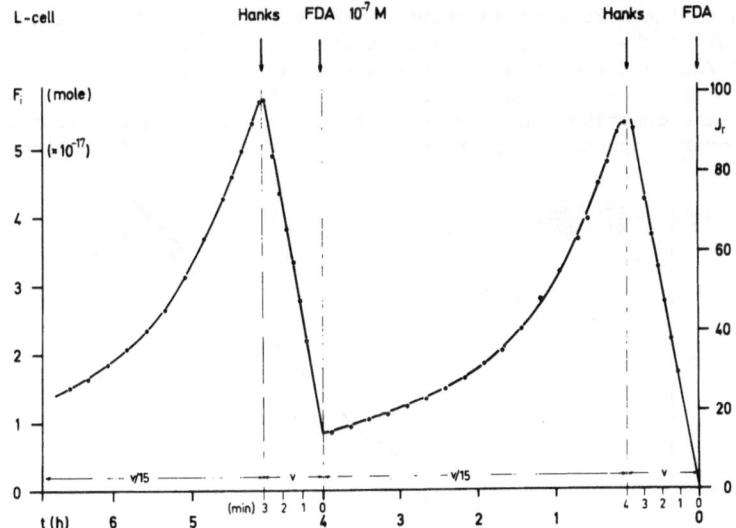

Figure 7. *Repeated incubation of a single L cell with 10^{-7} M FDA and washing with substrate-free medium for demonstrating the unchanged initial turnover rates and elimination characteristics even after keeping for hours in a flow chamber. For measuring the elimination period, the recorder speed was reduced 15 times. The left ordinate indicates intracellular mass of fluorescein (F_i), as calculated after calibration by means of the capillary technique*

free medium.

This fluorescence decrease indicates an elimination process which exactly obeys first-order kinetics $-dF_i/dt = k_2 \cdot F_i$ (see Figure 8). This fact has likewise been found by Rotman (1966). The elimination half-life periods, $t_{1/2}$, of L and HeLa cells did not differ significantly; on the basis of a log-normal distribution, as discussed below, a mean of $t_{1/2} = 31$ min was calculated (n = 50).

In order to find out whether at high intracellular concentrations a shift from a first-order to a zero-order process can be observed, indicating "saturation" of the elimination capacity by concentration-independent elimination rates, intracellular fluorescein concentrations up to 10^{-4} molar were produced. However, no significant deviations have been observed. Such high intracellular concentrations were avoided in all other experiments.

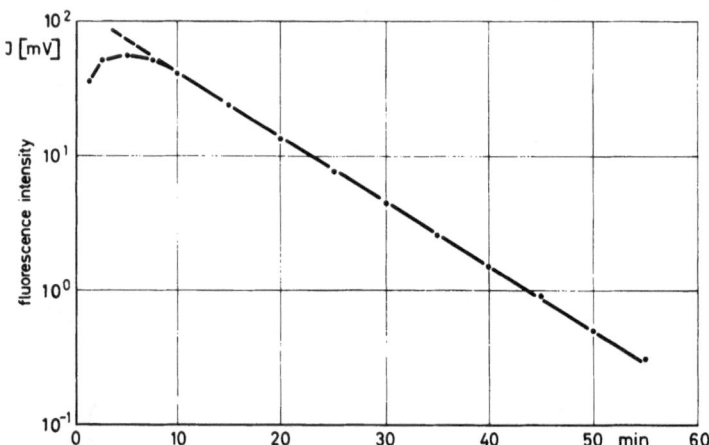

Figure 8. *Semilogarithmic plot of the fluorescence decrease in a single L cell after having passed the fluorescence maximum, indicating first-order elimination kinetics*

In addition to the direct measurement of the elimination from a single cell, a procedure which gives the reaction order even without absolute calibration, the elimination rates $-dF_i/dt$ have been plotted versus the intracellular mass, F_i, at a given time for single cells chosen at random (Figure 9). This statistical approach to the solution of the same question confirms conclusively the finding of first-order elimination kinetics over a concentration range of about two orders of magnitude.

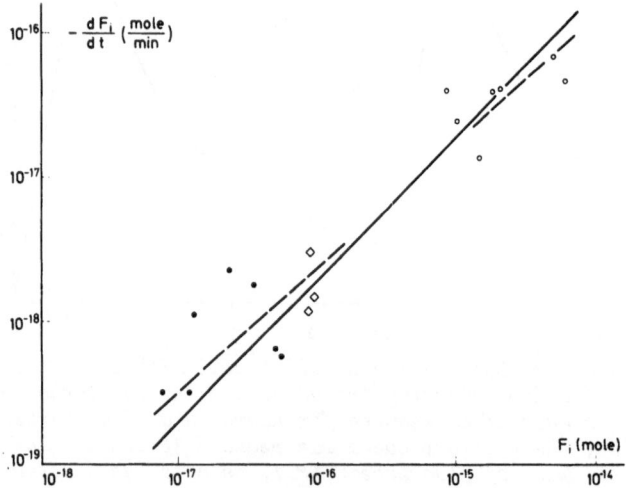

Figure 9. *Elimination rates, $-dF_i/dt$, plotted versus intracellular mass, F_i, of fluorescein for several L cells chosen at random, for statistical confirmation of first-order elimination kinetics. The symbols denote initial incubation with 10^{-7}, 10^{-6}, and 10^{-5} M FDA*

For evaluation of the initial substrate turnover rates as a function of the FDA concentration applied, single L cells have been incubated in stepwise-increased FDA concentrations in the flow chamber. The results of a typical experiment are depicted in Figure 10, which shows that, in order to maintain zero conditions, the measurement can be restricted to a very early phase and one that is remote from the maximum of fluorochromasia.

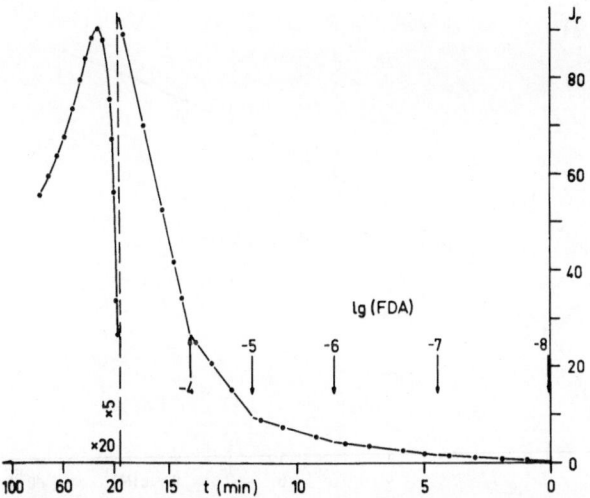

Figure 10. *Fluorescence intensity of a single L cell after incubation at rising concentrations of FDA. After the last exchange of medium, the recorder speed was reduced 20-fold and the recorder sensitivity 5-fold in order also to record the fluorescence maximum*

Rotman and Papermaster (1966) suggested previously that the intracellular FDA turnover rates obey simple Michaelis-Menten kinetics. Based on this assumption they derived a Michaelis constant for the intact cell in the range of $K_M \simeq 2.9 \times 10^{-6}$ M; however, this value is about 10 times lower than that of the cell-free extract with $K_M \simeq 3.6 \times 10^{-5}$ M.

Michaelis-Menten kinetics seem to be valid within a narrow range of FDA concentrations between 3×10^{-5} to 1×10^{-4} M; this leads to K_M values of about 3.6×10^{-5} M (Figure 11), which are similar to those for the cell-free extract of Rotman. However,

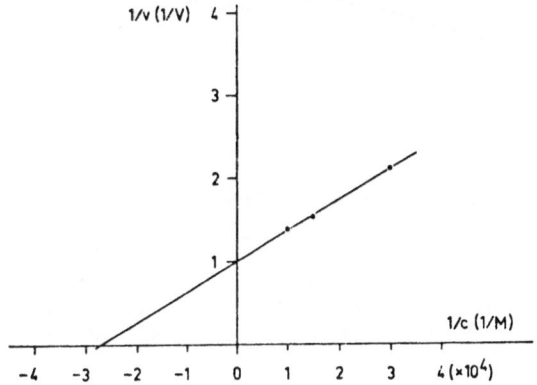

Figure 11. *Lineweaver-Burk plot of reciprocal relative FDA-turnover rates in a single cell versus reciprocal FDA concentrations. Linearity within one decade of FDA concentrations leads to an apparent 'K_M' of about 3.6 $\times 10^{-5}$ M*

when extending the substrate concentration range by more than one order of magnitude as indicated in Figure 10, the initial intracellular FDA turnover rates neither reached a maximum at 10^{-4} M (Figure 12), nor led to a linear Lineweaver-Burk plot (Figure 13).

For various reasons these deviations from simple Michaelis-Menten kinetics are not unexpected. The fact that the apparent K_M for FDA is as high as 3.6×10^{-5} M suggests that FDA is not a very specific esterase substrate and that saturation conditions cannot be reached. Moreover, several esterases may hydrolyze FDA concurrently with different specifity. In principle, this would have the same effect on a Lineweaver-Burk plot (Dixon and Webb 1966). Also, in view of the limitations of the substrate solubility and the requirement of maintaining cell integrity, experiments have to be performed at FDA concentrations in the range of and even much lower than the apparent K_M's. This, however, means that one must always expect to work in the first-order part of the

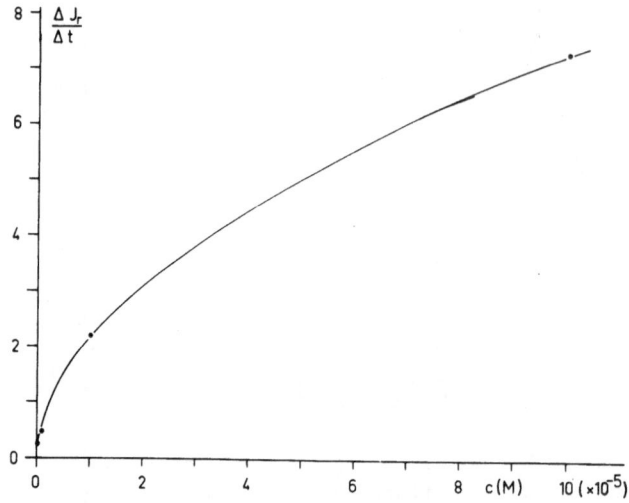

Figure 12. *Relative initial turnover rates, ΔJ/Δt, of FDA in a single L cell as a function of the FDA concentration applied*

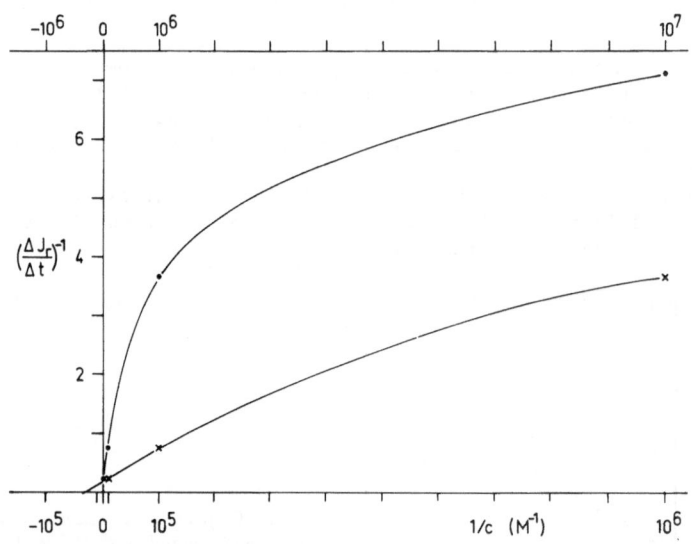

Figure 13. *Lineweaver-Burk plot of the reciprocal relative FDA turnover rates in a single cell versus reciprocal FDA concentrations. Upper curve at a 10-fold contracted abscissa. At a concentration range covering more than one order of magnitude, deviation from linearity is evident as compared with Figure 11*

enzyme kinetics. Thus, if initial FDA turnover rates are calculated, the FDA concentrations remain a parameter for the description of the reaction.

The rate of F_i production in the single cell is the result of at least two consecutive kinetic steps of the overall process: substrate permeation into the cell and enzymatic hydrolysis. For the permeation itself, a saturation function may be valid; therefore it has to be treated kinetically like the enzymatic reaction. Depending on which one of these reactions under saturation condition is the limiting one, the measurement of the initial velocities may serve to describe either the permeability of the cell for the substrate or the enzymatic activity of the intracellular esterases. The measurable turnover rates were shown to be far from saturation, but rather proportional to the inital FDA concentrations applied. Under these experimental conditions, first-order reaction kinetics may serve as a first approximation for the isolated permeation step as well. However, as long as both permeation and enzymatic turnover obey first-order kinetics, they may be combined to a single first-order reaction phenomenologically describing the 'invasion' of S_e to yield F_i as a counterpart of the 'elimination' of F_i to yield F_e.

As previously shown (Sernetz and Thaer 1972), the emission spectrum of intracellular fluorescein is shifted to longer wavelengths owing to protein binding of the product F_i (Figure 14). The decrease in the turnover rates as the intracellular accumulation of the product increases can likewise be interpreted in terms of product inhibition, as Burch (1955) demonstrated for the inhibition of horse liver esterase by Rhodamine B. Rapid spectroscopy of emission spectra during intracellular FDA turnover did not indicate any significant accumulation of *free* intracellular fluorescein during any phase of the turnover process.

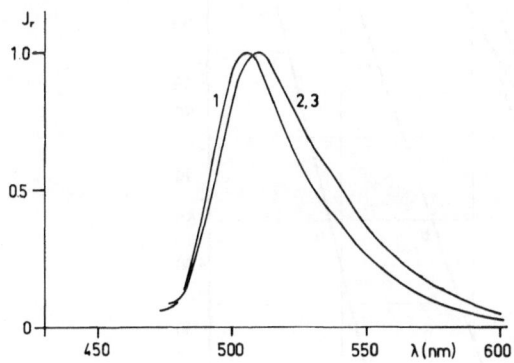

Figure 14. *Uncorrected relative fluorescence emission spectra of intracellular fluorescein liberated from FDA and fluorescein-dibutyrate FDB as compared with fluorescein (10^{-5} M in 0.06 M phosphate buffer ph 7.0) measured in capillary cuvettes under identical optical conditions (see also Sernetz and Thaer 1973)*

Therefore, the total mass of intracellular fluorescein and the intracellular turnover rates were calculated by taking into account a quench ratio of 2.1 between free and protein-bound fluorescein (Sernetz and Thaer 1972). This is true especially for the initial rates, as at zero time the fluorochrome concentration is low compared with the total cellular protein. Thus, the equilibrium is shifted to the side of the protein-fluorochrome complex.

Apart from the kinetic analysis, measurement of the initial FDA turnover rates at given FDA concentrations is of value for the description of specific properties of different cell types. As shown in Figures 15 and 16, at 10^{-5} M FDA, the mean turnover rates of L cells are about 10 times higher than those of HeLa cells. For each cell

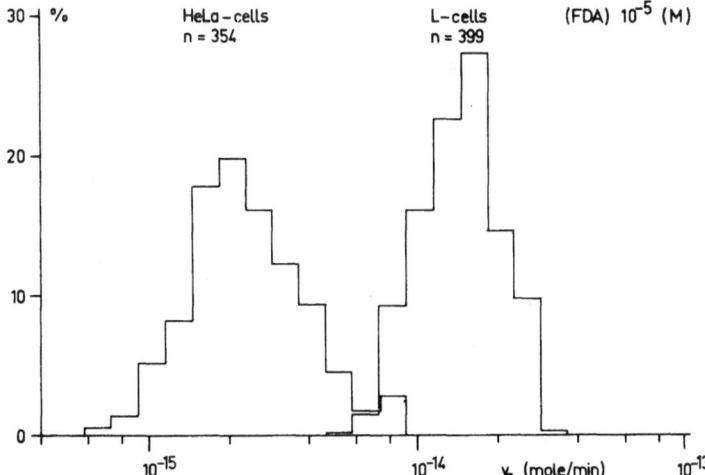

Figure 15. *Frequency distributions of FDA turnover rates, V_O, in individual L and HeLa cells, incubated in 10^{-5} M FDA*

population examined the rates show log-normal distribution. In Figure 16, the cumulative frequencies of the initial rates of several different cell populations at different FDA and FDB concentrations have been plotted on logarithmic probability paper (Aichinson and Brown 1969). From these plots the log mean and also the variance can be derived with-

out any difficulty; this is important to note, since experiment with solutions, cell homogenates or cell suspensions in a cuvette will only give a mean value (Löhr and Waller 1964).
The replacement of the "manual" microscope fluorometry by rapid automated flow-through techniques (see Rotman, this volume, part IV) would certainly lead to further reduction of the
variances given in Figure 16. Distinct differences also exist between the turnover rates
of a cell population exposed to different fluorogenic substrates of identical concentrations
(Figure 16 e and f). These findings offer promising new possibilities for cell diagnosis by
use of their typical enzyme equipment as studied by Löhr and Waller (1964), and in particular
for developing rapid flow methods (Dost 1968) for use in clinical diagnosis.

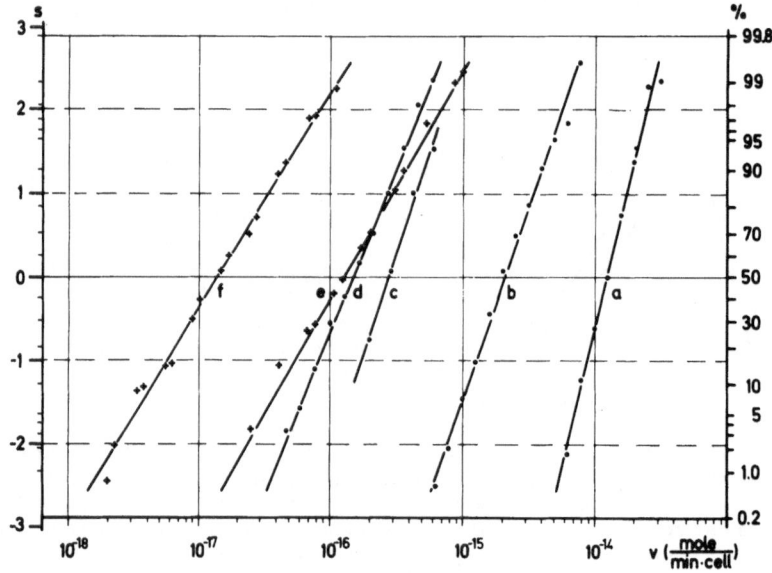

Figure 16. *Cumulative frequency distributions of FDA turnover rates in individual cells of different cell populations (logarithmic probability paper)*

		n	M FDA	$\log \bar{v}$	
a	L cells	399	1×10^{-5}	-13.92	± 0.15
b	HeLa cells	354	1×10^{-5}	-14.70	± 0.21
c	Dog kidney cells	65	5×10^{-7}	-15.57	± 0.21
d	Macrophages	342	5×10^{-7}	-15.80	± 0.25
e	BK 21	245	5×10^{-7}	-15.90	± 0.36
f	BK 21	273	FDB "	-16.86	± 0.39

Summarizing the kinetic features for the modification of the model introduced in
the beginning, a simplified version can be proposed for the phenomenological description of fluorochromasia at low substrate concentrations. The overall process may be
described in terms of two consecutive first-order reactions: the first reaction combining permeation and enzymatic hydrolysis to the process of "invasion", the second one
describing the "elimination". Thus the measurement of F_i, the fluorochromasia, is best
represented by a Bateman function (Dost 1968), of the type

$$F\ (t) = \frac{a \cdot k_1}{k_2 - k_1}\ (e^{-k_1 t} - e^{-k_2 t})$$

with k_1 and k_2 being the invasion and elimination constants, and a the initial concentration of the fluorogenic substrate. It should be noted that F_i denotes the fluore-
scein-protein complex (see above). Thus the 'elimination' constant, k_2, likewise represents a reaction resulting from a mere membrane permeability of the cell and the
liberation of the fluorescent product from the protein-bound form.
    In the case of L and HeLa cells, the elimination half-life periods do not differ,
whereas the invasion rates differ appreciably. At identical FDA concentrations this
accounts for the pronounced fluorochromasia of L cells as compared with that of HeLa
cells.

The analysis of turnover kinetics of fluorogenic substrates in cells is intimately related to similar cases of constraint enzymes or heterogeneous catalysts (Satterfield and Sherwood 1963). Therefore, some special analogous systems (Figure 17) will finally be discussed.

The histochemical techniques for enzyme localization in cells (Holt and O'Sullivan 1958)
differ from the fluorogenic substrate technique not only in that they are performed on fixed
cells, but also in that the intracellular product of the enzymatic reaction has to be
trapped by a fast accessory precipitation reaction $P + Q = PQ$ (Figure 17c) in order to
deposit a colored and absorbing product indicating the position of turnover. Therefore these reactions are useful only for localization and cannot be used for the simultaneous evaluation of enzyme kinetic data. Although up to now most chromogenic substrates used in histochemistry have been chosen merely for high absorbency, they are
in most cases also suitable as fluorogenic substrates. Moreover, rather than avoiding
diffusion of the product by precipitation, it is just this diffusion which provides an
additional criterion for the appraisal of cell reactions.

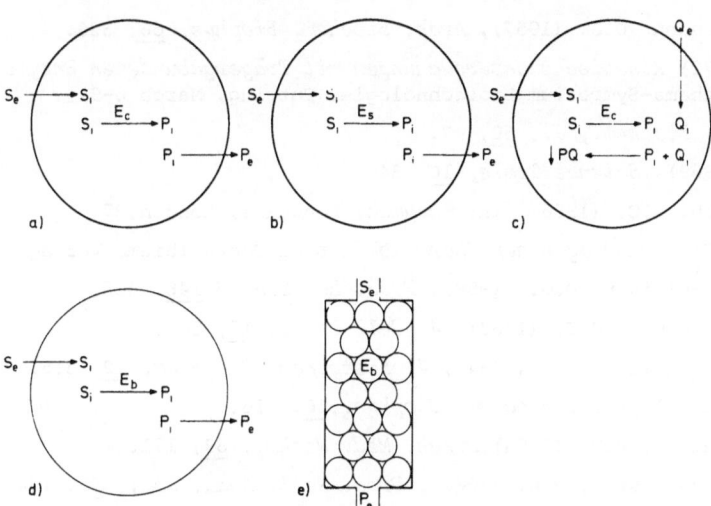

Figure 17. *Comparison of the cell model with several related experimental systems.
The circles refer to constrained enzyme systems, double contour stands for "membrane"
surface, single contour means without special membrane. S =substrate, P= product,
E = enzyme, Q =accessory chromogen, PQ =precipitate, Index i = intracellular or internal,
Index e =extracellular or external, E_c = cellular enzyme, E_s = solute, encapsulated
enzyme, E_b =bound, solid-phase enzyme.
(a) Fluorochromasia as described here, measuring parameter P_i.
(b) Encapsulated enzyme (Chang 1969), measuring parameter P_e
(c) Histochemical technique for localization of enzymes (Holt 1958; Hopsu 1963;1964)
measuring parameter PQ and its local distribution.
(d) Solid-phase enzyme bead, analogous to (a), but without assuming an enclosing membrane; measuring parameter P_i.
(e) Solid-phase enzyme column(Buchholz 1972); measuring parameter P_e*

A very similar system is that of encapsulated enzymes (Figure 17b) as described
by Chang (1969). The encapsulated enzymes may be soluble — as indicated by the index,
E_s — or insoluble derivatives. In both cases a distinct membrane is involved similar
to that in the cell models (Hopsu and Glenner 1963; Hopsu and McMillan 1964).

Solid-phase enzymes in the modification as beads, investigated either microfluoro-
metrically (Figure 17d) as described here or on columns (Figure 17e) (Buchholz 1972;
Katchalski 1957), likewise represent heterogeneous catalysts; but in contrast to the
other models, they lack particular "membrane".

The analytical treatment of a living cell as a heterogeneous catalyst, as a constraint enzyme system (Satterfield and Sherwood 1963; Blum and Jenden 1957), or as a compartment (Dost 1968), for pharmacokinetics at the cell level, compared with the analogous systems
referred to above, should provide further information on the metabolic reactions taking place
in single cells.

ACKNOWLEDGEMENTS

I wish to express my thanks to my colleague, A. Thaer, for helpful discussions and to Mrs. R. Menges and Mrs. K. Couwenbergs for skilful assistance. Parts of the experiments have been performed at Battelle Northwest Laboratories, Cellular and Molecular Biology Section. I am indebted to B. Wiley and E. Alpen for valuable support and discussions.

This investigation was supported by Battelle Memorial Institute, Columbus, Ohio.

REFERENCES

AITCHINSON, J. and Brown, J.A.C. (1969), *The Lognormal Distribution,* Cambridge University Press

BLUM, J.J. and Jenden, D.J. (1957), *Arch. Biochem. Biophys,* 66, 326.

BUCHHOLZ, K. (1972) *Kinetische Untersuchungen mit Trägergebundenen Enzymen in Reaktionssäulen*, Dechema-Symposium Biotechnologie, Tutzing, March 6-9.

BURCH, J. (1955), *Biochem., J.*, 59, 97.

CHANG, T.M.S. (1969), *Science Tools,* 16, 34

DIXON, M. and Webb, E.C. (1966), In: *Enzymes,* Longmans, London, 87.

DOST, F. H. (1968), *Grundlagen der Pharmakokinetik,* Georg Thieme Verlag, Stuttgart.

HOLT, S.J. and O'Sullivan, D.G. (1958), *Proc. Roy Soc. B* 148, 465.

HOPSU, V.K. and Glenner, G.G. (1963), *J. Cell Biol.*, 17, 503.

HOPSU, V.K. and McMillan, P.J. (1964), *J. Histochem. Cytochem.* 12, 315.

KATCHALSKI, E. (1957), *Arch. Biochem. Biophys.,* 66, 316.

LÖHR, G.W. and Waller, H.D. (1964), *Dtsch. Med. Wschr.*, 89, 171.

ROTMAN, B. and Papermaster, B.W. (1966), *Proc. Natl. Acad. Sci.*, 55, 134.

SATTERFIELD, C.K. and Sherwood, T.K. (1963), *The Role of Diffusion in Catalysis,* Addison-Wesley, Redding, Mass.

SERNETZ, M. and Thaer, A. (1970), *J. Microscopy,* 41, 43.

SERNETZ, M. and Thaer, A. (1972), *Analyt. Biochem.,* 50, 98.

SERNETZ, M. and Thaer, A. (1973), This Volume.

CHANGES IN THE MEMBRANE PERMEABILITY OF HUMAN LEUKOCYTES MEASURED BY FLUOROCHROMASIA IN A RAPID FLOW FLUOROMETER

Boris Rotman

Division of Biological and Medical Sciences
Brown University, Providence, Rhode Island

INTRODUCTION

Fluorochromasia is a property of cells to accumulate fluorescein intracellularly as a result of the enzymatic hydrolysis of a fluorogenic substrate, thereby becoming brightly fluorescent under blue light (Rotman and Papermaster 1966). The fluorochromatic reaction requires a nonpolar fluorogenic substrate, such as diesters of fluorescein and aliphatic acids. This type of substrate can penetrate the cell membrane and can be hydrolyzed intracellularly, thus liberating fluorescein, a negatively charged molecule, which does not diffuse readily across the cytoplasmic membrane of normal cells. If the rate of hydrolysis of the substrate exceeds the rate of fluorescein excretion, the intracellular concentration of fluorescein builds up until it reaches a steady state (in about seven minutes in our experiments) in which the rate of substrate hydrolysis is equal to the rate of fluorescein excretion.

Fluorochromasia has been observed in a variety of cells from different species including bacteria (Rotman and Papermaster 1966; Medzon and Brady 1969). Since the retention of intracellular fluorescein requires integrity of the cell membrane, fluorochromasia can be used to study processes affecting the membrane. For instance, the reaction of the membrane with specific cytotoxic antibodies and complement causes a total loss of fluorochromasia (Celada and Rotman 1967). This reaction has been used for detecting histocompatibility antibodies in human blood (Celada and Rotman 1967; Bodmer et al. 1967; Tosi et al. 1967; Douglas et al. 1971).

Another practical use of fluorochromasia is as a test for cell viability (Rotman 1966). After allowing the cells to hydrolyze fluorescein diacetate for five or ten minutes, the cell fluorescence is scored visually under a microscope equipped with a darkfield condenser and a broad band (440-480 nm) light filter. (A 6-volt tungsten lamp normally provided with a microscope is suitable for this test.) The cells are classified either as fluorescent or nonfluorescent, corresponding to living and dead, respectively. While this technique is adequate for routine evaluations of viability, it serves only to detect gross differences. For measuring discrete differences which may occur during growth, differentiation, or exposure to abnormal physiological conditions, one can determine the rate of fluorescein excretion across the cell membrane (Rotman 1966). However, these measurements are of value only when studying individual cells because of both time limitations and changes in the cell environment which occur during the assay. More meaningful data related to the distribution of fluorochromasia among a population of cells could be obtained by measuring a large number of individuals within a relatively short time. This has been accomplished by means of a rapid-flow fluorimeter recently developed in Van Dilla's laboratory (1969). The conditions and results of some experiments done with this instrument are described here.

MATERIALS AND METHODS

Freshly drawn human blood from a given individual was used routinely for comparison purposes. A drop of blood obtained by finger puncture was collected in 2 ml of phosphate-buffered saline (Dulbecco and Vogt 1954), and stored in ice. Other blood samples were procured at local hospitals and contained heparin as anticoagulant.

Fluorescein diacetate (five times recrystallized, purchased from Schwarz/Mann Corporation), was dissolved in reagent-grade acetone at a concentration of 5 mg per ml. The stock solution was stored at -20°C. Shortly before an experiment, the stock solution was diluted 1:400 with warm (about 35°C) phosphate-buffered saline. For this, 0.05 ml of the stock solution was rapidly mixed with 10 ml of the saline using a blow-out pipette. The milky suspension was further diluted 1:2 and was then used within minutes since it tends to flocculate with time. Blood samples were diluted with the

1:400 solution of fluorescein diacetate to give between 10^4 and 10^5 leukocytes per ml; they were then incubated at 37°C for 10 minutes. The cells were separated by centrifugation (2 minutes at 480 × g), resuspended in cold phosphate-buffered saline and kept in ice until the fluorometer assay[*].

The rapid-flow fluorometer was identical to that described by Van Dilla et al. (1969), except that instead of the laser we used a 100-watt mercury arc lamp and a broad band (440-480 nm) interference filter. The instrument was checked and standardized by measuring mulberry paper spores fixed with ethanol, stained with acridine orange, washed extensively with water by centrifugation and resuspended in water. These spore suspensions were stable for weeks when kept in the refrigerator. The standardization was made by adjusting the voltage of the photomultiplier to obtain a given average fluorescence per spore (Figure 1, see Plate I). The spores were also used to ascertain that the counting rate of the fluorometer corresponded to the total number of cells passing through the system. This was calculated by multiplying the volume per unit time passing through by the number of spores per ml measured in a hemocytometer.

RESULTS AND CONCLUSIONS

With the rapid-flow fluorometer used in our experiments, one determines quantitatively the fluorescence of individual cells. The principle of the instrument is to make a narrow stream of fluid, in which cells travel in single file, pass through a beam of exciting light. The pulses of light produced by fluorescent cells transversing the beam of exciting light are viewed by a photomultiplier positioned at right angles to both the light beam and the cell stream (Van Dilla et al. 1969). The resulting electrical pulses from the photomultiplier are amplified, analyzed for amplitude, and stored in a 400-channel pulse-height analyzer (Nuclear Chicago Corporation). A frequency-distribution histogram of fluorescent light emission per cell can be obtained from the memory of the analyzer when it is displayed on the screen of an oscilloscope. Integration in any portion of the histogram is provided by a scaler connected to the analyzer.

When the fluorochromasia of normal human blood cells was examined in the flow fluorometer, a characteristic bimodal distribution was observed (Figure 2, see Plate I, middle and lower curves). The peak near the ordinate axis corresponds mainly to lymphocytes, the other to granulocytes. Evidence for this conclusion comes from two types of observations. One was that lymphocytes separated in glass wool columns gave a single peak in the fluorometer that corresponded to the peak near the ordinate. The second line of evidence came from correlating a number of curves obtained with blood of people with different ratios of lymphocytes to granulocytes.

Control experiments with red blood cells separated from leukocytes showed that red cells did not interfere with the fluorometer measurements. This finding is consistent with previous observations that human red cells do not exhibit fluorochromasia (Rotman 1966).

The pattern of fluorochromasia distribution in fresh blood of a given person was constant for the entire span of this investigation (about eight months) except for one occasion which was associated with an abnormal condition of the blood donor. Similar bimodal distribution was observed in all the tested blood samples from normal persons (ca. 40). In contrast, several leukemic patients exhibited abnormal patterns, which sometimes consisted of only one peak (Figure 2, upper curve). The significance of this finding in terms of properties of the membrane of the leukemic cell cannot be evaluated without additional data concerning cell volume and rate of fluorescein excretion.

Storage of blood at room temperature for periods as brief as three or four hours resulted in patterns of fluorochromasia with lower average fluorescence per leukocyte. In contrast, the curves from blood stored at 0° C were constant for about 24 hours. After this time, significant changes in the patterns were observed resembling those from blood stored at room temperature.

The fast-flow fluorometer was also valuable for determining the rate at which

---

Serum proteins should be either excluded or kept below 0.5% during incubation with fluorescein diacetate because they have esterase activity. However, as indicated in the text, it may be advantageous to add fetal calf serum to fluorochromatic cells after the incubation with substrate.

PLATE I

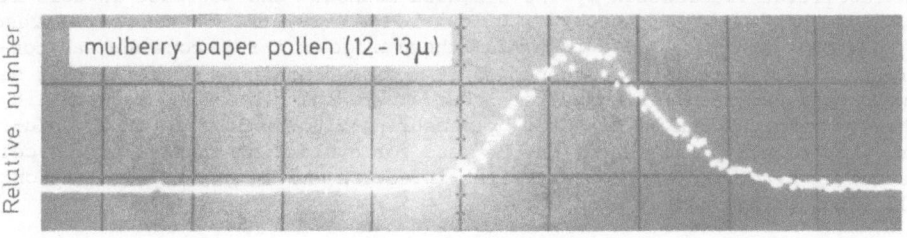

Figure 1. *Photograph of oscilloscope display showing the frequency distribution of fluorescent light emission per spore of mulberry paper stained with acridine orange. The curve represents a total of 10,000 spores*

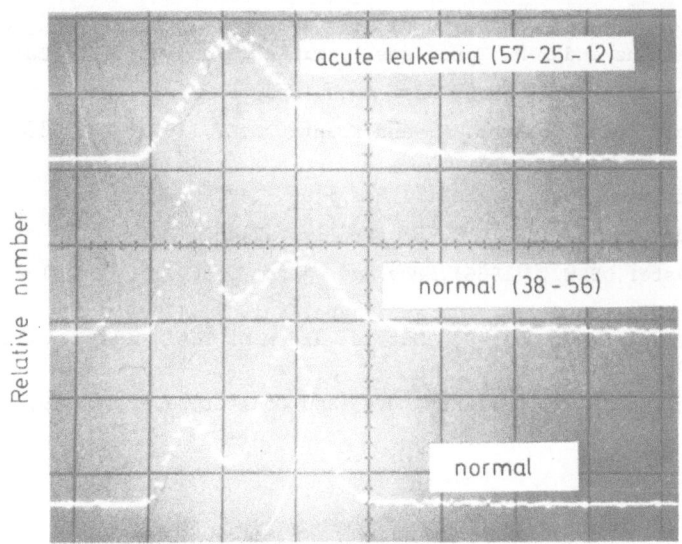

Figure 2. *Photograph of ocilloscope display showing the frequency distribution of fluorescent light emission per leukocyte. The cells had been made fluorochromatic as indicated in the text. Each curve represents a total of 10,000 readings.* **Upper** curve: *Sample from the blood of a patient with acute leukemia. The differential count was 57% blast, 25% lymphocytes and 12% granulocytes.* **Middle** curve: *Sample from a hospital patient with normal blood. Its count was 38% lymphocytes, 56% granulocytes.* **Lower** curve: *Standard sample from a normal person*

fluorescein is released into the medium. The procedure was simply to measure at intervals samples from a suspension of cells which had been previously allowed to accumulate intracellular fluorescein by the standard method. The decrease in cell fluorescence which occurred with time was used to calculate the exit-rate constant of fluorescein
(Rotman and Papermaster 1966). It was found that the average exit-rate constant was considerably reduced if the cells were suspended in medium containing 5% fetal calf serum.

In conclusion, our results demonstrate that subtle changes in the permeability
of the leukocyte membrane can be detected by measuring fluorochromasia in a rapid-flow
fluorometer. Accordingly, the system may be used for evaluating changes of the cell
membrane caused by cytotoxic antibodies, virus infection, malignant transformations,
cell-to-cell interactions and other agents.

ACKNOWLEDGEMENTS

I am very grateful to Dr. M. Van Dilla and Mr. J. Coulter for their help during
the construction and testing of the fast-flow fluorometer. Also, I am thankful to
Miss Gitte Lif-Ahlstrand for able assistance and to Dr. P. Calabresi for arranging
the supply of clinical material. Supported by NASA and USPHS GM-14198.

REFERENCES

BODMER, W., Tripp, M. and Bodmer, J., (1967), In: *Histocompatibility Testing*, (E.S.
    Curtoni, P.L. Mattiuz and R.M. Tosi, eds.), Williams and Wilkins, Baltimore, 341.

CELADA, F. and Rotman, B, (1967), *Proc. Nat. Acad. Sci. U.S.*, 57, 630.

DOUGLAS, K.S., Perkins, H.A., Cochrum, K. and Kountz, S.L.,(1971), *J. Clin. Invest.*,
    50, 274.

DULBECCO, R. and Vogt, M. (1954), *J. Exptl. Med.*, 99, 167.

MEDZON, E.L. and Brady, M.L., (1969), *J. Bacteriol.*, 97,402.

ROTMAN, B and Papermaster, B.W., (1966),*Proc.Nat. Acad. Sci. U.S.*, 55, 134.

TOSI, R.M., Pellegrino,M., Scudeller, G. and Cepellini, R.,(1967), In: *Histocompat-
    ibility Testing*, (E.S. Curtoni, P.L.Mattiuz and R.M. Tosi, eds), Williams and
    Wilkins, Baltimore, 351.

VAN DILLA, M.A., Trujillo, T.T., Mullaney, P.F. and Coulter, J.R., (1969), *Science*,
    163, 1213.

Part V
Fluorescent Molecular Probes for Complex Biological Molecules

PROBES FOR ENZYME CONFORMATION

G.K. Radda

Department of Biochemistry, University of Oxford, England

INTRODUCTION

Enzyme activity may be modulated in several ways. In particular, the interaction of small ligands with the enzyme, usually at a site distinct from the active site may alter the conformation of the protein molecule, a change often resulting in sigmoidal (cooperative) inhibition or activation curves. Alternatively the enzyme may be covalently modified (the process being catalyzed by another enzyme) which may again induce the appropriate conformational changes.

To understand the mechanism of enzyme regulation therefore, we have to be able to explore and define these conformational changes. Because such changes may be relatively small, we have developed or extended methods that rely on the introduction of extrinsic probes which, when attached to specific regions of a macromolecule, are able through their spectroscopic properties to give us some information about the nature of their binding environment and changes within it. Several forms of spectroscopy can be used as a method of detection, but the present paper will deal mainly with the use of fluorescent probes and how they can be used to detect conformational changes. Since, however, some of our conclusions are based on observations using additional probe methods, reference will be made to these.

CHEMICAL AND SPECTROSCOPIC PROPERTIES OF SOME FLUORESCENT PROBES

Fluorescent molecules can be attached to enzymes either noncovalently or covalently. Some examples of the former are compounds I, II, and III (Figure 1); and of the latter, 7-chloro-4-nitrobenzo-2-oxa-1,3-diazole (NBD-Cl) (IV), which is a reagent specific for SH-groups (Birkett et al. 1971) (though it can be forced to react with other nucleophiles). To illustrate some of the factors that influence the fluorescence properties of organic molecules, some of our studies on N,N-dimethylnaphtheurhodine (NE) and N-phenyl-1-naphthylamine (NPN) are discussed below.

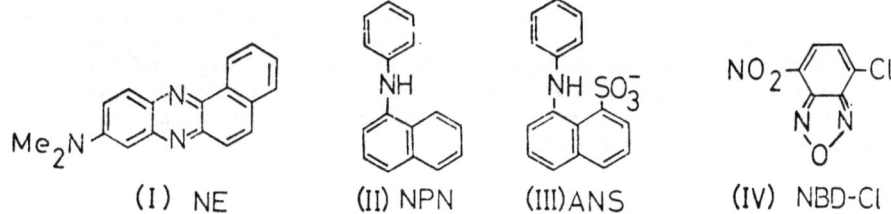

$$(I)\ NE \qquad (II)\ NPN \qquad (III)\ ANS \qquad (IV)\ NBD-Cl$$

Figure 1. *The structure of some fluorescent probes*

When light of appropriate wavelength is absorbed by a molecule, the Franck-Condon excited state is reached. Because of the time scale of this transition, the only changes that occur are those in electronic charge distribution of the chromophore and those in the polarization of the environment. Now before light emission occurs ($\sim 10^{-8}$ sec), several events take place. In particular because of the altered charge distribution in the chromophore, its geometry may change and a rearrangement of the solvent molecules around the chromophore may also lead to increased stabilization of the "equilibrium" excited state.

One consequence of this difference between the Franck-Condon and equilibrium excited states is that while the absorption spectrum will be a relatively faithful "indicator" of environmental *polarity* (which includes both dipole- and induced-dipole-type interactions), the fluorescence emission spectrum could be affected by *environmental constraint* as well. This constraint may be defined as any restriction imposed by the environment either on the ability of solvent molecules to rearrange in response to the excited-state dipole (*orientation constraint*) or on the ability of the chromophore to change its geometry (*packing constraint*).

Absorption Spectrum

The long wavelength-absorption band of NE, for example, is progressively blue shifted as the solvent polarity is decreased (Figure 2).

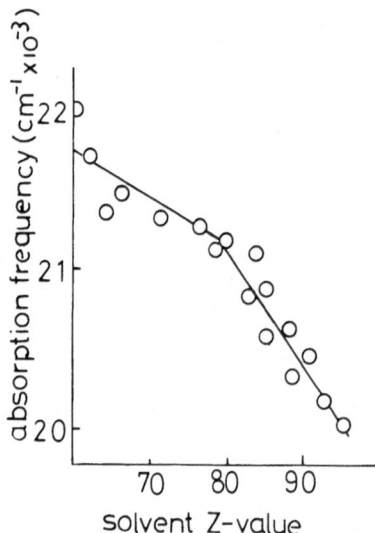

Figure 2. *The frequency of maximum absorption of the long wavelength band of NE in: 1) hexane; 2) benzene; 3) dichloromethane; 4) acetone; 5) 2-methylpropan-2-ol; 6) propan-2-ol; 7) propan-1-ol; 8) ethanol; 9) propandiol; 10) methanol; 11) 80% ethanol-water; 12) ethandiol; 13) 60% ethanol-water; 14) glycerol; 15) 40% ethanol-water; 16) 20% ethanol-water; 17) 5% ethanol-water*

In this figure an empirical solvent parameter (Z-value) introduced by Kosower (1958) is used to express solvent polarity. Even for the most viscous solvents used no deviation is found from the general relationship. The correlation of Z-value with absorption maximum does, therefore, provide in principle a method of measuring the polarity of the environment of the naphtheurhodine chromophore, *e.g.*, of the binding site on a macromolecule. In practice, however, the difficulty of accurately measuring the absorption maximum of a chromophore bound to a macromolecule would introduce a large uncertainty into such a determination.

Emission Spectra

As the excited singlet of NE has a large dipole moment, the fluorescence emission maximum is more sensitive to polarity than is the absorption maximum. However, the emission maximum might also be susceptible to environmental constraint. To examine this possibility, the emission of NE was studied in viscous and nonviscous solvents of known polarity. A comparison was made with the emission of N-phenyl-1-naphthylamine (NPN). As the emission from N-arylnaphthylamine dyes is known to be affected by environmental constraint (Radda 1971a and b; Gohlke and Brand 1971), this provided a useful check on the method used.

The fluorescence emission maxima of NE in solvents ranging in polarity from cyclohexane to water are progressively blue shifted as the solvent is made less polar. Plots of the uncorrected maximum against the solvent Z value are shown in Figures 3 and 4. The correlation between emission frequency and solvent polarity is good, but not linear. The viscous solvents, glycerol; 1,2-propandiol; and ethandiol cause blue shifts expected from their polarities. Any constraint of the equilibrium excited state, therefore, does not cause a blue shift detectable in this plot. However,

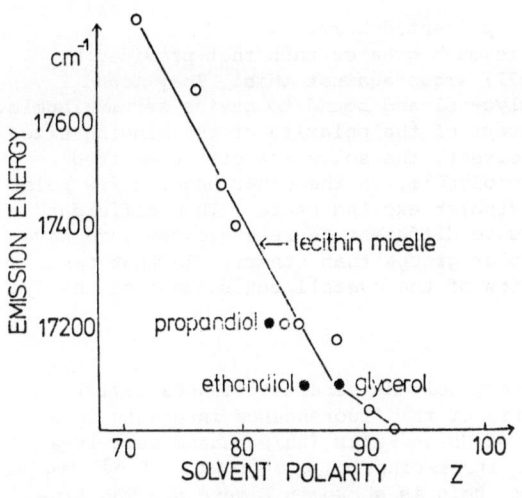

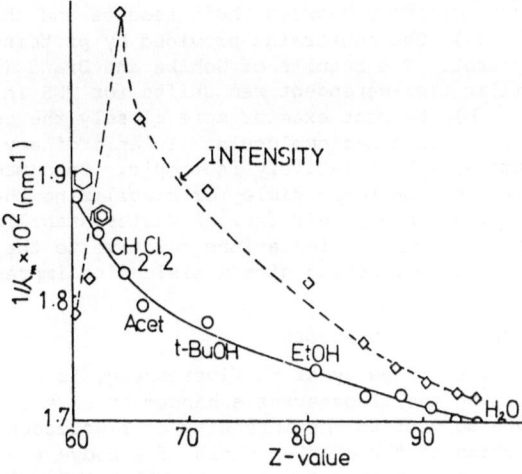

Figure 3. *The frequency of the fluorescence emission maximum of NE in different alcoholic solvents*

Figure 4. *The enhancement of NE fluorescence (relative to water) (upper curve) in different solvents*

environmental constraint can be detected in other measurements; this will be discussed later.

A similar result was found for NPN (Table I). This is perhaps surprising in view of the results of Gohlke and Brand (1971). They showed directly that the emission band of TNS in glycerol is shifted to the red during the lifetime of the excited state. The reason for this discrepancy is that the shift observed by Gohlke and Brand during the fluorescence decay is no greater than the average deviation of the points from the smooth curve constructed from the values given in Table I.

TABLE I

EMISSION MAXIMA OF NPN IN VARIOUS SOLVENTS

SOLVENT	EMISSION MAXIMUM	
	nm	cm^{-1}
Water	460	21,700
20% Ethanol	458	21,800
40% Ethanol	449	22,300
60% Ethanol	439	22,800
80% Ethanol	431	23,200
Methanol	426	23,500
Ethanol	419	23,900
Propan-1-ol	421	23,800
Propan-2-ol	418	23,900
2-methyl propan-2-ol	416	24,600
Acetone	416	24,600
Ethandiol	430	23,260
1,2-propandiol	426	23,470
Glycerol	432	23,100

These results suggest that the blue shift in emission does provide a valid measure of the polarity of the dye environment. Ainsworth and Flanagan (1969) came to the opposite conclusion. They showed that the emission from various probes, which competed for the same sites on bovine serum albumin, gave different values of binding-site polarity. They suggested that environmental constraint caused a considerable extra blue shift for the N-arylnaphthylamine dyes. Two explanations are possible for the discrepancy between their results and those presented here.

(a)   The constraint provided by proteins is much greater than that provided by glycerol.   The results of Gohlke and Brand (1971) argue against this.   They found similar time-dependent red shifts for TNS in glycerol and bound to bovine serum albumin.

(b)   We must examine more closely the concept of the polarity of the binding site of a dye on a macromolecule.   In an ordinary solvent, the solvation of the excited state will be relatively isotropic.   On a macromolecule, on the other hand, a few polar groups may be responsible for stabilizing the dipolar excited state.   Thus different dyes could have their excited states stabilized to different degrees because some have more favorable orientations relative to the polar groups than others.   In that case, the blue shift could give a misleading impression of the overall environment of the dye.

Fluorescence Intensity

The intensity of NE fluorescence is also very solvent sensitive (Radda 1971a). Defining the fluorescent enhancement as the ratio of the fluorescence intensity in a given solvent to the intensity of fluorescence of NE in water (this enhancement is a function of the quantum yield of the dye and of its extinction coefficient at 480 nm) yields a plot as a function of solvent Z value;   this is shown in Figure 4.   The fluorescence enhancement rises to a maximum as the solvent polarity is lowered from that of water to that of dichloromethane.   As the solvent polarity is further reduced, the enhancement falls again.   Viktorova et al. (1960) showed that this is due to a fall in quantum yield.   NPN shows a similar maximum (Ballard et al. 1971).

A possible explanation of this low enhancement in solvents with very low polarity, is that an n - π* transition overlaps the long wavelength transition of NE in these solvents.   n - π* absorptions are usually red shifted in apolar solvents, and the fluorescence in these transitions has a low quantum yield.

The quantum yields also fall on a smooth curve, and no deviation is found for NE in glycerol, 1,2-propandiol, or ethandiol (Figure 5).   Although viscous solvents do constrain the excited state, as shown below, the effect of constraint on the fluorescence quantum yield is too small to measure.

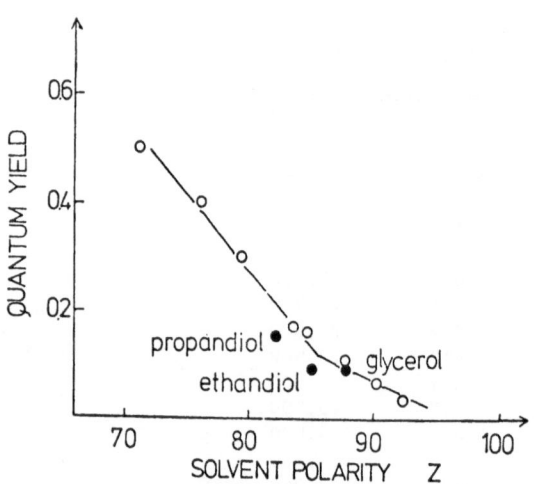

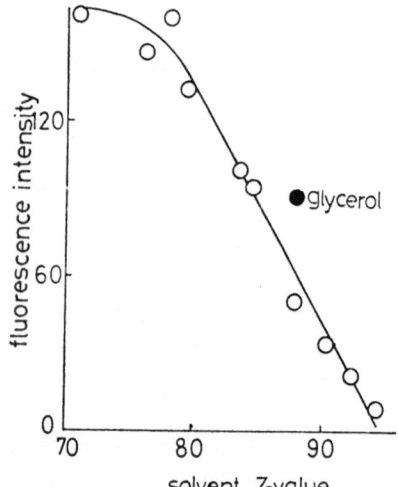

Figure 5.   *The quantum yield of NE flu-*
orescence in different alcoholic solvents
(cf. Figure 3)

Figure 6.   *Fluorescence intensity*
of NPN in different alcoholic solvents

This is in contrast to NPN, which shows an enhanced quantum yield in viscous solvents (Figure 6).   Both NE (Lippert 1955) and NPN (Radda 1971b) show an increase in dipole moment on excitation.   It is unlikely that they would not be equally sensitive to orientation constraint.   This suggests that the difference between NE and the N-arylnaphthylamine dyes is that the latter shows packing constraint.   This constraint is observed with chromophores whose geometries change on going from the ground (and

Franck-Condon excited) state to the equilibrium excited state. NE has a rigid poly-
cyclic structure: its geometry would not be expected to change much after excitation.
This is not the case with the N-Arylnaphthylamine dyes. If the hybridization of the
nitrogen changes after excitation, as suggested by both Kasha (1967) and El Bayoumi,
et al. (1970), the relative orientation of the two rings will change too. Hence these
dyes can be subject to packing constraint.

Absorption and Emission Overlap

A parameter that might be more sensitive to solvent rigidity than either quantum
yield or blue shift is the overlap between the absorption and emission bands of the
chromophore. This parameter was evaluated for NE in alcoholic and glycolic solvents
and was plotted as a function of solvent Z value in Figure 7. In the nonviscous solvents, a minimum in the overlap integral is found for NE in methanol. As the polarity
of the solvent is lowered from that of methanol, the overlap increases. This is because the emission band is blue shifted more rapidly than the absorption band.

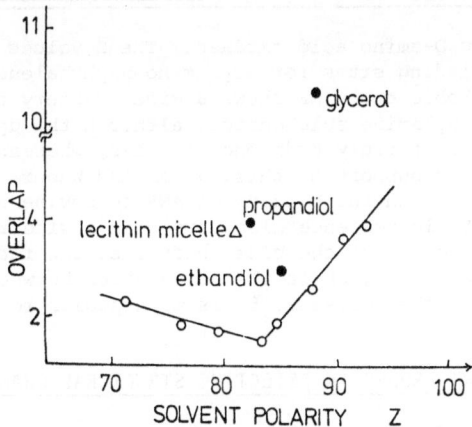

Figure 7. *The overlap between the absorption and emission bands of NE in alcoholic solvents*

Surprisingly, the overlap also rises as the solvent polarity is increased from
methanol. This is probably a reflection of the broadening of the emission and absorption bands which occurs in more polar solvents. The overlap for NE in the viscous
solvents is greater than that expected for solvents of comparable polarity. The extra
overlap contributed by viscosity rise in the order ethandiol<propandiol<glycerol.
This is the order of viscosity of the solvents.

This shows that the Franck-Condon excited state and its solvation shell have not
come to equilibrium before emission takes place. Nevertheless, the difference between
the emitting excited state and the equilibrium excited state is too small to have any
measurable effect on the quantum yield or blue shift of fluorescence. Thus by using
the overlap parameter, it is possible to measure separately the effects of constraint
and polarity. As the geometry of NE is not expected to change much after excitation,
the overlap probably measures orientation constraint.

The Interaction of NE with Proteins

NE interacts with a variety of proteins (and membranes)(Radda 1971a). The fluorescence of fully bound NE can be obtained by titrating NE with aliquots of concentrated
protein solutions. Linear plots of 1/corrected fluorescence (for scatter) against
1/protein concentration were extrapolated to infinite protein concentration to obtain
the fluorescence of fully bound dye (Table II).

As the blue shift in NE fluorescence is governed by polarity, it can be used to
measure the "polarity" of the binding sites on these proteins. Even this small selection of proteins shows a wide variety of shifts. The polarities of the sites, defined as the Z value of a solvent giving an equivalent shift, vary from 63 for bovine

TABLE II

EMISSION MAXIMA AND ENHANCEMENTS OF FLUORESCENCE OF NE BOUND TO PROTEINS

Protein	Emission Maximum		Enhancement
	nm	cm^{-1}	
Apo D-amino acid oxidase	580	17,240	4
Phosphorylase b	560	17,860	91
Glutamate dehydrogenase	560	17,860	12
α-chymotrypsinogen	548	18,250	115
Bovine serum albumin	546	18,310	110

serum albumin to 87 for apo D-amino acid oxidase. The Z values reported by Turner and Brand (1968) for protein binding sites for arylamino-naphthalene sulphonates vary from 74 to 86. Thus the hydrophobic probe NE shows a wider variety of site polarities than the amphiphilic N-arylnaphthylamine sulphonates, although the upper limit is similar. This may be because NE binds at truly hydrophobic sites, whereas ANS and TNS always bind near charged groups. In support of this, Joras and Weber (1971) have shown that arginine residues are involved in the binding of ANS to bovine serum albumin.

The enhancements of NE fluorescence also vary over a wide range, from 115 to 4. The enhancements follow the order of the blue shifts, so the factors governing blue shift also govern enhancement. But while the correlation between blue shift and enhancement is good for NE in all the solvents, it is much poorer for NE bound to proteins.

THE USE OF NONCOVALENT PROBES IN DETECTING STRUCTURAL CHANGES IN ENZYMES

Some years ago we showed that 1-anilino-naphthalene-8-sulphonate can be used to detect conformational changes in glutamate dehydrogenase (Dodd and Radda 1969; Brocklehurst and Radda 1971). In particular, the fluorescence of this probe when bound to glutamate dehydrogenase is enhanced by the regulatory ligand GTP (in the presence of the coenzyme NADH). The rate of this change is biphasic indicating at least two separate steps in the protein structure change.

The same probe can be used to study the rather more complex effect of other inhibitors such as diethylstilbestrol. Figure 8 shows that here too ANS detects at least two separate steps, and NADH exerts an additional enhancement. NADH alone has no effect. This contrasts with the behavior of NE, which also binds to glutamate dehydrogenase (Table II). A further enhancement is obtained on titrating the enzyme with NADH (Figure 9), and an additional effect of GTP can be ascribed to the tightening of NADH binding rather than to the GTP-induced conformational change detected by 1-anilino-naphthalene-8-sulphonate. The two probes clearly bind onto different regions of the enzyme molecule (the number of binding sites are two for NE and nine for 1-anilino-naphthalene-8-sulphonate per enzyme molecule) which undergo localized structural perturbations following ligand binding.

These observations also highlight the weakness of the approach, namely that the location of the probe molecules is usually not known.

COVALENT FLUORESCENT PROBES

To overcome the difficulties inherent in locating noncovalently linked probes we have studied the reaction of NBD-Cl with a series of model compounds and enzymes (Birkett et al. 1970). With the exception of cysteine (where a special mechanism is operative) the reagent is specific for SH groups around neutral pH, and yields a fluorescent product although the reagent itself is not fluorescent. The time course of the reaction can be easily followed by absorption spectroscopy (at 425 nm) and the extinction at this wavelength also gives a quantitative estimate of the number of SH groups that have reacted.

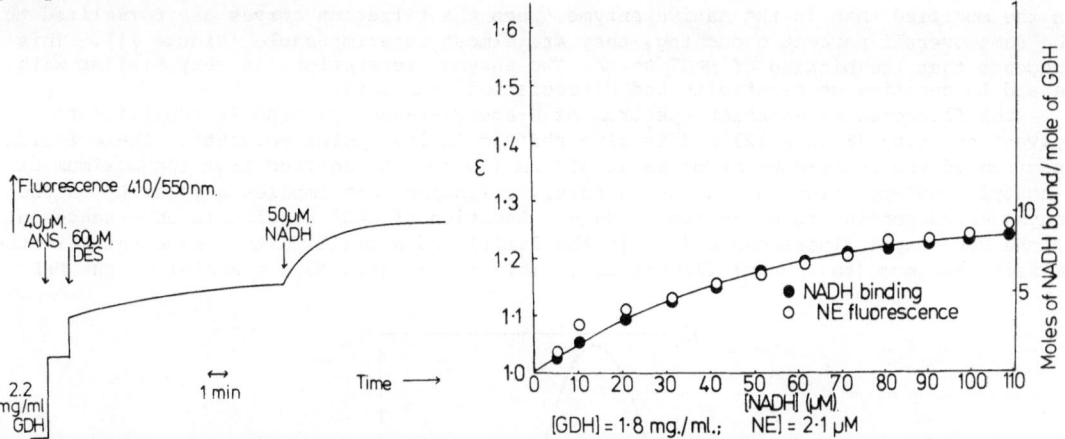

Figure 8. *Enhancement of 1-anilinonaphthalene-8-sulphonate (ANS) fluorescence on glutamate dehydrogenase (GDH) by diethylstilbestrol (DES)*

Figure 9. *Enhancement of NE fluorescence on NADH binding to glutamate dehydrogenase*

Glyceraldehyde-3-phosphate Dehydrogenase and Its Negative Cooperativity

In order to investigate whether the phenomenon of negative cooperativity in NAD^+ binding (Conway and Koshland 1968) is linked to conformational changes in the enzyme subunits we have prepared a fully active NBD-modified enzyme. The reaction of NBD-Cl with glyceraldehyde-3-phosphate dehydrogenase leads to the modification of eight SH groups per tetramer. The two groups in each subunit that react are cysteines 149 and 244, but the former react considerably faster than the latter. Modification of all eight groups completely inactivates the enzyme (Price 1971). However, addition of β-mercaptoethanol raises the activity to 95% in 15 min (Figure 10) with the removal of four NBD-groups from the active-site SH groups (cysteine 149). Binding of NAD^+ to both the native (Price and Radda 1971) and

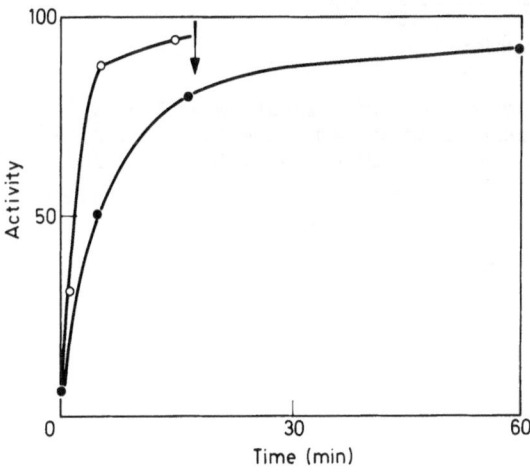

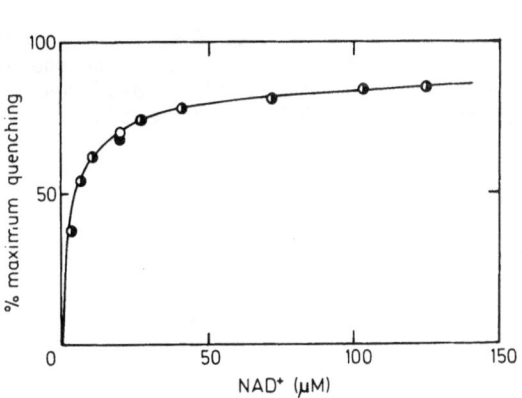

Figure 10. *Reactivation of NBD-G3PDH (approximately 7.0 -SH groups modified per mole) by β-mercaptoethanol. o——o 1 mM βME. ●——● 50 μM βME. The emzyme was at 0.3 mg/ ml in 0.1 M triethanolamine buffer at pH 7.6 (T = 0°C). At the time indicated (↓) the reactivated enzyme was dialysed or gel filtered (see text)*

Figure 11. *Comparison of native G3PDH (o) and tetra-NBD-active G3PDH (●) protein fluorescence quenching curves. The enzymes were at concentrations of 0.43 and 0.39 mg/ml respectively in 0.1 M triethanolamine buffer at pH 7.6 (T = 29°C)*

active tetra NBD-enzyme leads to the quenching of protein fluorescence (Price 1971). Although the extent of quenching and the inital protein fluorescence are somewhat lower in the modified than in the native enzyme, when the titration curves are normalized to the same overall percent quenching, they are almost superimposable (Figure 11). This suggests that the binding of NAD$^+$ to the two enzyme preparations is very similar with regard to negative cooperativity and dissociation constants.

The fluorescence emission spectrum of N-acetyl-S-NBD-cysteine is sensitive to solvent polarity (Figure 12) and is blue shifted in less polar solvents. The emission spectrum of the tetra-NBD-enzyme is at 503 nm (20 nm blue shifted from the maximum of N-acetyl-S-NBD-cysteine in aqueous buffers), a finding that implies a polarity on the enzyme corresponding to a Z-value of 84.4. Addition of NAD$^+$ results in an enhancement of the NBD-enzyme fluorescence (13% in the limit) and a small further blue shift in the emission maximum (to 500 nm) (Figure 13). This corresponds to a transfer of the NBD

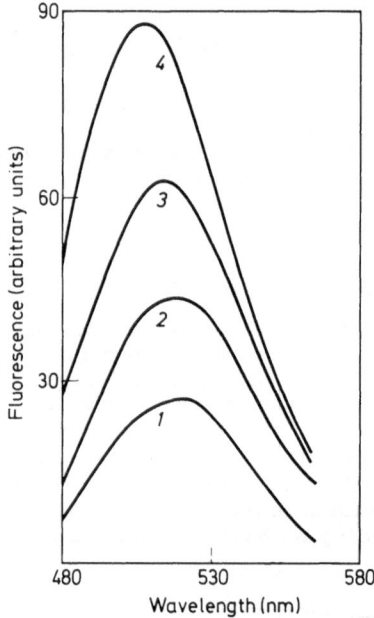

Figure 12. *The variation of the (uncorrected) emission spectrum of N-acetyl-S-NBD-cysteine with solvent polarity in the ethanol/water system. Spectra 1, 2, 3, 4 correspond to 0, 20, 40, 60% ethanol respectively. Excitation was at 420 nm. The N-acetyl-S-NBD-cysteine concentration was 12.2 μM (T = 25°)*

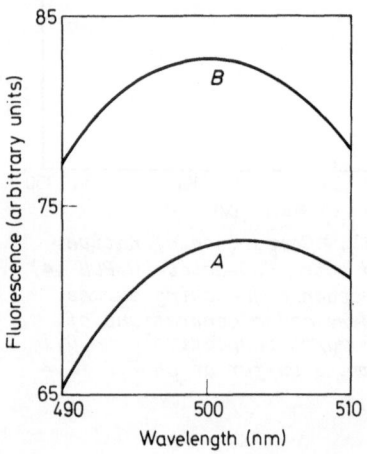

Figure 13. *Uncorrected NBD emission spectra of tetra-NBD-active G3PDH (0.39 mg/ml). (A) no NAD$^+$; (B) after addition of 700 μM NAD$^+$. Excitation was at 420 nm. For conditions see Figure 11*

moiety to a less polar environment (Z = 83.0) in the liganded state. Figure 14 shows the titration of the tetra-NBD-enzyme with NAD$^+$ and compares the increase in NBD fluorescence with the decrease in protein fluorescence (both expressed as a percent of the limiting changes). The two effects are clearly correlated. The fluorescence enhancement could arise from either a direct interaction between the bound NAD$^+$ and the NBD group, or an indirect interaction through a conformational change. The first alternative is unlikely for two reasons. First, the addition of large concentrations of NAD$^+$ (10mM) to N-acetyl-S-NBD-cysteine leads to 20% quenching of the NBD fluorescence. Second, the tetra-NBD-enzyme binds NAD$^+$ as well as does the native enzyme. An explanation for the enhancement of NBD fluorescence, based on the exclusion of neighboring water molecules by bound NAD$^+$, is difficult to exclude, but can be considered unlikely since NADH causes very little change in the NBD fluorescence. The most likely explanation therefore is that the probe is following the ligand-induced conformational change.

The problem then arises as to how one can use fluorescence to define the extent of the structural change. This can best be considered by relating fluorescence measurements to other observations, as will be illustrated below.

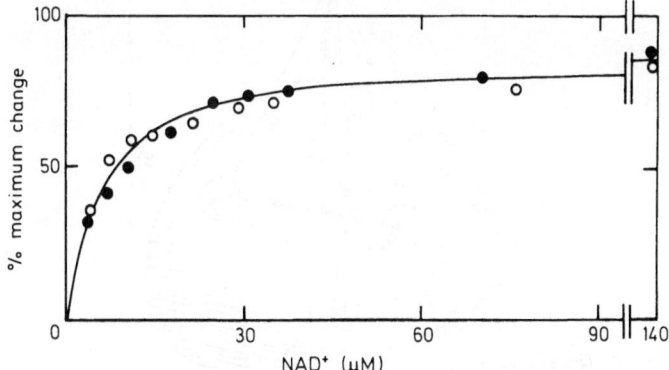

Figure 14. *Comparison of titration curves of tetra-NBD-active G3PDH with NAD$^+$. (o) protein fluorescence quenching; (●) NBD fluorescence enhancement. The titrations were performed in 0.1 M triethanolamine buffer, pH 7.6 (T = 29°C) and the enzyme concentration was 0.39 mg/ml. Protein fluorescence was excited at 300 nm and observed at 350 nm; NBD fluorescence was excited at 420 nm and observed at 500 nm*

POSITIVE HOMOTROPIC AND HETEROTROPIC INTERACTIONS IN PHOSPHORYLASE b

Glycogen phosphorylase b is activated by AMP. This activation is accompanied by a sigmoidal ligand-induced conformational change. This conformational change can be

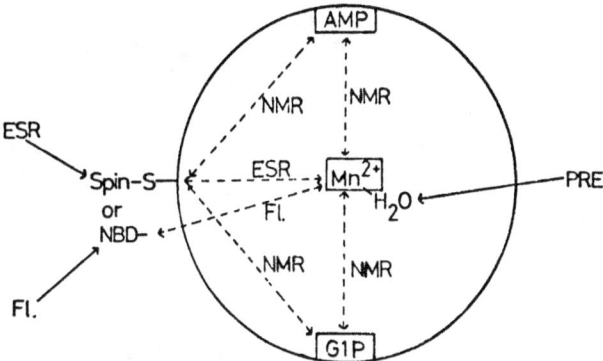

Figure 15. *A schematic representation of the probes and methods used in the study of phosphorylase b*

detected by a variety of methods (Birkett et al. 1971; Bennick et al. 1971) including chemical reactivity, fluorescence, proton relaxation enhancement, and electron spin resonance in a spin-labeled enzyme. Figure 15 summarizes the probes that we have been able to introduce without interference with enzymatic activity. Mn^{++} is bound by phosphorylase b (1 per subunit) and this site can be explored by studying the effect of the bound paramagnetic ion on the longitudinal nuclear magnetic relaxation rate of water molecules able to "visit" the Mn^{++} ions. One rapidly reacting SH group can be modified with the nitroxide spin label shown in the figure, and the same SH could also be labeled with the fluorescent NBD group. Both of these labels will detect the effect of AMP and of the substrate glucose-1-phosphate (or their combined influence) on the conformation of the enzyme (Birkett et al. 1971). By using the interaction among the various labels and those between the labels and the ligands (studied by fluorescence, electron spin resonance, and nuclear magnetic resonance), it has been possible to construct a stereochemical model for phosphorylase and its ligand binding sites shown in Figure 16 (Bennick et al. 1971).

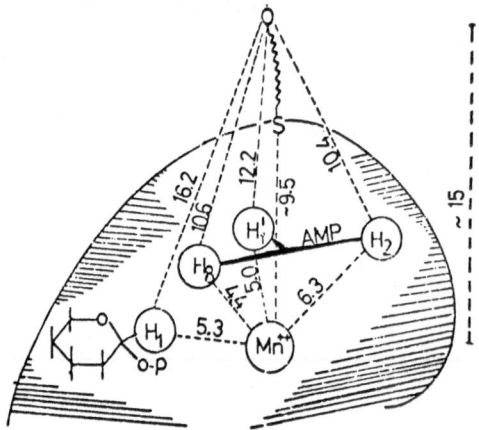

Figure 16. *The stereochemical relation of probe and ligand binding sites on phosphorylase b*

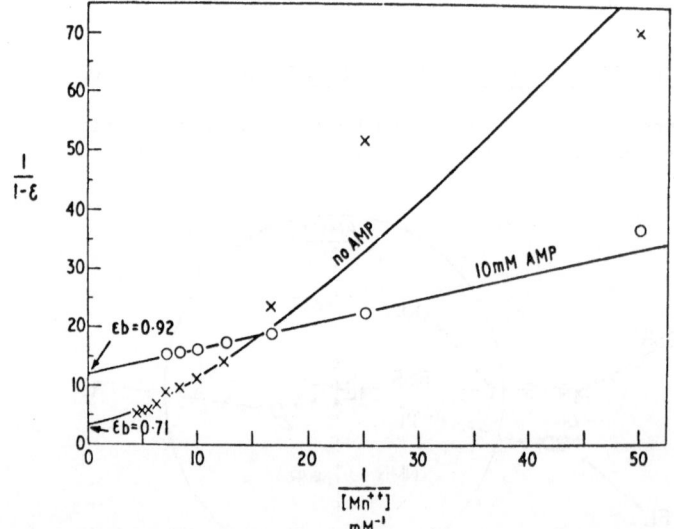

Figure 17. *Quenching of NBD-phosphorylase fluorescence on Mn^{2+} binding. NBD-phosphorylase in Tris-HCl (50 mM) buffer containing KCl (100 mM) at pH 8.5. Temperature was 25°C. Excitation was at 420 nm and emission at 515 nm. ε on the ordinate is the fluorescence intensity relative to one at $Mn^{2+}=0$. (From Bennick et al. 1971)*

For our discussion it is sufficient to consider the interaction between the bound Mn^{++} ions and the NBD label placed on the rapidly reacting SH group. The binding of Mn^{++} to the mono-NBD-enzyme results in the quenching the NBD fluorescence (30% quenching in the limit) (Figure 17). This quenching can be related to the distance between the two probe sites (Bennick et al. 1971). On the basis of some model studies (partly described before; Bennick et al. 1971) the observed 30% quenching corresponds to a Mn^{++}-NBD distance of 8 Å. Addition of AMP results in a diminution of this interaction (the limiting quenching is now only 10%, Figure 17) that corresponds to a lengthening of the Mn^{++}-NBD distance by approximately 1.5 Å. (Our preliminary model studies on labeled proline oligopeptides support a $1/(distance)^3$ dependence in the "paramagnetic" fluorescence quenching.) Thus it is clear (particularly since no gross conformational changes can be detected in phosphorylase b following ligand binding) that the extent of the changes is relatively small.

Phosphorylase can also be activated by covalently attaching a phosphate group to a serine residue using ATP and phosphorylase kinase (Fischer et al. 1970). Thus if the activation process and the 1.5 Å change in the Mn^{++}-NBD distance are related, one might expect a similar change when phosphorylase b is converted into the modified 'a' form. This is indeed the case, as is shown in Figure 18. The Mn^{++} quenching of the NBD enzyme fluorescence is again about 10% in the 'a' form, just as is observed for the 'b' form in the presence of AMP.

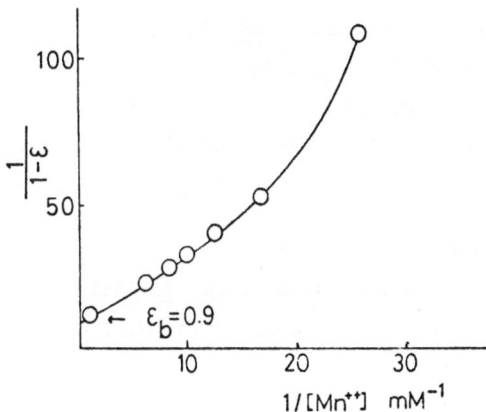

Figure 18. *Quenching of NBD-phosphorylase 'a' fluorescence on Mn^{2+} binding (cf. Figure 17)*

Thus fluorescence methods can be used in a variety of ways for probing and characterizing conformational changes in enzymes, furthermore the methods can be easily extended to the study of more complex systems such as membranes (Radda 1971a and b).

ACKNOWLEDGMENT

This work was supported by the Science Research Council.

REFERENCES

AINSWORTH, S. and Flanagan, M.T. (1968), *Biochim. Biophys. Acta.,* <u>194</u>, 213.

BALLARD, S.G., Barker, R.W., Barrett Bee, K.J., Dwek, R.A., RADDA, G.K., Smith, D.S., and Taylor, J.A., (1971), *Proceedings of the International Conference on Mitochondria,* Bressanone (1971).

BENNICK, A., Campbell, I.D., Dwek, R.A., Price, N.C., Radda, G.K., and Salmon, A.G., (1971), *Nature*, <u>234</u>, 140.

BIRKETT, D.J., Dwek, R.A., Radda, G.K., Richards, R.E., and Salmon, A.G. (1971), *Eur. J. Biochem.*, <u>20</u>, 494.

BIRKETT, D.J., Price, N.C., Radda, G.K., and Salmon, A.G., (1970), *FEBS Lett.*, <u>6</u>, 346.

BROCKLEHURST, J.R., and Radda, G.K., (1971), In: *Probes of Structure and Function of Macromolecules and Membranes*, (B. Chance, T. Yonetani, and A.S. Mildran, eds.), Vol. 2, Academic Press, New York, 59.

CONWAY, A., and Koshland, Jr., D.E., (1968), *Biochemistry*, <u>1</u>, 4011.

DODD, G.H., and Radda, G.K., (1969), *Biochem. J.*, <u>114</u>, 407.

EL-BAYOUMI, M.A., Dalle, J.P., and O'Dwyer, M.F., (1970), *J. Amer. Chem. Soc.*, <u>92</u>, 3494.

FISCHER, E.H., Pocker, A., and Saari, J.C., (1970), In: *Essays in Biochemistry*, (P.N. Campbell and F. Dickens, eds.), Vol. 6, Academic Press, London, 23.

GOHLKE, J.R., and Brand, L., (1971), *J. Biol. Chem.*, <u>246</u>, 2317.

JONAS, A., and Weber, G., (1971), *Biochemistry*, <u>10</u>, 1335.

KASHA, M., (1967), In: *Fluorescence — Theory, Instrumentation and Practice,"* (G.G. Guilbault, ed.), Edward Arnold, London, 201.

KOSOWER, E.M., (1958), *J. Amer. Chem. Soc.*, <u>80</u>, 3253.

LIPPERT, E., (1955), *Z.Naturforsch.*, <u>A 10</u>, 541.

PRICE, N.C., (1971), *Ph.D. Thesis*, Oxford.

PRICE, N.C., and Radda, G.K., (1971), *Biochim. Biophys. Acta.*, <u>235</u>, 27.

RADDA, G.K., (1971a), In: *Current Topics in Bioenergetics*, (D.R. Sanadi, ed.)Vol. 4, Academic Press, New York, 81.

RADDA, G.K., (1971b), *Biochem. J.*, <u>122</u>, 385.

TURNER, D.C., and Brand, L., (1968), *Biochemistry*, <u>7</u>, 3381.

VIKTOROVA, E.N., Kochemirovskii, A.S., Krasnitskaya, N.D, and Reznikova, I.I., (1960), *Opt. Spectrosc. (USSR)*, <u>9</u>, 288.

THE FLUORESCENCE OF BILIRUBIN-ALBUMIN COMPLEXES

Raymond F. Chen

Laboratory of Technical Development, National Heart and Lung Institute
Bethesda, Maryland

INTRODUCTION

In normal blood plasma, bilirubin is essentially completely bound to albumin (Ostrow and Schmid 1963), which thus can be considered the physiological transport protein for this bile pigment. Unbound bilirubin is thought to occur in plasmas of experimental animals and patients with hyperbilirubinemia such that the binding capacity of albumin is exceeded. In these cases, it is thought that the unbound bilirubin may produce toxic manifestations, especially to the central nervous system. The interaction of bilirubin with albumin therefore has the greatest importance and has attracted much study (Lester and Schmid 1964). Moreover, bilirubin-albumin complexes have recently drawn interest from optical rotatory dispersion (ORD) and circular dichroism (CD) studies on the mode of binding of the bilirubin. Blauer and King (1970) found that bilirubin-BSA* complexes had the largest molar rotation in the visible spectrum of any substance yet reported, and postulated the coiling up of bilirubin in the complex into the shape of a left-handed helix. Blauer et al. (1970) and Woolley and Hunter (1970) have extended these studies to the CD spectra of bilirubin complexes of HSA and BSA and have shown species-related differences.

The present studies utilize both the intrinsic fluorescence of the albumins, and an extrinsic fluorescence of the bilirubin to characterize the complexes. The binding constants for the bilirubin complexes with albumins of several species were obtained, and the extrinsic fluorescence was then able to be used to examine aspects of the conformation of bilirubin in the complexes.

MATERIALS AND METHODS

Crystalline HSA and rabbit albumin were obtained from Pentex, Inc. Crystalline HSA was purchased from Armour, Inc. Sheep, pig, and horse albumins were Fraction V preparations from Pentex. All albumins were treated with charcoal to remove impurities such as bound fatty acids (Chen 1967b). Bilirubin was obtained as the crystalline powder from both Sigma Chemical Company and Nutritional Biochemicals Corporation, and was repurified by recrystallization from chlorofrom-methanol mixtures according to the procedure of Ostrow, et al. (1961). In chloroform, the extinction of the purified bilirubin at 453 nm was found to be $5.9 \pm 0.1 \times 10^4$ cm^{-1} M^{-1}, corresponding to that of bilirubin of highest purity (Clinical Chemistry 1962). Stock solutions of bilirubin (10^{-4}M) were made up by dissolving the pigment with a small amount of 0.1 M KOH containing 10^{-3} M EDTA and diluting with water. Titrations were performed immediately after dissolving the bilirubin.

Fluorescence was measured with an Aminco-Bowman spectrofluorometer. Emission spectra were corrected for photodetector response (Chen 1967a). Quantum yields were determined relative to a quinine standard in 1 N H_2SO_4, assumed to have a yield of 0.55 (Melhuish 1961). Fluorescence decay times were measured with a TRW lifetime apparatus (Chen et al. 1967a). Absorption spectra were determined on a Cary spectrophotometer. Stopped-flow fluorescence measurements were made in the spectrofluorometer, using the propulsion unit described previously (Chen et al. 1969), and recording the phototube response on a Hewlett-Packard storage oscilloscope.

Molecular weights of 66,000 and 68,000 were assumed for BSA and HSA. All other albumins were assumed to have molecular weights of 67,000. The fluorescence intensity measurements used in the titrations of albumin with bilirubin were corrected for inner filter effects. Albumin fluorescence was linear with concentration under the conditions employed down to 2×10^{-8} M, although the titrations were done at concentrations of 10^{-7} M or higher.

---

* *Abbreviations: HSA, BSA, and RSA are human, bovine, and rabbit albumins.*

RESULTS

Instability of Bilirubin as Shown by the Absorption Spectra

The instability of bilirubin in aerated aqueous solutions is well known (With 1954; Gray 1953). Bilirubin appears to be even more unstable when excited with ultraviolet light (Ostrow and Branham 1970). The instability is a matter of concern in spectral and binding studies, since one must determine whether the rate of degradation will influence the results. Figure 1 shows absorption spectra of bilirubin and its complexes with BSA and HSA as measured initially and after 18 hours in the dark. Bilirubin solutions in

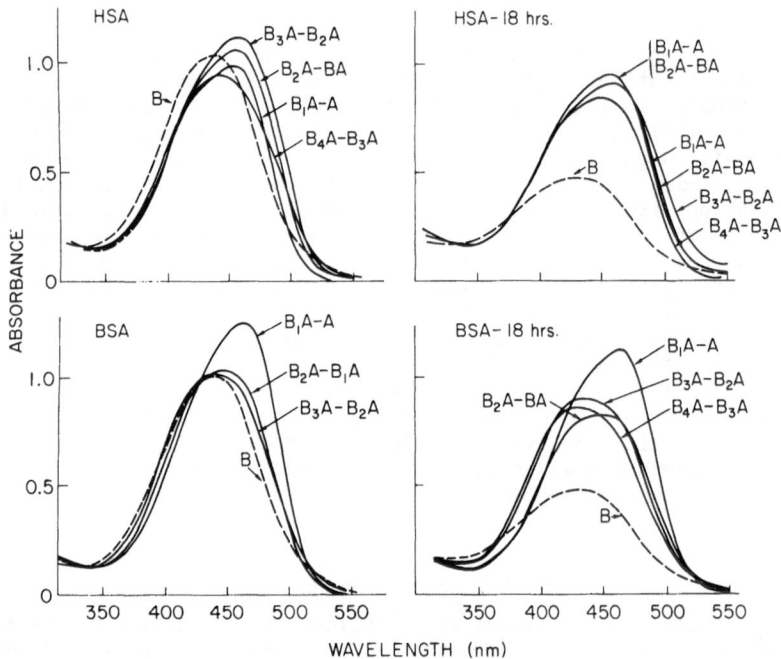

Figure 1. *Absorption spectra of free and albumin-bound bilirubin. The solid lines are differential spectra. The symbols are explained in the text. All solutions contained 0.1 M potassium phosphate buffer, pH 7.4. The spectra taken after 18 hours were obtained with solutions which had been kept in the dark but were not deaerated. Albumin concentration was 2×10^{-5} M*

phosphate buffer are seen to lose about half their optical absorption near 440 nm in 18 hours. In contrast, there is almost no change in absorption of the 1:1 bilirubin-HSA complex, and less than 10% decrease in absorbancy of the corresponding BSA complex. The absorption spectra of Figure 1 are differential spectra obtained with B_nA in the sample cell, where B = bilirubin, A = albumin, and n = mole ratio of added bilirubin to albumin; while the reference cell contained $B_{n-1}A$. The data show that the different bilirubin molecules have different spectra and are protected from degradation to different degrees by binding. Especially remarkable is the spectrum of the first bilirubin bound to BSA; subsequently bound bilirubins have much lower extinction. From the binding constants which were determined in this study and are reported below, the first two bilirubins are fully bound under the conditions used in these spectral determinations. The binding constants for the third and fourth bilirubins are not known, but the spectra of Figure 1 show that there is considerable protection against degradation of at least four bilirubins by each albumin molecule. This in turn suggests that at least four bilirubins per mole of albumin can be bound.

Figure 2 depicts absorption spectra for the complexes with sheep and rabbit albumins. The spectra here again illustrate the concepts that different bilirubins have different spectra, that the spectra obtained with albumins of different species are different, and that the spectra of bound and free bilirubin are easily distinguished.

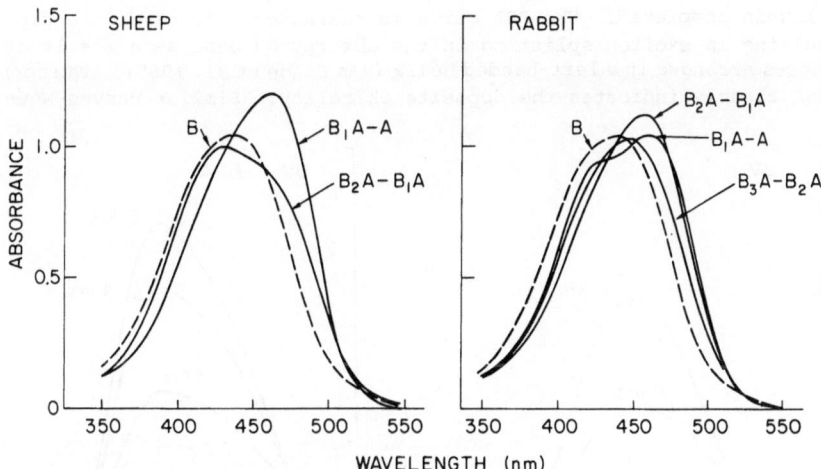

Figure 2. *Absorption spectra of sheep and rabbit albumin complexes containing bilirubin. The symbols are the same as in Figure 1*

TABLE I

SPECTRAL PROPERTIES OF BILIRUBIN COMPLEXES

1:1 bilirubin complex of:	Absorption: λ^{max} (nm)	ε (M^{-1} cm^{-1})	Emission: λ^{max} (nm)	Quantum Yield
HSA	458.8	46500	527	0.00102
BSA	462.5	57000	546	0.00085
RSA	460.3	52300	531	0.00084
Sheep Albumin	460.3	56600	528	0.00037
Free bilirubin in:				
0.1 M phosphate buffer, pH 7.4	438	52000	Negligible fluorescence in H_2O, dioxane, cyclohexane, ethanol, and	
$CHCl_3$	459	59800	benzene	

Table I indicates the absorption spectral differences shown in Figures 1 and 2.

Ostrow and Branham (1970) have studied the degradation of bilirubin by ultraviolet light and have concluded that binding to BSA actually accelerates photodegradation. Recent evidence suggests that the process in the absence of albumin involves singlet oxygen and is a self-sensitized photooxidation (McDonagh 1971). Our experience (Figure 3) appears contrary to that of Ostrow and Branham (1970) in that binding to BSA seems to stabilize against ultraviolet light.

Since the usual fluorescent room lighting contains a strong mercury line at 436 nm which is absorbed to a great extent by bilirubin, it seems wise to use incandescent lighting instead during bilirubin studies. Studies of bilirubin in the presence of albumin should not be invalidated due to degradative processes even if the studies require an hour or more, provided that adequate protection exists against strong ultraviolet light.

Fluorescence and CD of Bound Bilirubin

From the data of Woolley and Hunter (1970) it is known that the conformation of bilirubin in the HSA and BSA complexes must be very different. We have repeated many of the same measurements, and Figure 4 shows the ellipticity patterns obtained with BSA

and HSA bilirubin complexes. The BSA curve is characteristic of dipole-dipole interactions resulting in exciton splitting in the absorption band as a result of interact-
ing chromophores arranged in a left-handed helix (van Holde et al. 1965). The corresponding
curve for HSA clearly indicates the opposite chirality. Similar curves have been ob-

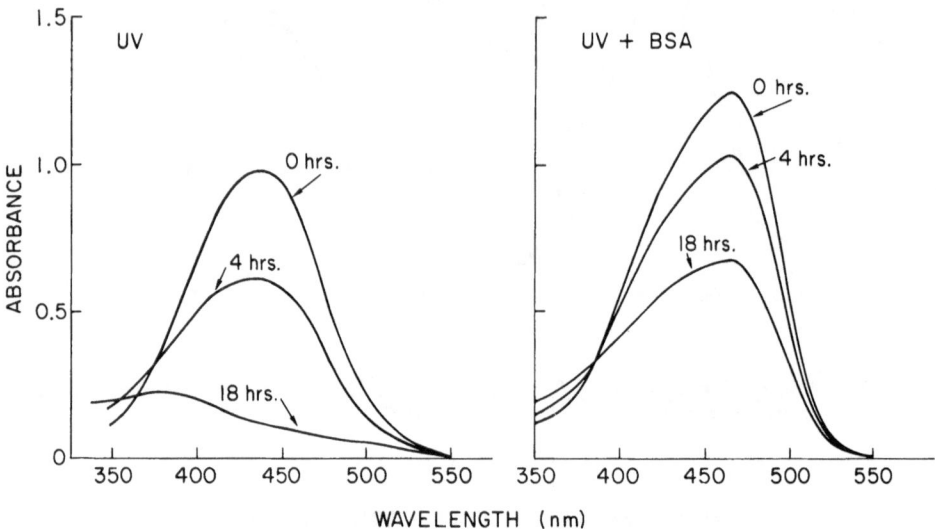

Figure 3. *The effect of ultraviolet light (UV) on the absorption spectrum of bilirubin. Solutions of bilirubin or bilirubin and albumin in 1:1 ratio were placed in
15 × 125 mm test tubes 6 inches from a GE F15TB "black light" with high output at 436
nm for the times shown, and their spectra were taken*

tained for bilirubin complexes of albumins from different species, and there are marked
differences. Those results will be reported elsewhere.

It was found that the bilirubin-albumin complexes had a faint fluorescence which
was visible to the naked eye when suitable solutions were held to an ultraviolet source.
The corrected emission spectra are shown in Figure 5. Measurement of the quantum
yields showed species differences, and the data are summarized in Table I. Unlike
other compounds which fluoresce when bound to albumin, bilirubin did not give similar
fluorescence when placed in apolar solvents such as cyclohexane or dioxane.

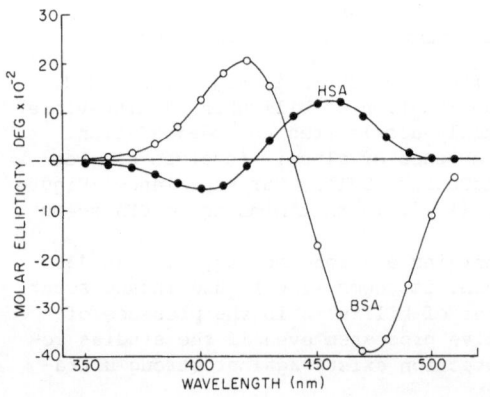

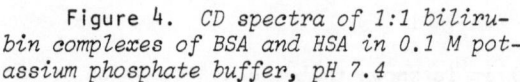

Figure 4. *CD spectra of 1:1 bilirubin complexes of BSA and HSA in 0.1 M potassium phosphate buffer, pH 7.4*

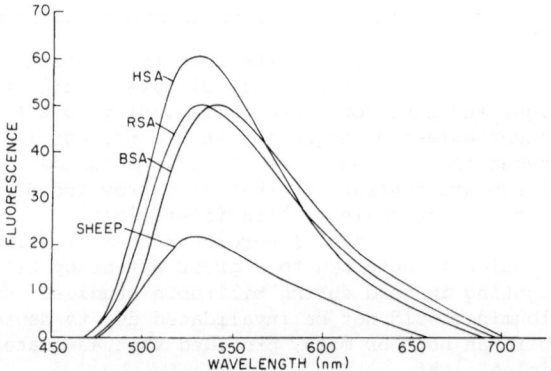

Figure 5. *Emission spectra of 1:1 bilirubin-albumin complexes of several species.
The spectra are corrected for detector response and normalized so that the areas
underneath the curves are proportional to
the quantum yields*

The fluorescence of bilirubin correlates with some features of the CD and absorption spectra. From Figure 1, it can be seen that the first bilirubin to bind to BSA has an anomalously high extinction, and a spectral shape distinct from subsequently bound bilirubins. Likewise, the data of Woolley and Hunter (1970) showed that only the first bilirubin gave a biphasic CD spectrum characteristic of a left-handed helix; additional bilirubins decreased the absolute magnitude of the observed ellipticity. The anomalous nature of the first bilirubin bound to BSA is also apparent in the fluorescence titration (Figure 6). It can be seen that the fluorescence yield is much higher from the initial bilirubin than from subsequent molecules bound to BSA. In contrast, the yield of fluorescence is about the same for the first three bilirubins to bind to HSA. In the case of the HSA complexes, Woolley and Hunter's (1970) data also show the CD effects to be additive for the first three bilirubins to bind.

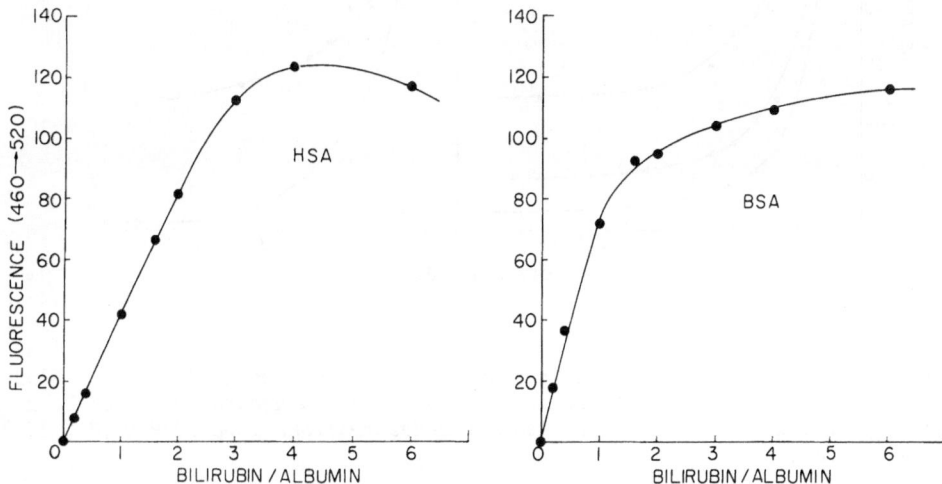

Figure 6. *Fluorescence titration of albumin with bilirubin. Albumin concentration was constant at 3 × 10⁻⁶ M. The bilirubin fluorescence was corrected for a solvent blank as well as inner filter effects at the exciting and emitting wavelengths*

Determination of the Binding Constants for the Bilirubin Complexes

The binding of bilirubin is accompanied by a loss in protein-intrinsic fluorescence. Because of the strong absorption of bilirubin in the region of protein emission, the most likely explanation of the quenching is resonance energy transfer from excited tryptophan to bilirubin. The phenomenon can be used to determine the binding constants; the method of Steiner et al. (1966), was employed.

Titrations of albumins with bilirubin were performed at different concentrations (Figure 7). In cases where relatively high concentrations of albumin were used, the titration curves were identical when plotted in terms of percent initial fluorescence vs. mole ratio of added bilirubin to albumin. These curves then represent the amount of fluorescence remaining when all the added bilirubin is bound. At lower concentrations, enough dissociation occurs so that the titration curves do not coincide with the titrations at high concentration. However, the amount of bilirubin binding can be determined at any concentration from the degree of fluorescence quenching, since the quenching-binding relation is known from the high-concentration titration.

Typical Scatchard plots derived from such titration data are shown in Figure 8. The data are clearly better for determination of the first association constant than for additional binding, since the first bilirubin causes 65%-80% quenching, leaving only a small amount of protein fluorescence whose quenching is utilized to measure further binding. The results are, for BSA, $k_1 = 2.0 \times 10^7$ M^{-1} and $k_2 = 0.2 \times 10^7$ M^{-1}; and for HSA, $k_1 = 7.0 \times 10^7$ M^{-1} and $k_2 = 0.6 \times 10^7$ M^{-1}. Table II gives the binding constants for the complexes of bilirubin with albumins of different species. It is of interest that the binding of bilirubin to HSA is tighter than to BSA, and that in both albumins the first bilirubin seems to bind about about an order of magnitude more tightly than the second.

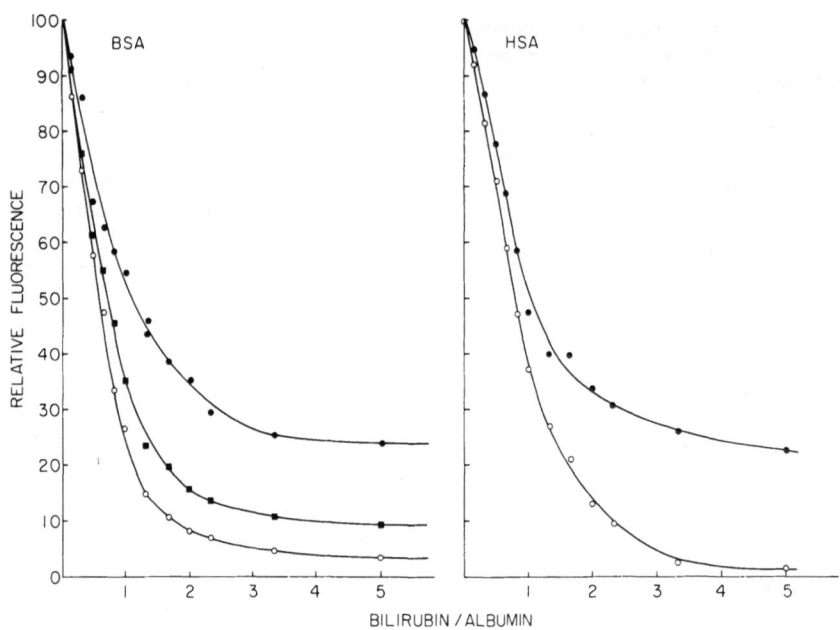

Figure 7. *Fluorescence quenching titration of albumins with bilirubin. Fluorescence excited at 290 nm and monitored at 340 nm is plotted against ratio of total concentrations of bilirubin and albumin. Albumin concentrations were 10^{-5} M (○), 10^{-6} M (■), and 10^{-7} M (●)*

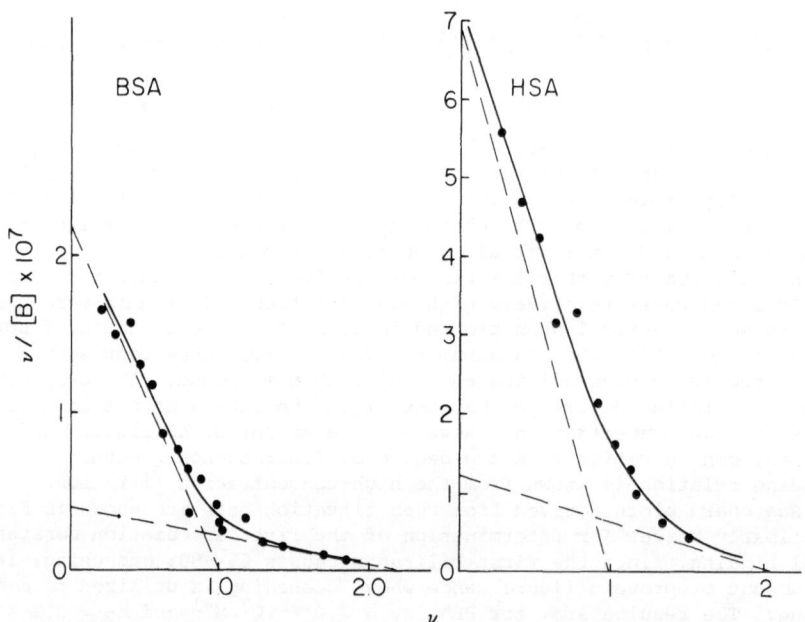

Figure 8. *Scatchard plots for the binding of bilirubin to BSA and HSA. ν here is the average number of moles of bilirubin bound to albumin, and [B] is the free bilirubin concentration*

TABLE II

ASSOCIATION CONSTANTS OF BILIRUBIN-ALBUMIN COMPLEXES AT 25°

	k_1 (M^{-1})	k_2 (M^{-1})
HSA	7.0×10^7	0.6×10^7
BSA	2.2×10^7	0.2×10^7
RSA	1.2×10^7	------
Horse albumin	1.7×10^7	------
Porcine albumin	1.8×10^7	------

Bilirubin-to-tryptophan Distance in the Complexes

The fluorescence quenching can be used to determine the approximate distance between the tryptophans and the bilirubin, using energy-transfer calculations. Figure 9 shows the overlap between the emission of albumin and the absorption spectrum of bilirubin necessary for such energy transfer to occur. Although the absorption spectrum of bound bilirubin has a peak in the range of 22,000 cm^{-1}, and a minimum near the peak of protein emission, the extinction coefficient is high even at the minimum, being of the order of 6×10^3 cm^{-1} M^{-1}.

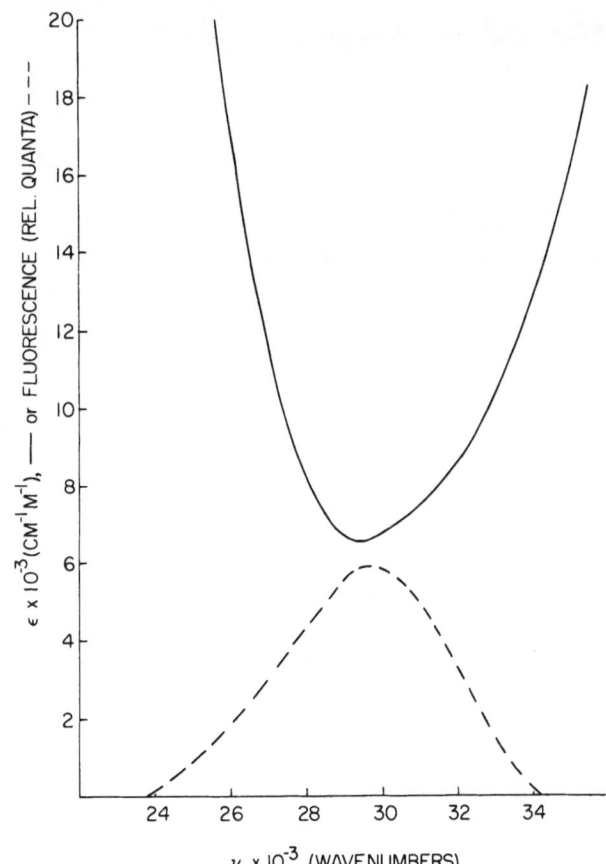

Figure 9. *Intrinsic HSA fluorescence spectrum and part of the bilirubin-HSA absorption spectrum, showing the spectral overlap necessary for energy transfer to occur. ν is the frequency, in wave numbers*

According to Förster's theory (1948), the critical transfer distance R_c, where the probabilities of energy transfer and of fluorescence from the donor are equal, is given by:

$$R_c = \sqrt[6]{\frac{1.66 \times 10^{-33} \, \tau J}{i^2 \times \bar{\nu}_0^{\,2}}}$$

PLATE I

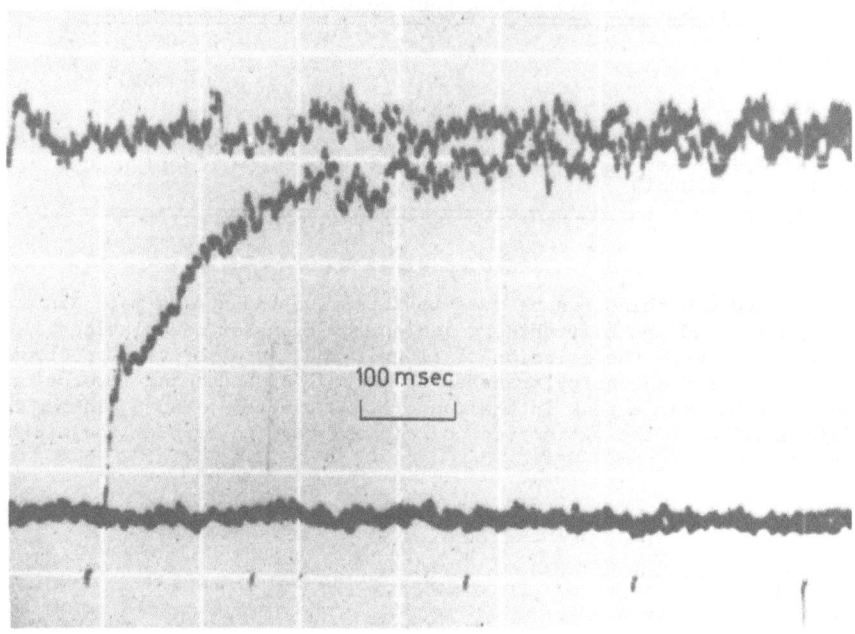

Figure 10. *Oscilloscope tracing showing the rate of appearance of bilirubin fluorescence following mixing of bilirubin and albumin in the stopped-flow apparatus. Final concentrations of bilirubin and albumin were both 1 x 10⁻⁵ M. The time base is 100 msec per major division*

where τ is the donor fluorescence decay time, $\bar{\nu}_0$ is the mean of the peak positions (in wave numbers) of the donor emission and lowest energy absorption bands, J is the overlap integral, and i the refractive index. The overlap integral, $J = \int \varepsilon(\nu) f(\nu) d\nu$ was evaluated graphically for the bilirubin-albumin system by plotting the absorption and fluorescence spectra in terms of frequency, i.e., ε_ν and f_ν vs. ν (in wave numbers) as shown in Figure 9 for bilirubin and HSA.

In calculating R_c, it was assumed that i = 1.6 and $\bar{\nu}_0 = 3.3 \times 10^4$ cm^{-1}. For the BSA-bilirubin system τ was measured to be 6.9 nsec, J was 2.15×10^9 M^{-2} cm^{-3}, and R_c was 31.9 Å. For the HSA-bilirubin system, τ was 5.0 nsec, J was 2.10×10^9 M^{-2} cm^{-3}, and R_c was 31.0 Å. These transfer distances are large compared with many other systems studied, and explain in part the large amount of quenching observed.

The bilirubin-tryptophan distance, R, is inferred from the fractional degree of energy transfer, X, from the relation (Förster 1948):

$$1 - X = \frac{1}{(R_0/R)^6 + 1}$$

From Figure 7, X is seen to be about 0.65 and 0.80 for HSA and BSA. R is then found to be 28 Å for the distance between the first bilirubin bound and the single tryptophan residue of HSA, and 25 Å for the average distance in BSA between bilirubin and the two tryptophans. These separation distances are so large as to exclude the possibility of tryptophan's being at the binding site.

The R_c calculations have assumed an orientation factor between donor emission and acceptor absorption dipoles that corresponds to random orientation. Such random orientation obviously does not occur, since the CD data show a fixed conformation of the bilirubin. However, one can exclude a perpendicular orientation of the dipoles since this would have prevented any energy transfer, a situation contrary to the observed quenching. Polarization spectra (Chen 1971) also show that several bilirubin absorption oscillators are present in the region of protein emission, and these oscillators are at large angles to each other. The CD data also show that the bilirubin molecule is probably nonplanar. In view of these considerations, the assumption of random orientation probably gives the best estimate of R_c.

Also, the calculation of R has utilized the degree of quenching as an indicator of degree of energy transfer, although these two quantities may not coincide, as we have emphasized previously (Chen and Kernohan 1967). Errors may occur when energy transfer occurs from less fluorescent tryptophans; i.e., transfer is competitive mainly with radiationless processes rather than radiative transitions. However, in HSA, there is only one tryptophan; while in BSA, the two tryptophans are apparently approximately equally fluorescent (Luk 1971). The use of quenching as a measure of energy transfer is thus indicated in these special cases.

Kinetic Measurements of Bilirubin Binding

Utilizing stopped-flow fluorometry, one may study the binding process by monitoring the quenching of protein emission at 340 nm, or the emission of the bound bilirubin at 540 nm. Our measurements have shown that the quenching of protein emission occurs very rapidly. The apparatus used has a dead time of about 20 msec, and the quenching is already at least 95% complete before the first observation can be made. On the other hand, the appearance of bilirubin fluorescence follows a slower course. Figure 10 (Plate I) shows an oscilloscope trace of the bilirubin fluorescence after mixing bilirubin with BSA. The time required for half the bilirubin fluorescence to appear is about 100 msec. The course of the process is exponential and independent of the final concentration of bilirubin and albumin, indicating a first order process with a time constant of about 7 sec^{-1}.

Species differences in the kinetic course of binding and bilirubin fluorescence appearance have been observed and will be reported elsewhere.

DISCUSSION

Fluorescence measurements were used in this study to contribute information concerning the physical mode of binding of bilirubin to albumins from different species. Fluorescence is a multifaceted optical parameter that is amenable to use in kinetic studies. The data shown here indicate that the binding of bilirubin to albumin is very rapid and tight, but the bound bilirubin undergoes a measurably slow transformation resulting in fluorescence.

Investigators have attempted to determine the binding constants for bilirubin-albumin complexes, but none of these studies can be considered reliable. The major problems are the instability of the bilirubin and the very tight binding; these factors invalidate studies utilizing equilibrium dialysis or other time-consuming techniques. The binding constants obtained in this study show that bilirubin is among the substances most tightly bound by albumin, rivalling the fatty acids in the size of the association constants.

Bilirubin is essentially nonfluorescent in organic solvents, so it is tempting to attribute the fluorescence from the complexes to a constrained conformation of bilirubin. A restricted shape of the bilirubin molecule is already indicated by the CD data, a fact that suggests that the tetrapyrrole structure is coiled in helical form.

The shape change that occurs on binding of bilirubin must involve a considerable entropy change. Preliminary studies at different temperatures have permitted evaluation of thermodynamic parameters and have largely confirmed the prediction of a large entropy change associated with binding.

These studies have illustrated the use of fluorescence to study a difficult binding problem. Conformational changes in the ligand as well as in the binding protein can be detected by fluorescence spectroscopy.

REFERENCES

BLAUER, G., Harmatz, D., and Naparstek, A., (1970), *FEBS Lett.*, 9, 53.

BLAUER, G., and King, T.E., (1970), *J. Biol. Chem.*, 245, 372.

CHEN, R.F., (1967a), *Anal. Biochem.*, 20, 339.

CHEN, R.F., (1967b), *J. Biol. Chem.*, 242, 173.

CHEN, R.F., (1971), *Anal. Lett.*, 4, 459.

CHEN, R.F., Alexander, N., and Vurek, G.G., (1967), *Science*, 156, 949.

CHEN, R.F., and Kernohan, J.C., (1967), *J. Biol. Chem.*, 242, 5813.

CHEN, R.F., Schechter, A.N., and Berger, R.L., (1969), *Anal. Biochem.*, 29, 68.

FÖRSTER, T., (1948), *Ann. Phys.*, Leipzig, 2, 55.

GRAY, C.H., (1953), *The Bile Pigments*, Methuen, London.

LESTER, R., and Schmid, R., (1964), *N. Eng. J. Med.*, 270, 779.

LUK, C.K., (1971), *Biopolymers*, 10, 1229.

MCDONAGH, A.F., (1971), *Biochem. Biophys. Res. Comm.*, 44, 1306.

MELHUISH, W.H., (1961), *J. Phys. Chem.*, 65, 129.

OSTROW, J.D., and Branham, R.V., (1970), *Gastroenterology*, 58, 15.

OSTROW, J.D., Hammaker, L., and Schmid, R., (1961), *J. Clin. Invest.*, 40, 1442.

OSTROW, J.D., and Schmid, R., (1963), *J. Clin. Invest.*, 42, 1286.

RECOMMENDATION on a Uniform Bilirubin Standard, (1962), *Clin. Chem.*, 8, 405.

STEINER, R.F., Roth, J., and Robbins, J., (1966), *J. Biol. Chem.*, 241, 560.

VAN HOLDE, K.E., Brahma, J., and Michelson, A.M., (1965), *J. Mol. Biol.*, 12, 726.

WITH, T.K., (1954), *Biology of Bile Pigments Including a Review of Their Chemistry and a Discussion of Analytical Methods*, Arne-Frost-Hansen, Copenhagen.

WOOLLEY, P.V., III, and Hunter, M.J., (1970), *Arch. Biochem. Biophys.*, 140, 197.

THE EFFECT OF THE EXCITATION WAVELENGTH ON THE FLUORESCENCE DEPOLARIZATION OF PROTEIN-DYE COMPLEXES OR CONJUGATES

Bernard Witholt[*] and Ludwig Brand[**]

* Department of Chemistry, University of California, San Diego, Calif.
Present Address: Department of Chemistry, University of Groningen, Groningen, The Netherlands
** Department of Biochemistry, Johns Hopkins University, Baltimore, Md.

The depolarization of fluorescence of protein-dye complexes in solution has usually been interpreted in terms of Perrin's equation for the depolarization to be expected for spheres (Perrin 1926):

$$\frac{A_0}{A} = \frac{\frac{1}{p} - \frac{1}{3}}{\frac{1}{p_0} - \frac{1}{3}} = 1 + \frac{3\tau}{\rho_0} = 1 + \frac{R\tau}{V_M} \cdot \frac{T}{\eta} \qquad (1)$$

where:

A is the emission anisotropy defined by Jablonski (1960) as $A = 1/(\frac{1}{p} - \frac{1}{3})$,

$A_0 = 1/(\frac{1}{p_0} - \frac{1}{3})$,

p is the experimental polarization, defined as $p = \frac{I_V - I_H}{I_V + I_H}$,

p_0 is the polarization when T/η approaches 0,

τ is the lifetime of the excited state of the chromophore in seconds,

ρ_0 is the rotational relaxation time of the sphere in seconds, at a specific T/η,

T is the solution temperature (°K),

η is the solution viscosity (poises),

V_M is the molar volume (ml/mole),

R is the gas constant (8.317×10^7 ergs°K^{-1} mole^{-1}).

This equation is usually plotted as $1/p$ or $(1/p-1/3)$ versus T/η, to produce a Perrin plot. It applies strictly to spheres. To illustrate this, let us consider a chromophore bound to a spherical macromolecule. For simplicity we will assume that the chromophore contains only a single oscillator; that is, the absorption and emission oscillators are parallel. This oscillator is therefore fixed to a sphere which is free to rotate in solution. Between the time of excitation and the time of emission (usually 5 to 20 nanoseconds) there is a reorientation of the oscillator due to rotation of the sphere in solution. The amount of the reorientation prior to emission depends on the lifetime of the excited state, and on the rotational relaxation time of the sphere. Both of these variables are expressed explicitly in Perrin's equation.

Since all axes through a sphere have identical rotational relaxation times (ρ_0), the specific orientation of the chromophore oscillator relative to the sphere is irrelevant. Thus, Perrin's equation will always apply to such a sphere-dye complex, regardless of the precise orientation of the dye relative to the sphere.

Furthermore, although we have given an example involving parallel absorption and emission oscillators, Perrin's equation applies equally well to cases where the oscillators are nonparallel. This is easily seen from equation (1): since the right side of equation (1) is independent of orientation parameters, all plots of A_0/A versus T/η should therefore be identical regardless of the relative orientations of the absorption and emission oscillators. Plots of A_0/A versus T/η will be referred to as normalized Perrin plots.

Equation (1) does not apply to nonspherical complexes except under special circumstances. This is so because different axes within a nonspherical molecule have different rotational relaxation times. Since depolarization depends on the reorientation of that particular molecular axis, which contains the chromophore oscillator AE (we are still referring to a single oscillator or parallel absorption and emission oscillators), the precise orientation of the chromatophore relative to a nonspherical macromolecule will influence the extent of depolarization to be expected in any given case.

This can be illustrated by fixing a chromophore (or a single oscillator AE) relative to a prolate ellipsoid with a long axis of revolution i and short equatorial axes j and k, as shown in Figure 1 (left panel). When the oscillator AE is parallel to the long axis i of the prolate, for example, it reorients very slowly, because the rotational relaxation time of the axis of revolution i is long. A normalized Perrin plot of such a complex would show a small slope, indicative of a long rotational relaxation time ρ_i. When the oscillator AE is parallel to one of the short axes j or k, it reorients rapidly as a result of the fast rotation of the ellipsoid about the axis of revolution. Hence, there will be rapid depolarization, reflected in a normalized Perrin plot with a steep slope, as shown in Figure 1 (right panel). The rotational relaxation times ρ_i and ρ_j, calculated from such plots would in fact be the rotational relaxation times of the i and j axis.

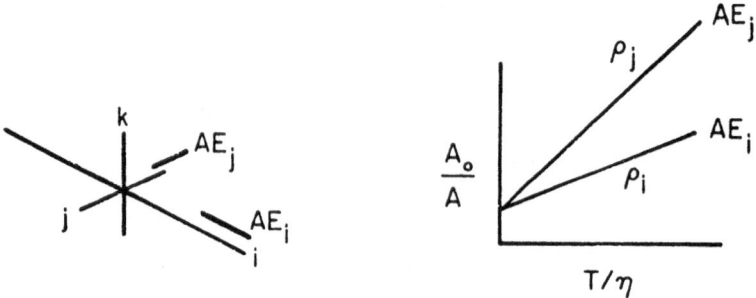

Figure 1. *Left: Axial system* ijk *for a prolate ellipsoid of revolution. Right: Normalized Perrin plots to be expected when a dye binds with its oscillators (AE) parallel to either the* i *axis or the* j *axis*

Thus, the orientation of a chromophore relative to a nonspherical molecule has a profound influence on the normalized Perrin plots to be obtained from such a complex; it is therefore difficult to interpret a Perrin plot for a nonspherical complex.

So far, we have only considered the relatively simple case in which the absorption and emission oscillators are parallel or identical. In practice, however, this need not be so. Often the absorption and emission oscillators are not parallel. In that case it becomes extremely difficult to see intuitively what effect the oscillator orientations relative to a prolate will have on the depolarization of fluorescence.

Fortunately, Perrin (1936) derived a general equation, relating depolarization to the ellipsoid dimensions, the direction cosines of the chromophore oscillators, the lifetime of the excited state, and the solvent conditions. This equation has been examined in detail by several investigators.

Memming (1961) rederived Perrin's equation for ellipsoids of revolution; this equation is a specific case of Perrin's general equation for ellipsoids with three different axes. Gottlieb and Wahl (1963) derived equations to predict the polarization to be expected when there is some free rotation of the chromophore relative to the protein, which phenomenon had also been discussed by Weber in 1952. Weber and Anderson (1969) examined the effect of energy transfer among chromophores bound to the same prolate.

We have rewritten Perrin's general equation in terms of a specific set of parameters illustrated in Figure 2 (Witholt 1969; Witholt and Brand 1970). A protein-chromophore complex is treated as a prolate ellipsoid of revolution (with an axis of revolution i and two identical shorter axes j and k) which contains an absorption oscillator A and an emission oscillator E, positioned relative to the prolate by means of angles A_1, A_2 and A_3. Figure 3 shows an example of the normalized Perrin plots which are obtained when the orientation of the oscillators is varied within the ellipsoid.

These curves represent the polarization behavior of a molecule with a molar volume of 50,000 ml/mole, to which is attached a dye with a lifetime τ of 5 nanoseconds. When the molecule is a sphere, all normalized Perrin plots will follow curve a, with an apparent rotational relaxation time at 20°C of 61.6 nanoseconds, corresponding to the true rotational relaxation time for such a sphere. The remaining curves are all for a prolate, still at 50,000 ml/mole, but with an axial ratio of 5. The only variables are the orientations of the absorption and emission oscillators relative to the ellipsoid. When $A_2 = 0$, the emission oscillator is parallel to the i axis, and curve

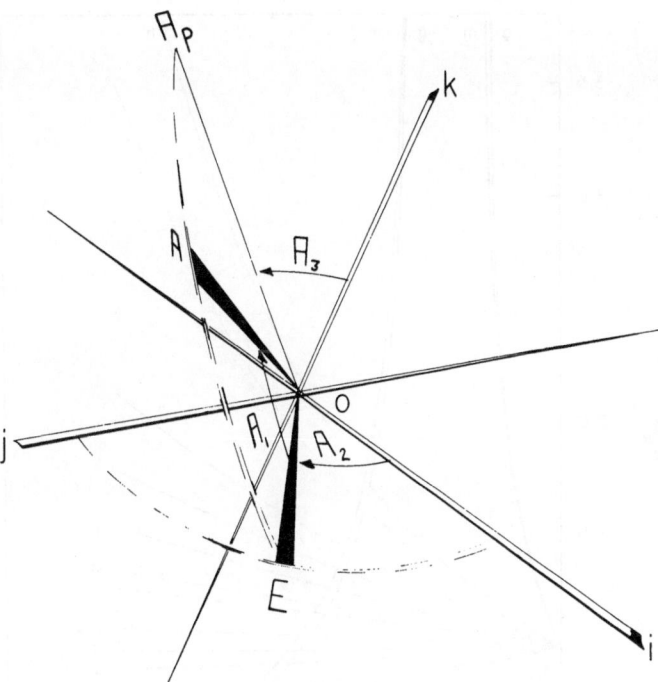

Figure 2. *Oscillator orientations relative to an ellipsoid of revolution around axis* i. *Axes* j *and* k *are identical.* E *is the emission oscillator in the* ij *plane, forming an angle* A_2 *with* i. *A is the absorption oscillator in the* A_pOE *plane, forming an angle* A_1 *with* E, *while* A_p *makes an angle* A_3 *with axis* k. A_p *is perpendicular to* E

b is obtained, yielding an apparent rotational relaxation time of 285.8 nanoseconds, which is that of the long axis of this particular prolate. Interestingly enough, the orientation of the absorption oscillator is irrevelant; all orientations lead to curve b.

When $A_2 = 90°$, the emission oscillator coincides with the j axis. A_3 is maintained at 0°, which means that the absorption oscillator lies in the jk plane. As A_1 increases, curves c through j are generated. Clearly, a wide range of apparent relaxation times could be calculated from such plots, starting with 78 nanoseconds for curve c, where both oscillators are parallel to the j axis, and continuing to 40 nanoseconds for curve j when the two oscillators are normal to each other and in the equatorial plane. It is interesting that at certain angles A_1, the polarization actually increases rather than decreases as T/η increases. This is true of curves d and e which represent transitions with $p_0 = .097$ and $.049$, respectively. Obviously no meaningful rotational relaxation times can be calculated from such curves.

When the absorption oscillator lies in the ij plane ($A_3 = 90°$), another set of curves is obtained as A_1 increases. These are curves k through q. Again, a variety of slopes are obtained, including negative slopes for curves n, o, and p, corresponding to transitions with $p_0 = -.0101$, $-.0510$, and $-.1037$ respectively.

The main point of this exercise is that when a chromophore with several absorption dipoles is fixed relative to a prolate ellipsoid, an infinite variety of normalized Perrin plots may be obtained depending on the precise angles A_1, A_2, and A_3. In addition, most of these curves are essentially straight lines and it would be impossible to determine from any single curve what the orientation of the oscillators relative to the protein axes might be.

Figure 3 confirms our intuitive predictions based on parallel oscillators: the depolarization of fluorescence from chromophores, fixed relative to prolate ellipsoids, results in highly variable normalized Perrin plots. While the rotational relaxation times calculated from such plots reflect the actual rotational relaxation times of specific molecular axes when the oscillators are parallel, this is not so when the oscillators are not parallel. In that case several unexpected effects may occur, including extremely fast depolarization and polarization instead of depolarization

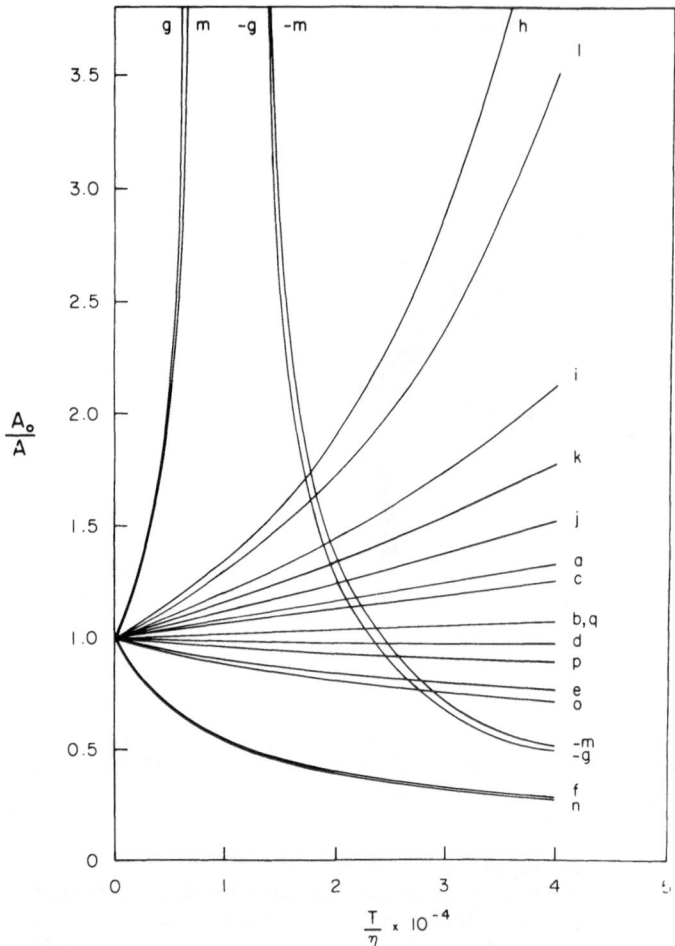

Figure 3. *Normalized Perrin plots for a sphere (curve a) and a prolate (curves b-q). A volume of 50,000 ml/mole is assumed for both prolate and sphere, which have axial ratios of 5 and 1, respectively. The associated chromophore has a lifetime of 5 nsec and the specific oscillator orientations relative to the ellipsoid are discussed in the text. Note that curves g and m disappear toward + ∞ and then reappear from - ∞. The negative portions of curves g and m have been multiplied by -1 and are plotted as positive curves. (Witholt and Brand, 1970)*

with increased rotation.

From what has been said before it is clear that when an adduct, involving a dye fixed relative to a prolate, is excited at different wavelengths, a series of nonidentical normalized Perrin plots will be obtained, provided that the lifetime of the excited state remains constant at all exciting wavelengths. On the other hand, when there is random binding, all normalized Perrin plots should be identical, since Perrin's simple formula applies. Thus, an examination of the normalized Perrin plots for different adducts should allow a classification as to the type of binding involved.

We have measured the depolarization due to several protein-dye conjugates and complexes (Witholt and Brand 1970). Figure 4 shows normalized Perrin plots for a bovine serum albumin (BSA)-anthracene conjugate containing fewer than 1.3 moles of anthracene per mole of BSA. The deviations in the normalized Perrin plots suggested a specific orientation of anthracene relative to BSA. It was likely that there was energy transfer from the BSA tryptophans to the bound anthracene which complicated the interpretation of the normalized Perrin plots obtained at 265 and 280 nanometers.

The polarization data reported here were corrected for several instrumental and experimental artifacts, the most important of which was due to the rotation of the incident polarized beam by high sucrose concentrations. Depending on the exciting wavelength, the true polarization was lowered by as much as 15% at sucrose concentrations of 60%.

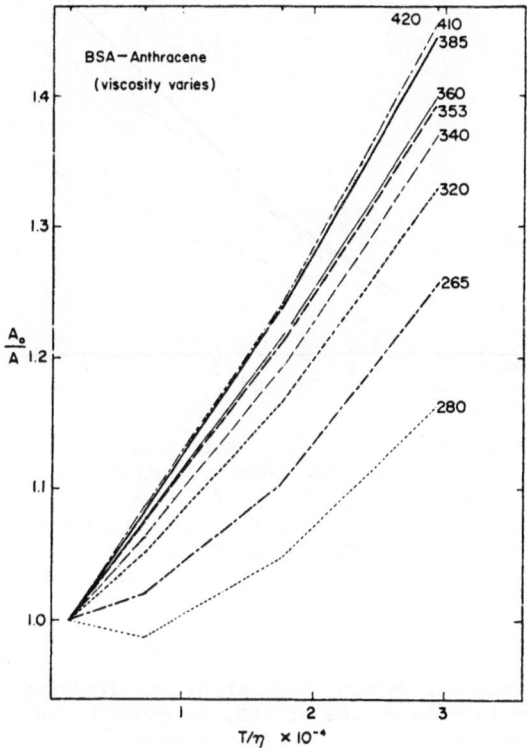

Figure 4. *Normalized Perrin plots for a bovine serum albumin-anthracene conjugate. The dye-to-protein ratio was less than 1.3 and the concentration of conjugate was 8.7 µM in 0.1 M Tris (pH 7.3) at 20.0±0.2°.*
Data were obtained at four sucrose concentrations equivalent to T/η = 1355, 7118, 17,566 and 29,260°K per P (poise). The excitation band width was 4.8 nm and the emission band width was 33 nm. The emission monochromator was set at 450 nm. The excitation wavelength is indicated for each curve. (Witholt and Brand, 1970)

Figure 5 shows the normalized Perrin plots for a BSA-fluorescein conjugate, containing fewer than 0.3 fluorescein molecules per BSA molecule. Here too, deviations in the normalized Perrin plots suggested a fixed orientation of the chromophore relative to the protein.

In each of the two preceding examples, it was not surprising that there was a preferred orientation of the dye molecules relative to the protein molecules. The dyes may have bound noncovalently to hydrophobic regions followed by covalent linkage to nearby reactive groups, resulting in orientationally specific binding.

Figure 6 shows a noncovalent complex of 1-anilinonaphthalene-8-sulfonate (1,8-ANS) and BSA, with $\bar{n}$ (the ratio of moles of ANS per mole of BSA) equal to 0.64. Under these conditions more than 99% of the total emission was due to complexed ANS. The normalized Perrin plots for this complex indicated that the binding was orientationally nonspecific; this surprised us, since ANS binds BSA with a dissociation constant of 3.2×10^{-6} (Weber and Young 1964). We would have expected such relatively tight binding to require a considerable degree of orientational specificity, to allow for the necessary alignment between the ligand and the binding site on the protein.

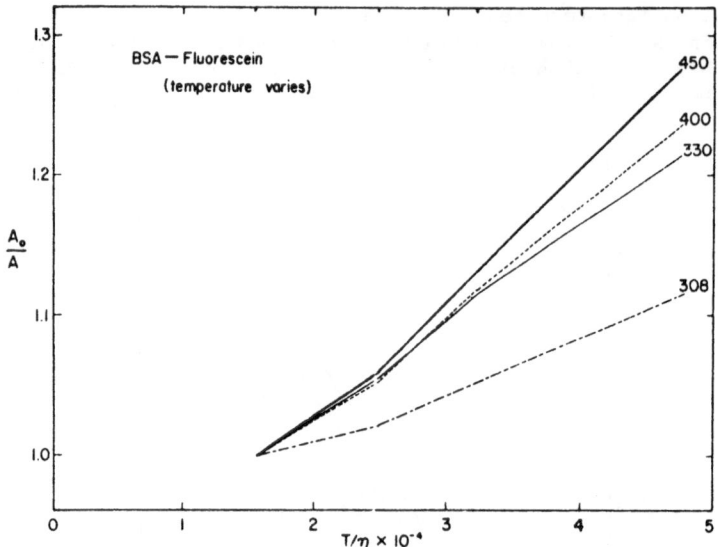

Figure 5. *Normalized Perrin plots for a bovine serum albumin-fluorescein conjugate. The dye-to-protein ratio was less than 0.3 and the concentration of the conjugate was 5.9 x 10⁻⁵M in 0.1 M Na phosphate, pH 7.4. Data were obtained at T/η = 15,400, 24,900, 32,300 and 47,400°K/poise by varying the temperature of the solution. The excitation band width was 1.7 nm and the emission band width was 37 nm. The emission monochromator was set at 530 nm. The excitation wavelength is indicated for each curve*

We also obtained normalized Perrin plots at higher dye/protein ratios since BSA was capable of binding up to 5 molecules of ANS, as shown by Weber and Young (1964). Figure 7 shows results obtained at $\bar{n}$ = 1.93 (lower segment) and $\bar{n}$ = 4.55 (upper segment). It was clear that as the number of ANS molecules per BSA molecule increased, there was an increasing deviation among the normalized Perrin plots. This was further shown in Figure 8, where the deviation among the normalized Perrin plots was plotted as a function of the number of ANS molecules per BSA molecule. σ was simply a measure of the deviations among the normalized Perrin plots. As was the case in Figure 7, these differences increased as the number of ANS molecules per BSA molecule increased.

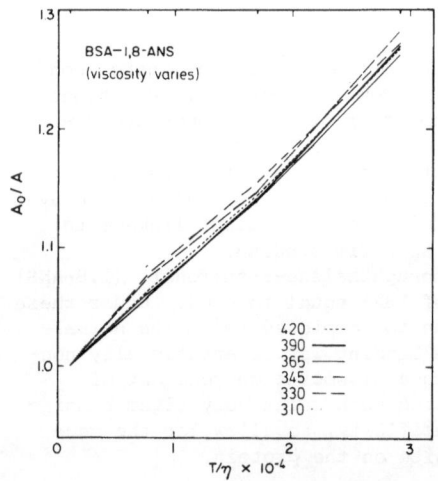

Figure 6. *Normalized Perrin plots for a 15 μM bovine serum albumin-anilinonaphthalenesulfonate complex. The protein had 0.64 mole of dye absorbed per mole of protein. The solvent was 0.05 M carbonate-bicarbonate buffer at pH 9.0 at 20.0 ± 0.2°. Data were obtained at four sucrose concentrations equivalent to T/η = 1000, 7672, 16,990 and 29,260°K per P. The excitation band width was 4.8 nm at the wavelengths indicated for each curve. The emission monochromator was set at 480 nm. (Witholt and Brand, 1970)*

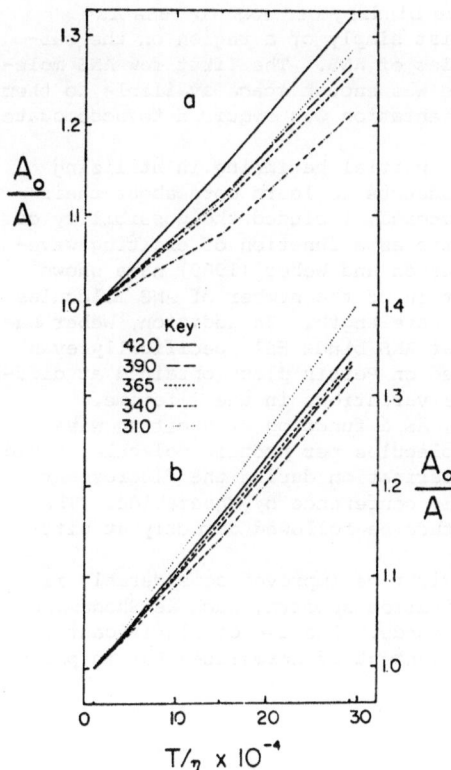

Figure 7. *Normalized Perrin plots for two bovine serum albumin-anilinonaphthalenesulfonate complexes. The solvent was 0.1 M sodium phosphate (pH 6.8) at 20.0 ± 0.3°. The excitation band width was 4.8 nm at the excitation wavelengths indicated for each curve. The emission monochromator was set at 480 nm. Data were obtained at three sucrose concentrations. For set a, $\bar{n}$ = 4.55 and T/η = 1142°, 14,970° and 29,260° K per P; for set b, $\bar{n}$ = 1.93 and T/η = 1087°, 14,870° and 29,260° K per P. (Witholt and Brand, 1970)*

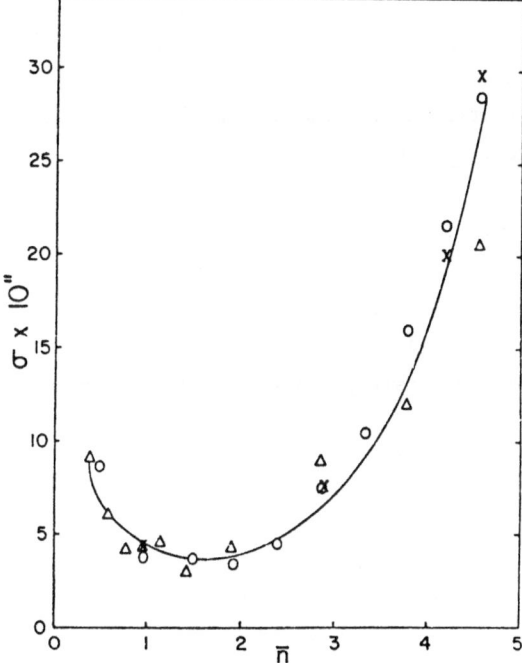

Figure 8. *Change of σ, the degree of orientational anisotropy with $\bar{n}$. Titration of 50 μM bovine serum albumin with 1,8-anilinonaphthalenesulfonate in 0.1 M sodium phosphate (pH 6.8) at 20.0 ± 0.2°. Each point on this figure is based on a set of normalized Perrin plots such as those shown in Figure 7. $\bar{n}$ was varied by adding 5-μl increments of 1,8-anilinonaphthalenesulfonate. The circles and triangles represent duplicate experiments carried out under slightly different optical conditions. The crosses are exact duplicate of the corresponding circles (Witholt and Brand 1970)*

One interpretation of these results is that the binding of ANS to BSA is rather general. The binding site or sites may consist simply of a region on the surface of BSA, which can accomodate up to five molecules of ANS. The first few ANS molecules could orient relatively randomly because there was enough space available to them but as more ANS molecules bound, a more specific orientation was required to accomodate all of the bound ANS.

It should be clear that this approach is only a partial beginning in utilizing the complex depolarization behavior of protein-dye adducts to learn more about their mutual interaction. Other factors, which we have ignored, included the possibility of variations in the lifetime and the protein axial ratio as a function of exciting wavelength, sucrose concentration, or temperature. Anderson and Weber (1969) have shown that variations in the lifetime may occur as a function of the number of ANS molecules per BSA molecule, and as a function of the exciting wavelength. In addition, Weber has presented data during this conference suggesting that ANS binds BSA specifically even at low ratios of ANS to BSA. Weber's data were based on Perrin plots obtained at different exciting wavelengths, which were corrected for variations in the lifetime.

Problems imposed by variations in the lifetime, as a function of exciting wavelength, solvent conditions, and the number of dye molecules per protein molecule, could be avoided altogether, however, by measuring the polarization during the fluorescent decay time after excitation, as described during this conference by Yguerabide. The actual orientation of various molecular axes could then be followed directly at different exciting wavelengths.

Finally, the intrinsic power of this approach might be improved considerably if it were possible to develop dyes with complex polarization spectra, such as Rhodamine B, but with lifetimes of the order of 15 to 30 nanoseconds. The use of significantly different transitions would enhance the information content of normalized Perrin plots.

ACKNOWLEDGEMENTS

Figures 2-4, 6-8 are reprinted from *Biochemistry*, (1970), 9, 1948. Copyright 1970 by the American Chemical Society. Reprinted by permission of the copyright owner.

REFERENCES

ANDERSON, S.R. and Weber, G. (1969), *Biochemistry*, 8, 371.

GOTTLIEB, Y.Y. and Wahl, P. (1963), *J. Chim. Phys.*, 60, 849.

JABLONSKI, A. (1960), *Bull. Polish Acad. Sci. (Math. Phys. Series)*, 8, 259.

MEMMING, R. (1961), *Z. Physik. Chem. (Frankfurt)*, 28, 168.

PERRIN, F. (1926), *J. Phys.*, (6), 7, 390.

PERRIN, F. (1936), *J. Phys. Radium*, Serie VII, T. VII, 1,1.

WEBER, G. (1952), *Biochem. J.*, 51, 145.

WEBER, G. and Anderson, S.R. (1969), *Biochemistry*, 8, 361.

WEBER, G. and Young, L. (1964), *J. Biol. Chem.*, 239, 1415.

WITHOLT, B. (1969), *Ph. D. Thesis*, Johns Hopkins University, University Microfilms No. 69-13,500.

WITHOLT, B. and Brand, L. (1970), *Biochemistry*, 9, 1948.

YGUERABIDE, J. (1973), This Volume.

FLUORESCENCE, ABSORPTION, AND OPTICAL ROTATORY BEHAVIOR OF ACRIDINE ORANGE:POLY-L-GLUTAMIC ACID COMPLEXES

Brian Myhr* and John Foss**

* *National Cancer Institute, Bethesda, Maryland.*
** *Department of Biochemistry and Biophysics, Iowa State University, Ames, Iowa*

INTRODUCTION

The work described below was actually carried out a few years ago at Iowa State University, but for the most part, it has not yet been published (Myhr 1968). This work was concerned with the use of metachromatic dyes as probes of polyanion conformation under test tube conditions, yet it now seems that the results and experience obtained may in some sense be useful to those who are interested in utilizing the optical properties of intracellular acridine orange, or another metachromatic dye, to derive polymer structural information or to follow regulatory changes.

The acridine orange-polyglutamic acid (AO:PGA) complex is a simple and appropriate noncellular model system for study. There is essentially only one type of binding site, the γ-carboxylate group. Furthermore, as the negatively charged carboxylate groups are protonated, PGA undergoes a well-known conformational transition from a coil form to an α-helical structure. Other biological polyanions either bind these dyes by additional modes, such as intercalation in the case of nucleic acids, or do not possess regular structures and do not undergo well-defined conformational changes (i.e., mucopolysaccharides). Thus, the AO:PGA complex appeared to be the system of choice for observing in what ways the optical properties of metachromatic dyes may depend on the conformational state of the polymer. An example of such an interdependence was shown a number of years ago by Stryer and Blout (1961), who discovered that AO became strongly optically active on binding to helical PGA but not to the coiled form. Moreover, the sign of the induced optical activity inverted upon replacing poly-L-glutamic acid (a right-handed α-helix) with poly-D-glutamic acid (a left-handed helix).

GENERAL BEHAVIOR OF THE COMPLEX

These results led us to further studies of this model system, including the absorption and fluorescence behavior of bound dye. We had not been involved in this work very long, however, before it became apparent that the optical properties of the complex often depended dramatically both on the method of preparation and on how soon the complex was observed (Myhr and Foss 1966). This variability will be described later, but for the moment it should be emphasized that this phenomenon is probably the most important message of this paper to histologists. In essence, the complicated behavior of the AO:PGA model system serves as a warning to hasty interpretations of the optical behavior of intracellularly bound AO and its changes as a function of experimental variables.

The optical variability is probably due in large part to the formation of a number of metastable states, each with its own peculiar optical behavior. Freshly prepared complexes were generally clear, but within seconds in some cases, particularly at low polymer-to-dye ratios and in salt solutions, suspended particulate matter would appear and aggregate to a flocculent precipitate. Under other conditions, such as higher polymer-to-dye ratios, several minutes or even days would be required to observe aggregate formation. Obviously, the highly substituted PGA molecules lose their water solubility and the observed optical behavior may be as much or more a function of aggregates as it is of single polymer molecules containing bound dye. It therefore becomes very important to pay particular attention to how a complex is prepared and how soon optical mearurements are made.

We found that reproducible absorption and fluorescence measurements could be obtained by adopting a standard method of complex formation (which often resulted in measurements of some nonequilibrium state of the complex). In this method, the complexes were prepared directly in a 1 × 1 cm cuvette, and rapid mixing was accomplished with a looped wire attached to an electromagnetic vibrator. With the use of micro-burets, the PGA solution was first prepared at the desired concentration, pH, and NaCl concentration. Then 1×10^{-3}M AO stock was added quickly from another microburet, the solution mixed until just homogeneous, and the optical measurement taken immediate-ly. Absorption and fluorescence scans were obtained within one minute. The complex was then discarded unless the effects of aging were to be noted.

Using this methodology, we have already shown that AO:PGA complexes exhibit the well-known metachromatic behavior of other dye:chromotrope systems (Myhr and Foss 1971). Near the equivalence of γ-carboxylate groups and dye molecules, the AO fluorescence band at 535 nm is highly quenched and the visible absorption area is reduced, with the ab-sorption maximum shifting from 492 nm to as far as 450 nm for the coil complex in the absence of added salt. These spectral changes arise from the electronic coupling of adjacently bound dyes and can be reversed with a larger excess of polymer, which in-creases the proportion of noninteracting, bound dye molecules. It was shown by fluore-scence, absorption, and acid-base titrations that one dye molecule will bind to each carboxylate and that the attendant metachromasy is not reversed until very high glu-tamyl residue/dye ratios, R/D, are reached, as shown in Figure 1. In fact, the α-hel-ical and extended coil conformations could not be distinguished on the basis of the tendency of AO to bind as dimers, indicating that the spacing of binding sites on both conformers may be close to optimal. The value of Bradley's stacking coefficient (Lamm et al. 1965), K, was about 1,500, which corresponds to a free energy of interaction of -4.4 kcal/mole of dye-pairs.

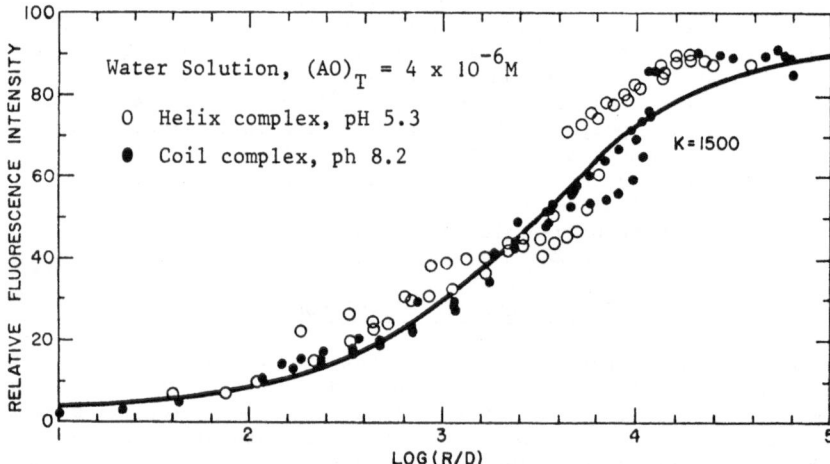

Figure 1. *The relative fluorescence intensity of AO:PGA at 535 nm as a function of high R/D ratios. Total dye concentration is 4×10^{-6}M. (From Myhr and Foss (1971). Reproduced with permission.)*

The degree of metachromasy, however, was found to vary, depending on the polymer conformation and the concentration of added salt. As a practical convenience, the de-gree of metachromasy is loosely defined as the extent of the spectral changes measured either as the *loss* in fluorescence at 535nm or as the *loss* in absorption at the dye monomer maximum at 492nm. Thus, measurements at these single wavelengths reached a minimum (maximal metachromasy) near the equivalence of sites and dye. The magnitude of this minimum level was dependent on the solution conditions (pH, NaCl concentration), and it remained constant to R/D ratios of at least 50, except in the case of the coil without added salt. In this last case, a redistribution of dye from larger stacks to dimers occurs soon after equivalence, causing the fluorescence and absorption levels to rise about 10% in the R/D range of 1 to 50. Thus, it was possible to prepare AO: PGA complexes at R/D ratios around 15-20, where all of the dye in the system is bound to the polymer, and to follow the changes in the degree of maximal metachromasy as functions of pH and salt concentration.

OPTICAL BEHAVIOR VERSUS HELIX-COIL TRANSITION

In the lower half of Figure 2, the helix-coil transition of PGA is shown as a function of pH in aqueous solution with no added salt. This transition was monitored in the usual manner by the optical rotation at 233 nm, which is the dispersion trough of the α-helix. The transition midpoint is seen to occur at pH 6.0, with virtually the entire transition taking place in the pH range 5.5 to 6.5. This pH range will be marked by a crosshatched bar in the subsequent figures as an aid to the comparison of meta-chromasy changes with PGA conformation. Although it is not shown, the transition mid-point shifts to pH 6.16 in 0.1M NaCl, and the pH transition range of 4.5 to 5.75 will be marked for this solution condition.

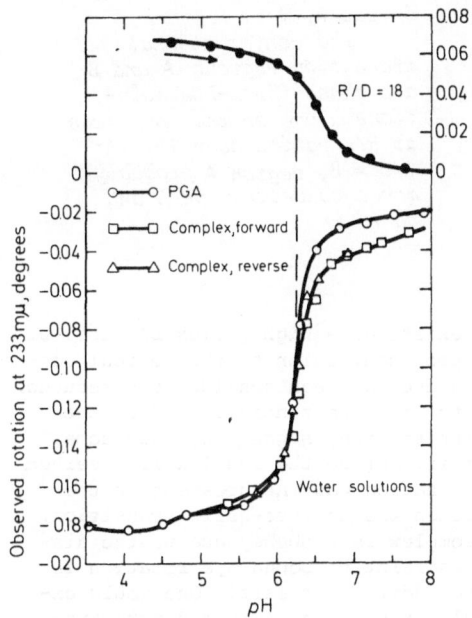

Figure 2. *Observed rotations for water solutions of L-PGA at 535 nm as a function of pH. Forward refers to a titration from high to low pH, reverse to the opposite direction. The total dye concentration is 4 × 10⁻⁵M and the complex was prepared at R/D = 18*

Also shown in Figure 2 is the lack of an effect by bound AO on the position and pH range of this transition for complexes prepared at R/D = 18. It is important to show this to be sure that any observed changes in metachromasy as a function of pH are correctly compared with the PGA transition observed by the amide interactions. Thus, a complex prepared at pH 8.0 was titrated with acid into the helical pH region and then reverse titrated with base. These two curves were coincident with each other and essentially so with the PGA curve. (At the lower R/D ratio of 4, however, bound AO did interfere markedly with the PGA conformational change, causing a major transition at pH 7 [Myhr 1968].) Therefore, the presence of bound dye at R/D = 18 has a minor effect on the conformational mobility of PGA; furthermore, any lag in the responsiveness of bound dye was minimized by adding AO to pretitrated PGA rather than by titrating the complex.

The PGA transition can also be followed by the optical activity induced in the dye absorption bands upon binding to helical PGA. As shown in the upper half of Figure 2, this optical activity disappears as the helical complex is titrated with base, in agreement with the results of Stryer and Blout (1961). However, the transition observed in this manner appeared to be broadened and to lag slightly the titration midpoint of helical PGA alone.

Before discussing the metachromatic response as a function of pH, it is instructive to consider the pH dependency of the fluorescence depolarization of the complexes, both in water and in 0.1M NaCl (Figure 3). In this experiment the AO:PGA complex was excited in a number of its UV absorption bands, so the purpose was to look for changes in the degree of polarization rather than in its value. As is well known, the extent of depolarization will reflect the degree of rotational freedom of the bound dye, which in turn is some function of the size, shape, and flexibility of the entire polymer. It therefore seemed possible that an increase in the degree of polarization might be observed in the transition from a random, flexible coil complex to the helical complex. As is shown in Figure 3, this was in fact observed for the complex in 0.1M NaCl. In the basic pH region, the degree of polarization was almost zero, in agreement with the

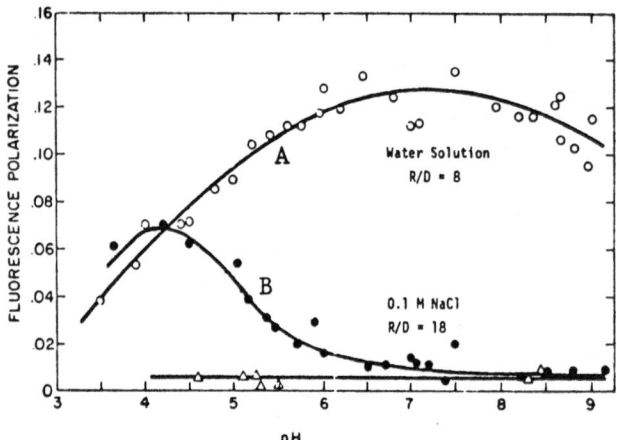

Figure 3. *The fluorescence polarization of the complex as a function of pH in water and 0.1M NaCl. Emission observed at 535 nm. Free AO in water,* (Δ - Δ).

The conformational transition regions A *and* B *are those for L-PGA solutions alone or the complexes at R/D ratios near 18. At R/D = 8, region* A *probably moves to between pH 6 and pH 7*

conclusion of other physical measurements that PGA exists as a highly flexible coil at this salt concentration. The rise in polarization upon conversion to the helical complex shows that the rotational mobility of the bound dye has been considerably reduced. It may be significant that the polarization begins to rise substantially before the coil-to-helix transition is observed by the amide interaction, suggesting that some restriction occurs in the polymer or sidechain mobility before the α-helix is observed.

In water solution, however, the degree of polarization was the greatest in the coil pH region and its magnitude did not change through the coil-to-helix transition. The decrease observed at lower pH for the helical complex is probably due to the displacement of dye by protons. Thus, a significant restraint to bound dye rotation is present in the coil complex that is destroyed by the addition of salt. One would expect from electrostatic arguments that the coil would exist as an extended structure with much less internal rotational movement than in the case where added counterions are present. The fluorescence polarization results also suggest that such an extended coil essentially twists and tightens into an α-helix (with an increasing degree of protonation) without much change in the rotational mobility of the γ-carboxylate groups. In the coil region, moreover, the loss of fluorescence polarization upon addition of salt suggests an explanation for the loss in the degree of metachromasy caused by salt.

The degree of metachromasy, as monitored by the fluorescent intensity at 535 nm, is shown in Figure 4 as a function of pH for complexes in water and 0.1M NaCl solutions. (A similar set of curves was obtained for absorption measurements at 490 nm, Figure 5, except that the curve in water clearly showed the attainment of maximal metachromasy at pH 6.3.) The 535 nm fluorescent band is highly quenched for the coil complex in water and does not indicate the coil-to-helix transition. The rise in intensity in the helical pH region has been interpreted as being the result of both an increasing proportion of bound monomers and the displacement of dye below pH 4.

In striking contrast, however, is the behavior of the complex in 0.1M NaCl. Our optical titration curves for the coil complex have fairly sharp equivalence points and then remain absolutely flat in the R/D range from 1 to above 50. This strongly suggests that all of the dye remains bound in the presence of 0.1M NaCl. The high level of fluorescence (low degree of metachromasy) in the coil region therefore appears to be due to structural changes in the complex and not to the presence of free dye. It is also immediately obvious that the 535 nm band is responsive to the random coil-to-helix transition, but not in any simple way. A large drop in the fluorescence intensity (and in the absorption at 490 nm) occurred before the helical content reached 4 to 5%, and a second large drop began near the middle of the coil-to-helix transition region.

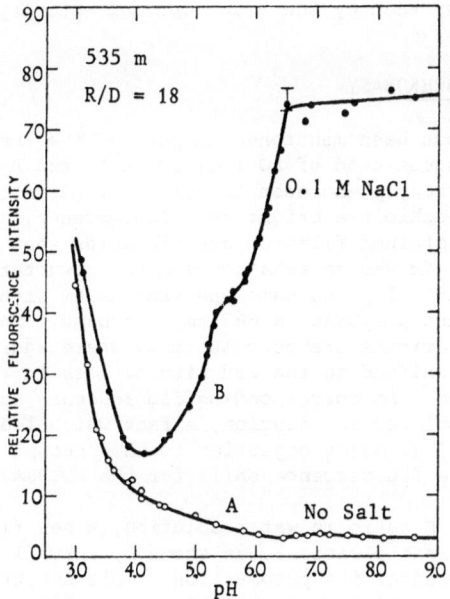

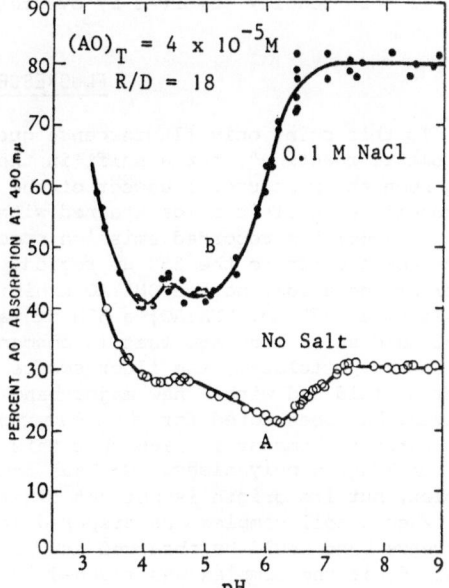

Figure 4. *The fluorescence at 535 nm of the AO:PGA complex relative to AO as a function of pH. Total dye concentration is 4 × 10⁻⁵M.*
Region A: Helix-coil transition region in water.
Region B: Helix-coil transition region in 0.1M NaCl

Figure 5. *The absorption at 490 nm of the AO:PGA complex relative to AO as a function of pH. Conditions same as in Figure 4*

An interpretation of the fluorescence behavior in 0.1M NaCl versus water, as suggested by the fluorescence polarization results, is that the dye is primarily sensing changes in the flexibility of the PGA molecule. In other words, the development of metachromasy may depend upon a restriction in the relative motion of adjacent γ-carboxylate binding sites. The addition of salt to the helical complex does not change the degree of metachromasy very much, because the polymer remains in the comparatively rigid, α-helical conformation in both cases. The addition of salt to the coil form of PGA, however, causes a conformational change from an extended structure with restricted side chain motion to a randomly coiled polymer; this increase in flexibility destroys the effectiveness of PGA as a stacking template. The kind and degree of spectral shifts also depends, of course, on the exact geometry (distance, angle, number) of associated dye molecules, so that an effective chromotrope (or stacking template) must have appropriately spaced sites. Underlying this, however, is the additional requirement of sufficient rigidity. Obviously, in a hypothetical case where the sites would be as free to rotate with respect to each other as if the polymer subunits were not covalently linked, there would be no more tendency for bound dye to associate than there would for free dye. In the case of PGA, the γ-carboxylates are properly spaced to result in a high tendency of AO to associate, regardless of the chain conformation; so large changes in metachromasy will primarily reflect changes in polymer flexibility.

If this argument is correct, the fluorescence (and absorption) behavior of the AO:PGA complex signals the existance of a sharp change in PGA flexibility at pH 6.5 in 0.1M NaCl (Figures 4 and 5). The sudden decrease in fluorescence suggests that side-chain mobility is significantly decreased prior to the pH where α-helical formation is indicated by the amide Cotton effect. Such a physical change was also suggested by the results of Doty et al. (1957), who described an "anomalous" drop in intrinsic viscosity just prior to the coil-to-helix transition of PGA in 0.2M NaCl-dioxane (2:1). From a theoretical standpoint, Ptitsyn (1967) has shown that the root mean square of the end-to-end distance of PGA should pass through a minimum in the early stages of the coil-to-helix transition. Perhaps, therefore, such an early increase in compactness is more readily observed by dye metachromasy than by the amide optical activity.

FLUORESCENCE METACHROMASY

To this point only fluorescence quenching has been mentioned as part of the "metachromatic reaction", but a shift in the fluorescent band of AO from green to red has long been the most useful aspect of the metachromasy phenomenon to cell biologists. Intracellular nucleic acids stained with AO may exhibit a bright red fluorescence, and Rigler (1966) has recorded emission spectra for stained films of nucleic acids that also show a shift to the 650 nm region. This shift was correlated with the structural order of the acids: native DNA:AO exhibited essentially the same spectrum as AO with a maximum at 535 nm; RNA:AO, a 650 nm maximum; and polyU:AO, a 667 nm main band. Zanker (1952) had shown long ago that in concentrated acridine orange solutions, where aggregation was postulated, the fluorescence maximum shifted to the red with no loss in quantum yield and with a new major band at 650 nm. No corresponding fluorescence shift, however, has been noted for dye:chromotrope complexes in solution, a fact which led Van Duuren to comment in 1966 that this was the last major objection to the concept of dye stacking on polyanions. We have found such a fluorescence shift for the AO:PGA complex, but its origin is not yet clear.

When a coil complex was prepared at a low R/D ratio in water solution, a new fluorescence band could be observed at 595 nm (apparent position), as shown by curve 1 in Figure 6, if the complex was scanned immediately after its preparation. This spectrum is highly distorted at the red end because the sensitivity of the 1P28 photomultiplier tube used falls off rapidly above 550 nm. Therefore, the intensity of the new band relative to the 535 nm intensity is much greater than shown, and its real position lies above 600 nm, perhaps near the 650 nm position found by Rigler (1966) for AO:RNA complexes. In spite of this distortion by our measuring system, it is clear that fluorescence metachromasy does occur with the AO:PGA complex prepared in solution, at least under certain conditions.

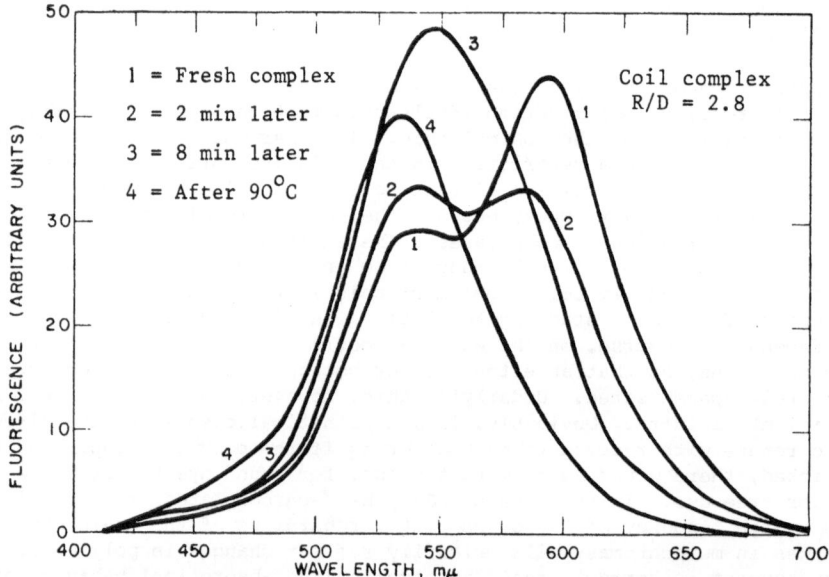

Figure 6. *The effect of age and heat on the fluorescence spectrum (uncorrected) of the coil complex in water at a low R/D ratio. Total dye concentration is 4×10^{-5} M, pH 8.5*

Our reason for qualifying the previous statement is that the red band appears to be characteristic of some nonequilibrium state of the complex that happens to be kinetically favored. Thus, with the passage of time, or upon heating and cooling, the fluorescence spectrum shifts back toward the original band at 535 nm. Immediately after mixing, the 590 nm intensity first decreases slightly, then increases to a maximum value in about 30 sec. Within the next two minutes it falls rapidly; thereafter, at much slower rates. Curve 2 in Figure 6 shows a typical scan obtained two minutes after complex preparation; curve 3 is at eight minutes. If either this "aged complex" or a freshly prepared one is heated to 90°C and then cooled to room temperature, an emission scan like curve 4 is obtained. All trace of the 595 nm band is now lost. The extended coil complex at a higher R/D ratio (ca. 20) and also the helical complex show the 595 nm band only as a prominent shoulder, but in these cases, the shoulder is not destroyed by the heat treatment.

These fluorescence changes are not accompanied by significant changes in the absorption spectrum, which is not, as a general rule, a particularly sensitive optical parameter to structural or environmental changes. It may be that the fluorescence behavior is more sensitive to the state of aggregation of the complex, for example, and that some particular aggregate is kinetically favored in the case of the fully charged polymer at low R/D. This possibility must be considered because visually obvious aggregation occurs soon after preparation of an initially clear solution. Heating causes redissolution, yet the complex will precipitate again upon standing at room temperature. On the other hand, it is possible that the red fluorescence is a property of bound dimers that is quenched by higher stack formation. If the formation of dimers is a kinetically favored process, then the initially formed fluorescent dimers at R/D $\cong$ 3 could be imagined as rearranging into nonfluorescent stacks of three or more dyes either with the passage of time or upon heating. Such a rearrangement would not occur either at higher R/D ratios, where the dimer is energetically favored, or with the helix complex, where protonation of the sites prevents the formation of stacks greater than dimers. This interpretation would explain the observed stability of the 595 nm band for the helix complex and the extended coil complex at higher R/D, and would be in qualitative agreement with the prediction of a red band by the theory of exciton splitting (Cundall et al. 1970; Bradley et al. 1963).

It must be stated at this point, however, that we really do not understand the fluorescent behavior of the AO:PGA complex in terms of any theoretical model. This is largely due to the fact that the 535 nm band is always present under conditions where dye monomers are not expected. Thus, in the previous discussion, if the dyes rearrange to larger stacks on the coil at low R/D, why should the 535 nm band become enhanced? Do the dimers and higher stacks emit weakly at 535 nm or is there always a small percentage of noninteracting dyes present and emitting at this wavelength?

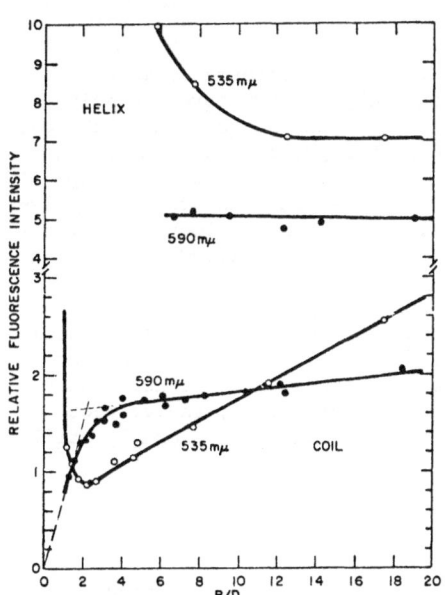

Figure 7. *The AO:PGA complex fluorescence intensities at 535 nm and 590 nm relative to free AO emission at 535 nm as a function of R/D. Total dye concentration is 4 × 10⁻⁵M. The helix complexes were prepared at pH 4.3 to 4.5, the coil complexes at pH 8.1 to 8.7*

The behavior of the 535 nm and 590 nm intensities as functions of the R/D ratio are shown in Figure 7. In the helix case, where absorption spectra indicate that only dimers are formed, the 590 nm intensity is seen to remain constant from the point where most of the dye is bound to R/D ratios above 20. Thus, dimers in this case appear to be characterized by both a 535 nm band at 7% and a 595 nm band at 5% of the unbound dye fluorescent intensity. (Recall that dimers do not begin to dissociate into monomers until R/D values exceed 100, (See Figure 1.) In contrast, in the coil case both bands are minimal at the equivalence point of R/D = 1 (intersection of 535 nm limbs) and then rise immediately with an increasing number of available sites. The rise in 535 nm fluorescence parallels a rise in absorption at 490 nm and is probably due to a redistribution of dye toward dimers. At R/D = 20, the 535 nm intensity of 3% is still below the helix complex level, but at very high R/D, the two complexes follow the same unstacking curve (Figure 1). Since the 590 nm intensity is kinetically controlled at low R/D, no useful interpretations of its shape can yet be made. It is only obvious that it arises as a result of complex formation and that its intensity also appears to increase with a higher percentage of dimers. Thus, these titration results have suggested that large stacks of AO are nonfluorescent and that either dimers emit at both wavelengths or a small percentage of bound dye always behaves as monomers whenever excess sites are present.

OPTICAL ACTIVITY

The optical activity behavior of the AO:PAG complex will not be discussed at length here. Suffice it to say that the induced optical activity of the dye is highly sensitive to the structural details of the complex and to the formation of aggregates, both of which depend on the method of complex preparation. Induced Cotton effects as large as those observed for the helix complex can be reproducibly obtained for coil complexes in water solution (Myhr and Foss 1966). In dilute buffered solutions on the other hand, highly variable optical activity, including complete sign reversals, were obtained, and we never were able to devise a method of complex preparation which would yield reproducible behavior. Heating always destroyed the optical activity of the coil complexes and also converted the slightly cloudy, optically active coil complex solutions into clear ones. Certain mixing procedures that yielded clear solutions also never produced an optically active complex. These and other observations strongly suggested that complex aggregate formation may be responsible for the remarkable optical activity of the coil complexes.

CONCLUDING REMARKS

In conclusion, it should be noted that the absorption and fluorescence metachromasy of AO bound to PGA reflects primarily the local polymer structure and only incidentally the overall polymer conformation. The extent of the metachromatic changes depends, of course, on the stacking geometries allowed by the polymer structure. In addition, a certain restriction in the relative motion of adjacent binding sites appears necessary for the development of metachromasy. Thus, metachromasy reflects polymer primary structure and flexibility in a very general way. When one looks, however, at optical parameters more sensitive to the precise structure of the AO:PGA complex in water solution — $i.e.$, the new red fluorescence band and induced optical activity — one begins to realize the importance of kinetically controlled processes and the state of aggregation in determining the optical behavior of the complex. These factors will operate to a greater degree in intracellular staining. It would therefore seem that while some interpretations of intracellular chromotrope structures are possible, such as those by Rigler (1966) on the degree of randomness of nucleic acids, the correct interpretation of the details of intracellular metachromatic changes is probably still a long way off.

ACKNOWLEDGEMENT

This work was supported by Grant GM-15540 to J.F. from the National Institutes of Health, U.S. Public Health Service.

REFERENCES

BRADLEY, D.F., Tinoco, I., Jr., and Woody, R.W., (1963), *Biopolymers,* 1, 239.

CUNDALL, R.B., Lewis, C., Llewellen, P.J., and Phillips, G.O., (1970), *J. Phys. Chem.,* 74, 4172.

DOTY, P, Wada, A., Yang, J.T., and Blout, E.R., (1957), *J. Polymer. Sci.,* 23, 851.

LAMM, M.E., Childers, L., and Wolf, M.K.,(1965), *J. Cell Biol.,* 27, 313.

MYHR, B.C., (1968), *Ph.D. Thesis,* Iowa State University, Ames, Iowa.

MYHR, B.C., and Foss, J.G., (1966), *Biopolymers,* 4, 949.

MYHR, B.C. and Foss, J.G., (1971), *Biopolymers,* 10, 425.

PTITSYN, O.B., (1967), In: *Conformation of Biopolymers,* (G.N. Ramachandran, ed.), Vol. I, Academic Press, New York, 381.

RIGLER, R.,Jr., (1966), *Acta Physiol. Scand.,* 67, Suppl. 267, 32.

STRYER, L. and Blout, E.R.,(1961), *J. Am. Chem Soc.,* 83, 1411.

VAN DUUREN, B.L., (1966), In: *Fluorescence and Phosphorescence Analysis,* (D.M. Hercules, ed.), Interscience Publishers, New York, Chapter 7.

ZANKER, V.Z., (1952), *Physik. Chem.,* Leipzig, 199, 225.

ETHIDIUM BROMIDE: A FLUORESCENT PROBE OF NUCLEIC ACID STRUCTURE AND ITS POTENTIAL FOR IN-VIVO STUDIES

Jean-Bernard Le Pecq

Laboratoire de Physico-chimie Macromoleculaire,
Institut Gustave-Roussy, France

INTRODUCTION

Owing to the fact that tryptophan is a strongly fluorescent amino acid which is present in most of the proteins, nature has provided the biochemist and biophysicist with a very powerful tool. This conference gave a clear illustration of this situation. Unfortunately, nature was not so kind to the nucleic acids investigators. Except in very interesting but exceptional cases, like base Y in phenylalanine t-RNA of yeast (Beardsley et al. 1970), nucleic acids generally have no "built-in" fluorescent probe. Introducing artificially such probes into nucleic acids gives rise to many difficulties especially if the native and original conformation is to be maintained. Ideally one would like to introduce a fluorescent base analogue which presumably would not distort the nucleic acid structure. This very promising approach has been attempted successfully after the discovery of the antibiotic formycin (Ward et al. 1969a, 1969b). Another approach has been introduced. Lerman (1961) discovered the possibility of intercalating a dye between two base pairs in double stranded nucleic acids. In principle, this is very similar to the introduction of an additional fluorescent base-pair into the double helix, since the helix is lengthened accordingly. Of course, this gives rise to a distortion in the helix, but as will be shown later, additional information on the conformation of the nucleic acids can be obtained in some cases if the nature of the distortion is accurately known. One of the main advantages of this approach is that many different dyes are known which are able to intercalate in double stranded DNA and RNA. Thus, in principle, a dye of suitable fluorescent properties can be selected. While searching for such a dye we studied a few years ago (Le Pecq et al. 1964), the compound ethidium bromide (Figure 1) which seemed to have many of the properties required for a suitable fluorescent probe of the nucleic acid conformation.

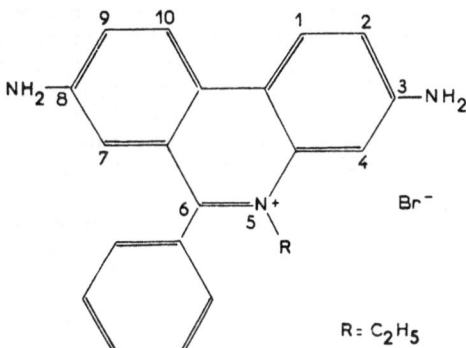

Figure 1. *Ethidium Bromide*

Further studies (LePecq and Paoletti 1965b; 1967; Paoletti et al. 1971: Paoletti and Le Pecq 1971b), confirmed that ethidium bromide may serve as a useful fluorescent probe. At the present stage of these studies two different questions may be raised:

(a) How can the present *in-vitro* studies be developed further and what information may be expected?

(b) How can the present knowledge of the in-vitro systems be applied to the study of the conformation of nucleic acids at the *in-vivo* level?

This last question is of considerable interest, since fluorescence is one of the few techniques which can be applied directly to living systems. In this paper we wish to comment briefly on these two questions. In the first part the results of the *in-vitro* studies are reviewed and discussed briefly, while the second part deals with the problems of the application of this methodology to living systems. To fully understand the possibilities of this approach it is obviously necessary to know accurately the mechanisms of the binding of ethidium bromide to nucleic acids.

MECHANISM OF THE BINDING OF ETHIDIUM BROMIDE TO NUCLEIC ACIDS

As this subject has recently been reviewed in detail (Le Pecq 1971) we would like to comment on this topic here very briefly. Ethidium bromide (EthBr) binds to nucleic acids at two different sites:

A Fluorescent Site

This site is an intercalation site similar to that described by Lerman (1961), for proflavine. This fact has been clearly demonstrated by a variety of techniques, for instance, X-ray diffraction (Fuller and Waring 1964), flow dichroism, singlet-singlet energy transfer from DNA to EthBr (Le Pecq and Paoletti 1967) and viscosimetry applied to linear and circular DNA (Saucier et al. 1971). At this site the dye is bound between two base pairs, its plane being parallel to the plane of the base pairs. There is an almost complete overlap of the dye and of the adjacent base pairs in the complex. Therefore, the name "sandwich complex" which has often been used for this type of complex is a good illustration of the intercalation complex. When intercalated EthBr shows a fluorescence quantum efficiency of 0.14 (Paoletti and Le Pecq 1971b), i.e., about 20 times that of free EthBr in aqueous solution. The fluorescence lifetime is close to 24 ns (Wahl et al. 1970). Such a long lifetime is not surprising since the absorption coefficient of the dye is relatively small in the visible spectrum and since the emission occurs at long wavelengths. The absorption spectrum of the dye is shifted to longer wavelengths after binding to nucleic acids. Therefore, selective excitation of the fluorescence of the bound dye can be achieved by exciting its fluorescence at a wavelength where the bound dye has an extinction coefficient much larger than that of the free dye. Under optimum conditions, the fluorescence intensity of the dye can be increased by a factor of almost 100 if it is bound to nucleic acid. The fact that the fluorescent site is an intercalation site is specific for double stranded nucleic acids, either DNA, RNA or DNA-RNA hybrid. Figure 2 clearly shows that the increase in fluorescence emitted from the dye is negligible in the presence of single stranded structures and proportional to the amount of double stranded structures available for binding. The reactivity of a triple stranded structure is intermediate between that of single- and double-stranded nucleic acids and may be the result of its imperfect structure. There seems to be a mixture of double and triple stranded structures. The increase in fluorescence quantum efficiency of the dye on binding is most likely the result of the immersion of the dye in the hydrophobic medium of the intercalation site. This conclusion is drawn because of the similar properties of the dye when intercalated in nucleic acids or when dissolved in organic solvents. As a consequence, this fluorescence site is almost exclusively specific for nucleic acids and no fluorescence increase of the dye is observed with other compounds (Le Pecq and Paoletti 1966).

The binding constant is of the order of $1 \times 10^6 M$ at low ionic strength and is only slightly reduced at high ionic strength. The dye binds to nucleic acids in highly concentrated cesium chloride solutions. Since the dye-nucleic acid complex has a smaller density than free DNA or RNA (Le Pecq and Paoletti 1967), the selective reactivity of the dye for different nucleic acids can be utilized for separating them in a density gradient of cesium chloride in the ultracentrifuge. According to this principle, double stranded RNA can be separated from single stranded RNA (Kelly 1967) and covalently closed circular DNA from other DNA's (Radloff et al. 1967). In the case of double stranded DNA, the binding situation is best interpreted by the so-called excluded site model (Crothers 1968) where each inter-base pair of the double helix represents a potential site; however when one site is occupied the neighboring sites are excluded. For structural reasons which have been explained elsewhere (Paoletti and Le Pecq 1971a) saturation of the double helix is finally obtained if one dye molecule is bound for every four nucleotides. Single stranded DNA or RNA of random sequences are able to fold thus forming short double stranded regions in a hair-pin like structure. EthBr binds to these structures as it does to double stranded nucleic acids. Therefore, the amount of dye which binds to these structures is a measure of the amount of these double stranded regions. In general, the binding constant for RNA is higher than that

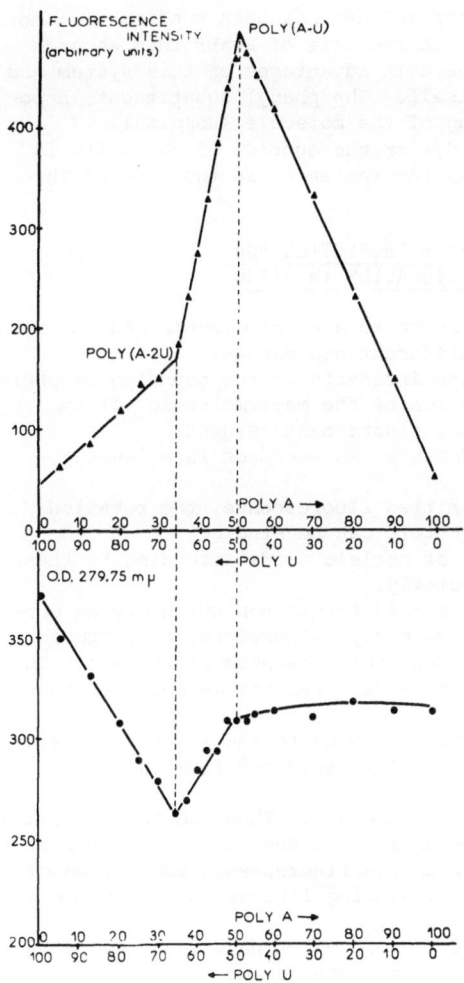

Figure 2. *Illustration of the specificity of the fluorescence increase of ethidium bromide on binding to double-stranded polynucleotides.*

Poly (A) and Poly (U) are mixed at different ratios. The total nucleotide concentration is kept constant. Part of the solution is used for measuring the UV absorption and for controlling the formation of the double and triple stranded structure poly (A:U) and poly (A:2U). Then an excess of ethidium bromide necessary to saturate all available binding sites is added and its fluorescence intensity is measured. (For experimental details see Le Pecq and Paoletti, 1965a). (Figure from LePecq and Paoletti 1965a)

for DNA. On the other hand binding is independent of base composition and molecular weight (Le Pecq and Paoletti 1967).

Second Binding Site

This site is mainly populated at low ionic strength and preferentially in the case of stranded polynucleotides. In contrast to the properties of the first site, the fluorescence quantum efficiency of the dye bound to this second site is small. The absorption spectra of EthBr bound to this second site and to the first fluorescent site are almost identical since, in general, three well-defined isobestic points are observed whatever the conditions are. Therefore, spectrofluorometry is the only way of distinguishing between the two different sites. However, quenching by energy transfer between the dye molecules bound to the fluorescent site and those bound to this second site may become very efficient.

Since for studying the conformation of the nucleic acids it is necessary to measure the binding exclusively to the first fluorescent site, it follows that it is almost impossible to make reliable measurements if this second site is populated. Fortunately, the binding constant to this site is very much dependent on the ionic strength of the medium and can generally be eliminated completely by increasing the salt concentration. This is an important fact to be considered when using EthBr fluorescence as a probe of the nucleic acid structure. Results of some experiments carried out without such precautions would probably have to be reassessed. This is probably true also for the measurement of EthBr binding to t-RNA (Bittman 1969). The number of sites found in that study for t-RNA is obviously too large and has not been confirmed in more detailed studies (Thomas,personal communication). In the case of EthBr this second site can be readily eliminated. This is one of the main advantages of this system and is certainly inherent in the structure of EthBr itself. The phenyl substituent in position 6 is oriented perpendicular to the main ring of the molecule (Hospital and Busseta 1969). This explains why stacking of the dye on the outside of the helix is not favored, in contrast to the situation observed, for instance, in the case of the acridine derivatives.

USE OF ETHIDIUM BROMIDE FLUORESCENCE PROPERTIES FOR STUDYING THE CONFORMATION OF NUCLEIC ACIDS IN VITRO

The fluorescence properties of a chromophore bound to a macromolecule can in principle be used to study its conformation by three different approaches:

(a) It is possible to analyze the fluorescence intensity of the bound chromophore and its changes due to alterations of the conformation of the macromolecule, or to measure the reactivity of a macromolecule toward the fluorescent reagent.

(b) The bound chromophore can be used as a donor or an acceptor in an energy transfer experiment.

(c) By measuring the depolarization of the emitted fluorescence, the rotation properties of the macromolecule or its dynamic structure can be studied.

EthBr has been used to study the conformation of nucleic acids according to these three principles. The results are here reviewed briefly.

In the case of proteins, for example, the amino acid tryptophan which may be considered as a built-in fluorescent probe is bound covalently. Therefore, a fluorescent change can immediately be interpreted in terms of a change in quantum efficiency. In the case of an EthBr-nucleic acid complex the fluorescence intensity measured is the resultant of two independent factors:

(a) The number of dye molecules which are actually bound to the fluorescent site resulting both from the affinity of the dye for the nucleic acid and from the total number of available sites.

(b) The fluorescence quantum efficiency of the bound dye. Thus, to interpret any fluorescence change observed, it is necessary to determine the number of molecules of EthBr bound to the first and second site together with the fluorescence properties of the bound dye. This information can be obtained by measuring fluorescence lifetime and fluorescence intensity simultaneously.

The fluorescent properties of the EthBr-nucleic acid complex have been used mainly for the following investigations into polynucleotides conformation.

Study of the Reactions of Hybridization Between Polynucleotides

In this case it is the number of available fluorescent binding sites which varies. The number reaches a maximum for double stranded structures. Any change from single to double or triple stranded structures can readily be observed, since in the presence of EthBr it is accompanied by a significant change in fluorescence intensity. An example of this reaction is shown in Figure 2. The formation of a double stranded structure in solution between poly (A) and poly (I) can only be detected with this technique (Le Pecq and Paoletti 1965b; Le Pecq 1971).

Study of Displacement Reactions Between Polynucleotides

In general, reactions of this type are accompanied by a change from single or triple stranded structures to a double stranded structure or vise versa. In the presence of EthBr, these reactions result in major changes in fluorescence intensity (Le Pecq and Paoletti 1965a).

Study of Polymerization reactions

When DNA-dependent RNA polymerase transcribes a single stranded DNA, a DNA-RNA hybride is formed. The number of fluorescent sites available for EthBr increases at the same rate as the formation of double stranded DNA-RNA hybride proceeds during the enzymatic reaction. This principle could also be applied to the study of DNA polymerase,

reverse transcriptase or any enzyme which causes the formation of double stranded polynucleotides (Le Pecq et al. 1966).

Study of Circular DNA and its Associated Enzymes

The principles on which the study of covalently closed DNA using EthBr is based
are somewhat more complex (see Figure 3). When EthBr binds to a naturally supercoiled
DNA, the superhelix is progressively released because of the change in torsion of the
double helix induced by the intercalation process. Owing to this superhelix the binding constant of EthBr for the naturally supercoiled DNA is larger than for the nicked
or linear DNA (Bauer and Vinograd 1968). Let us consider two samples, one containing a given
concentration of naturally supercoiled DNA (A in Figure 3), the other containing exactly
the same concentration of nicked DNA (A' in Figure 3). If EthBr is added in increasing
amounts to the two samples of identical concentrations and if the fluorescence intensities of the two samples are measured under identical conditions, the fluorescence intensity measured for the sample containing the supercoiled DNA will be higher than the
fluorescence intensity measured for the sample containing the nicked DNA; this clearly
demonstrates the difference in affinity of the dye to the two different DNA's. When
the natural superhelix is completely released (B in Figure 3) the affinity of the dye
is the same for both DNA species, and the fluorescence intensities of the two samples
are identical. The quantity of dye bound at this step is proportional to the number of super-
turns (Bauer and Vinograd 1968). If EthBr is added beyond this point, a superhelix of opposite
sign compared to that of the original superhelix is formed in covalently closed circular

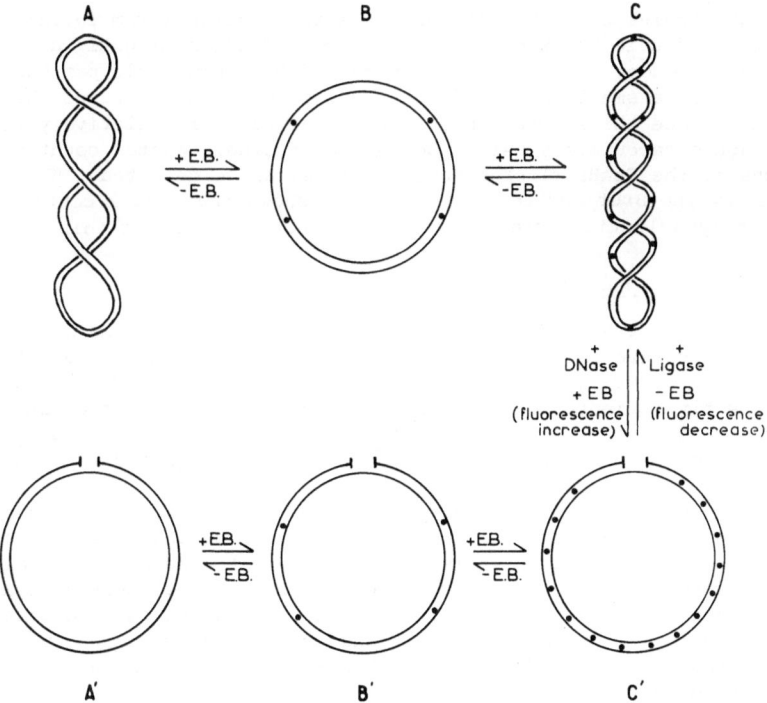

Figure 3. *Schematic illustration of the specific reactivity of covalently closed
circular DNA with ethidium bromide.*
*When ethidium bromide is added to naturally supercoiled DNA (A), it first releases
the superhelix and produces a relaxed circle (B). Further addition of ethidium bromide
causes the formation of a superhelix of opposite sign (C). In the presence of excess
ethidium bromide the supercoiled DNA binds less ethidium bromide than the nicked circle and therefore is less fluorescent. The conversion of one form of circular DNA into
the other (C ⇌ C') is accompanied by a drastic change in the fluorescence intensity if
ethidium bromide is present in excess. Details are discussed in the text*

DNA. This superhelix of opposite sign now decreases the affinity of the dye so that the sample containing the supercoiled DNA becomes less fluorescent than the sample containing the nicked DNA. Therefore, a well-defined point is observed where the fluorescence intensity of the two samples reaches equality. The measured fluorescence intensity then permits the number of dye molecules bound per molecule of circular DNA to be calculated and, provided that the change of torsion associated with the intercalation process is known, it is possible to determine the sign of the superhelix and to measure the degree of superhelicity of the original supercoiled DNA. Owing to the sensitivity of fluorescence measurements this determination can be carried out on very small quantities of DNA, possibly no more than a few tenths of a microgram.

Let us now consider the structure of supercoiled and nicked DNA in the presence of a large excess of EthBr (C and C' in Figure 3). The number of dye molecules bound to the supercoiled DNA will be smaller than the number of EthBr molecules bound to the nicked circular DNA owing to the difference in affinity of EthBr for these two conformational forms of DNA. If a single nick is made on a supercoiled DNA, the fluorescence intensity measured in the presence of an excess of dye will increase, since the topological constraint limiting the binding of the dye will be released. The reverse reaction which is represented by the repair of the nicked circular DNA will be associated with a decrease of fluorescence intensity in the presence of an excess of dye. This principle provides the basis for a very sensitive method of investigating these enzymatic reactions (Paoletti et al. 1971).

Measurement of the Fluorescence Lifetime Changes of Ethidium Bromide Associated With Changes of the Double Helix Structure

Recent X-ray studies of double stranded DNA in solution (Bram 1971a; 1971b; 1971c) have shown that the double helix can change its conformation slightly as a result of changes of its environment and of a minor change of the turn angle per base pair; under these conditions only a small change of the fluorescence properties of the bound dye may be expected. Since the fluorescence lifetime of EthBr is relatively long and can therefore be measured accurately using the method of single photon counting (Wahl 1970), minor variations of the EthBr fluorescence lifetime can be detected. Work in progress at this laboratory (Le Bret and Le Pecq 1972) has shown that small conformational changes of the polynucleotides can actually be detected by use of this method.

Use of EthBr for Energy Transfer Experiments

The rate of energy transfer between a donor and an acceptor has been calculated by Förster (1959). This rate is inversely proportional to the sixth power of the distance between acceptor and donor and proportional to an orientation factor which takes into account the respective orientations of donor and acceptor. Since the rate of energy transfer is so much dependent on distance, the latter can under certain conditions be accurately determined. Therefore, this method has been designated as a spectroscopic ruler (Stryer and Haughland 1967).

Thus, it would be possible to detect a conformational change causing a minor variation of the distance between two chromophores by measuring the variation of the energy transfer rate. This possibility has clearly been demonstrated by Tao et al. (1970) who measured the rate of energy transfer between the Y base of phenylalanine t-RNA of yeast and EthBr bound to the strongest binding site. When the salt concentration was changed in the solution the rate of energy transfer changed dramatically. The authors interpreted this result as a conformational change of the t-RNA molecule due to the change in the environment.

The energy transfer between EthBr bound to DNA has likewise been measured (Paoletti and Le Pecq 1971b). Assuming the dye molecules to be randomly distributed along the helix, the distance between adjacent molecules of EthBr on the helix can be calculated. It follows that in this case the orientation factor is the only unknown quantity. This orientation factor is simply the angle between the transition moments of two adjacent EthBr molecules. Thus, the angle between two adjacent molecules can be measured and the change of torsion of the double helix caused by the intercalation of one EthBr molecule may be deduced. The result of this determination is controversial, since it predicts that intercalation further winds the double helix and that the sign of the superhelix of naturally supercoiled DNA is positive. Up to now it was generally admitted that the opposite is true (Fuller and Waring 1964; Crawford and Waring 1967; Vinograd et al. 1968). This problem should, therefore, be reinvestigated.

The Dynamic Structure of DNA

EthBr is rigidly bound when intercalated in polynucleotides and cannot move relative to the helical axis. Furthermore, the direction of the transition moments of absorption and emission which are in the plane of the molecule has a known direction relative to the helix axis. On the other hand, the fluorescence lifetime of intercalated EthBr is of relatively long duration. It is well known that a longer lifetime of the fluorescence label considerably facilitates the determination of the rotational relaxation time of a macromolecule by measuring the fluorescence depolarization (for a review of this method see Weber 1953). As a first approximation it may be assumed that the kinetic rotational relaxation time of a sphere is about 1 ns per 3,000 Daltons. Thus, EthBr is a convenient fluorescent probe for studying polynucleotides up to a molecular weight of 10^5 Daltons on the basis of this method. This range includes t-RNA, 5 S-RNA and other species of low molecular-weight nuclear RNA. EthBr has been used as such a probe for investigating the t-RNA conformation (Tao et al. 1970). Since DNA is a large molecule, the assumption seems to be justified that its rotation during the lifetime of the EthBr excited state can be neglected. Measurements of the fluorescence depolarization of EthBr bound to large molecules of DNA show that this assumption is not correct and that DNA has a dynamic structure (Kelly 1967). Measurement of the decay of fluorescence anisotropy of EthBr bound to DNA permits the conclusion that DNA molecules in solution are subject to an internal oscillatory Brownian motion. The amplitude of this oscillation was found to be 35° and the relaxation time of this motion was 28 ns.

POSSIBILITIES OF STUDYING THE CONFORMATION OF NUCLEIC ACIDS IN ORGANIZED STRUCTURES AND AT THE "IN VIVO" LEVEL

From the results of these *in vitro* studies it becomes evident that the use of ethidium bromide as a fluorescent probe offers valuable information as to the conformation of nucleic acids. Hence, it is tempting to apply the same methodology to more complicated systems like organized particles containing nucleic acids or even directly to living cells using fluorescence microscopy. Both approaches have been tried and are briefly discussed with respect to their prospects and limitations.

From the results of several experiments it may be concluded that the presence of proteins at the surface of the nucleic acids limits their reactivity for EthBr. One of the most demonstrative results was obtained by an investigation into the reconstitution of the ribosome out of its protein and nucleic acid constituents (Bollen et al. 1970). Ribosomes do not bind EthBr, but ribosomal RNA binds EthBr in its double stranded regions and the fluorescent intensity of EthBr in the presence of ribosomal RNA can easily be measured. During the reconstitution experiment when proteins are progressively added to the RNA, EthBr is expelled from the macromolecular complex and fluorescence intensity decreases drastically. Thus, the reconstitution kinetics can be followed easily by fluorometry. It is interesting to note that the reconstituted ribosome exhibits activity in the protein synthesis assay *in-vitro*, but in contrast to the native ribosome binds some EthBr (Bollen et al. 1970). The same conclusions concerning the effect of proteins on the binding behavior are drawn from dissociation and reconstitution experiments of nucleohistones (Angerer and Moudrianakis 1972; Riou 1967), and from the study of the binding of EthBr to intact trypanosoma by fluorescence microscopy. It is evident from these experiments that EthBr readily binds to the kinetoplast where the DNA is free of proteins, in contrast to the binding of EthBr to the nucleus where DNA is complexed with proteins (Riou 1968; Caspersson et al. 1969). Nevertheless, it seems that in all these cases the situation is more complex. In addition to the easily understandable limitation of the binding for instance by chromosomal proteins, a limitation of the EthBr binding due to the folding of DNA in such a structure or due to an unknown DNA conformation must certainly also be taken into account. Such limitations of EthBr binding have been made evident by the work of Caspersson et al (1969). These authors observed for instance, that the dye quinacrine mustard, which covalently binds to DNA and presumably reacts with any protein-free DNA, gives fluorescence patterns in chromosomes which are quite different from those obtained by use of EthBr and other intercalating dyes.

Limitations to EthBr binding due to the folding of DNA can experimentally be studied in intact phages and viruses by measuring the fluorescence characteristics of the bound dye in these systems. The constraints imposed by DNA folding can be varied by studying dye binding to different deletion mutants of these phages. These mutants have the same head volume but contain a smaller amount of DNA. Experimental work based on this approach is in progress at our laboratory (Paoletti and Le Pecq Unpublished). Preliminary results show that the structure of the DNA helix is changed due to the constraint imposed by the folding. The EthBr fluorescence characteristics are modified and progressively return to the fluorescent characteristics of the dye bound to DNA in solution, when the folding constraint is progressively released by decreasing the amount of DNA to be packed as in the case of the deletion mutants. It is also observed that the second binding site becomes favored if the intercalation is limited by the folding constraint. This makes quantitative fluorescence measurements much more difficult.

A better understanding of the binding limitations imposed by such folding is obviously necessary before any meaningful measurement can be made by microscope fluorometry, for instance, on chromosomes. This would be of considerable interest especially in view of Crick's hypothesis (1971) concerning the role of DNA folding in regulation processes of eucaryotes. Crick also stresses the importance of superhelical DNA. It seems that the present state of development of fluorescence microscopy and microscope spectrofluorometry justifies an attempt to map superhelices in chromosomes using EthBr. This attempt could profit by different procedures selected according to the characteristic property of the superhelix required for the experiment. One could for instance release the constraint of the superhelix by nicking the DNA. A simple procedure would be the irradiation of chromosomes with X-rays of the right dose. It would be necessary to compare the fluorescence pattern before and after nicking of the DNA. Possibly an even easier way of detecting a superhelix by this method would be through energy transfer experiments. Thus, the EthBr fluorescence could be quenched by energy transfer, e.g. by additionally using non-fluorescent ethidium bromide derivative. In a superhelix the two dyes will be further apart and no quenching of the EthBr fluorescence may be observed; it is hoped that the superhelix will be specifically labeled under these conditions. Energy transfer experiments may also be considered for measuring inter-helix spacing and for a direct study of DNA packing in the chromosomes.

REFERENCES

ANGERER, L.M., and Moudrianakis, E.M. (1972) *J. Mol. Biol.*, 63, 505.

BAUER, W., and Vinograd, J. (1968) *J. Mol. Biol.*, 33, 141.

BEARDSLEY, K., Tao, T., Cantor, C.R., (1970) *Biochemistry*, 9, 3524.

BITTMAN, R., (1969) *J. Mol. Biol.*, 46, 251.

BOLLEN, A., Herzog, A., Favre, A., Thibault, J., Gros, F., (1970) *FEBS Letters*, 11, 49.

BRAM, S., (1971a) *J. Mol. Biol.*, 58, 277.

BRAM, S., (1971b) *Nature New Biology*, 232, 174.

BRAM, S., (1971c) *Nature New Biology*, 233, 161.

CASPERSSON, T., Zech, L., Modest, E.J., Foley, G.E., Wagh, V., Simonsson, E.,(1969) *Exp. Cell Res.*, 58, 141.

CRAWFORD, L.V., and Waring, M.J., (1967) *J. Mol. Biol.*, 25, 23.

CRICK, F., (1971) *Nature*, 234, 25.

CROTHERS, D.M., (1968) *Biopolymers*, 6, 575.

FÖRSTER, T., (1959) *Disc. Faraday Soc.*, 27, 7.

FULLER, W., and Waring, M.J., (1964) *Ber. Bunsenges, Physik. Chem.*, 68, 805.

HOSPITAL, M., and Busseta, B., (1969), *Comptes Rendus série C. Paris*, 268, 1232.

KELLY,R., and Sinsheimer, R.L., (1967) *J. Mol. Biol.*, 29, 229.

LE BRET, M., and Le Pecq, J.B., (1972) Unpublished results.

LERMAN, L.S., (1961) *J. Mol. Biol.*, 3, 18.

LE PECQ, J.B.,(1971) In: *Methods of Biochemical Analysis*, (Glick, D.,ed.) J. Wiley and Sons publishers, Vol. 20, 41.

LE PECQ, J.B., and Paoletti, C. (1965a), *Comptes Rendus Paris*, 261, 838.

LE PECQ, J.B., and Paoletti, C.,(1965b), *Comptes Rendus Paris*, 260, 7033.

LE PECQ, J.B., and Paoletti, C.,(1966), *Analyt. Biochem.*, 17, 100.

LE PECQ, J.B., and Paoletti, C. (1967), *J. Mol. Biol.*, 27, 87.

LE PECQ, J.B., and Jeanteur, Ph., Ravicovitch, R., Paoletti, C., (1966), *Biochim. Biophys. Acta*, 119, 442.

LE PECQ, J.B., Yot, P., Paoletti, C.,(1964), *Comptes Rendus Paris*, 259, 1786.

OLINS, D.E. (1969) *J. Mol. Biol.*, 43, 439.

PAOLETTI, J., and Le Pecq, J.B., Unpublished results.

PAOLETTI, J., and LE PECQ, J.B., (1971a),*Biochimie*, 53, 969.

PAOLETTI, C., and Le Pecq, J.B., Lehman, I.R., (1971), *J. Mol. Biol.*, 55, 75.

PAOLETTI, J., and Le Pecq, J.B., (1971b), *J. Mol. Biol.*, 59, 43.

RADLOFF, R., Bauer, W.R., Vinograd, J.,(1967), *Proc. Nat l. Acad. Sci. USA*, 57, 1514.

RIOU, G., (1967), *Comptes rendus Paris Série D*, 265, 2004.

RIOU, G., (1968), *Comptes rendus Paris Série D*, 266, 250.

SAUCIER, J.M., Festy, B., Le Pecq, J.B.,(1971), *Biochimie*, 53, 973.

STRYER, L., and Haughland, R.P., (1967), *Proc. Natl. Acad. Sci. USA*, 58, 719.

TAO, T., Nelson, J.H., Cantor, Ch. R. (1970), *Biochemistry*, 9, 3514.

THOMAS, G., Personal Communication.

VINOGRAD, J., LeBowitz, J., Watson, R., (1968),*J. Mol. Biol.*, 33, 173.

WAHL, Ph., Paoletti, J., Le Pecq, J.B., (1970), *Proc. Natl. Acad. Sci. USA*, 65, 417.

WARD, D.L., Reich, E., Stryer, L, (1969a),*J. Biol. Chem.*, 244, 1228.

WARD, D.L., Cerami, A., Reich, E., Acs, G. and Altwerger, L., (1969b), *J. Biol. Chem.*, 244, 3243.

WEBER, G., (1953), *Adv. Protein Chem.*, 8, 415.

NANOSECOND FLUORESCENCE SPECTROSCOPY OF BIOLOGICAL MACROMOLECULES AND MEMBRANES

Juan Yguerabide*

Department of Molecular Biophysics and Biochemistry,
Yale University

I. INTRODUCTION

Fluorescence spectroscopy is a very sensitive technique capable of giving information on various aspects of the structure and dynamics of biological macromolecules and membranes. Its applicability in these areas has been greatly enhanced in recent years by improvements in techniques both for the generation and detection of nanosecond light pulses and for numerical analysis of time-dependent emission data. This article reviews the basic concepts used in these applications and presents results obtained in our laboratory with proteins, nucleic acids, and membranes.

II. BASIC CONCEPTS

There are three basic uses of fluorescence spectroscopy in the study of the conformation and dynamics of biological systems. In one, the dependence of the emission properties of a chromophore on its environment is used to study the polarity of interesting sites and to detect conformational changes which may occur when conditions are changed or when the system interacts with small molecules. In a second application, electronic excitation-energy transfer between donor and acceptor chromophores positioned at suitable sites is used to measure the distance between the sites and to detect conformational changes which occur when the system is perturbed. In a third application, polarized fluorescence is used to establish the size, shape, and flexibility of macromolecules in solutions and to determine orientation and motion of molecules in membranes. Recent developments allow the use of fluorescence spectroscopy to study the electrical potential at the surface of membranes and to investigate the effects of this potential on ionic transport mechanisms.

A. *Fluorescent Probes and Polarity*

The application of fluorescence techniques to a biological system requires the presence of a fluorescent chromophore at a suitable site. The chromophore may be either intrinsic to the system, such as tryptophan in proteins, or extrinsic, in which case it is called a fluorescent probe. Extrinsic chromophores with well-defined properties are the ones most generally used since intrinsic chromophores a) may not be present in systems of interest, b) may not be positioned at interesting sites, or c) may have emission properties which are difficult to interpret. Fluorescent probes can be covalently or noncovalently attached to a system and may be highly specific, as in the case of fluorescent analogs of substrates and receptor agonists. For membranes, the chromophore may be designed to probe specific regions of these structures. Tables I and II give a list of some fluorescent probes and their uses.

Chromophores whose fluorescent properties are strongly dependent on the polarity of the environment have been especially useful as sensitive indicators of polarity and conformational changes (Stryer 1968; Edelman and McClure 1968). These chromophores are exemplified by ANS and dansyl, which are used in the study of proteins and membranes and by ethidium bromide, which is used in the study of nucleic acids. The effect of environment on the emission properties of ANS is shown in Table III. As the polarity decreases, the wavelength of maximum emission λ_{max} shifts to shorter wavelengths, while the quantum yield ϕ_f and lifetime τ increase. The effects are most pronounced for ϕ_f and τ.

Ever since the discovery of polarity-sensitive fluorescence probes (Weber and

---

* *Present address: Department of Biology, University of California, San Diego, California.*

TABLE I

FLUORESCENT PROBES

Probes	Investigation
1 - Anilino - 8 - naphthalene sulfonate (ANS).	Polarity of heme-binding sites of apomyoglobin and apohemoglobin (Stryer 1965). Conformational changes in muscle, nerve, and nerve endings during action potential (Tasaki, et al., 1968; Carnay and Benn 1969; Patrick et al. 1971). Conformation, dynamics and surface electrical potential of erythrocyte membranes (Gitler et al. 1969; Fortes et al. 1972). Structural changes in mitochrondrial membranes on activation with substrates (Azzi et al. 1969).
2 - p - Toluidinylnaphthalene - 6 - sulfonate (TNS).	Conformational changes on activation of chymotrypsinogen (McClure and Edelman 1967). Conformation of nerve membrane (Tasaki et al. 1971).
1 - dimethylaminonaphthalene 5 - sulfonyl chloride (dansyl chloride) and derivatives.	Conformation of active site of carboxypeptidase A (Latt et al. 1970). Polarity of sulfonamide site of carbonic anhydrase (Chen and Kernohan 1967). Polarity of hapten binding site and flexibility of IgG (Parker et al. 1967; Yguerabide et al. 1970).
2,7 - diamino - 10 - ethyl - 9 - phenylphenanthridinium bromide (ethidium bromide).	Secondary structure of DNA and RNA (Le Pecq and Paoletti 1967).
1 - Pyrenebutyric acid.	Accessibility of binding site to O_2 (Vaughen and Weber 1970).
12 - (9 - Anthroyl) - stearic acid (AS).	Conformation and dynamics of membrane s (Waggoner and Stryer 1970; Yguerabide 1971).
N,N' - di (octadecyl) oxacarbocyanine	Conformation and dynamics of membrane s (Yguerabide and Stryer 1971).

Laurence 1954), it has been of interest to determine the mechanism responsible for the striking effects of environment on the emission properties of these chromophores. Various experiments now indicate that the sensitivity results from the unusually large dipole moments which the probes have in the excited state. The dipole moment of TNS, for example, increases by 10 debyes upon excitation (Edelman and McClure 1968). Because of the difference in moments, interaction with solvent dipoles is greater in the excited state than in the ground state, thereby decreasing the energy separation of the two states and shifting the fluorescence spectrum to the red, as is observed experimentally. The increased interaction in the excited state, however, requires that the solvent dipoles in the vicinity of the chromophore rearrange themselves within the lifetime of the excited molecule. The shift in fluorescence spectrum is thus dependent on the mobility of the solvent dipoles. Indeed, the effect of a polar solvent on emission properties decreases with decreasing temperature (Edelman and McClure 1968). Because of the dependence of emission properties on mobility, ANS and similar chromophores are now called probes of dynamic polarity.

The shifts in energy levels discussed above are also indirectly responsible for the effects of polarity on τ and ϕ_f through mechanisms which will not be described here. τ and ϕ_f are directly related to the rates of emission k_e and internal quenching k_i by the expressions

$$\tau = 1/(k_e + k_i) \tag{1}$$

$$\phi_f = k_e/(k_e + k_i) \tag{2}$$

k_i is in turn determined by a variety of internal quenching processes, including the process by which excited molecules are converted from the emitting singlet state to the nonemitting triplet state (intersystem crossing). Polarity can effect τ and ϕ_f by influencing either k_e or k_i. Recent measurements in various solvents reveal that the main influence is on k_i. As shown in Table III, k_i increases with increasing polarity. So far, however, experiments have not shown which of the several internal quenching processes is being affected. It is believed, nevertheless, that intersystem crossing is the one involved. This contention is based on experiments which indicate that for ANS the singlet-triplet energy separation ΔE decreases with increasing polarity (Seliskar and Brand 1971), and on theoretical calculations which indicate that the rate of intersystem crossing increases rapidly with decrease in ΔE (Robinson and Frosch 1963). On these bases, it is expected that k_i will have a steep dependence on ΔE, which acts as an amplification factor so that relatively small changes in ΔE will have pronounced effects on τ and ϕ_f, as observed experimentally.

Intersystem crossing, however, is a complex process that depends not only on polarity, through its effects on ΔE, but also on more specific properties of the environment. In particular, theoretical and experimental considerations indicate that the rate of intersystem crossing will depend on whether the environment is deuterated or not (Seliskar and Brand 1971; Robinson and Frosch 1963). Experiments with ANS have indeed revealed that ϕ_f is twice as large in D_2O as in H_2O, although λ_{max} is the same in both solvents (Stryer 1966). The isotope dependence results from the difference in mass of hydrogen and deuterium atoms, a difference which affects the rate at which excitation energy can be transferred to the various vibrational modes of the solvent. k_i is also substantially increased by relatively low concentrations of paramagnetic substances, such as O_2, or heavy-atom-containing substances, such as bromobenzene, which increase spin-orbit coupling without affecting λ_{max}. Thus ϕ_f and τ, are not simply related to the polarity of the environment.

In many applications of fluorescent probes, however, deuterated compounds or heavy-atom quenchers are not present in the system under study; in these cases ϕ_f and τ will be chiefly dependent on dynamic polarity as determined, for example, by measuring the emission properties in a series of straight-chain alcohols. On the other hand, deuterium and heavy-atom effects may be useful in determining whether the site in which the probe is located is accessible to the solvent or to small molecules. In any case, the high sensitivity of τ and ϕ_f on environment make these parameters very sensitive indicators of conformational changes.

Some of the uses of fluorescent probes are shown in Table I. A detailed discussion of the use of dansyl to determine the polarity and conformational changes of the active site of antibodies is presented in Section III, on Advantages of Nanosecond Fluorescence Spectroscopy, below.

Excitation Transfer

It has been known since 1922 (Cario and Franck) that electronic excitation energy can be transferred from a singlet excited molecule (donor) to an unexcited molecule (acceptor) over distances as large as 100 Å (singlet-singlet transfer). According to Förster (1947), this transfer occurs by a transition dipole-dipole moment interaction, and its specific rate, k_{DA}, is related to spectroscopic and molecular parameters by the equation (Haugland et al. 1969):

$$k_{DA} = 8.7 \times 10^{23} \, JK^2 N^{-4} k_e R^{-6} sec^{-1} \tag{3}$$

where R is the distance (in Å) between the centers of the transition moments of the donor D and acceptor A, N is the refractive index of the medium between A and D, and k_e (sec^{-1}) is the rate of donor emission. K is an orientation factor defined by the equation:

$$K = \hat{a}.\hat{d} - 3 \, (\hat{a}.\hat{r}) \, (\hat{d}.\hat{r}) \tag{4}$$

where $\hat{a}$ and $\hat{d}$ are, respectively, unit vectors along the transition moments of A and D, and $\hat{r}$ is a unit vector along the line joining the center of A and D. J is a spectral overlap parameter defined by the equation:

TABLE II

STRUCTURAL FORMULAS FOR FLUORESCENT PROBES

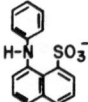

1-Anilino-8-naphthalene sulfonate (ANS)

1-Pyrenebutyric acid

2-p-Toluidinylnaphthalene-6-
sulfonate (TNS).

2,7 diamino-10-ethyl-9-phenylphenanthri-
dinium bromide (ethidium bromide).

1-dimethylaminonaphthalene-5-sulfonyl
chloride (dansyl chloride).

12-(9-Anthroyl)-stearic acid (AS).

N,N'-di(octadecyl)oxacarbocyanine

TABLE III

FLUORESCENCE PARAMETERS OF ANS IN A SERIES OF SOLVENTS[a,b]

Solvent	λmax (nm)[c]	ϕ_f	τ(nsec)	k_e (sec^{-1}) × 10^7	k_i (sec^{-1}) × 10^7
Octanol	464	0.646	12.3	5.23	2.86
Butanol	464	0.516	10.9	4.73	4.44
Propanol	466	0.476	10.2	4.65	5.12
Ethanol	468	0.361	8.85	4.08	7.22
Methanol	476	0.216	6.05	3.57	12.9
Dioxane		0.505	11.8	4.28	4.19
Water	515	0.004[c]	0.55	0.727	181
Bound to Apomyoglobin	454	0.98[c]	16.4[d]	5.98	.122

[a]Results are for magnesium salt of ANS. The parameters ϕ_f, τ, k_e and k_i are defined in text.

[b] Yguerabide 1972.

[c] Stryer 1965.

[d] Tao 1969.

$$J = \frac{\int F_D(\lambda) \varepsilon_A(\lambda) \lambda^4 d\lambda}{\int F_D(\lambda) d\lambda} \tag{5}$$

where $F_D(\lambda)$ is the fluorescence intensity of the donor at wavelength λ and $\varepsilon_A(\lambda)$ is the molar decadic absorption coefficient (in $cm^{-1}\ M^{-1}$) of the acceptor at λ. Integration is over the region where the fluorescence spectrum of D overlaps the absorption spectrum of A. The validity of Equation (5) with respect to the dependence of k_{DA} on J and R has been verfied experimentally (Haugland et al. 1969; Latt et al. 1965; Bücher et al. 1967; Stryer and Haugland 1967).

Because of the dependence of k_{DA} on R, excitation transfer between chromophores attached to sites S_1 and S_2 on a macromolecule or membrane can be used to determine the distance between the sites. In this determination, the parameters k_{DA}, k_e, N, J, and K are evaluated for the system of interest, and R is then calculated from Equation (3). Methods for evaluating these parameters are as follows (Haugland et al. 1969).

a) The rate of energy transfer k_{DA} can be evaluated by exciting the donor with a fast pulse of light and then measuring the decay of fluorescence intensity of either the donor or the acceptor. It is, however, usually simpler to use the donor fluorescence because the acceptor emission kinetics are more complex. Thus, by measuring the lifetime of the donor in the absence, τ_D^o and presence, τ_D, of the acceptor, k_{DA} can then be evaluated with the expression:

$$k_{DA} = \frac{1}{\tau_D} - \frac{1}{\tau_D^o} \tag{6}$$

With modern techniques of nanosecond fluorescence spectroscopy, k_{DA} can be determined with very high accuracy.

b) To evaluate k_e, the fluorescence quantum yield of the donor ϕ_D^o in the absence of acceptor is measured by standard techniques and k_e is then calculated with the equation:

$$k_e = \phi_D^o\ \tau_D^o \tag{7}$$

c) The value of the refractive index N is usually estimated from values given in the literature for similar systems. For proteins, a value of 1.4 is normally used. Errors in the calculated value of R introduced by uncertainties in N are usually not greater than 10%.

d) The spectral overlap integral J is evaluated with Equation (5), using the fluorescence spectrum of D and absorption spectrum of A as measured in the system of interest.

e) The orientation factor K^2 cannot be measured with available techniques, and its evaluation presents the major problem in the use of excitation transfer to measure distances. Its value ranges from 0 to 4, and in the absence of any knowledge concerning K^2, only an upper limit, corresponding to $K^2 = 4$, can be given for the value of R. The method usually used in practice to estimate the value of K^2 is based on the following observations. The transition moments $\hat{a}$ and $\hat{d}$ do not usually have a fixed orientation with respect to each other, as implicitly assumed above, but instead rotate through relatively large angles in times less than one nsec. This rotation occurs by mechanisms involving internal molecular processes as well as rotational motion of the donor and acceptor chromophores at the site of attachment. The rotational motions cause K^2 to be averaged over several orientations, and in the case where $\hat{d}$ and $\hat{a}$ are randomly distributed and rapidly rotating (complete dynamic random case), K^2 becomes equal to 2/3. If the conditions of this case can be shown to apply in a given experiment, then $K^2 = 2/3$, and the orientation ambiguity is removed.

For proteins, experiments indicate that $\hat{d}$ and $\hat{a}$ usually rotate rapidly over angles ranging from 10° to 30°. Although these ranges do not satisfy the conditions of the complete dynamic random case, it has nevertheless, become common practice to assume that $K^2 = 2/3$ in the calculation of R. The errors which may result from this assumption can be easily calculated. Thus, if in a given situation, R_r and K_r^2 are the true values of the distance and orientation factor, and R_c is the distance that is calculated using $K^2 = 2/3$, then the error α in the calculated distance is given by the expression:

$$\alpha = R_r/R_c = \sqrt[6]{\frac{3}{2}\ K_r^2} \quad . \tag{8}$$

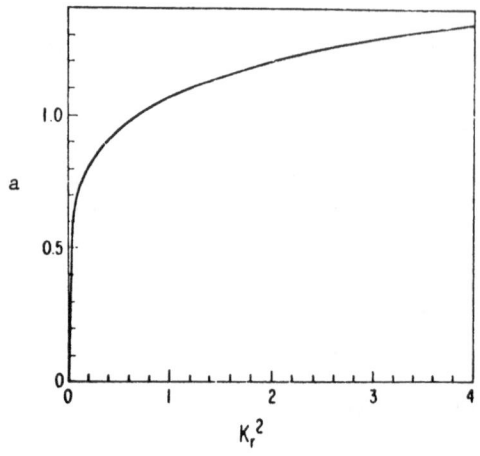

Figure 1. *Plot of the error α in the distance R_c calculated from excitation energy transfer measurements assuming $K^2 = 2/3$ versus the true orientation factor K_r^2. See Equation (8)*

Figure 1 shows a plot of α vs K_r^2 covering all possible values of K_r^2. For K_r^2 greater than 2/3, R_c is less than R_r, while for K_r^2 less than 2/3, R_c is greater than R_r. The error in R_c is not greater than 10% when K_r^2 is in the range of 0.4 to 1. For K_r^2 greater than 1, α increases slowly with increasing K_r^2, and the error approaches a value of about 35% for $K_r^2 = 4$. For values of K_r^2 less than 0.4, however, α increases rapidly with decreasing K_r^2 and the error in distance becomes greater than 30% for $K_r^2 < 0.1$. It is evident from the graph that the most serious errors will occur when K_r^2 is small, although large values of K_r^2 also cause significant errors.

The important question, then, is how probable are those values of K_r^2 which give large errors in R_c? That is, what fraction of the total number of possible orientations of $\vec{d}$ and $\hat{a}$ correspond to those values of K_r^2? Furthermore, how much angular motion is necessary to make the value of K_r^2 sufficiently close to 2/3 that the error in R_c is acceptable. These questions are presently being investigated theoretically (Yguerabide and Stryer) and their answers should provide a quantitative basis for evaluating the uncertainties involved in determining distances by excitation transfer.

Table IV gives a list of pertinent studies related to electronic excitation energy transfer.

Nanosecond Fluorescence Polarization Spectroscopy

Fluorescence polarization techniques can be used to measure the rotational motion of macromolecules in solution. The interest in making such measurements is that rotational motion is related to the size, shape and flexibility of the macromolecules and can, therefore, be used to evaluate these important parameters.

Until recently, fluorescence polarization studies were restricted to steady state measurements. Such measurements, however, have limited scope because they require determination of polarized intensities at different temperatures T or viscosites η before structural parameters can be evaluated from the polarization data. Changes in conformation which may occur when T and η are changed complicate the interpretation of the results. On the other hand, time-dependent nanosecond fluorescence-polarization measurements give all the necessary information with a single solution at constant temperature. Furthermore, these measurements provide a more detailed picture of rotational motion. In essence, steady-state measurements give the time average of the nanosecond data.

In the application of nanosecond fluorescence polarization techniques, a sample of the fluorescent-labeled macromolecules is excited with fast pulses of light, and the resulting fluorescent pulses are detected through a polarizer oriented first in a vertical and then in a horizontal direction. These measurements thus yield two graphs of fluorescence intensity F vs. time. The graphs are designated $F_V(t)$ and $F_H(t)$; the subscripts refer to the direction of the detecting polarizer. $F_V(t)$ and $F_H(t)$ are complex functions of time and are dependent on both the lifetime τ of the chromophore and the rotational motion of the macromolecule. It is, however, possible to separate the dependence on τ from that on rotational motion by defining two new functions in terms

TABLE IV

INVESTIGATIONS: USING EXCITATION ENERGY TRANSFER

Investigation	Donor	Acceptor	Reference
Verified that the rate of excitation transfer depends on R^{-6}.	Naphthyl	Dansyl	Stryer and Haugland 1967
Verified that the rate of excitation transfer depends on spectral overlap J as predicted by Förster equations.	N-methylindole	Ketone group	Haugland et al. 1969
Distance between anticodon and amino acid acceptor for phenylalanine transfer RNA	Y base	Acridine dyes	Beardsley and Cantor 1970
Distance between cobalt and dansyl group of dansyl peptides bound to active site of cobalt, carboxypeptidase	Dansyl peptide	Cobalt atom	Latt et al. 1970
Distance between transferrin Tb^{+++} and Fe^{+++} binding sites	Tb^{+++}	Fe^{+++}	Luk 1971
Shape of rhodopsin molecule	Tryptophan and Tyrosine	N-retinyl	Ebrey 1971
Thickness of red-cell ghost	Cyanine dye	Gold layer	Peters 1971
Shape of rhodopsin molecule	Iodoacetamide and Acridine derivatives	11-Cis retinal	Wu and Stryer 1972

of $F_V(t)$ and $F_H(t)$. One of these functions is called the sum function $S(t)$ and is defined by the expression:

$$S(t) = F_V(t) + 2F_H(t) \qquad (9)$$

$S(t)$ depends on the decay characteristics of the chromophore but not on its rotational motion. In cases where the chromophore is situated in only one type of site and thus has a single lifetime τ, $S(t)$ obeys the expression:

$$S(t) = S_0 e^{-t/\tau} \qquad (10)$$

In more complex situations, $S(t)$ is given by a sum of exponential functions. Because of its relation to τ, $S(t)$ gives information on the polarity of the chromophore environment.

The second function is called anisotropy $A(t)$ and is defined by the expression:

$$A(t) = \frac{F_V(t) - F_H(t)}{S(t)} \qquad (11)$$

This function depends only on rotational motion and is the one normally used to obtain structural information from polarized fluorescence data. The usual analysis consists of comparing the experimental graph of $A(t)$ with graphs calculated theoretically for rigid particles of different shapes. This comparison yields information on the size, shape, and flexibility of the macromolecules. In the case where the molecules are rigid spheres and the chromophore is firmly attached, the decay of anisotropy is described by the expression:

$$A(t) = A_0 e^{-t/\phi} \tag{12}$$

where ϕ is the rotational correlation time of the sphere. ϕ is related to the size (volume) of the sphere by the equation:

$$\phi(sec) = \frac{\eta V}{kT} \tag{13}$$

$$= \frac{1}{6D} \tag{14}$$

where V (in cm^3) is the hydrated volume of the sphere, η (in poise) is the viscosity, and k (in ergs deg^{-1} $molecule^{-1}$) is the Boltzman constant (1.35×10^{-16}). D (in sec^{-1}) is the rotational diffusion coefficient. V is related to the molecular weight M, specific volume v (in cm^3/g), and hydration h (cm^3 of water per gram weight of macromolecular substance) by the equation:

$$V = \frac{M (v + h)}{N} \tag{15}$$

where N is Avogadro's number. Thus, if the experimental graph of log $A(t)$ vs. t is linear, as for Equation (12), it indicates that the macromolecule is spherical or approximately spherical, and the rotational correlation time ϕ can be evaluated from the slope of the graph. This value of ϕ can then be used in Equation (13) to calculate V.

If a graph of log $A(t)$ vs. t shows curvature, the macromolecule is either highly asymmetric or flexible. In this case, the experimental data are usually first analyzed in terms of rigid particles having the shapes of prolate ellipsoids of revolution. These ellipsoids are characterized by their volume V and axial ratio, which is the ratio of the lengths of the major to the minor axes. The decay of anisotropy for an ellipsoid is described by the expression (Tao 1969):

$$A(t) = A_o \sum_{i=1}^{3} f_i e^{-t/\phi_i} \tag{16}$$

The three rotational correlation times ϕ_i are related to the rotational diffusion coefficients about the major, $D_{||}$, and minor, $D_{\perp}$, axes by the expressions:

$$\phi_1 = (6D_{\perp})^{-1} \tag{17}$$

$$\phi_2 = (5D_{\perp} + D_{||})^{-1} \tag{18}$$

$$\phi_3 = (2D_{\perp} + 4D_{||})^{-1} \tag{19}$$

$D_{||}$ and $D_{\perp}$ are in turn related to γ and the rotational diffusion coefficient D of a rigid sphere having a volume equal to the volume V of the ellipsoid by the equations:

$$\frac{D_{\perp}}{D} = \frac{3}{2} \frac{\gamma[(2\gamma^2-1)\beta-\gamma]}{(\gamma^4-1)} \tag{20}$$

$$\frac{D_{||}}{D} = \frac{3}{2} \frac{\gamma(\gamma-\beta)}{\gamma^2-1} \tag{21}$$

$$\beta = \frac{1}{\sqrt{\gamma^2-1}} \ln (\gamma+\sqrt{\gamma^2-1}) \tag{22}$$

The coefficients f_i depend on θ (the angle between the emission transition moment and the major axis of the ellipsoid) as shown by the expressions:

$$f_1 = \left(\frac{3}{2} \cos^2\theta - \frac{1}{2}\right)^2 \tag{23}$$

$$f_2 = 3\cos^2\theta \sin^2 \theta \tag{24}$$

$$f_3 = \frac{3}{4} \sin^4 \theta \tag{25}$$

Equations (16) to (25) show that the decay of anisotropy A(t) for an ellipsoid of revolution is determined by three independent variables, namely, D (which is determined by V through Equations (13) and (14), γ, and θ. The interesting parameters are V and γ, which contain information on the size and shape of the ellipsoid.

If the fluorescent probe is randomly, but rigidly, attached to the ellipsoid so that the emission moments are randomly oriented with respect to the major axis, Equation (16) reduces to:

$$A(t) = \frac{A_o}{5} (e^{\frac{-t}{\phi_1}} + 2e^{\frac{-t}{\phi_2}} + 2e^{\frac{-t}{\phi_3}}) \tag{26}$$

The most general approach for analyzing the experimental anisotropy data in terms of a rigid ellipsoid is to fit Equation (16) to the experimental graph of A(t). In principle, this procedure would yield values of f_i and ϕ_i from which $D_{||}$, $D_{\perp}$, θ, and γ could be calculated with Equations (17) to (25). This approach, however, cannot be applied in practice because $A(t)$ vs. t can normally be measured experimentally over only one decade of $A(t)$. Unique values for f_i and ϕ_i cannot be obtained from such data. An alternative approach which is simple to apply has been introduced by Yguerabide, Epstein, and Stryer (1970). This approach is based on the fact that the parameters necessary to calculate V with Equation (15) are usually known in an anisotropy experiment. In this case there are only two independent parameters in Equation (16), namely γ and θ. The known information can be introduced into the analysis of A(t) by using Equation (13) to calculate ϕ, the rotational correlation time of a rigid sphere with a volume equal to that of the ellipsoid. It is then possible to plot the experimental data as $A(t)/A_0$ vs. t/ϕ on a semilogarithmic scale. The meaning and usefulness of such a plot can be seen from Equation (16). This equation can be rearranged so that $A(t)/A_0$ is explicitly dependent on t/ϕ instead of simply on t. The rearranged equation can be written as:

$$\frac{A(t)}{A_0} = \sum_{i=1}^{3} f_i e^{-(t/\phi)\Omega_i} \tag{27}$$

$$\text{where } \Omega_1 = D_\perp/D \tag{28}$$

$$\Omega_2 = (5D_\perp + D_{||})/6D \tag{29}$$

$$\Omega_3 = (2D_\perp + 4D_{||})/6D \tag{30}$$

Since the parameters Ω_i depend only on γ (see Equations (20) and (21), whereas f_i depend only on θ, a plot of $A(t)/A_0$ vs. t/ϕ depends on γ and θ, but not on V. This method of plotting the experimental data thus reduces the problem of curve fitting to evaluating two parameters. Figure 2 shows semilogarithmic plots of $A(t)/A_0$ vs. t/ϕ for $\gamma=2.5$

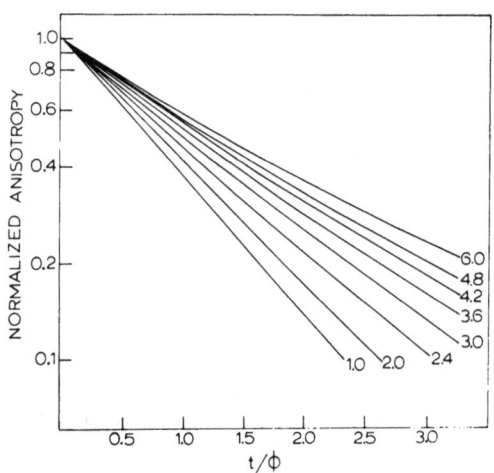

Figure 2. *Time dependence of the normalized anisotropy for rigid prolate ellipsoids ranging in axial ratio from 1.0 to 6.0. The chromophore is assumed to be randomly orientated relative to the ellipsoid axes. (Yguerabide, Epstein and Stryer 1970)*

and different values of θ. These plots are called normalized anisotropy plots. A series of such plots for different values of γ give a complete representation of Equation (16), i.e., these plots show all the possible shapes which the graph of $A(t)/A_0$ vs. t/ϕ can assume for an ellipsoid of revolution. It should be especially noted, that the shape of a plot of $A(t)$ vs. t is determined essentially by the values of γ and θ, whereas V merely compresses or expands the time scale.

From the discussion given above, it can be seen that the comparison of Equation (16) with experimental data can be made easily by superimposing and comparing visually the experimental normalized anisotropy plot with the theoretical plots. This comparison allows one to determine if the macromolecule is flexible or rigid. If the molecule is flexible in times of nanoseconds, the experimental plot will have a component which is decaying faster than that demonstrated by any ellipsoid having the same volume V as the macromolecule. On the other hand, if the molecule behaves as a rigid particle, γ can be determined from the theoretical graph which best fits the experimental data. The shape of the molecule can then be calculated from the value of γ. An application of these techniques is given in Section IV on protein applications, below.

The expressions presented above for A(t) apply only to systems where the macromolecules are randomly oriented in a homogenous environment, as is indeed the case when the molecules are dissolved in an ordinary solvent. These expressions were derived by Perrin (1934; 1936) and others (Tao 1969; Ehrenberg and Rigler 1972) from the hydrodynamic equations of motion for rigid ellipsoids. In a biological membrane, however, the environment of a probe is highly asymmetric, and the above expressions for A(t) do not apply. The solution of the hydrodynamic equations of motion in a bimolecular membrane is complex, and expressions for A(t) applicable to such oriented structures have not yet been derived. The problem may be somewhat simplified by a new theoretical approach recently introduced by Weber (1971) in which he uses phenomenological rate equations to describe the time dependence of polarized emission. The phenomenological equations are only approximately correct (Ehrenberg and Rigler 1972; Belford et al. 1972), but they are much easier to use and the deviations from the correct expressions are probably within experimental error. Nanosecond fluorescence measurements on oriented bimolecular membranes should give information on how much and how fast small molecules can move in different directions in the membrane and also how these motions are affected by specific lipids and proteins.

III. ADVANTAGES OF NANOSECOND FLUORESCENCE SPECTROSCOPY

Nanosecond fluorescence techniques have definite advantages over steady-state measurements in the study of the conformation and dynamics of macromolecules. Some of these advantages are listed below.

(1) In the application of fluorescent probes to determine polarity and to detect conformational changes, nanosecond measurements are more informative and direct. Thus, for example, a probe that is attached to a protein or membrane may reside at more than one kind of site, or may reside on a site which has different conformations in equilibrium with each other. In these frequently occurring cases, the fluorescent probe will have different lifetimes corresponding to the different dynamic polarities of the sites, and the decay of fluorescence intensity F(t) will be described by a sum of exponential terms,

$$F(t) = \Sigma \; a_i e^{-t/\tau_i} \tag{31}$$

where a_i is related to the number of probe molecules at the ith site and τ_i is the lifetime of the probe at that site (It should be noted that unbound chromophore does not usually contribute to $F(t)$ since its fluorescence efficiency in water is very low.) The existance of different binding sites can be readily detected by nanosecond meassurements since, in this case, a semilogarithmic plot of $F(t)$ vs. t will show curvature. Furthermore, by fitting Equation (31) to the nanosecond emission data, one can obtain values of a_i and τ_i which provide information on the relative number of probe molecules and polarity at each site. On the other hand, the presence of different binding sites cannot be detected from a measurement of quantum yield alone and may actually be difficult to detect by any steady-state measurement. Even in cases where there is only one binding site, nanosecond measurements are still advantageous, because they are more direct. Thus, in experiments where the chromophore is noncovalently bound — as is often the case, for example, in the study of membranes — it is necessary to establish

the fraction of molecules which are bound in order to evaluate the fluorescence efficiency ϕ_f. The lifetime of a chromophore, however, is independent of the amount that
is bound (as long as the fluorescence of the unbound chromophore is low, so that it
does not contribute to $F(t)$) and can be evaluated directly from the slope of $F(t)$ vs. t.

Finally, in experiments where conformational changes are to be detected with fluorescent probes that are noncovalently bound, a change in steady-state fluorescence
intensity when the system is perturbed can be due to changes in both the quantum yield
and the binding constant. Binding studies are necessary to separate the two effects.
Such a separation, however, becomes difficult when there are several binding sites.
The lifetime of the chromophore, on the other hand, depends only on the environment and
can thus be used directly to follow conformational changes. In cases where there is
more than one site, fitting Equation (31) to $F(t)$ before and after the perturbation
allows one to determine the effects both on conformation, through changes in τ_i, and on
relative binding, through changes in α_i.

(2) In the use of excitation transfer to determine distance, nanosecond fluorescence measurements have several advantages. First, they yield directly the value of
k_{DA}. Secondly, they readily allow one to determine, for example in the case of proteins, whether the donor and acceptor chromophores are positioned at unique sites. If
the sites are indeed unique, each chromophore will have a single lifetime. Thirdly,
nanosecond measurements give information on the relative orientation of donor and acceptor transition moments. This information is valuable in estimating a value of K^2.
Possible types of orientations and their nanosecond emission characteristics are listed
below. (It is assumed in this list that donor and acceptor are separated by a unique
distance.)

a) The donor and acceptor have a unique and fixed orientation with respect to
each other. In this case, the excited donor in the presence of acceptor will have a
single lifetime τ_D, and the decay of polarized fluorescence will be determined by the
rate of rotation of the macromolecule.

b) There is a distribution of orientations of the transition moments, but the
moments are rigidly fixed to the macromolecule. In this case, there will be a distribution of values of k_{DA} so that $\log F(t)$ vs. t will show curvature, but the decay of anisotropy will again be determined by the rotational motion of the macromolecule.

c) The donor and acceptor transition moments have a distribution of orientations
and are rapidly moving. In this case, there will be a single average value of k_{DA} and,
consequently, a single value of τ, but $A(t)$ will have a fast component corresponding to
to the rapid motion of $\hat{a}$.

(3) The advantages of nanosecond measurements in the use of fluorescence polarization to study the dynamics of macromolecules and membranes have already been described in the section on nanosecond fluorescence polarization spectroscopy, above.

The full potential of nanosecond fluorescence spectroscopy, as described above,
can be realized only if good techniques are available for measuring $F(t)$ vs. t and for
analyzing time-dependent emission data. In particular, the nanosecond techniques must
have very high sensitivity so as to be able to detect the small amounts of biological
material that are often encountered in fluorescence experiments. Furthermore, they must
be able to record $F(t)$ vs. t over a range of several decades of fluorescence intensity,
since the amount of information which can be extracted from this record depends strongly on the intensity range.

Various techniques have been used in the past to measure $F(t)$ vs. t (for a review,
see Yguerabide 1972). At present, the time-resolved single-photon counting technique
appears to be the one most suitable for biological studies. It has a very high sensitivity and can record $F(t)$ over at least three decades of light intensity. A variety
of numerical techniques have also been used to analyze time-dependent emission data.
To understand these techniques it must first be noted that $F(t)$ represent the emission
kinetics of a system which has been excited by an infinitely fast pulse of light, often
called a delta pulse of excitation. Indeed, theoretical expressions given in the literature for the decay of fluorescence intensity from particular systems are usually derived for this type of excitation. In practice, however, the sample is excited with
a flash of finite duration, and the emitted light is recorded with a detector having
finite response-time characteristics. The recorded nanosecond emission, therefore,
does not represent $F(t)$, but instead represents a function $R(t)$ which depends on $F(t)$,
on the duration of the light flash, and on the response time of the detector. It can
be shown from theory that $R(t)$ and $F(t)$ are related by the convolution integral:

$$R(t) = \int_0^t L(T)\ F(t-T)\ dT \qquad (32)$$

where $L(t)$ represents the light flash as recorded with the same detector used to record $R(t)$; i.e., $L(t)$ represents the combined response-time characteristics of the light flash and detector.

The analysis of nanosecond emission data usually begins with the evaluation of $I_F(t)$ from records of $R(t)$ and $L(t)$. This evaluation is usually called deconvolution. The method of moments (Yguerabide 1972; Isenberg and Dyson 1969) has proved to be a useful and precise technique for this evaluation whenever $F(t)$ can be represented by a sum of exponential functions such as is shown for example, in Equation (31). In this case, the method of moments yields values of a_i and τ_i. More general techniques of analysis — Fourier and Laplace transforms, for example — are presently being investigated. A complete description of the nanosecond lamps, the single-photon technique, and the method of moments has recently been given by the present author (Yguerabide 1972).

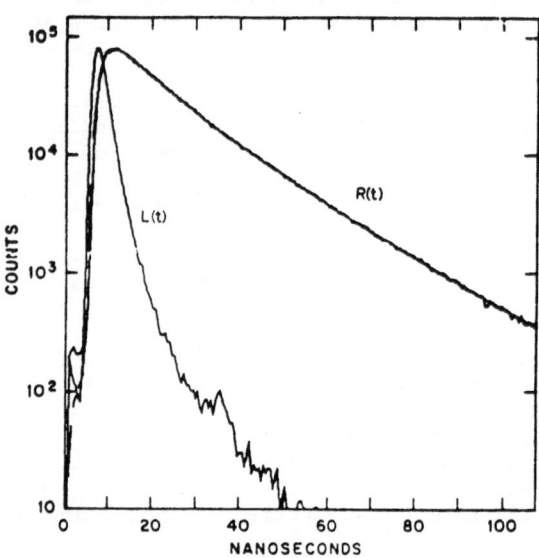

Figure 3. *Nanosecond emission $R(t)$ obtained by adding in the computer the emission data from a solution of $5 \times 10^{-5}M$ 1-anilino-8-naphthalene sulfonate in ethanol to the data from a solution of $10^{-6}M$ quinine sulfate in $0.1\,N\,H_2SO_4$. The dashed curve superimposed on $R(t)$ is a theoretical fit $C(t)$ obtained by convoluting a two-exponential function with the lamp pulse $L(t)$. $C(t)$ and $R(t)$ are almost indistinguishable from each other because of the good fit. (Yguerabide 1972)*

Figure 3 demonstrates the type of information which can be obtained with the single-photon counting technique and the method of moments. The graph $L(t)$ represents the exciting light pulse as measured with a detector operating in the single-photon mode. The graph $R(t)$ was obtained by adding the nanosecond emission data (i.e., $R(t)$ graph) of a solution of $5 \times 10^{-5}M$ ANS in ethanol to the nanosecond data for a solution of $10^{-6}M$ quinine sulfate in $0.1N\ H_2SO_4$. The graph thus corresponds to a system whose emission $F(t)$ is a two-exponential function:

$$F(t) = a_1 e^{-t/\tau_1} + a_2 e^{-t/\tau_2} \qquad (33)$$

where τ_1 and τ_2 are the lifetimes of ANS and quinine sulfate, respectively. Analysis of $R(t)$ by the method of moments yielded the values $a_1 = 0.30$, $a_2 = 0.17$, $\tau_1 = 9.4$, and $\tau_2 = 20.3$. To determine how well these values represent the experimental data, we used them to calculate with Equation (32) a graph of $R(t)$, designated as $C(t)$, which could then be compared with the experimental data. This calculation was done by integrating Equation (32), using $L(t)$ of Figure 3 and Equation (33) with the values a_1, and a_2, τ_1, and τ_2 given above. The graph of $C(t)$ is shown in Figure 3 as a dashed curve superimposed on $R(t)$. The agreement between $C(t)$ and $R(t)$ is sufficiently high that the two graphs are barely distinguishable from each other. A variety of tests indicates that the method of moments gives unique values of a_i and τ_i when the emission $F(t)$ does not involve more than two exponential functions. However, for systems where $F(t)$ contains three or more exponential functions, the values of a_i and τ_i given by the method of moments are not unique; i.e., there are sets of significantly different values of a_i and τ_i which equally well fit the experimental data. The problem here is that three decades of intensity data, containing noise, are not sufficient to define uniquely a_i and τ_i when more than two exponential functions are involved.

IV. APPLICATION

Proteins

The use of nanosecond fluorescence spectrocopy in the study of the structure and dynamics of proteins will be illustrated here by experiments done in our laboratory with antibodies of the IgG type. According to the Y model of Valentine and Green (1967), the four polypeptide chains of an IgG molecule are folded in such a manner that the overall conformation of the molecule consists of three units or segments loosely bound to each other through small flexible lengths of polypeptide chains (Figure 4). Two of the segments, called Fab segments, are identical, and each contains one binding site. The third unit is called the Fc segment. The model predicts that the IgG molecule is flexible; i.e., the three segments are able to move with respect to each other. It has been postulated that this segmental flexibility facilitates the binding of antigen. An additional property of the IgG molecule is that it can be split by pepsin into two fragments. One of these fragments, called the $F_{(ab')_2}$ fragment, contains the two active sites. IgG can also be split into three fragments by papain. Two of these fragments, called the F_{ab} fragments, are identical and each contains one active site. The ability of IgG to bind a specific antigen is retained almost in full by the $F_{(ab')_2}$ and F_{ab} fragments.

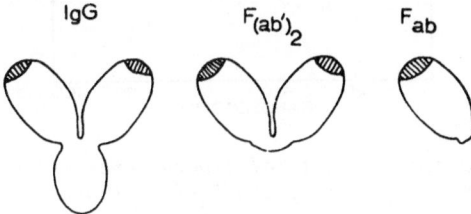

Figure 4. *Schematic diagram of* IgG *according to the* Y *model of Valentine and Green (1967) and of the* $F_{(ab')_2}$ *and* F_{ab} *fragments produced by enzyme digestion. The combining sites are shown by shaded areas. (Yguerabide et al. 1970)*

Our studies were initiated to determine whether the F_{ab} segments are capable of segmental motion in the nanosecond time range. In these studies, the fluorescent hapten, ε-dansyl-L-lysine, was used to label the active site of antidansyl IgG. The fluorescence efficiency of the hapten increases 25-fold when it binds to the active site of IgG. Measurements of the polarized fluorescence intensities $F_v(t)$ and $F_H(t)$ were made with fluorescent labeled IgG, as well as with labeled $F_{(ab')_2}$ and F_{ab} fragments. The functions S(t) and A(t) were then calculated as described in Section II, under Nanosecond Fluorescence Polarization Spectroscopy, (See Equations (9), (11)).

Figure 5 shows the graphs of S(t) for the three species of antibody. It should be recalled that S(t) contains information on the lifetime characteristics of the probe and thus reflects the dynamic polarity of the environment of the chromophore. There are three important features which are to be noted from Figure 5. First, the S(t) graphs for the three particles are almost identical, a fact which indicates that the active sites of these particles are almost the same; i.e., splitting $F_{(ab')_2}$ and F_{ab} from the IgG molecule does not significantly affect the conformation of the active site. This result is consistent with the idea of flexible joints between the three segments in the intact IgG molecule. Secondly, the semilogarithmic plot of S(t) shows curvature, which indicates that the active site has a distribution of conformations. This result is consistent with the well-known heterogeneity of antibodies. Thirdly, the average lifetime of the bound dansyl hapten is about 24 nsec, indicating that the active site is highly nonpolar.

Plots of A(t) for the fluorescent-labeled IgG, $F_{(ab')_2}$, and F_{ab} particles are shown in Figure 6. (The anisotropy data in this figure are raw data that have not been deconvoluted. In the present case, however, the average lifetime and rotational correlation times of the labeled antibody particles are much longer than the response-time characteristics of the nanosecond spectrofluorimeter and deconvolution is not necessary.) In contrast to S(t), the A(t) graphs are quite different for the three particles, as is to be expected from their different hydrodynamic properties. For the F_{ab} fragment, the semilogarithmic plot of A(t) vs. t is linear, yielding a single rotational

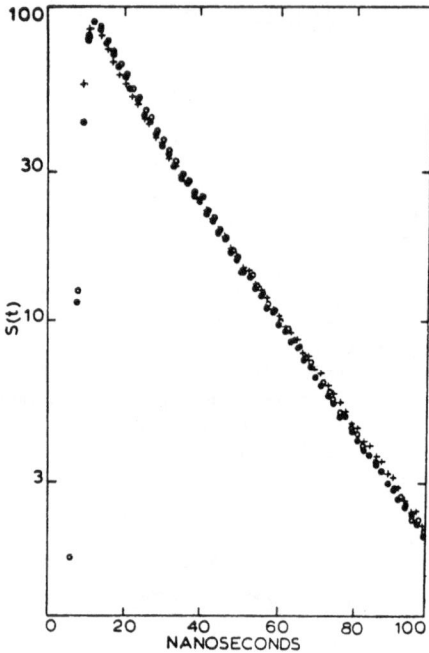

Figure 5. *Time dependence of the total fluorescence intensity* S(t) *emitted by dansyllysine bound to IgG(●),* $F_{(ab')_2}$(○), *and* F_{ab}(+). *(Yguerabide et al. 1970)*

correlation time of 33 nsec. The rotational correlation time expected for the F_{ab} fragment if it were a rigid sphere has a value of 20 nsec. This value is calculated from Equation (13) and Equation (15) with M = 50,000, v = 0.73 ml/g, η = 0.94 cP at 23°C, and h = 0.32 ml/g. The fact that the measured correlation time is larger than that expected for a rigid sphere indicates that the Fab fragment is indeed rigid and also elongated. The elongation of the fragment, expressed in terms of the axial ratio γ, was estimated to be 2.5. This estimation was made by plotting A(t) vs t/φ and comparing the results with theoretical normalized anisotropy plots for rigid ellipsoids with different elongations (See Figure 6).

The semilogarithmic plot of A(t) versus time for the labeled IgG molecule is not linear, a fact which shows that this particle has more than one rotational correlation time. This result indicates that IgG is either (a) rigid but highly asymmetric, or (b) not rigid. To determine which of these alternatives applies, the experimental data

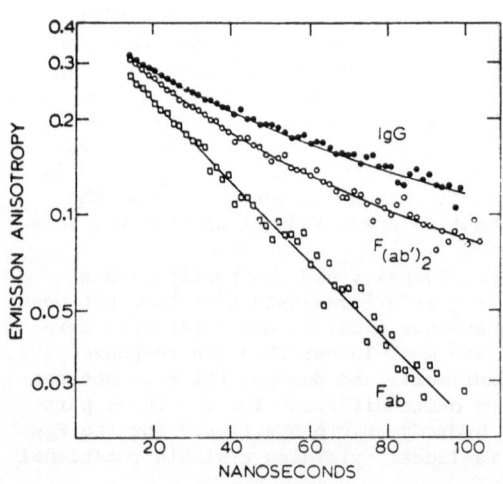

Figure 6. *Time dependence of the emission anisotropy* A(t) *of dansyllysine bound to* IgG(●), $F_{(ab')_2}$(○), *and* F_{ab}(□). *(Yguerabide et al. 1970)*

were compared with the theoretical normalized anisotropy plots of rigid ellipsoids. An example of this comparison is shown in Figure 7 for the case in which the fluorescent probe is assumed to have a random orientation at the active site. The value of ϕ used to normalize the experimental A(t) data was 53 nsec. This value was calculated from Equations (13) and (15), using M = 150,000 and h = 0.2 ml/g at 23°C. The experimental graph deviates markedly from the theoretical plots for the rigid ellipsoid. Similar deviations were observed when the experimental data were compared with graphs for ellipsoids having the chromophore at a unique orientation. In general, the comparisons show that the experimental plot of A(t) has an initial component that decays faster than that shown by the graph of any rigid ellipsoid. The results thus indicate that IgG is not rigid. Further, theoretical analysis in terms of a nonrigid model showed that the F_{ab} segments in the intact IgG molecule can move with respect to each other over an angle of 30° in about 30nsec.

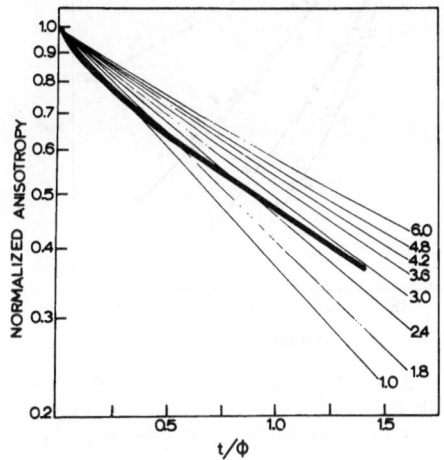

Figure 7. *Comparison of observed and calculated anisotropy plots. The heavy line is the normalized observed emission anisotropy data for IgG. This curve was obtained using ϕ = 53 nsec and A_o = 0.32. The light lines are emission anisotropy curves for rigid prolate ellipsoids of axial ratio ranging from 1.0 to 6.0. The observed emission anisotropy data are clearly incompatible with any of these rigid structures. (Yguerabide et al. 1970)*

Nucleic Acids

Nanosecond fluorescence polarization measurements have given valuable information on the flexibility of double-stranded DNA. Wahl, Paoletti and Le Pecq (1970) have found, and we have confirmed, that the nanosecond emission anisotropy A(t) of double-stranded calf thymus DNA (M=7×10^6) labeled with ethidium bromide has a fast initial component. This result indicates that small segments of the DNA chain can perform oscillatory motions with a rotational correlation of about 30 nsec. To elucidate the nature of these motions and to gain insight into the factors which govern the structure of DNA, we (Yguerabide, Elson and Sherwood) have initiated studies with dAT oligomers having the generic structure d(pTpA)$_i$ (Scheffler et al. 1968). Ethidium bromide, EB, which intercalates between adjacent base pairs of the double helix, is used as a fluorescent probe. The lifetime of the intercalated EB is around 24 nsec; this indicates a highly nonpolar site.

It is expected that in a solution of dAT oligomers, there will be an equilibrium among different types of oligomer helices (Scheffler et al. 1968). The two predominating forms will be (a) single hairpin helices formed by internal base pairing of single dAT oligomers, and (b) two-chain helices formed by base pairing between two oligomer molecules. The equilibrium between these two forms of the oligomer is depicted in Figure 8. Theory indicates that the single hairpin helices will be favored at high temperatures.

Figure 9 shows semilogarithmic plots of emission anisotropy vs. time for a solution of d(pTpA)$_{16}$ at different temperatures. At each temperature, the graphs are highly curved. Such curvature can arise from (a) segmental motion, (b) high asymmetry, or (c) equilibrium among species of different shapes. Consideration of theoretical normalized anisotropy plots, however, indicates that none of the helical species expected for the d(AT)$_{16}$ oligomer has sufficient asymmetry to account for the curvature in the anisotropy plots. In fact, the normalized plots show that the nanosecond emission anisotropy of rigid single hairpin helices H and of rigid two-chain helices D can be represented by single exponential functions. These conclusions are based on axial ratios of 2.8 and 5.5 for H and D, respectively, and on a value of 90° for the angle between the emission moment of ethidium bromide and the long axis of the helix. High asymmetry,therefore, cannot explain the anisotropy data.

On the other hand, the results of Figure 9 can be qualitatively understood in terms of an equilibrium between single-chain hairpin and two-chain helices. The two-chain

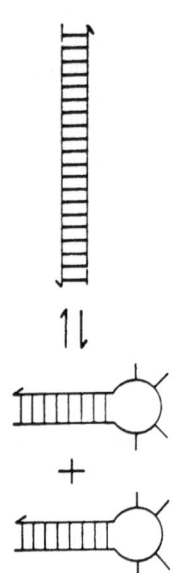

Figure 8. *Schematic representation of the equilibrium between single hairpin helices formed by internal base pairing of a dAT oligomer and two-chain helices formed by base pairing between two oligomer molecules*

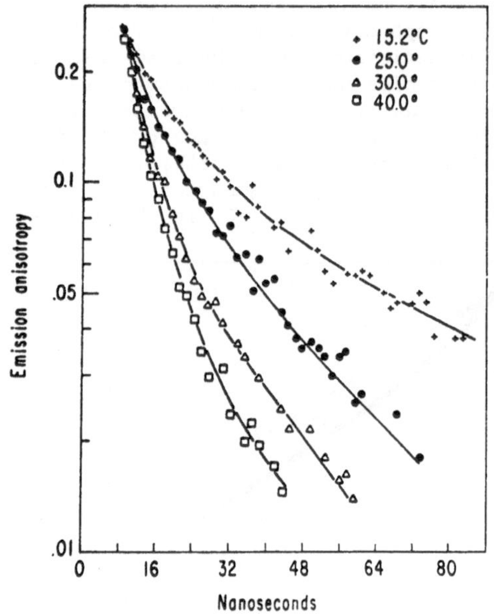

Figure 9. *Time dependence of the emission anisotropy A(t) at different temperatures for ethidium bromide bound to d(AT)$_{16}$ oligomer*

helices are expected to have a rotational correlation time about ten times greater than that of the hairpins because of their higher axial ratio and molecular weight. The slow component in the decay of emission anisotropy may, therefore, be associated with D. The rotational correlation time calculated for this component from the graph at 15.2° is 60 nsec, which agrees very well with the value expected for a rigid two-chain helix. Figure 9 also shows that the contribution of the slow component to the decay of anisotropy increases with decreasing temperature. This indicates that the single hairpin helices are favored at higher temperatures, as expected from thermodynamic theory.

In order to make a quantitative assessment of the equilibrium mechanism mentioned above, it is necessary to deconvolute the data of Figure 9 so as to eliminate the response-time characteristics of the nanosecond spectrofluorimeter. The instrument response will especially affect the fast component of anisotropy decay which is associated with H, since the rotational correlation time of this helix is around 5nsec. The deconvolution of anisotropy data, however, is more difficult than the deconvolution of simple fluorescence emission R(t). Small errors which enter the deconvolution of $R_V(t)$ and $R_H(t)$ result in large errors in the deconvoluted anisotropy plots. We are presently investigating various approaches which we hope will yield reliable deconvoluted anisotropy data. From these data we should be able to study the equilibrium of D and H as well as the segmental motion which may appear with increasing chain length.

Membranes

Interest in the use of fluorescence spectroscopy to study the structure, dynamics and function of biological membranes has increased considerably in recent years. Some of the uses are listed in Table I. This section presents the results of fluorescence spectroscopic experiments done in our laboratory to study mechanisms of ionic permeability in red-cell membranes.

A striking feature of the red-cell membrane is its high permeability for anions.

It is about a million times more permeable for anions than for cations. During changes in CO_2 concentrations, this selective permeability allows red cells to maintain constant plasma pH by rapidly exchanging Cl^- for HCO_3^- between the plasma fluid and the inside of the cell. In a mechanism proposed by Passow (1969), the anions are postulated to cross the membrane through narrow aqueous channels lined with positive charges. These charges keep positive ions from entering the channels and endow the membrane with its selective anion permeability. A prediction of the Passow mechanism is that the negative ions should carry current when an external electrical potential is applied across the membrane. Recent experiments indicate, however, that this prediction is not correct (Harris and Pressman 1967; Hoffman and Lassen 1971). Instead, the anions appear to cross the membrane as neutral species, an effect probably produced by the association of the anions with positively charged carriers. There is at present considerable effort in various laboratories to characterize the mechanism of anion permeability in greater detail, and especially to determine the nature of the postulated carriers.

In our fluorescence spectroscopic studies of the red-cell membrane (Fortes, Yguerabide et al. 1972; Fortes 1972), ANS, which has a negative charge, appeared to be a choice fluorescent probe for studying the mechanism of anion permeability (Fortes and Hoffman 1971) because (a) it binds to the red-cell membrane with increase in fluorescence intensity, and (b) it readily penetrates the membrane through pathways which may be the same as those used by the regular anions (e.g., Cl^-, HCO_3^- and $SO_4^=$). The possibility that the pathways are the same is suggested by the result that ANS at low concentrations decreases (inhibits) the permeability of the regular anions. The object of our experiments was to determine if the inhibition of anion permeability by ANS is really due to competition for the same pathways, and if so, to use ANS to study these paths.

Our experimental results consist of nanosecond emission data for ANS bound to red-cell ghost membranes at different ANS and salt concentrations. Ghost membranes were used because the hemoglobin of the intact red cell absorbs the exciting and emitted light and thus prevents the measurement of fluorescence intensity. Samples were prepared by adding ANS to a suspension of ghost membranes in buffered electrolyte. ANS binds reversibly to the membrane so that at each ANS concentration there is an equilibrium between bound and unbound chromophore.

Analysis of the nanosecond emission data by means of the method of moments showed that at each ANS concentration, the emission kinetics of the bound ANS can be represented by the equation:

$$F(t) = a_1 e^{-t/\tau_1} + a_2 e^{-t/\tau_2} \tag{34}$$

This result suggests that ANS binds to at least two different sites, X_1 and X_2, characterized by the ANS emission lifetimes $\tau_1 = 7$ nsec and $\tau_2 = 19$ nsec. (The lifetimes are only weakly dependent on concentration.) Some information on the nature of these sites was obtained from experiments with synthetic bilayer lecithin and lecithin-cholesterol membranes. In these systems, the bound ANS was found to have a single lifetime of about 6.7 nsec. The similarity of this value of the lifetime with that of τ_1 suggests that the site X_1 is formed by the bilayer lipid region of the ghost membrane. Furthermore, the fact that there is only one lifetime in the pure lipid bilayer suggests that the highly hydrophobic site X_2 is probably on a protein or in the region where lipid and protein interact. The possibility that X_1 or X_2 is in the path used by regular anions to cross the membrane is discussed below.

The most interesting conclusions concerning mechanisms which affect anion permeability come from measurements of a_1 and a_2 as a function of ANS concentration. It was indicated above that a_1 and a_2 are proportional to the amounts of chromophore that are bound to sites X_1 and X_2, respectively. By measuring the absolute amount of ANS that is unbound at each ANS concentration, we were able to convert a_1 and a_2 into the concentrations [ANSX$_1$] and [ANSX$_2$] of ANS bound to the two sites.

The results show that at low concentrations of ANS, [ANSX$_1$] and [ANSX$_2$] increase with increasing probe concentration, but that both became approximately constant at higher concentrations of ANS, an indication that the sites have become saturated. Close examination of our results, however, suggests that the saturation is only apparent; i.e., the sites are not actually saturated. Instead, it appears that the bound ANS anions create a negative electrostatic surface potential which at high probe concentrations becomes sufficiently negative to prevent additional binding of ANS ions.

The surface potential thus limits the amount of ANS that can be bound, and this gives rise to an apparent saturation effect. The surface-potential hypothesis is supported by the following observations. First, the ratio $\gamma = [ANSX_2]/[ANSX_1]$ does not change significantly with increasing ANS concentration. If true saturation of the binding sites was responsible for the effects found in our experiments, then the constancy in γ would imply that the two sites saturate at the same rate. This can occur only if the binding constants for the different sites are very similar. Such similarity, however, would be rather fortuitous since the sites appear to be quite different, as can be inferred from the values of τ_1 and τ_2 and the synthetic-membrane experiments. It seems more likely that the apparent saturation is caused by some factor which is common to both sites, such as the membrane surface potential. Secondly, increasing the ionic strength, by increasing the salt concentration of the suspension electrolyte, increased both $[ANSX_1]$ and $[ANSX_2]$. The simplest explanation of this result is that the added cations shield the negative surface charges, thus reducing the surface potential and increasing the amount of ANS that is bound.

To test further the hypothesis that the apparent saturation of ANS binding sites is caused by a surface potential, Ψ, we derived the theoretical expressions which relate Ψ to $[ANSX_1]$ and $[ANSX_2]$ and compared them with the experimental data. We assumed the equilibrium:

$$ANS + X_1 \rightleftarrows ANSX_1 \qquad (35)$$

$$ANS + X_2 \rightleftarrows ANSX_2 \qquad (36)$$

for which one can write

$$K_1^{\sim} = \frac{[ANSX_1]}{[ANS]\,[X_1]} \qquad (37)$$

$$K_2^{\sim} = \frac{[ANSX_2]}{[ANS]\,[X_2]} \qquad (38)$$

where $K_1^{\sim}$ and $K_2^{\sim}$ are equilibrium binding constants. Because of the negative charge on ANS, $K_1^{\sim}$ and $K_2^{\sim}$ not only depend on chemical interactions, but also on ψ (in mV) as shown by the expressions

$$K_1^{\sim} = k_1^{\sim}\, e^{\,F\psi/RT} \qquad (39)$$

$$K_2^{\sim} = k_2^{\sim}\, e^{\,F\psi/RT} \qquad (40)$$

where $k_1^{\sim}$ and $k_2^{\sim}$ are the equilibrium constants related to the chemical potential, and $RT/F = 25.3$ mV at 22°C. The potential ψ is related (a) to the surface charge density σ_0 (ions/Å^2) normally present on the membrane, and (b) to the surface charge density σ_{ANS} (ions/Å^2) of the bound ANS ions as shown by the equation:

$$sinh\,\frac{F\psi}{2RT} = (\sigma_0 + \sigma_{ANS}) \left(\frac{\pi}{2DRTC}\right)^{1/2} \qquad (41)$$

where C is the concentration of monovalent electrolyte in moles/ liter and D is the dielectric constant of water. Note that since the bound ANS ions have a negative charge, the charge density σ_{ANS} has a negative value.

The equations presented above can be simplified by introducing the following assumptions. First, it is assumed that the binding sites are far from saturation, so that $[X_1]$ and $[X_2]$ are virtually independent of ANS concentration. Thus, assuming that $[X_1]$ and $[X_2]$ are constant and combining them with $K_1^{\sim}$ and $K_2^{\sim}$ reduces Equations (35) to (40) to:

$$K_1 = \frac{[ANSX_1]}{[ANS]} \qquad (42)$$

$$K_2 = \frac{[ANSX_2]}{[ANS]} \qquad (43)$$

$$K_1 = k_1 e^{F\psi/RT} \tag{44}$$

$$K_2 = k_2 e^{F\psi/RT} \tag{45}$$

We further assume that the potential ψ is not very large so that we can expand Equation (41) in a power series and retain only first-order terms. The potential is then given by the expression:

$$\frac{F\psi}{RT} = (\sigma_0 + \sigma_{ANS}) \left(\frac{\pi}{2DRTC}\right)^{1/2} \tag{46}$$

This equation is applicable for ψ <50mV. Under the conditions for which Equations (46) applies, the potential ψ can be separated into the potentials ψ_0 and ψ_{ANS} produced by σ_0 and σ_{ANS}, respectively. This separation is shown explicitly by the expression (which follows from Equation (46)):

$$\psi_{ANS} = (\psi_0 + \psi_{ANS}) \tag{47}$$

where

$$\psi_{ANS} = \frac{\sigma_{ANS}RT}{F} \left(\frac{\pi}{2DRTC}\right)^{1/2} \tag{48}$$

and

$$\sigma_{ANS} = - \frac{([ANSX_1] + [ANSX_2])}{A} \tag{49}$$

A is the total area (in Å^2) of the ghost membranes in one liter of the ghost cell suspension. Note that since σ_{ANS} is negative, ψ_{ANS} is also negative.

It can now be shown that Equations (42) to (49) account qualitatively for the main features of our experimental data. Thus, according to Equations (48) and (49), ψ_{ANS} becomes more negative with increasing ANS concentration. This causes K_1 and K_2 (Equations (42) and (43)) to decrease and eventually approach zero, producing an apparent saturation effect. Furthermore, the ratio $\gamma = [ANSX_2]/[ANSX_1]$ is independent of ANS concentration ($\gamma = k_1/k_2$ from Equations (42) and (43)), so that both sites saturate at the same rate. Finally, according to Equation (48), ψ becomes less negative when the salt concentration C is increased. This increases the amount of ANS bound by increasing K_1 and K_2, as found experimentally.

To determine whether the surface-potential hypothesis also accounts quantitatively for the experimental results, we used a special curve-fitting procedure to compare Equations (42) to (49) with the experimental data. We found that these equations indeed give an excellent fit to the experimental graphs of $[ANSX_1]$ and $[ANSX_2]$ vs. total ANS concentration. Values obtained from the curve fits, for the membrane surface potentials were -6.7, -13 and -24 mV for total ANS concentrations of 6.6, 16.6, 33.3 μM, respectively. We also found that the experimental values of ψ had a $1/\sqrt{c}$ dependence on salt concentration as predicted by Equation (48). Thus there is excellent quantitative agreement between the experimental results and prediction of the surface-potential model.

The results presented above indicate that the surface-potential created by the ANS ions bound to the membrane strongly influence additional binding of the chromophore. Since the regular permeant anions also sense this potential, our results further suggest that ANS decreases anion permeability mainly by creating a negative potential on the membrane surface. Indeed, theoretical calculations show that the values of the surface potentials calculated in our experiments can account quantitatively for the effects of ANS on, say, the permeability of Cl^-. The ability of ANS and similar anions to produce significantly large surface potentials at low concentrations is to a large extent dependent on their amphipathic properties. The nonpolar part of these anions allows them to bind very efficiently to nonpolar regions on proteins and the lipid bilayer of the membrane.

It is, of course, possible that a small fraction of the decrease in anion permeability may be due to direct competition between ANS and regular anions for carriers (or channels). If this is so, then part of the ANS fluorescence must come from the ANS ions bound to these structures. It is indeed tempting to suggest that the highly hydrophobic site X_2 is related to the carriers, while the site X_1 is related to the bilayer regions of the membrane. Were this the case, a_2 and τ_2 could be used to study the conformation and dynamics of the carriers. This possibility could be evaluated by using specific inhibitors of anion permeability. These inhibitors would preferentially affect that component of the emission associated with the carriers. Experiments, however, have shown that the commonly available inhibitors, are not sufficiently specific and have almost the same effects on a_1 and a_2. Most of these inhibitors are negatively charged and probably exert their effects by influencing the surface electrostatic potential. More specific inhibitors must therefore be found in order to decide whether ANS emission can be used directly to study the postulated anion carriers.

ACKNOWLEDGEMENTS

This work was supported by grants from the National Institute of Health (GM — 16708) and the National Science Foundation (GB — 27408).

REFERENCES

AZZI, A., Chance, B., Radda, G.K., and Lee, C.P., (1969), *Proc. Natl. Acad. Sci., U.S.A,* 62, 612.

BEARDSLEY, K, and Cantor, C.R., (1970), *Proc. Natl. Acad. Sci., U.S.A.,* 65, 39.

BELFORD, G.G., Belford, R.L., and Weber, G., (1972), *Proc. Natl. Acad. Sci., U.S.A.,* 69, 1392.

BÜCHER,H., Drexhage, K.H., Fleck, M., Kuhn, H., Möbius, D.,Schäfer, F.P., Sondermann,J., Sperling, W., Tillman, P., and Wiegand, J., (1967),*Mol. Cryst.,* 2, 199.

CARIO, G., and Franck, J., (1922), *Z. Physik.,* 11, 161.

CARNAY, L. D., and Barry, W.H., (1969), *Science,* 165, 608.

CHEN, R. F., and Kernohan,J.C., (1967), *J. Biol. Chem.,*242, 5813.

EBREY, T.G., (1971),*Proc. Natl. Sci., U.S.A.,* 68, 713.

EDELMAN, G.M., and McClure, W.O., (1968), *Accounts Chem. Res.,* 1, 65.

EHRENBERG, M., and Rigler, R., (1972), *J. Chem Phys.,* 14, 539.

FÖRSTER, T.,(1947), *Ann. Phys.,* 2,55.

FORTES, P.A.G., (1972), Ph. D. thesis, Yale University.

FORTES, P.A.G., and Hoffman, J.F.,(1971), *J. Membrane Biol.,*5, 154.

FORTES, P.A.G., Yguerabide, J. and Hoffman, J.F., (1972), *Biophysical Soc. Abstracts,* Sixteenth Annual Meeting, 255a.

GITLER, C., Rubalcava, B., and Caswell, A., (1969), *Biochim. Biophys. Acta,* 193, 479.

HARRIS, E.V., and Pressman, B.C., (1967), *Nature,* 216, 918.

HAUGLAND, R.P., Yguerabide, J., and Stryer, L.,(1969), *Proc. Natl. Acad. Sci., U.S.A.,* 63, 23.

HOFFMAN, J.F., and Lassen, U.V.,(1971), *Abstracts 25th International Physiol. Congress,* 253.

ISENBERG, I., and Dyson, R.D., (1969), *Biophys. J.,*9, 1337.

LATT, S.A., Auld, D.S., and Vallee, B.L., (1970), *Proc. Natl. Acad. Sci., U.S.A.,* 67, 1383.

LATT, S.A., Cheung, H.T., and Blout, E.R., (1965), *J. Am. Chem. Soc.,* 87, 995.

LE PECQ, J.B., and Paoletti, C., (1967), *J. Mol. Biol.,* 27, 87.

LUK, C.K., (1971), *Biochemistry,* 10, 2838.

MC CLURE, W.O., and Edelman, G.M., (1967), *Biochemistry,* 6, 567.

PASSOW, H., (1969), *Prog. Biophys. Mol. Biol.,* 19, 425.

PARKER, C.W., Yoo, T, Johnson, M.C., and Godt, S.M., (1967), *Biochemistry*, 6, 3408.

PATRICK, J., Valeur, B., Monnerie, L., and Changeau, J., (1971), *J. Mem. Bio.*, 5,102.

PERRIN, J., (1934), *J. Phys.*, 5, 497.

PERRIN, F., (1936), *J. Phys.*1, 1.

PETERS, R.,(1971),*Biochem. Biophys. Acta,* 233, 465

ROBINSON, G.W., (1961), *J. Mol. Spectrosc.*, 6, 58.

ROBINSON, G.W., and Frosch, R.P., (1963), *J. Chem. Phys.*,38, 1187.

SCHEFFLER, I.E., Elson, E.L., and Baldwin, R.L., (1968), *J. Mol. Biol.*, 36, 291.

SELISKAR, C.J., and Brand, L. (1971), *Science*, 171, 799.

STRYER, L., (1965), *J. Mol. Biol.*, 13, 482.

STRYER, L., (1966), *J. Am. Chem. Soc.*, 88, 5708.

STRYER, L., (1968), *Science* , 162, 526.

Stryer, L., and Haugland, R.P. (1967), *Proc. Natl. Acad. Sci., U.S.A.*, 58, 719.

TAO, T., (1969), *Biopolymers*, 8, 609.

TASAKI, I., Watanabe, A., and Hallett, M., (1971), *Proc. Natl. Acad. Sci., U.S.A.*, 68, 938.

TASAKI, I., Watanabe, A., Sandlin, A., and Carnay, L., (1968), *Proc. Natl. Acad. Sci., U.S.A.*, 61,883.

VALENTINE, R.C., and Green, N.M., (1967), *J. Mol. Biol.*, 27, 615.

VAUGHEN, W.M., and Weber, G., (1970),*Biochemistry*, 9, 464.

WAGGONER, A.S., and Stryer, L., (1970), *Proc. Natl. Acad. Sci., U.S.A.*, 67, 579.

WAHL, P., Paoletti, J., and Le Pecq, J.B.,(1970), *Proc. Natl. Acad. Sci., U.S.A.*,65, 417.

WEBER, G., (1971), *J. Chem. Phys.*,55,2399.

WEBER, G., and Laurence, D.J.R., (1954), *Biochem. J.*, 56, xxxi.

WU,C.W., and Stryer, L., (1972), *Proc. Natl. Acad. Sci., U.S.A.*, 69, 1104.

YGUERABIDE, J., (1972), *Methods Enzymol.*, Vol. 26, 798.

YGUERABIDE, J., Elson, E., and Sherwood, D., unpublished results.

YGUERABIDE, J., Epstein, H.F., and Stryer, L., (1970), *J. Mol. Biol.*, 51, 573.

YGUERABIDE, J., and Stryer, L., (1971), *Proc. Natl. Acad. Sci., U.S.A.*, 68, 1217.

YGUERABIDE, J., and Stryer, L., unpublished results.

MEASUREMENTS OF SINGLE MOLECULES OF ANTIBODY BY THEIR ABILITY TO ACTIVATE A DEFECTIVE ENZYME

Boris Rotman

Division of Biological and Medical Sciences, Brown University, Providence, Rhode Island

INTRODUCTION

Certain defective enzymes produced by genetically altered bacteria show a substantial and specific activation in the presence of antibodies directed against the corresponding normal enzyme, Pollock et al. 1967; Rotman and Celada 1968; Arnon and Cinader 1971). This type of response has inherent capabilities for measuring antibodies with ease and sensitivity since it transduces the antigen-antibody binding into a catalytic reaction. The increases in enzymatic activity mediated by antibodies are generally modest except for those of a penicillinase (Pollock 1967) and a β-D-galactosidase (Rotman and Cinader 1968; Messer and Melchers 1970) which are 70- and 550-fold, respectively. The β-D-galactosidase reaction was chosen for our studies not only because of its high activation factor, but also for the fluorogenic assay of this enzyme, which measures quantitatively the activity of a single β-D-galactosidase molecule, (Rotman 1961;1970). The expectation was that the sensitivity of the fluorogenic assay would serve to measure individual molecules of activating antibody. The results of the experiments presented here fulfilled this expectation and, furthermore, demonstrated that one molecule of monovalent antibody (Fab) is sufficient for activating a molecule of a defective β-D-galactosidase (AMEF). The latter conclusion could not be reached *a priori*, because AMEF is a tetramer with four identical subunits (Rotman and Celada 1968) and, therefore, it could have required binding to each of the subunits for activation. Although conventional experiments using dilution kinetics may indicate whether one, two, or more molecules of antibody are required for activation (Celada et al. 1970), they do not exclude the possibility that the activity of a molecule of AMEF is proportional to the number of antibody molecules bound to the AMEF. In the single molecule experiment this dilemma is not present, since it is possible to establish by direct measurements the activity of AMEF molecules at different AMEF-antibody ratios.

The assay of individual antibody molecules has additional utility for comparing immunoglobulin preparations which may differ in ability to activate AMEF. The molecular assay can answer the question: does a given antiserum have less activating capacity than others because it has fewer molecules of antibody with normal activating ability or because it has antibody molecules incapable of eliciting normal activation? Our results favor the latter alternative since they show that molecules of activating antibody may differ from each other with respect to activating ability. It should be noted that the dissociation constant of the activating antibody does not influence the results of these experiments because the single molecule assay is done in the presence of an excess of AMEF; therefore, any antibody dissociation would be followed by rapid reassociation to another AMEF molecule.

MATERIALS AND METHODS

Monovalent fragments (Fab) of antibodies were prepared by papain digestion of the specific immunoglobulin (Porter 1959) and subsequent purification in a Sephadex G-100 column. Preparations of Fab have been shown previously to activate AMEF to the same extent as intact antibodies (Celada et al. 1970).

The procedures for preparing purified β-galactosidase and AMEF, and for obtaining and assaying antisera directed against β-D-galactosidase were similar to those described earlier (Rotman and Celada 1968; Celada et al. 1971).

For the assay at the single molecule level, a given dilution of a Fab preparation was mixed with an excess of purified AMEF, and the mixture was incubated 30 minutes at 37°C and then at 4°C for several hours until no further increase in enzymatic activity was detectable. Dilutions of the reaction mixture were mixed with the fluorogenic substrate, fluorescein-di-(β-D-galactopyranoside) (purchased from Schwarz/Mann Inc.), and dispersed in silicone oil, as previously reported for the assay of single molecules of β-D-galactosidase (Rotman 1961; 1970). After 16 hours of incubation at 35°C, the fluorescence of selected individual microdroplets of 20.3 μm diameter was measured using a standard microscope equipped with a dark field condenser, appropiate light filters and a photomultiplier attachment (Rotman 1961; 1970). The performance of the microscope fluorometer was monitored by several methods. The light output of a constant radio-phosphorescent source was used to calibrate the photomultiplier and the amplifier. A fluorescent glass was used to monitor the intensity of the microscope lamp (an ordinary six-volt tungsten lamp supplied with the microscope) which was connected to a constant voltage power supply. The ultimate performance of the microscope fluorometer was ascertained by measurements of droplets containing a given concentration of fluorescein and by the results of controls with droplets which had single molecules of β-D-galactosidase.

The data obtained in using the microscope fluorometer were converted into a frequency distribution histogram (Figure 1) by a computer, and from this histogram, the frequency of microdroplets *without* enzymatic activity was measured. This frequency

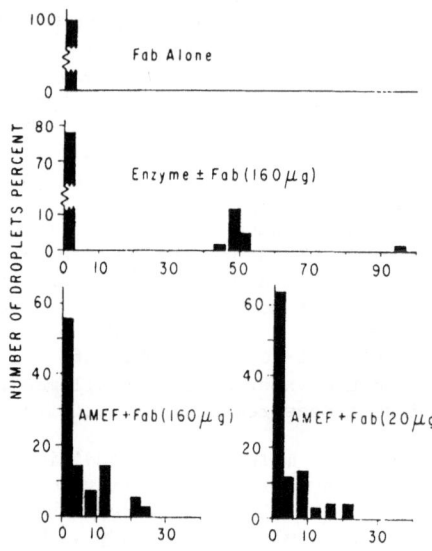

Figure 1. *Distribution of fluorescence among microdroplets containing the fluorogenic substrate, fluorescein-di-(β-D-galactopyranoside), and the indicated additions.*
Upper histogram: *controls with either Fab alone at 160 μg per ml, AMEF alone at 100 μg per ml, or buffer.* Middle histogram: *normal β-D-galactosidase at 19 units per liter (equal to 0.03 μg per liter) mixed with 160 μg per ml of Fab.* Lower histograms: *genetically defective β-D-galactosidase (AMEF) mixed with either 20 (right figure) or 160 μg per ml of Fab (left figure).*
A minimum of 80 microdroplets were quantitatively measured for each histogram. Arbitrary units of fluorescence are used in the abscissa

serves to calculate the *average number* of enzymatically active centers distributed per droplet according to Poisson's equation (Rotman 1961). The average number, divided by the volume of a microdroplet (4.4 × 10^{-9} ml) and corrected for the dilutions made before spraying the droplets, yields the number of enzymatically active AMEF-Fab complexes per ml of antiserum. No corrections were necessary, since neither Fab alone nor AMEF alone gave any fluorescent droplets (Figure 1). From the same data, one can obtain the *average activity* of an AMEF-Fab complex by dividing the number of units per ml of the assayed mixture by the number of molecules of complex per ml.

RESULTS AND CONCLUSIONS

The sensitivity of the method for measuring single molecules of enzyme stems from carrying the reaction between the enzyme and its fluorogenic substrate in a droplet of microscopic size. This is accomplished by dispersing the reaction mixture into a micro-chamber filled with silicone oil by means of a sprayer (Collins et al. 1964). Using adequate dilutions of enzyme, it is possible to obtain a random all-or-none distribution of enzyme molecules among droplets of a given diameter. If very dilute solutions of enzyme are dispersed so that statistically the chance of two molecules being in one droplet is minimized, direct measurements of the hydrolysis rate of an individual enzyme molecule can be made (Rotman 1961; 1970). In order to use this method for the antibody-mediated activation of a defective enzyme, the problem of polymolecular complexes and precipi-tates had to be circumvented. This was done by using monovalent fragments of antibodies (Fab) which do not precipitate but activate AMEF (Celada et al. 1970).

As shown in Figure 1, appropriate dilutions of a mixture of AMEF and activating Fab assayed at the single-molecule level yielded a proportion of microdroplets with enzymatic activity, while neither of the components of the mixture assayed alone yield-ed such droplets. Additional controls (not shown here) in which AMEF was mixed with either normal rabbit serum or unspecific antiserum Fab gave no fluorescent microdrop-lets. The activity of the normal enzyme is not significantly affected by the presence of anti-β-D-galactosidase Fab, however, it is presented as a reference in Figure 1 (middle histogram). The activity of a normal molecule of enzyme produces about 50 units of fluorescence, as shown by the cluster of discrete number of active droplets. The droplets with 96 units on the right hand of the histogram most likely represent a class of droplets with two molecules, since both the fluorescence and the frequency correspond to the expected values.

From the data of the experiment shown in Figure 1, one can calculate that the ave-rage AMEF-Fab complex with enzymatic activity produced 7 fluorescence units equivalent to 1/7 of the activity of a normal enzyme molecule. This value was characteristic of the given antiserum used and was shown to be independent of the ratio AMEF:Fab over a wide range covering excess of both antigen and antibody. Therefore, we have taken this value as a measurement of the ability to induce conformational changes in AMEF which cause activation. Anti-β-D-galactosidase antibodies from different sources were found to vary considerably in this conformational ability. For instance, the best mouse antisera had one tenth of the conformational ability of the best rabbit antisera.

Among the best rabbit antisera, there was one showing an unusual degree of homo-geneity. It was estimated that 60% of the Fab molecules were similar in conformational ability with an average value of 12 fluorescence units for the AMEF-Fab complex. This activity corresponds to one fourth of the normal enzyme activity. The question of whether this is or is not the maximal attainable value of activated AMEF can not be answered conclusively since, at present, one can not rule out that there are antibodies in the antiserum which can block the reaction of activating antibodies. If purified activating antibodies were available, this question could be easily answered.

The results described above show that activating Fab combines with AMEF forming a complex with discrete enzymatic activity; however, the results do not provide the answer to the question of how many molecules of Fab are necessary for activation of an AMEF molecule. For this, we measured the number of AMEF-Fab complexes formed in the pre-sence of excess AMEF as a function of the Fab concentration. Under this condition, if one Fab molecule can activate a molecule of AMEF, all AMEF-Fab complexes should be act-ive and the number of them should be equal to the number of activating Fab molecules present in the reaction mixture. In excess of AMEF, the rate of dissociation of the complex can be neglected, since a ratio AMEF:Fab larger than 700 was used and, thus, there was a good chance for reassociation. Accordingly, if one Fab molecule can act-ivate, the single-molecule assay should show a frequency of droplets with enzymatic activity equal to $1-e^{-m}$. The term m represents the average number of Fab molecules distributed among the droplets; and e^{-m}, the frequency of droplets without enzymatic activity (i.e., without Fab) given by Poisson's equation. The alternative assumption that at least two molecules of antibody are necessary for activation predicts that the frequency of droplets with active AMEF-Fab complexes will be equal to the probability of two or more Fab molecules being present in a droplet. This is equal to $1-e^{-m}-me^{-m}$, since the frequency of droplets with one Fab molecule is me^{-m}. Similarly, equations for the frequency of droplets with activity can be calculated, assuming that three, four or more molecules of Fab are required for activation. Theoretical curves constructed according to these equations are presented in Figure 2 together with experimental data. It is clear that the results fit only the theoretical curve assuming one molecule of Fab. This conclusion confirms and extends the results obtained by conventional dilution kinetics (Celada et al. 1970). It extends because the conventional method leaves open the alternative that activation is directly proportional to the number of Fab molecules bound; i.e., a complex made of one molecule of AMEF and n molecules of Fab would have n times the enzymatic activity of a complex with one molecule. This alternative is ruled out in experiments with single molecules since direct observations of the enzymatic activity of individual AMEF-Fab complexes show that the activity remains constant throughout the range of Fab concentration.

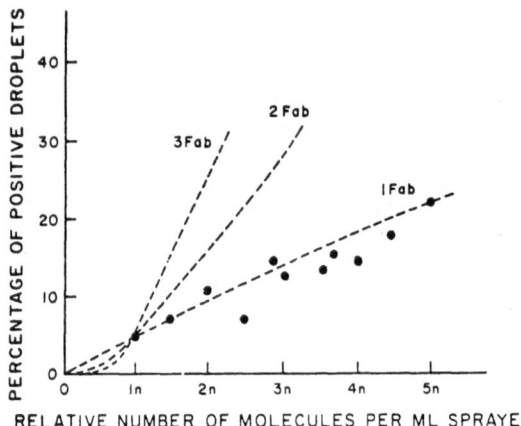

Figure 2. *Frequency of microdroplets with enzymatic activity as a function of the relative concentration of activating Fab antibodies in the sprayed solution. The concentration of AMEF was constant in all the assays. Dotted lines are theoretical curves calculated according to the assumption that either one, two or three molecules of Fab are necessary for the activation of a molecule of AMEF. The circles show experimental values, each one of which represents more than 300 droplets measured for the lower dilutions, and 500 for the higher dilutions. See text for details*

In conclusion, it was established that the fluorogenic assay for single molecules of β-D-galactosidase can be used to study the antibody-mediated activation of AMEF, a defective β-D-galactosidase. The assay was used to show that one molecule of Fab antibody is enough to activate one molecule of AMEF, and that the maximal activity obtained is equal to one fourth of that of the normal enzyme. The method is also useful for comparative measurements of antibodies in terms of ability to activate AMEF.

ACKNOWLEDGEMENTS

The invaluable collaboration of Miss J. Grosvenor and Miss R. Guzman is gratefully acknowledged. Supported by grants GM - 14198 from the National Institutes of Health and GB - 12615 from the National Science Foundation.

REFERENCES

ARNON,R., and Cinader, B., (1971), In: *Progress in Immunology*, (Amos, B., ed.) Academic Press, New York, 1199.

CELADA, F., Ellis, J., Bodlund, K., and Rotman, B., (1971), *J. Exp. Med.*, **134**, 751.

CELADA, F., Strom, R., and Bodlund, K., (1970) In: *The Lactose Operon*, (Beckwith, J.R. and Zipser, D., eds), Cold Spring Harbor Laboratories Press, 291.

COLLINS, J.F., Mason, D.B., and Perkins, W.F., (1964), *J. Gen. Microbiol.*, **34**, 353.

MESSER, W., and Melchers, F., (1970), In: *The Lactose Operon*, (Beckwith, J.R. and Zipser, D., eds.) Cold Spring Harbor Laboratories Press, 305.

POLLOCK, M.R., Fleming, J., and Petrie, S., (1967), In: *Antibodies to Biologically Active Molecules*, (Cinader, B., ed.), Pergamon Press Ltd., Oxford, 139.

PORTER, R.R., (1959), *Biochem. J.*, 73, 119.

ROTMAN, B., (1961), *Proc. Natl. Acad. Sci. U.S.A.*, 47, 1981.

ROTMAN, M.B., (1970), In: *The Lactose Operon*, (Beckwith, J.R., and Zipser, D., eds.) Cold Spring Harbor Laboratories Press, 279.

ROTMAN, M.B., and Celada, F., (1968), *Proc. Natl. Acad. Sci. U.S.A.*, 60, 660.

NANOSECOND TIME-RESOLVED FLUORESCENCE SPECTROSCOPY IN MOLECULAR BIOLOGY

M. R. Loken, J. R. Gohlke and L. Brand

The Johns Hopkins University, Baltimore, Maryland

INTRODUCTION

Fluorescent dyes are capable of providing specific information concerning the physical and chemical character of macromolecules in biological systems. The unique feature of fluorescence spectroscopy is the creation of a new, short-lived chemical entity, the excited state. The lifetime of this species is several nanoseconds. This is sufficient time for the excited probe to interact with its immediate environment.

Some of the interactions which may occur during the lifetime of the excited chromophore are depicted in Figure 1. Following the absorption of light, the excited state may be depopulated either by emission of light or by means of a variety of nonradiative quenching processes. The excited molecule may also reorient due to brownian motion, transfer its energy to another chromophore, cause its solvent shell to reorient, or undergo a chemical reaction such as proton transfer. These events must occur in the nanosecond time domain if they are to compete with radiative emission.

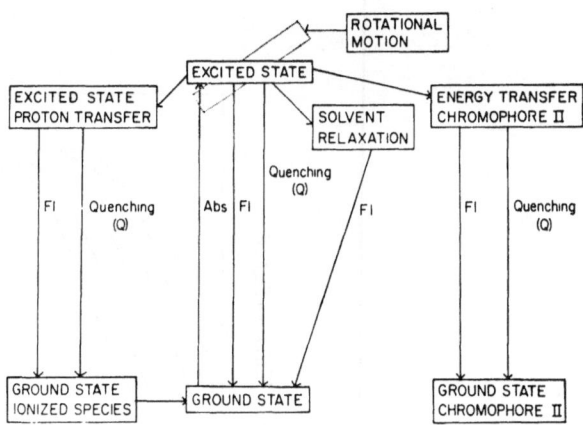

Figure 1. *Interactions which take place during the excited state of a molecule can provide information about molecular size and shape (rotational motion), intermolecular distance (energy transfer), and molecular microenvironments (solvent relaxation and excited state proton transfer)*

The well-established techniques of steady-state fluorescence provide some information about each of these processes. Since steady-state measurements represent a time-averaged signal obtained under photostationary conditions, *direct* information regarding the time dependency of the events cited above is lost.

In contrast, nanosecond time-resolved emission spectroscopy provides a means for the detection of excited-state reactions by direct examination of the fluorescence decay (Brand and Gohlke 1971, Deluca et al. 1971).

Several instrumental techniques are available for the measurement of fluorescence decay times. These include direct intensity decay measurements, phase fluorometry, and the single-photon counting method for obtaining the probability of radiative transitions (Ware 1971). A single-photon counting instrument has been described in detail by Schuyler and Isenberg (1971), and similar equipment has been used in our laboratory. Decay curves are obtained at all wavelengths of interest and are either analyzed directly for exponential character by the "method of moments" (Isenberg and Dyson 1969), or normalized to constant intensity and converted to nanosecond time-resolved emission spectra with the aid of a digital computer.

SOLVENT RELAXATION

It has been known for some time that the fluorescence of N-arylaminonaphthalene sulfonates is highly sensitive to the polarity and viscosity of the solvent (Brand and Gohlke 1972). In general, the emission maxima shift from blue to red, the band width increases, and the quantum yield decreases as the polarity of the solvent increases (Edelman and McClure 1968). These molecules have a much higher dipole moment in the excited state than in the ground state, and it has been proposed that these effects are due to reorientation or relaxation of the solvent shell around the excited molecule (Brand and Gohlke 1972). The theory of excited-state solvent relaxation has been investigated in detail by Lippert (1961), Mataga (1963), and Bakshiev (Mazurenko and Bakshiev 1970).

If excited-state solvent relaxation takes place on the same time scale as the fluorescence decay (i.e., on the order of nanoseconds), a shift in emission spectra from blue to red should be observed with increasing time during the fluorescence decay. The fluorescence decay of a pure chromophore which does not undergo an excited-state reaction should be independent of emission wavelength. In contrast, a nanosecond time-resolved spectral shift will also be reflected by changes in the decay profile as a function of wavelength. Depending on the mechanism of the solvent interaction with the excited molecule, a wavelength-dependent multiexponential or nonexponential decay may be observed.

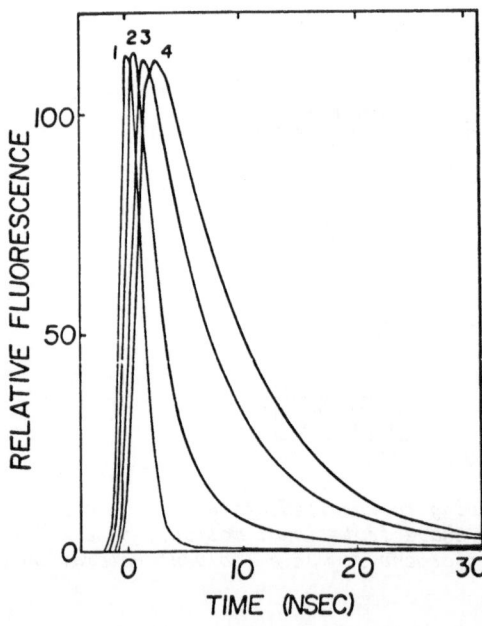

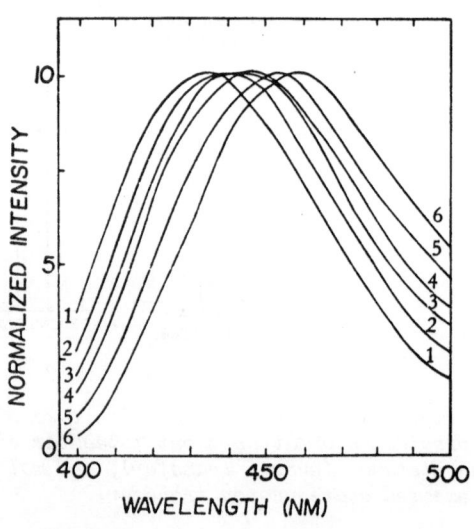

Figure 2. *Wavelength dependence of the fluorescence decay of 2,6-p-TNS in glycerol:water (95:5) at 3°C. #1, exciting lamp profile; #2, 400 nm; #3, 455 nm; #4, 500 nm. Excitation wavelength, 340 nm*

Figure 3. *Time-resolved emission spectra of 2,6-p-TNS in glycerol:water (95:5). Time 0 is defined as the peak of the lamp profile. #1, -0.2 nsec; #2, 0.7 nsec; #3, 1.5 nsec; #4, 3.1 nsec; #5, 6.0 nsec; #6, 16.6 nsec. before or after the peak of the lamp flash*

Fluorescence decay curves of 2-p-toluidinonaphthalene-6-sulfonate (2,6 p-TNS) in glycerol:water (95:5 v/v) at 3°C at three emission wavelengths are shown in Figure 2 (curves 2, 3 and 4). Clearly the shape of the decay curve changes with emission wavelength. The decay in the blue region of the emission is more rapid than that at the red region.

Figure 3 depicts the nanosecond time-resolved emission spectra generated from decay curves obtained at five-nanometer intervals with the solution indicated above. Curve 1 represents the fluorescence emission spectrum 0.2 nanoseconds before the peak of the lamp flash (curve 1 in Figure 2). Curve 6 shows the emission spectrum 16.8 nanoseconds after the peak of the lamp flash. This time-dependent red shift is interpreted to be due to the excited-state interaction of the water and/or glycerol with the chromophore. The rate at which the emission maxima shift to the red increases as the temperature is increased.

Time-resolved spectral shifts are not observed in nonviscous solvents since solvent relaxation, if it occurs, is too rapid to be detected with nanosecond instrumentation. The decay curves of 2,6-p-TNS in ethanol are not wavelength dependent and do not show time-dependent spectral shifts.

Time dependent spectral shifts similar to those seen in glycerol have been observed with 2,6-p-TNS adsorbed both to some proteins (Brand and Gohlke 1971) and to phospholipid vesicles. This indicates that the dye-binding region contains polar residues capable of relaxing about the excited dye, yet the microviscosity is such that the relaxation takes place on the nanosecond time scale. If the polar groups in question are water molecules, they are not completely free to move since their movement would result in rapid relaxation which could not be detected on this time scale. It appears that this approach will be useful in providing new information regarding solvation characteristics and possibly microviscosity or microconstraint measurements at sites of interest on macromolecules or biological membranes.

PROTON TRANSFER

Aromatic alcohols represent another class of fluorophores whose emission properties are sensitive to the environment. It is well known from the work of Weller (1961) that ionization constants in the excited states may differ by 5 or 6 pH units from those in the ground state.

One well-studied system which undergoes excited-state proton transfer is that of 2-naphthol. The ground-state pKa is 9.4, while the pKa in the excited state is 2.8. The emission maximum of the protonated species, A, is near 340 nm; while the emission maximum of the ionized species, B, is near 425 nm. The basic reaction scheme for this system is shown in Figure 4. In the pH region between the ground-state and excited-state pKa, the chromophore is in the protonated form, A. After absorption of a photon, A* is formed. This can emit light with a rate constant K_{F1A}, it can return to the ground state by nonradiative mechanisms with a rate constant K_{QA}, or it can lose a proton to form B* with a rate constant K_F. Any B* which is formed must come from this excited-state reaction. B* can emit light with a rate constant K_{F1B}. It can undergo nonradiative transitions to the ground state with a rate constant K_{QB}, or if the hydrogen ion concentration is high enough, it can return to A* with a rate constant K_R. Determination of the rate constants K_F and K_R for reactions of this type provides a quan-

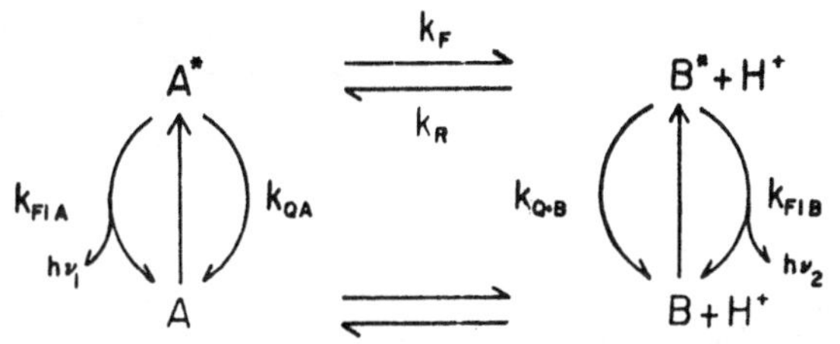

Figure 4. *Reaction scheme for excited-state proton transfer. Protonated and ionized forms are represented by A and B respectively. From Loken et al., Biochemistry,* **11,** *4779 (1972)*

titative measure of the proton-donating and -accepting properties of the environment around the chromophore.

The fluorescence decay curves for 2-naphthol at pH 2.7 are shown in Figure 5. Since an excited-state reaction is taking place while the emission is occuring, the fluorescence decay curves will not be described by single exponentials. The sharp spike is a profile of the lamp flash, the dashed curve is the decay of naphthol (collected at 360 nm), while the solid curve represents the decay of naphtholate (collected at 450 nm). The delay in naphtholate emission due to its formation from naphthol rather than its direct excitation is clearly evident.

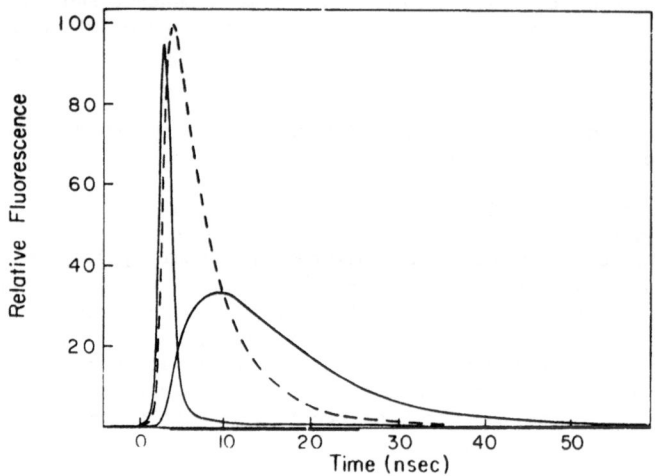

Figure 5. *Fluorescence decay curves of 2-naphthol, pH 2.7. The sharp curve is the profile of the exciting lamp. The dashed line is fluorescence from the protonated species (360 nm). The smaller curve is the decay of the ionized species (450 nm) expanded 2.5 times. Excitation, 340 nm. From Loken et al., Biochemistry, 11, 4779 (1972)*

Analysis of such decay curves can yield both the forward and reverse rate constants. One method of analysis makes use of the nonintegrated rate equations. These equations can be written:

$$- \frac{d\ [A^*]}{dt} = (K_{F1A} + K_{QA} + K_F)\ [A^*] - K_R\ [H^+]\ [B^*] \tag{1}$$

$$- \frac{d\ [B^*]}{dt} = (K_{F1B} + K_{QB} + K_R\ [H^+])\ [B^*] - K_F\ [A^*] \tag{2}$$

By dividing equation 1 by $[A^*]$ and equation 2 by $[B^*]$ the linear equations result:

$$- d\ [A^*]/dt/\ [A^*] = -K_R\ [H^+]\ \frac{[B^*]}{[A^*]} + (K_{F1A} + K_{QA} + K_F) \tag{3}$$

$$- d\ [B^*]/dt/\ [B^*] = -K_F\ \frac{[A^*]}{[B^*]} + (K_{F1B} + K_{QB} + K_R\ [H^+]) \tag{4}$$

These equations can be solved directly using the data from two decay curves and appropriate proportionality factors relating concentration to fluorescence intensities. Decay curves are collected at wavelengths where only one form emits. The rate constants can be determined from a graph in which the ratio of the two intensities at a specific time is plotted versus the ratio of the instantaneous derivative and the intensity at that time.

Figure 6 shows an example of a derivative-intensity plot, as does Figure 7. The slope of the plot in Figure 6 is equal to $K_R\ [H^+]$. The initial part of the graph is not used because the decay function is convolved with a lamp of finite width. The slope of the plot in Figure 7 is K_F. Convolution is not a major problem in this case. If the intercepts of plots, such as those shown in Figure 7, are replotted as a function of $[H^+]$, a value of K_R is obtained (Figure 8). K_F as determined for 2-naphthol in water is equal to $5.1 \times 10^7\ \text{sec}^{-1}$, while K_R is limited by the rate of hydrogen-ion diffusion and is equal to $5.5 \times 10^{10}\ \text{M}^{-1}\ \text{sec}^{-1}$.

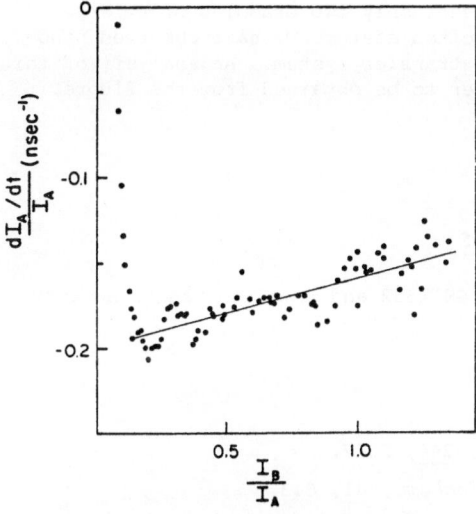

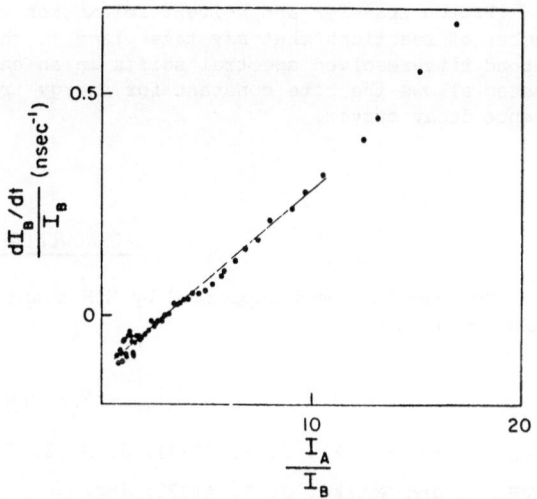

Figure 6. *Derivative-intensity plot of 2-naphthol, pH 3.43, having a slope equal to K_R [H^+]. The initial points deviate from linearity as a result of convolution with a lamp of finite width. From Loken et al., Biochemistry, 11, 4779 (1972)*

Figure 7. *Derivative-intensity plot of 2-naphthol, pH 3.43, with a slope equal to K_F. From Loken et al., Biochemistry, 11, 4779, (1972)*

In mixtures of ethanol and water the change in the hydrogen-ion accepting character of the solvent media is reflected by changes in the rate constants. The rate of excited-state proton transfer can be greatly increased by addition of a proton acceptor such as acetate or imidazol. When 2,6-naphtolsulfonate is adsorbed to macromolecules such as bovine serum albumin, the proton-donating and -accepting character to which the chromophore is exposed changes, and this change is reflected by changes in the rate constants for proton transfer.

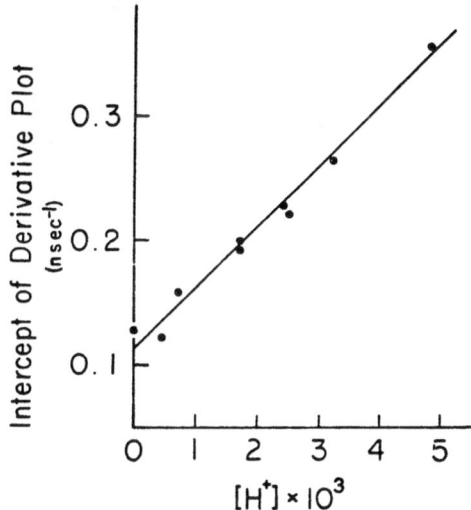

Figure 8. *Replotting the intercepts of the derivative-intensity graphs, similar to that shown in Figure 7, as a function of [H^+] yields a straight line with a slope equal to K_R. From Loken et al., Biochemistry, 11, 4779, (1972)*

Proton transfer and solvent relaxation represent only two examples of a large number of reactions that may take place in the excited state. We have observed nanosecond time-resolved spectral shifts in an energy transfer system. An analysis of this system allows the rate constant for energy transfer to be obtained from the fluorescence decay curves.

ACKNOWLEDGEMENTS

This research was supported by NIH grant No. GM11632 and American Cancer Society grant No. P-610.

REFERENCES

BRAND, L. and Gohlke, J. R. (1971) *J. Biol. Chem.*, **246**, 2317.

BRAND, L. and Gohlke, J. R. (1972) *Ann. Rev. of Biochem.*, **41**, 843.

DELUCA, M., Brand, L., Cebula, T.A., Seliger, H. H., and Makula, A.F. (1971) *J. Biol. Chem.*, **246**, 6702.

EDELMAN, G. M. and McClure, W. O., (1968) *Accounts Chem. Res.*, **1**, 65.

ISENBERG, I., and Dyson, R. D. (1969), *Biophys. J.*, **9**, 1337.

LIPPERT, E. (1961), *Angew. Chem.*, **73**, 695.

MATAGA, N., (1963), *J. Chem. Soc.,Japan*, **36**, 654.

MAZURENKO, Y. T. and Bakshiev, N. G., (1970), *Opt. Specktrosk.*, **28**, 905.

SCHUYLER, R., and Isenberg, I. (1971), *Rev. Sci. Instrum.*, **42**, 813.

WARE, W. (1971), In: *Creation and Detection of the Excited State*, (Lamola,A.A., ed.), Marcel Dekker, New York, 213.

WELLER, A. (1961), *Progr. React-Kinet.*, **1**, 187.

FLUORESCENCE SPECTROSCOPY AS A TOOL FOR STUDYING DRUG INTERACTIONS WITH BIOLOGICAL SYSTEMS

Colin F. Chignell

*Laboratory of Chemical Pharmacology, National Heart and Lung Institute,
Bethesda, Maryland*

INTRODUCTION

The first commercially available spectrophotofluorometers were based on a design by Dr. R. L. Bowman (Bowman et al. 1955). The instrument which he developed was built in response to a request by Dr. B. B. Brodie, a pharmacologist, who needed more sensitive instrumentation for the detection and assay of drugs and their metabolites in body tissues and fluids. While pharmacologists were slow to use this technique to study the molecular basis of drug interactions with biological systems, biochemists and biophysicists quickly added this new tool to an evergrowing array of physical instrumentation. Recently, however, the great potential of fluorescence spectroscopy as a tool for studying how drugs interact with macromolecules has been realized and exploited (Chen 1972; Chignell 1972).

The fluorescence of any molecule may be characterized by several parameters, including the wavelengths of maximal activation and emission, quantum yield, fluorescence lifetime, and degree of polarization (Chen 1972). Since these parameters are often very sensitive to changes in the microenvironment of the fluorophore, it is not surprising that the fluorescence of a drug molecule can be drastically altered when it binds to a macromolecule (Chignell 1970a; 1970b; 1972). Similarly, the fluorescence characteristics of a macromolecule can be modified by the binding of a drug. When such changes occur, it is often possible to gain valuable information about the nature of the drug-macromolecule interaction (Chignell 1970a; 1970b; 1972). If neither the drug nor the macromolecule possesses suitable fluorescence characteristics, the interaction can often be studied by complexing (covalently or otherwise) the macromolecule with a fluorescent label (Chignell 1970a; 1970b; 1972).

The interaction of a drug with a cellular component, or "receptor", initiates a chain of events which eventually leads to the expression of its pharmacological activity (Goldstein et al. 1968a). It should be emphasized, however, that drug molecules also interact with other cellular components, and although such interactions do not elicit a pharmacological response, they are nevertheless important. Thus the binding of a drug to serum and tissue proteins may profoundly affect its availability at the site of action (Brodie 1965). The covalent reaction of a drug with serum proteins may produce allergic reactions (Goldstein et al. 1968b), while drug combination with other cellular components, such as nucleic acids, may result in other toxic side reactions (Schumacher et al. 1968).

Although the number of receptor macromolecules that are available in pure form is small, there is one drug-receptor system that has been studied in great detail by fluorescence spectroscopy. Erythrocyte carbonic anhydrase is a globular zinc-containing metalloprotein that reversibly catalyzes the hydration of CO_2 (Maren 1969). This enzyme is specifically inhibited by aromatic sulfonamides that have the general structure $ArSO_2NH_2$, where Ar can be homocyclic or heterocyclic (Maren 1969). Since carbonic anhydrase is the site of action for diuretic sulfonamide drugs, such as acetazolamide and ethoxyzolamide, this system will be used as one example to illustrate the usefulness of fluorescence spectroscopy in the study of drug-receptor interactions. Membrane systems from red-cell ghosts, microsomes, mitochondria, and the eel electroplax will be used as examples of the fluorescent-probe technique for monitoring drug-receptor interactions. The binding of acidic drugs to serum albumin will be used to illustrate the application of fluorescence techniques to the interaction of a drug with a macromolecule that is not its cellular receptor. The importance of drug binding to plasma proteins has been discussed elsewhere (Brodie 1965).

CHANGES IN THE FLUORESCENCE QUANTUM YIELD AND POLARIZATION OF DRUGS
ON BINDING TO MACROMOLECULES

Serum Albumin

When the anticoagulant drug, warfarin (I), binds to human serum albumin (HSA), the flu-

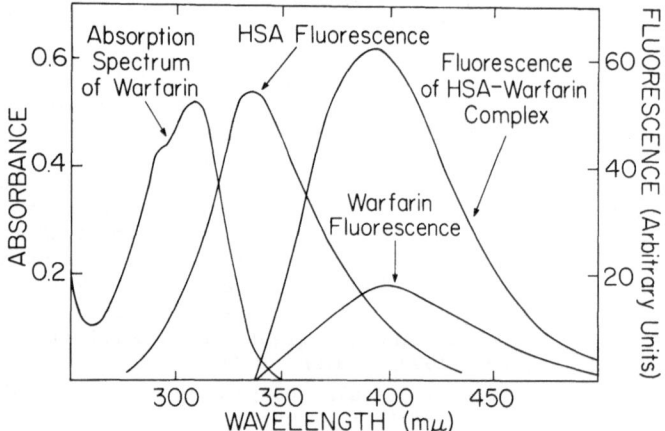

I
Warfarin

orescence quantum yield of the drug increases eightfold while its fluorescence emission maximum moves to shorter wavelengths (Figure 1, Table I) (Chignell, 1970c).

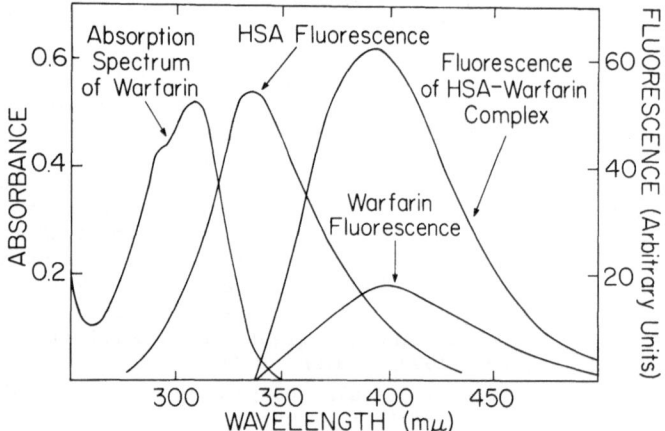

Figure 1. *Spectra of warfarin and HSA. The absorption spectrum of warfarin was measured in a 1 cm quartz cell, using a drug concentration of 4 × 10⁻⁴M in the presence of 0.1M sodium phosphate (pH 7.4). Fluorescence emission spectra were recorded using microcells (0.29 × 0.29 cm) and are displayed in terms of relative quanta. The concentrations were: HSA, 1.55 × 10⁻⁵M; warfarin alone, 4 × 10⁻⁵M; warfarin, 4 × 10⁻⁵M, + HSA, 1 × 10⁻⁵M. The HSA solution was activated at 290 nm, while those containing warfarin were activated at 320 nm. Bandwidths of excitation and emission were 12 nm. From Chignell (1970c), courtesy of Academic Press*

The fluorescence quantum yield of warfarin also increases when the drug is dissolved in certain solvents such as dimethylformamide and glycerol, but is relatively unaffected by other solvents such as methanol and ethyl acetate (Table II). There is little correlation between the dielectric constant of the solvent and its effect on the fluorescence of warfarin (Table II). The fluorescence quantum yield of warfarin also increases in the presence of the cationic detergent, cetyltrimethylammonium bromide (Table II), but only when the detergent is present at a concentration in excess of its critical micellar concentration (Chignell 1970c). While the solvents either produce little effect on the fluorescence emission maximum of warfarin or shift it to longer wavelengths (Table II), the fluorescence emission maximum of warfarin moves to shorter wavelengths when the drug binds to human serum albumin. Rat and canine serum albumins induce even greater blue shifts in the fluorescence emission maximum of warfarin (Table III). Rat serum albumin also causes a 13-fold increase in the quantum yield of bound warfarin. One of the complicating factors that makes interpretation of these results difficult is that the long wavelength band of warfarin that is being excited at 320 nm may contain both $n-\pi^*$ and $\pi-\pi^*$ transitions. Furthermore, it has been found that ion-

TABLE I

TABLE I

QUANTUM YIELDS OF WARFARIN, HSA, AND THE WARFARIN-HSA COMPLEX

Compound	Exciting Wavelength	λ_{max} of Emission	Quantum Yield*
	nm	nm	
HSA	290	335	0.063
Warfarin	320	400	0.012
Warfarin-HSA	290	390	0.015[a]
Warfarin-HSA	290	335	0.013[a]
Warfarin-HSA	320	390	0.090

The quantum yield of warfarin-HSA complex was determined using a solution containing 1×10^{-5}M HSA, 4×10^{-5}M warfarin, and 0.1M sodium phosphate buffer (pH 7.4). [a]Quantum yield*calculated on the basis of the tryptophan absorption.

TABLE II

QUANTUM YIELD OF WARFARIN DISSOLVED IN VARIOUS SOLVENTS

Solvent di-electric constant	Solvent[a]	Quantum Yield	Emission λ_{max}
Debye units			nm
78.5	Water (0.05M sodium phosphate, pH 7.4)	0.013	400
42.5	Glycerol	0.104	392
37.0	Ethylene glycol	0.047	400
36.7	Dimethylformamide	0.154	415
32.0	Propylene glycol	0.061	400
31.2	Methanol	0.016	400
25.8	Ethanol	0.033	400
19.2	1-Butanol	0.039	400
6.1	Ethyl acetate	0.010	400
3.0	Dioxane	0.022	412
	Cetrimide (0.05% in 0.05M sodium phosphate, pH 7.4)	0.041	400

[a]Solutions were made by adding 0.1 ml of a stock solution (made up in 0.01 N NaOH) to 9.9 ml of solvent. Fluorescence was excited at 320 nm.

TABLE III

THE QUANTUM YIELD AND FLUORESCENCE EMISSION MAXIMUM OF WARFARIN BOUND TO DIFFERENT SERUM ALBUMINS

Serum Albumin	Fluorescence* Quantum yield	Fluorescence* λ_{max}
		nm
None	0.012	400
Human	0.070	390
Bovine	0.069	380
Rat	0.153	380
Porcine	0.080	385
Canine	0.083	380
Ovine	0.076	385
Equine	0.088	390
Rabbit	0.047	390

*Activation wavelength was 320 nm using 12 nm band width.

ization of warfarin can alter the fluorescence quantum yield of the drug (Corn and Berberich 1967). Nevertheless, even though the nature of fluorescence increase that is observed when warfarin binds to albumin is not known, this phenomenon can still be used to measure the interaction. If the fluorescence of warfarin is monitored during titration of rat serum albumin by the drug, a fluorescence titration curve is obtained (Figure 2). From this curve, it is possible to derive (Chignell 1972) a Scatchard plot (Figure 3) which shows that warfarin has one binding site on rat plasma albumin with an association constant of $8.2 \times 10^5 M^{-1}$.

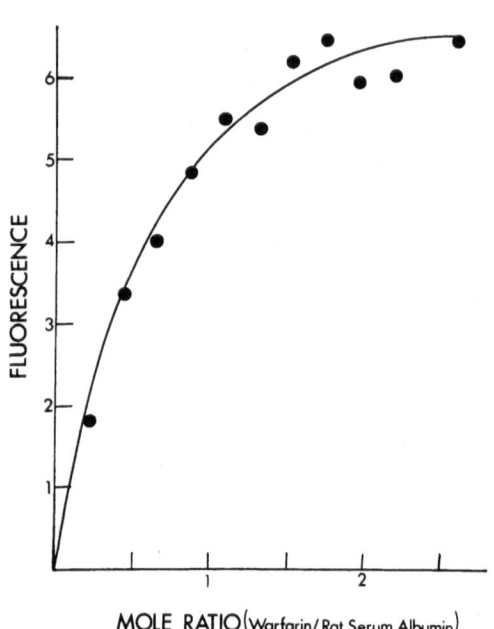

Figure 2. *Fluorometric titration of rat serum albumin with warfarin. The excitation and emission wavelengths were 320 nm and 390 nm, respectively. Bandwidths of excitation and emission were 12 nm*

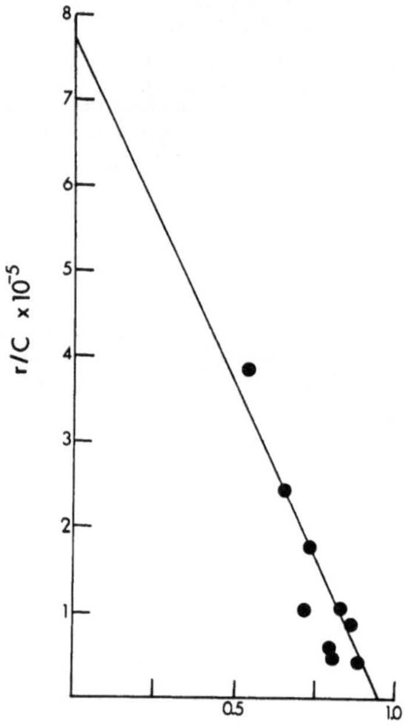

Figure 3. *A Scatchard plot of warfarin binding to human plasma albumin. This curve was calculated from the data in Figure 2. r = number of moles of warfarin bound per mole of rat serum albumin; C = molar concentration of free warfarin*

Light energy absorbed by the single tryptophan group of human serum albumin is nonradiatively transferred to bound warfarin which then reemits it as fluorescence (Figure 4) (Chignell 1970c). The critical transfer distance, R_0, for which resonance transfer between tryptophan and warfarin is 50% complete can be calculated from equation (1) to be 26.2 Å (Chignell 1970c; Förster 1948).

$$R_0 = \left(\frac{1.66 \cdot 10^{-33} \cdot \tau \cdot J_{\bar{\nu}}}{n^2 \cdot \bar{\nu}_o^2} \right)^{1/6} \tag{1}$$

τ = donor fluorescence decay time

$\bar{\nu}_o$ = mean peak positions (in wave numbers) of donor emission and lowest energy absorption bands

$J_{\bar{\nu}}$ = overlap integral

n = refractive index of the solvent.

The actual distance, R, between the tryptophan and the bound warfarin may be calculated from equation (2). Since the value of x, (the fraction of photons transferred from

$$1 - x = \frac{1}{(R_0/R)^6 + 1} \tag{2}$$

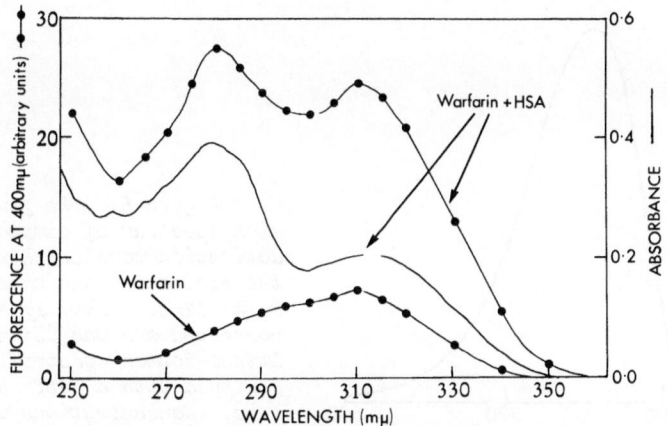

Figure 4. *Spectra of warfarin and the warfarin-HSA complex. The activation spectra (●—●) of warfarin and the warfarin-HSA complex have been corrected for the intensity of the activating light. Concentrations were: warfarin alone, $4 \times 10^{-5}M$; warfarin, $4 \times 10^{-5}M$,+ HSA, $1 \times 10^{-5}M$. The absorption spectrum of the warfarin-HSA complex (——) was measured in a 1 cm quartz cell, using a solution containing $1 \times 10^{-5}M$ HSA and $4 \times 10^{-5}M$ warfarin. A blank containing $2.3 \times 10^{-5}M$ warfarin was used to compensate for absorption due to the free drug. All solutions contained 0.1M sodium phosphate buffer (pH 7.4). From (Chignell 1970c), courtesy of Academic Press*

tryptophan to warfarin), is 0.166 (Table I, 0.015/0.09), the distance between the tryptophan in human plasma albumin and bound warfarin must be 26.2 Å. This value should be compared with a calculated diameter of 56 Å for human plasma albumin.

Another parameter which can change when a drug binds to plasma albumin is the fluorescence polarization of the drug. Camptothecin (III) is a quinoline alkaloid that exhibits antileukemic and antitumor activities in animals. Camptothecin is highly fluorescent

III   Camptothecin

with an emission maximum at 445 nm (Figure 5). While the fluorescence quantum yield and emission maximum of camptothecin remain unaltered when the drug binds human plasma albumin, there is a dramatic change in the fluorescence polarization of the drug molecule. When a fixed amount of camptothecin is titrated with increments of human serum albumin, the fluorescence polarization of the drug increases, reaching a maximum of +0.25 in the presence of a large excess of protein (Figure 6). The fraction of drug bound, $\bar{s}$, at each point in the titration can be calculated according to the method of Deranleau and Neurath (1966) from the following formula;

$$\bar{s} = (p - p_{min})/(p_{max} - p_{min}) \qquad 0 \leqslant \bar{s} \leqslant 1 \qquad (3)$$

where p = observed polarization, p_{min} = polarization at zero plasma albumin concentration, and p_{max} = polarization when all the drug is bound. The modified Scatchard plot which is obtained (Figure 7) from the polarization data (Figure 6) indicates that camptothecin has an affinity constant of 1.08×10^6 M^{-1} for human serum albumin. Since many drugs are highly fluorescent, this technique should be extremely useful for studying their interaction with biologically important macromolecules.

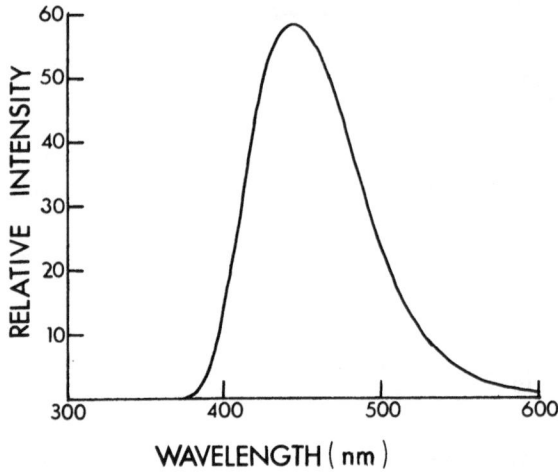

Figure 5. *The fluorescence emission spectrum of camptothecin. The activation wavelength was 370 nm, while the excitation and emission bandwidths were 12nm. The spectrum is uncorrected and was obtained with an Aminco-Bowman spectrofluorometer equipped with a R456 photomultiplier tube. Concentrations were camptothecin, $1 \times 10^{-5}M$; sodium phosphate (pH 7.4) 0.1 M*

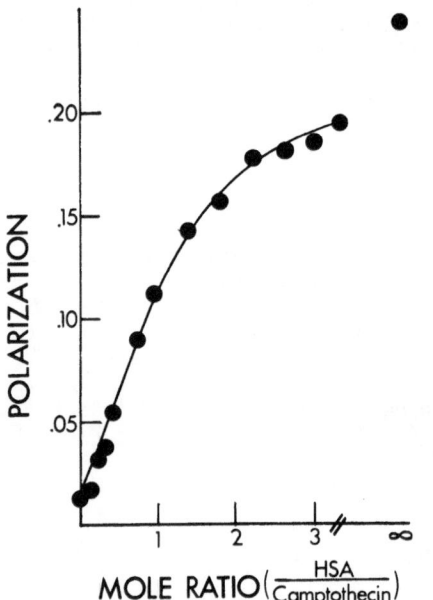

Figure 6. *The effect of human serum albumin on the fluorescence polarization of camptothecin. The concentration of camptothecin was $2 \times 10^{-6}M$, while sodium phosphate (pH 7.4) was 0.1M*

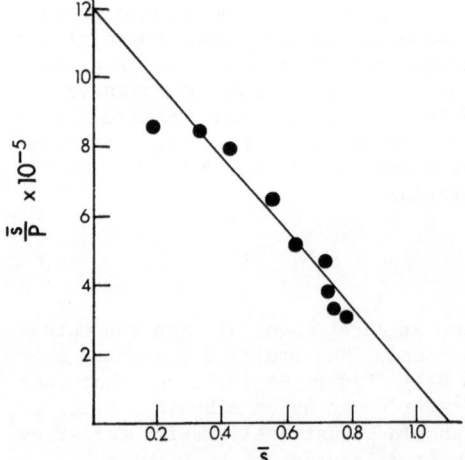

Figure 7. *Modified Scatchard plot of the binding of camptothecin to human plasma albumin. The points were calculated from the data shown in Figure 6, according to the method described in the text. $\bar{s}$ = fraction of drug bound; [p] = molar concentration of free HSA*

Carbonic Anhydrase

In 1967 Chen and Kernohan reported that 5-dimethylaminonaphtha-
lene-1-sulfonamide (DNSA)(IV) formed a highly fluorescent complex with bovine erythrocyte

DNSA

IV

carbonic anhydrase B (BCAB). The fluorescence of free DNSA in water had a peak emission at 580 nm and a quantum yield of only 0.055, while DNSA bound to BCAB had an emission maximum at 468 nm and a quantum yield of 0.84. Chen and Kernohan (1967) suggested
that the large emission blue shift was due to the hydrophobicity of the binding site
and the dissociation of the sulfonamide group. They also found that 85% of the photons
absorbed by the seven tryptophans of BCAB were transferred to the single bound DNSA
molecule. The effective average distance between DNSA and tryptophan was calculated
from equations (1) and (2) to be 16 Å.

Chen and Kernohan (1967) were able to estimate a fluorescence decay time of 22.1
nsec for the BCAB-DNSA complex by monitoring the fluorescence polarization of the bound
ligand as a function of temperature. Since they calculated a value of 24.3 nsec for
the rotational relaxation time of BCAB by assuming that the protein was an anhydrous
sphere of 30,000 molecular weight, they suggested that the enzyme had a low degree of
asymmetry.

CHANGES IN THE FLUORESCENCE OF A MACROMOLECULE ON INTERACTION WITH A DRUG MOLECULE

Serum Albumin

Förster (1948) has shown that when the fluorescence emission spectrum of compound
A (donor) overlaps the absorption spectrum of compound B (acceptor), resonance transfer
of energy from A to B can occur and the fluorescence of A may be quenched. Human plasma
albumin contains a single tryptophan residue which when activated at 290 nm emits fluorescence with a maximum at 335 nm (Figure 8). The tryptophan fluorescence emission
band of human serum albumin overlaps extensively (Figure 8) the absorption maximum of
4-butyl-1-(p-nitrophenyl)-2-phenyl-3,5-pyrazolidinedione VI, an analog of the antiinflam-
matory drug, phenylbutazone V, so that, according to Förster, resonance transfer of energy
could take place between the drug and the protein tryptophan. When VI binds to human

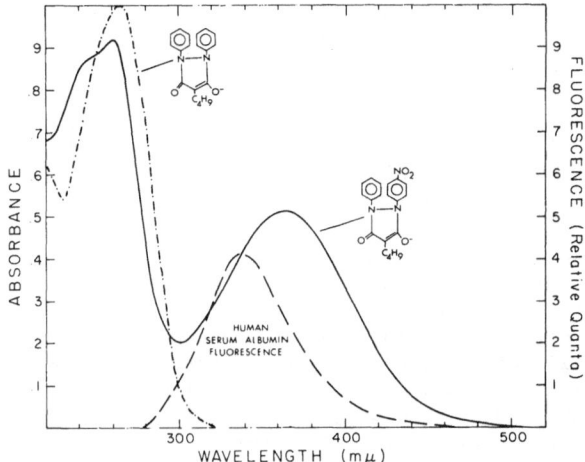

Figure 8. The ab-
sorption spectra of
phenylbutazone (V) and
4-butyl-1-(p-nitrophe-
nyl)-2-phenyl-3,5-pyra-
zolidinedione (VI) and
the fluorescence
emission spectrum of
human serum albumin.
The fluorescence activ-
ation wavelength was
290 nm. All solutions
contained 0.1 M sodium
phosphate buffer (pH
7.4)

plasma albumin, the tryptophan fluorescence of the protein is indeed quenched (Figure 9). When the tryptophan fluorescence of HSA was plotted as a function of the drug-al-

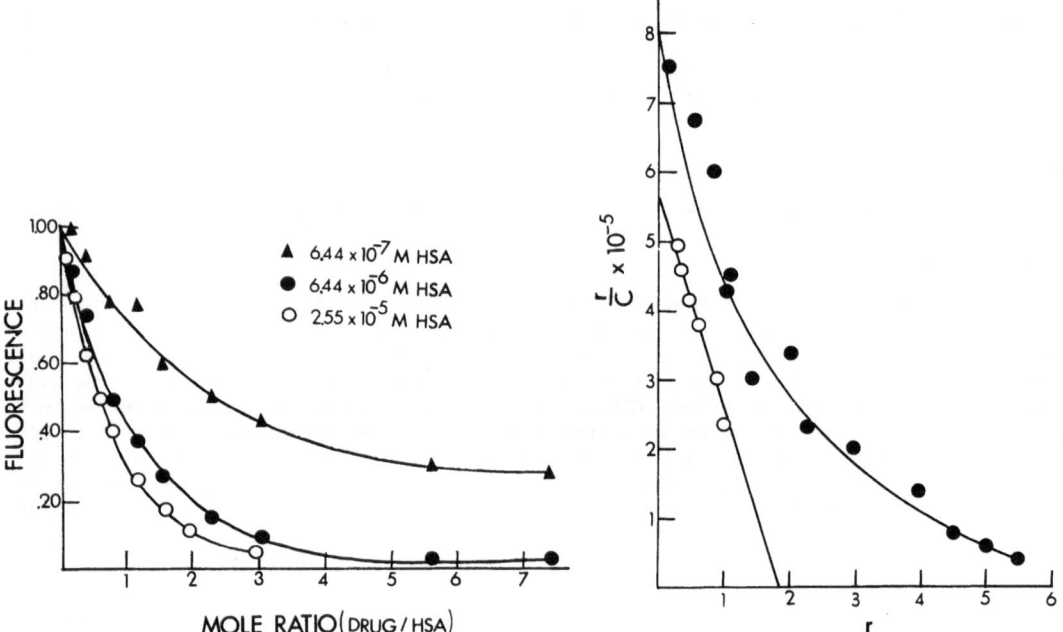

Phenylbutazone	R = H	V
Analog	R = NO$_2$	VI

bumin ratio, the degree of quenching was found to be dependent upon the protein concentration (Figure 9). When the concentration of HSA was 2.55 × 10^{-5}M or above, the quenching curves were superimposable, indicating that all the drug was bound. From

Figure 9. *The quenching of the fluorescence of human serum albumin by the binding of 4-butyl-1-(p-nitrophenyl)-2-phenyl-3,5-pyrazolidinedione (VI). The activation and emission wavelengths were 290 nm and 335 nm. Band widths were 12 nm. All solutions contained 0.1M sodium phosphate (pH 7.4)*

Figure 10. *A Scatchard plot of the binding of VI to human serum albumin. Results were obtained either from equilibrium dialysis experiments (—●—●—) or from the fluorescence titration curve in Figure 9 (o—o). All solutions contained 0.1M sodium phosphate buffer. C = molar concentration of free VI; r = the number of moles of VI bound per mole of HSA*

this curve it was possible to calculate the amount of bound drug during the titration of a lower concentration (6.44 × 10^{-7}M) of albumin (Chignell 1972). These data gave the Scatchard plot shown in Figure 10. It is of interest that the fluorescence-quenching experiments indicate the presence of only two binding sites on human plasma albumin for the phenylbutazone derivative (VI), while equilibrium dialysis measurements show that there are at least six sites which bind the drug. One possible explanation for these contradictory observations is that only two of the drug binding sites are sufficiently close to the tryptophan group of human plasma albumin to permit energy transfer to take place.

Carbonic Anhydrase

Chen and Kernohan (1967) have reported that binding of DNSA to BCAB quenches the tryptophan fluorescence of the enzyme. The association constant for this interaction was $2.4 \times 10^{-7}M$. Taylor and co-workers (1970) have studied the binding of a series of sulfonamides to human erythrocyte carbonic anhydrases B and C (HCAB, HCAC) by measuring the quenching of tryptophan fluorescence. They studied those sulfonamides that lacked the necessary spectral requirements for energy transfer by measuring their ability to displace competitively a sulfonamide that did quench the fluorescence of HCAB and HCAC. These workers also measured the kinetics of complex formation between the human carbonic anhydrases and the sulfonamides by means of a stopped-flow apparatus which monitored the fluorescence quenching of HCAB and HCAC. Their experiments showed that the association rate was not diffusion controlled, but that the formation of the sulfonamide-enzyme complex required a distinct activation energy. They also concluded that the complex was stabilized primarily through a favorable enthalpy change (Taylor et al. 1970).

CHANGES IN THE FLUORESCENCE OF LABELS ATTACHED TO MACROMOLECULES ON THE BINDING OF DRUGS

Serum Albumin

When dansyl glycine VII binds to human serum albumin, the quantum yield of the ligand increases fivefold while its fluorescence emission maximum shifts from 580 nm to 480 nm

Dansyl Glycine

VII

$N(CH_3)_2$

SO_2NHCH_2COOH

(Table IV) (Chignell 1969a). Fluorescence titration indicates that binding occurs at a single site for which dansyl glycine has an association constant of 4.6×10^5 M^{-1} (Figure 11). It is of interest that only one of the dansyl glycine binding sites on human serum albumin is capable of increasing the quantum fluorescence yield of the

TABLE IV

QUANTUM YIELDS OF DANSYL GLYCINE, HSA, AND THE DANSYL GLYCINE-HSA COMPLEX

Compound	Exciting Wave- length	λ_{max} of Emission	Quantum Yield
	nm	nm	
HSA	290	335	0.077
Dansyl glycine	350	580	0.051
Dansylglycine-HSA	290	480	0.235
Dansyl glycine-HSA	350	480	0.443
Dansyl glycine-HSA	290	335	0.046

The quantum yield of the dansyl glycine-HSA complex was determined using a solution containing $1 \times 10^{-5}M$ HSA, $1 \times 10^{-5}M$ dansyl glycine, and 0.1M sodium phosphate buffer (pH 7.4). Under these conditions, ultrafiltration indicated that the concentration of free dansyl glycine was $0.36 \times 10^{-5}M$.

ligand (Figure 11). This suggests that dansyl glycine has only one hydrophobic binding site on human serum albumin. Several anionic drugs such as phenylbutazone (Figure 12) (Chignell 1969a), flufenamic acid (Figure 13) (Chignell 1969b), and dicoumarol (Figure 14) (Chignell 1970c) can competitively displace dansyl glycine from its binding site on human serum albumin. This technique not only provides a convenient method for monitoring drug interactions with human serum albumin, but also gives information on the hydrophobic nature of the binding sites.

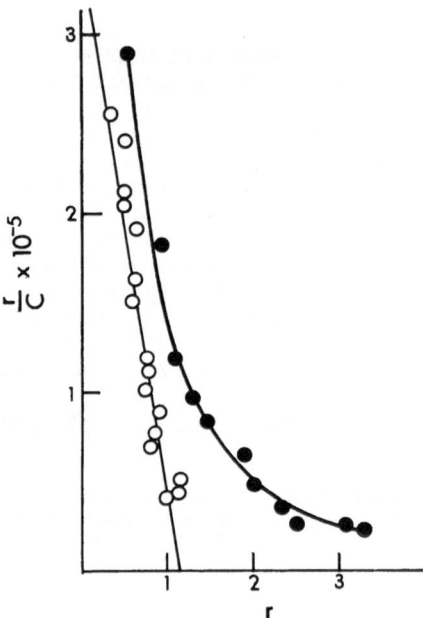

Figure 11. *A Scatchard plot of the binding of dansyl glycine to human serum al-bumin. Results were obtained either from equilibrium dialysis experiments (●—●) or from fluorescence titration measurements (o—o). Activation and emission wavelengths were 350 nm and 480 nm respectively. Band widths were 12 nm. All solutions contained 0.1M sodium phosphate buffer (pH 7.4). r = number of moles of dansyl glycine bound per mole of HSA; C = molar concentration of free dansyl glycine*

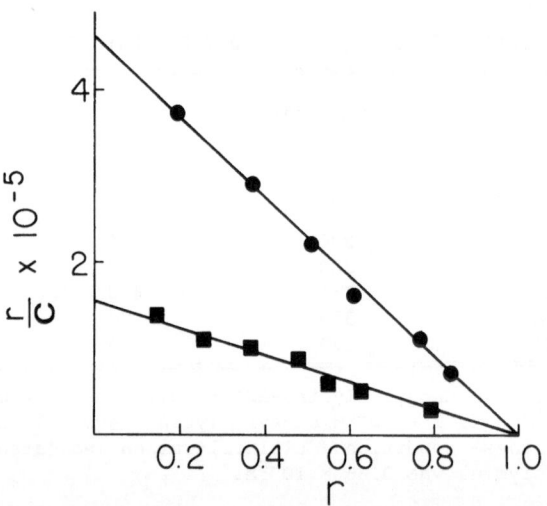

Figure 12. *Scatchard plot of the binding of dansyl glycine to human serum albu-min in the presence of phenylbutazone (V). C = molar concentration of free dansyl glycine, n = number of moles of dansyl glycine bound per mole of HSA. Binding was measured by monitoring the increase in dansyl glycine fluorescence at 480 nm while activating at 350 nm. Human serum albumin alone (1 × 10^{-5}M) ●—●; human serum albu-min (1 × 10^{-5}M) + phenylbutazone (2.5 × 10^{-5} M)—■—■. From Chignell (1969a), courtesy of Academic Press*

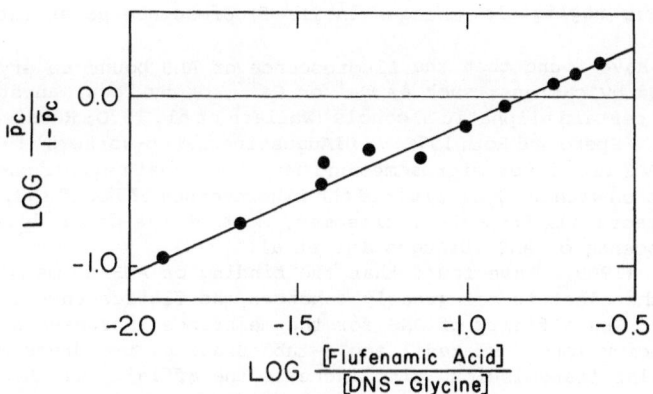

Figure 13. *Hill plot for the replacement of dansyl glycine bound to HSA by flu-fenamic acid. Activation and emission wavelengths were 350nm and 480 nm respectively. Band widths for excitation and emission were 12 nm. $\bar{p}_c$ = fractional occupation of dansyl glycine binding sites by flufenamic acid. From Chignell (1969b), courtesy of Academic Press*

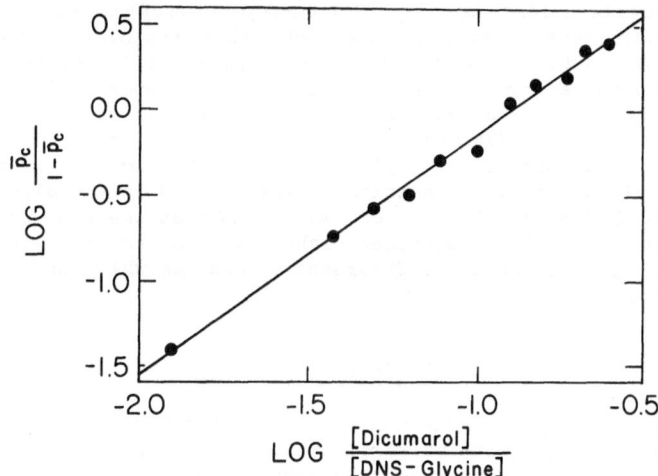

Figure 14. *Hill plot for the replacement of dansyl glycine by dicoumarol. The concentrations were: human plasma albumin, $1 \times 10^{-5}M$, dansyl glycine, $2 \times 10^{-4}M$; sodium phosphate buffer (pH 7.4) 0.1 M. Activation and emission wavelengths were 350 nm and 480 nm respectively. Bandwidths for excitation and emission were 12 nm. $\bar{p}_c$ = fractional occupation of dansyl glycine binding sites by dicoumarol. From Chignell (1970c), courtesy of Academic Press*

Membrane Systems

The fluorescent probe, 1-anilino-8-naphthalenesulfonic acid (ANS), has been used by several groups to study the interaction of drugs, cations, and other ligands with membrane systems. It has been shown, for example, that the binding of ANS to intact mitochondria or isolated mitochondrial membranes is accompanied by both a marked shift in the fluorescence emission maximum of the dye to shorter wavelengths and an increase in fluorescence yield (Azzi et al. 1969; Chance et al. 1969; Chance and Lee 1969; Packer et al. 1969). The addition of oligomycin, succinate, uncouplers of oxidative phosphorylation, or ATP produces changes in the fluorescence of membrane-bound ANS, a fact which suggests that conformational changes precede the utilization of the intermediates of energy production by the mitochondrion. The addition of either butacaine (a local anesthetic) or Ca^{++} to ANS-labeled rat liver mitochondria causes an increase in the fluorescence yield of the dye, but negligible changes in its fluorescence polarization (Chance et al. 1969).

Other workers have found that the fluorescence of ANS bound to erythrocyte membranes is markedly affected by cations, such as Na^+ or Ca^{++}, or by local anesthetic drugs, such as butacaine or certain aliphatic alcohols (Wallach et al. 1970; Rubalcalva et al. 1969; Feinstein et al. 1970; Spero and Roth 1970). DiAugustine and co-workers (1970) have studied the interaction of ANS with liver microsomes and have found that neither phenobarbital pretreatment nor reduction with $Na_2S_2O_4$ altered the fluorescence of bound ANS. While warfarin competitively displaced ANS from the microsomes, most of the other drugs tested increased the fluorescence of ANS (DiAugustine et al. 1970).

Kasai et al. (1969), have found that the binding of ANS to membranes from the electric organ of the electric eel greatly enhanced the fluorescence intensity and polarization of the dye. The affinity of ANS for the membranes increased significantly in the presence of calcium ions. Flaxedil and d-tubocurarine, two drugs which are inhibitors of neuromuscular transmission, also increase the affinity of the membranes for ANS. While it is at present unclear whether these fluorescence changes are directly related to the pharmacological effects of these agents, it does appear that the use of more specific fluorescence probes may permit the monitoring of the molecular events occurring during the transmission of impulses at the neuromuscular junction.

One of the difficulties in interpreting data from membrane systems containing ANS is that the precise location of the fluorescent label is unknown. Several groups have suggested, however, that ANS is bound to membrane phospholipids (Feinstein et al. 1970; Spero and Roth 1970; DiAugustine et al. 1970). Thus it appears probable that changes in ANS fluorescence may represent alterations not only in the conformation of membrane protein but also in the phospholipid environment of the dye. This problem has been overcome by Waggoner and Stryer (1970), who have synthesized several probes specifically for membrane studies. Three of these compounds, anthroyl stearic acid, dansyl phosphatidylethanolamine, and octadecylnaphthylamine sulfonic acid, were found to be specifically incorporated into phospholipid bilayer vesicles (Vanderkooi 1973). The emission spectra of these probes indicated that the chromophore of anthroyl stearic acid was located in the hydrocarbon region, that of dansyl phosphatidylethanolamine was located in the glycerol layer, and that of octadecylnaphthylamine sulfonic acid was located at the aqueous interface of the bilayer. When such specific fluorescent labels are incorporated into membranes, it will become easier to interpret the fluorescence changes which occur when drugs and other molecules perturb such systems.

PROGNOSIS

Fluorescence spectroscopy is undoubtedly one of the most versatile techniques for studying drug interactions with macromolecules. Even drugs which do not fluoresce can be studied by this technique provided that they quench the native fluorescence of the macromolecule or can displace a fluorescent probe. Fluorescence spectroscopy is especially useful for studying drug binding to proteins, since it is more rapid than equilibrium dialysis and far more sensitive than many other techniques such as ultraviolet and visible absorption spectroscopy. In addition to providing a useful tool for measuring drug interaction with macromolecules, fluorescence spectroscopy can also furnish valuable information on the structure of the drug binding site and the nature of binding forces involved. Although fluorescence spectroscopy has been most frequently used as an analytical tool in pharmacology, the future should see a significant increase in the use of this technique to study drug interactions with biological systems.

REFERENCES

AZZI, A., Chance, B., Radda, G.K. and Lee, C. P., (1969), *Proc. Nat. Acad. Sci.*, **62**, 612.

BOWMAN, R.L., Caulfield, P.A. and Udenfriend, S., (1955), *Science*, **122**, 32.

BRODIE, B.B., (1965) In: *Transport Function of Plasma Proteins*, (Desgrez, P. and Traverse, P.M., eds.), Elsevier, Amsterdam, 137.

CHANCE, B., Azzi, A., Mela, L., Radda, G.K., and Vainio,H., (1969), *FEBS Letters*, **3**,10.

CHANCE, B., and Lee, C., (1969) *FEBS Letters*, **4**, 181.

CHEN, R.F.,(1972), In: *Methods in Pharmacology*, (Chignell, C.F. ed.) Appleton-Century-Crofts, New York, Vol. 2, 1.

CHEN, R. F. and Kernohan, J.C., (1967), *J. Biol. Chem.*, 242, 5813.

CHIGNELL, C.F. (1969a), *Molec. Pharmac.*, **5**, 244.

CHIGNELL, C.F., (1969b), *Molec. Pharmac.*, **5**, 455.

CHIGNELL, C.F. (1970a), *Advan. Drug Res.*, **5**, 55.

CHIGNELL, C.F., (1970b), *Fluorescence News*, American Instrument Company, Silver Spring, Md., **5**, 1.

CHIGNELL, C.F.,(1970c). *Molec. Pharmac.*, **6**, 1.

CHIGNELL, C.F., (1972), In: *Methods in Pharmacology*, (Chignell, C.F., ed.) Appleton-Century-Crofts, New York, Vol. 2, 33.

CORN, M. and Berberich, R., (1967), *Clin. Chem*, **13**, 126.

DERANLEAU, D.A. and Neurath, H., (1966), *Biochemistry*, **5**, 1413.

DIAUGUSTINE, R.P., Eling, T.E. and Fouts, J.R., (1970), *Fed. Proc.*, **29**, 738.

FEINSTEIN, M.B., Spero, L. and Felsenfeld, H. (1970),*FEBS Letters*, **6**, 245.

FÖRSTER, T. (1948), *Ann. Phys. Leipzig*, **2**, 55.

GOLDSTEIN, A., Aronow, L. and Kalman, S.M., (1968a), In: *Principles of Drug Action*, Harper and Row, New York, 1.

GOLDSTEIN, A., Aronow, L., and Kalman,S.M.,(1968b), In: *Principles of Drug Action*, Harper and Row., New York, 476.

KASSAI, M., Changeux, J.P., and Monnerie, L., (1969), *Biochem. Biophys. Res. Commun.*, **36**, 420.

MAREN, T.H., (1969),*Physiol. Rev.*, **47**, 595.

PACKER, L., Donovan, M.P., and Wrigglesworth, J.M., (1969), *Biochim. Biophys. Acta*, **35**, 832.

RUBALCALVA, B., Munoz, D.M. and Gitler, C., (1969), *Biochemistry*, **8**, 2742.

SCHUMACHER, H., Blake, D.A. and Gillette, J.R., (1968), *J. Pharmac. Exp. Ther.*, **160**, 201.

SPERO, L., and Roth, S.,(1970), *Fed. Proc.* **29**, 474.

TAYLOR, P.W., King, R.W. and Bergen, A.S.V., (1970), *Biochemistry*, **9**, 2638.

VANDERKOOI, J. (1973), This volume.

WAGGONER, A.S., and Stryer, L., (1970), *Proc. Nat. Acad. Sci.*, **67**, 579.

WALLACH, D.F.H., Ferber, E., Selin, D., Weikekamm, E., and Fischer, H., (1970), *Biochim. Biophys. Acta*, **203**,67.

TEMPERATURE SENSITIVITY OF FLUORESCENT PROBES IN THE PRESENCE OF MODEL MEMBRANES AND MITOCHONDRIA

Jane Vanderkooi

Johnson Research Foundation, University of Pennsylvania, Philadelphia, Pennsylvania

INTRODUCTION

Although there is no dirth of models describing the structure of membranes, a universally accepted description of membrane structure with respect to function eludes us, a situation due in part to the lack of techniques with which to attack the problem. In recent years fluorescent compounds have been used as "probes" of membrane structure; the usefulness of this technique is characterized by the ease in obtaining information not only on the probe environment, mobility, location, but also on kinetics of probe response to changes in membrane structure.

A description of the response of various fluorescent dyes to changes in structure of artificial and natural membranes is presented here. Fluorescence intensity, excitation and emission maxima, and fluorescence polarization values are sensitive parameters of the environment, and can be used to monitor phase transitions occuring in artifical membranes.

MATERIALS AND METHODS

A Perkin-Elmer Spectrophotofluorometer equipped with 105 PB Polacoat polarizers was used for this work. Polarization P was defined as

$$P = \frac{I_{||} - TI_{\perp}}{I_{||} + TI_{\perp}}$$

where $I_{||}$ and $I_{\perp}$ are light intensity parallel and perpendicular to the exciting beam, and T is a constant which corrects for the inability of the monochromator to transmit the two components of polarized light with equal facility. Phospholipids obtained commercially were sonicated with a Branson sonifier for 5 min immediately prior to use.

Fluorescent compounds determined by X-ray analysis to be located in specific regions of artificial membranes were chosen as probes (Lesslauer et al. 1972). Dansylated phosphatidyl ethanolamine (DPE) and octadecyl naphthalene sulfonate (ONS) are probes of the polar head-group region, while 10-anthroyl stearic acid (AS) is located in the hydrocarbon-core region. The aqueous interface is thought to be the location of 8-anilino-1-naphthalene sulfonate (ANS). The structures of the compounds are given in Figure 1. DPE, ONS and AS were gifts of Dr. L. Stryer (Waggoner and Stryer 1970).

RESULTS AND DISCUSSION

Synthetic membranes made from pure phospholipids were used to calibrate the fluorescence response of dyes observed in biological membranes. As seen in Figure 2, the fluorescence intensity of AS in the presence of egg lecithin micelles decreases with increasing temperature, and a plot of reciprocal polarization as a function of temperature divided by the viscosity of water is linear between 10 - 50°C. In contrast, the fluorescence intensity of AS incorporated into dipalmitoyl lecithin micelles is less sensitive to changes in temperature, and a discontinuity in the Perrin plot occurs at around 40°C, the melting point of dipalmitoyl lecithin (Figure 2, center).

The results from the polarization values can be interpreted to mean that below the melting point of dipalmitoyl lecithin, AS is rigidly held in the crystalline lattice, while above the melting point, the decrease in polarization values as a function of increasing temperature is due to increased mobility of the dye. Over the temperature range studied, egg lecithin which is composed of a mixture of saturated and unsaturated fatty acids,

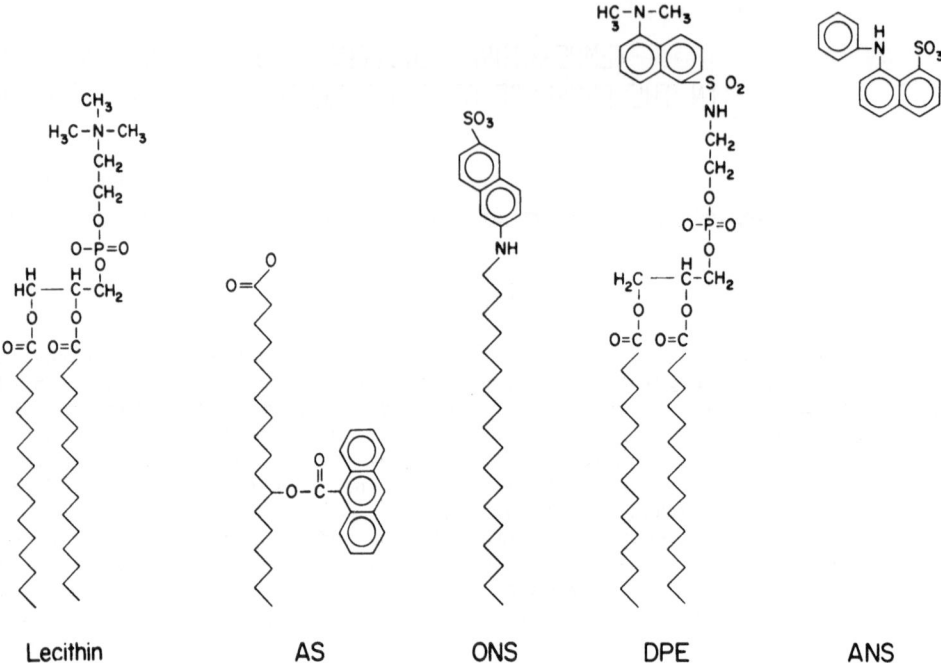

Figure 1. *Structure of fluorescent compounds.* AS: *10-anthroyl stearic acid;* ONS: *octadecyl naphthalene sulfonate:* DPE: *dansylated phosphatidyl ethanolamine;* ANS: *8-an-ilino-1-naphthalene sulfonate*

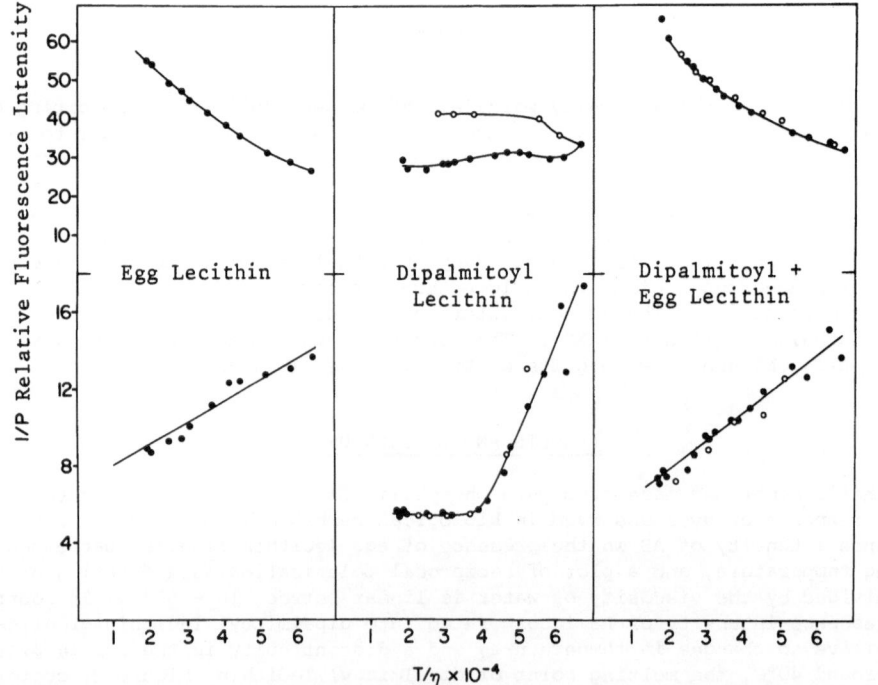

Figure 2. *Perrin plot of AS fluorescence polarization in the presence of phospholipids. All samples were sonicated in the presence of 10 µM AS and 0.1 mM morpholinopropane sulfonate, pH 7.2. Left: 1 mg egg lecithin/ml. Center: 1 mg L-α dipalmitoyl lecithin/ml. Right: 1 mg egg lecithin/ml and 1 mg dipalmitoyl lecithin/ml*

remains in a fluid state; consequently, a linear Perrin plot is obtained. Incorporation of egg lecithin into the dipalmitoyl lecithin micelles results in a disappearance
of the phase transition, and the fluorescence-intensity profile resembles that of pure
lecithin (Figure 2, right).

In addition to the marked sensitivity of fluorescence intensity and polarization
to temperature, the position of AS fluorescence excitation and emission maxima in the
presence of phospholipid micelles are functions of the hydrocarbon chain mobility.
At 24°C (below the melting point of dipalmitoyl lecithin) the fluorescence excitation
is at 384 nm, and the fluorescence emission is at 430 nm, compared with 382nm excitation maximum and 442 nm emission maximum observed in egg lecithin. At 48°C (above the
melting point of dipalmitoyl lecithin), the excitation and emission spectra of AS-di-
palmitoyl lecithin micelles resemble that of the AS-egg lecithin spectra at 24°C.

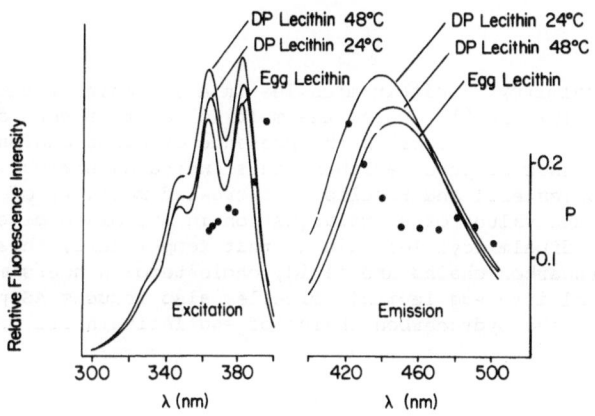

Figure 3. *Excitation and emission spectra of AS in the presence of lecithin
micelles. Lecithin (1 mg/ml) was sonicated in the presence of 10 μM AS. Excitation
was at 385 nm; emission, at 440 nm. Temperature of the egg lecithin was 24°C. The
points refer to polarization values of AS in the presence of dipalmitoyl lecithin
at 24°C*

It was of interest to determine whether the fluorescence parameters of DPE and
ONS, considered to be probes of the polar head-group region, and ANS, which is at the
aqueous interface, would also indicate the melting of the hydrocarbon chains of dipalmitoyl lecithin micelles. The Perrin plots of DPE or ONS incorporated into dipalmitoyl lecithin micelles are qualitatively similar to that observed using AS as a fluorescent probe, in that a distinct increase in probe mobility occurs at temperatures a-
bove the melting point of dipalmitoyl lecithin (Figure 4). Similar results were observed using ANS as the fluorescent probe (Vanderkooi and Martonosi 1969), demonstrating
that probes located at the surface of the micelle respond to phase transitions occurring in
artificial membranes in a similar manner as AS, a probe of the hydrocarbon region.

The interaction between the protein and lipid components of membranes is of great
biological significance and can easily be studied in model membranes using the fluorescent probes described here. Although changing the medium viscosity by addition of
sucrose or D$_2$0 had little or no effect on the observed melting of dipalmitoyl lecithin,
addition of various lipids profoundly changed the melting point. Incorporation of

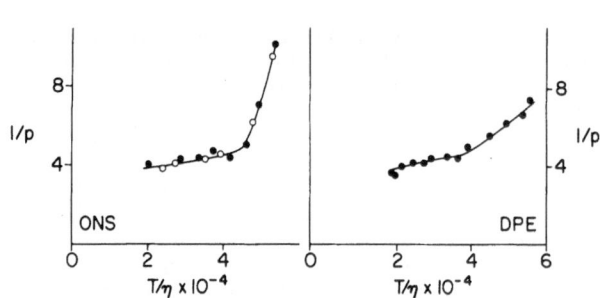

Figure 4. *Polarization of ONS and DPE incorporated into dipalmitoyl lecithin micelles. Dipalmitoyl lecithin (1 mg/ml) was sonicated with 10 μM ANS or DPE in the presence of 0.5 mM phosphate buffer, pH 7.2. Excitation for ONS (left) 350 nm; emission, 430 nm. Excitation for DPE (right) 340 nm; emission, 490 nm. T/η was varied by increasing (closed symbols) or decreasing (open symbols) the temperature*

cholesterol into the dipalmitoyl lecithin micelles in a 1:1 molar ratio completely abolished the melting point (Figure 5). Below the melting point of pure dipalmitoyl lecithin, the polarization of AS is lower in the presence of mixed cholesterol-lecithin micelles than in the presence of pure lecithin; this indicates scrambling of the rigid hydrocarbon chains by cholesterol and results in increased mobility of the probe. Above the melting point, the values of AS polarization in the mixed micelles are higher than those of AS in pure dipalmitoyl lecithin at that temperature; this finding suggests that when the hydrocarbon chains are fluid, cholesterol hinders movement. Incorporation of cholesterol into egg lecithin micelles also reduces AS polarization in a temperature range where the hydrocarbon chains of egg lecithin are melted.

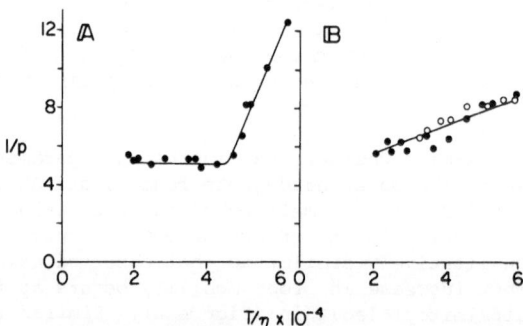

Figure 5. *Comparison of AS polarization in the presence of dipalmitoyl lecithin micelles and mixed cholesterol lecithin micelles. Samples were sonicated in the presence of 10 μM AS and 1 mM PO₄. Excitation was at 380 nm; emission at 450 nm. Left: 1mg dipalmitoyl lecithin/ml. Right: 1mg dipalmitoyl lecithin and 0.5mg cholesterol/ml*

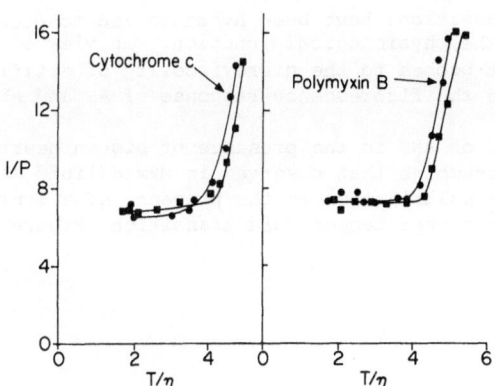

Figure 6. *Effect of protein and polypeptides on the polarization of AS in the presence of dipalmitoyl lecithin. The mixture contained 10 μM AS, 0.1 mM morpholino-propane sulfonate and 1 mg dipalmitoyl lecithin/ml. Left: 0.1 mg oxidized cytochrome* c/ml. *Right: 1.0 mg polymyxin B/ml. Excitation, 385 nm; emission, 440 nm*

Addition of proteins or polypeptides to dipalmitoyl lecithin results in a shift of the melting point to lower temperatures. In the presence of oxidized cytochrome c, the melting point shifts from 41° to 39°C, as indicated by AS polarization (Figure 6, left). A similar decrease in melting temperature is observed in the presence of Poly-myxin B sulfate, an antibiotic which acts as a membrane surfactant (Figure 6, right).

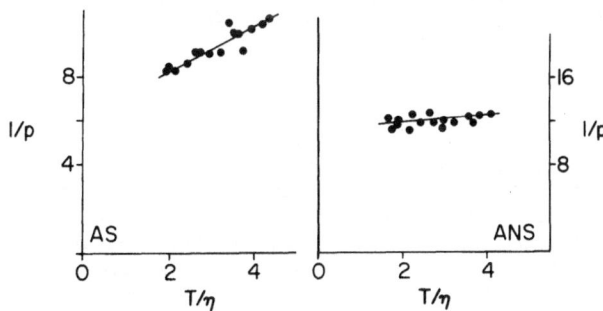

Figure 7. *Polarization of AS and ANS in the presence of pigeon heart mitochon-dria. Left: 10 μM AS was incubated in the presence of 0.225 M mannitol, 0.075 M su-crose, and 0.05 M morpholinopropane sulfonate, pH 7.2, for one hour. Excitation, 380 nm; emission, 450 nm. Right: mixture contained 30 μM ANS, 0.225 M mannitol, 0.075 M sucrose, 0.05 morpholinopropane sulfonate, and 0.2 mM KCN. Excitation, 360 nm; emission, 470 nm*

Temperature-phase transitions have been hypothesized to occur in biological systems as a requirement of the physiological function. In view of the well-established sensitivity of fluorescent probes to the microviscosity of artificial membranes, it was of interest to examine the fluorescence response of AS and ANS in the presence of natural membranes.

The Perrin plot of AS or ANS in the presence of pigeon heart mitochondria is linear from 6° to 40°C and resembles that observed in mixed lipid artificial membranes (Figure 7). Similarly, AS polarization in the presence of a lipid extract of mitochondria does not reveal a marked temperature transition (Figure 8).

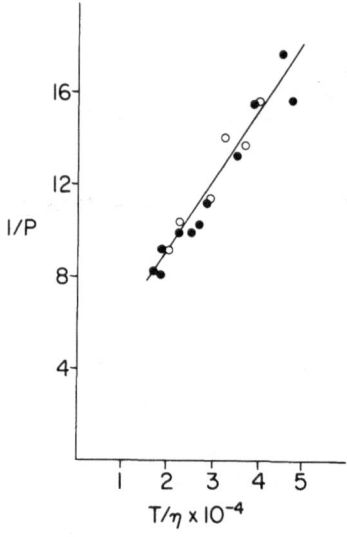

Figure 8. *Polarization of AS fluorescence in the presence of mitochondrial lipid. Mitchondrial lipid (4 mg/ml) was sonicated in the presence of 10 μM AS and 5 mM PO_4 buffer, pH 7.4. Excitation, 380 nm; emission, 450 nm*

Although there is no real evidence for a phase transition's occurring in mitochondria, as indicated by ANS or AS polarization profiles, the spinlabeled compound, 12-nitroxide stearate has been reported from a variety of sources to be sensitive to phase transitions in mitochondria (Raison et al. 1971). Reasons for the discrepancy between the results observed with AS or ANS and the spinlabeled probe include: the likelihood of different binding properties of the various molecules to the membrane; the possible perturbation of the system by any of the probes, resulting in alteration of membrane structure; or in the case of the fluorescent probes, the obscuring of small but significant deviations from linearity in the Perrin plots as a result of scatter of the experimental points. It is demonstrated that in micelles composed of mixed components, including lipids and proteins, the melting point of dipalmitoyl lecithin is significantly altered. In complex natural membranes such as mitochondria, the presence of a wide variety of phospholipids and proteins would be expected to shift the phase transition to temperatures considerably lower than those observed in the pure components, and perhaps lower than the temperature used in this study.

ACKNOWLEDGEMENTS

The author is grateful to Dr. B. Chance for help and support in all phases of this work. Supported in part by GM-12202-08.

REFERENCES

LESSLAUER, W., Cain, J., and Blaisie, J. K., (1972), *Proc. Natl. Acad. Sci., USA,* 69, 1499-503.

RAISON, J.K., Lyers, J.M., Mehlhorn, R.J., and Keith, A.D., (1971), *J. Biol. Chem.,* 246, 4036-40.

VANDERKOOI, J., and Martonosi, A., (1969), *Arch. Biochem. Biophys.,* 133, 153-63.

WAGGONER, A.S., and Stryer, L., (1970), *Proc. Natl. Acad. Sci. USA,* 67, 579-89.

CONFORMATIONAL DYNAMICS OF MODEL MEMBRANES

Juan Yguerabide*

Department of Molecular Biophysics and Biochemistry, Yale University

INTRODUCTION

Interest in the study of biological membranes has increased considerably in recent years as a result of the realization that these structures are complex systems involved in a great variety of functions. Electron microscopy has revealed that the membrane is not just a simple cell envelope; on the contrary, it is a highly convoluted structure which permeates much of the cell and encloses all cell organelles. Among its many functions are replication, energy transduction, nerve impulse propagation, sensory reception, and hormonal integration.

The main components of the membrane are proteins, lipids, and carbohydrates. The lipids form a bimolecular structure, except possibly where they meet proteins and carbohydrates. The polar heads of the lipids reside at the membrane surface, while their nonpolar hydrocarbon chains form the interior of the membrane. The exact conformation of the proteins and carbohydrates and their interactions with the lipid bilayer are not presently well understood.

The unraveling of the function, conformation, and dynamics of complex systems is greatly facilitated when individual components can be isolated and reassembled into simpler structures. The functions of the individual components and their combinations can then be more easily studied, while conformation and dynamics can be probed by techniques, such as fluorescence spectroscopy, which are often difficult to apply to systems with complex compositions.

Until recently, there were no known methods for disassembling a membrane and reassembling it into a simpler structure with specific functions. In 1962, however, Mueller, Rudin, and colleagues developed a technique which is beginning to make this reconstruction of the membrane possible (Mueller and Rudin 1969; Henn and Thompson 1969). In their technique, lipids isolated from natural sources are used to form a one mm-diameter bilayer membrane between two compartments filled with electrolyte (details are discussed in the section on formation of model membranes). The electrolyte and electrical potential in the two compartments can be adjusted to simulate physiological conditions and the electrical properties of the membrane can be measured by regular techniques of electrophysiology. The electrical functions of proteins and other membrane components can be tested by incorporating them into the bilayer.

Electron microscope, light reflectance, and electrical capacitance measurements have shown that the pure lipid synthetic membrane has a thickness and capacitance similar to that of natural membranes (Mueller and Rudin 1969; Henn and Thompson 1969). Its resistance (10^{10} Ω/cm^2), however, is much higher than that of the natural structures (10^3-10^4 Ω/cm^2), and it does not display any of the usual membrane functions. These functions must therefore be performed by other membrane components. Indeed, addition to the bilayer of certain cyclic polypeptides decreases the membrane resistance to 10^3 Ω/cm^2 and makes the membrane selectively permeable to cations. Action potentials can be induced by incorporating a protein called EIM that can be obtained from several sources, including nerve membranes (Mueller and Rudin 1969; Henn and Thompson 1969).

The Mueller-Rudin technique thus provides a pertinent method for reassembling membranes that are biologically significant. The use of these model membranes, however, has so far been limited in two important respects. First, the small amounts of material in the membrane are beyond the limits of detectability of most techniques for studying structure and dynamics, thus the studies undertaken have been mostly restricted to testing electrical functions. Second, it has been difficult to incorporate interesting proteins, such as rhodopsin, into the membrane. This problem is partly due to the limited knowledge presently available concerning the kinds of protein-lipid interactions that are involved in the incorporation of proteins in natural membranes. The most direct method for introducing a protein into the synthetic membrane is to

* *Present address:* Department of Biology, University of California at San Diego,
    La Jolla, California.

solubilize it in the membrane-forming solution prior to forming the membrane. This procedure, however, often denatures the protein or prevents the membrane from thinning into a bilayer structure. To avoid these problems, the membrane is usually formed first and a solubilized preparation of the protein is then added to the bathing electrolyte. However, when the protein added in this manner does not change the electrical properties of the membrane, it is difficult to decide whether the protein has not incorporated into the membrane or whether it has no electrical functions. Further progress thus requires the availability of techniques which can establish, independent of function, whether a protein added to the bathing electrolyte has entered the bilayer and, if so, whether it has the same orientation as in the natural membrane.

This article describes a simple method which we recently developed to apply fluorescence techniques to the Mueller-Rudin model membrane (Yguerabide and Stryer 1971). Fluorescence spectroscopy is promising in this application because of its very high sensitivity and its ability to give information on various aspects of conformation and dynamics. In particular, the incorporation of small amounts of fluorescent-labeled proteins in the membrane can be detected independent of function, while orientation, mobility, and proximity relations can be determined by the use of special protein and lipid fluorescent probes. The discussion presented here will be limited to studies with lipid probes, which can be used to demonstrate the main aspects of the technique. Studies with proteins are to be presented elsewhere (Yguerabide, in preparation).

FORMATION OF MODEL MEMBRANES, MICROSPECTROFLUOROMETER

Techniques of Membrane Formation

The synthetic membrane is usually formed across a one mm hole on the side of a teflon cup that is immersed in and filled with electrolyte. The formation of a membrane is initiated by depositing across the hole a small amount of membrane-forming solution consisting of 4% (w/v) lipid in a hydrocarbon solvent such as octane. Within a few minutes after depositing the solution, oil and excess lipid spontaneously migrate towards the walls of the hole and the membrane thins into a bilayer structure that remains connected to the cup through a torus of oil. For the fluorescence spectroscopic measurements, we added to the membrane-forming solution 0.04% (w/v) of a lipid-soluble fluorescent probe.

We experienced difficulties measuring fluorescence from the planar membranes described above because of light scattering from the cup and intense emission from the thick torus of oil. These difficulties, however, were eliminated by using a spherical form of the membrane (Pagano and Thompson 1967). These membranes can be readily formed from a solution of 4% oxidized cholesterol in octane. In our initial experiments, we used this lipid in preference to phospholipids because of the ease with which it forms stable membranes at room temperature (20 to 23° C). The technique which we used to generate the spherical membranes is as follows: a platinum wire with a 2-mm-diameter loop at one end is dipped into the membrane-forming solution. The loop, containing some membrane-forming solution, is then brought to the bottom of a $1 \times 1 \times 4.5$ cm quartz cuvette containing a KCl salt gradient ranging from 1.01 to 1.11 g/ml. The electrolyte is buffered at a pH between 6 and 6.2. The membrane-forming solution tends to emulsify at pH > 6.2. When the loop is abruptly moved upwards, several "bubles" — i.e., spherical membranes — are formed along the height of the cuvette. The bubbles contain electrolyte and have walls made of oil and lipid. Each spherical membrane remains indefinitely suspended at that height in the gradient corresponding to the concentration of KCl solution that is trapped inside the sphere. Within a few minutes after forming a bubble, oil and excess lipid spontaneously migrate to the top of the sphere, and the membrane eventually thins to a bilayer, except for the top which has a cap of oil. The thinned membranes usually last between 30 to 60 minutes on the average.

Figure 1 (Plates I - III) shows some spherical membranes in the process of thinning to a bilayer. The membrane spheres are 3 to 5 mm in diameter and can easily be seen with the eye. When first formed, the spheres show a bright, uniform fluorescence. As oil and excess lipid migrate, three sharply demarcated regions appear: a highly fluorescent cap, a strongly fluorescent middle region, and a weakly fluorescent lower region. The weakly fluorescent region spreads until it covers the whole sphere except for the intensely fluorescent cap.

PLATE   I

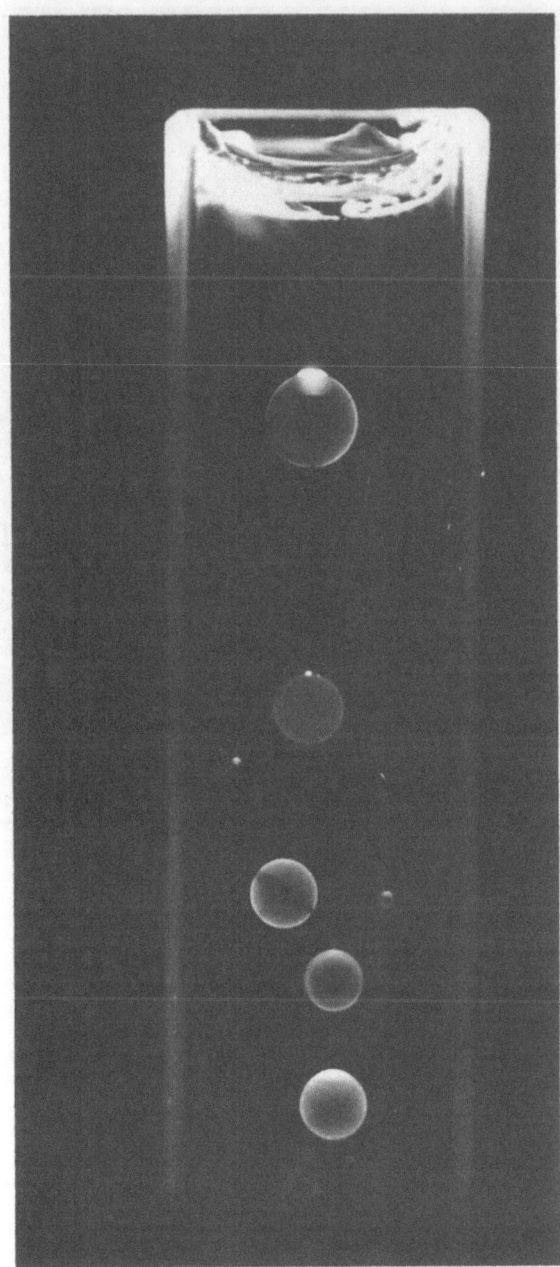

Figure 1a. *Formation of fluorescent spherical membranes in a salt gradient. The fluorescent probe is anthroyl stearic acid. The exciting wavelength was 365 nm, and the emission was viewed through a 3-71 Corning filter. The width of the cuvette is 1 cm.   Fluorescent spherical membranes are shown suspended in salt gradient along height of cuvette*

PLATE II

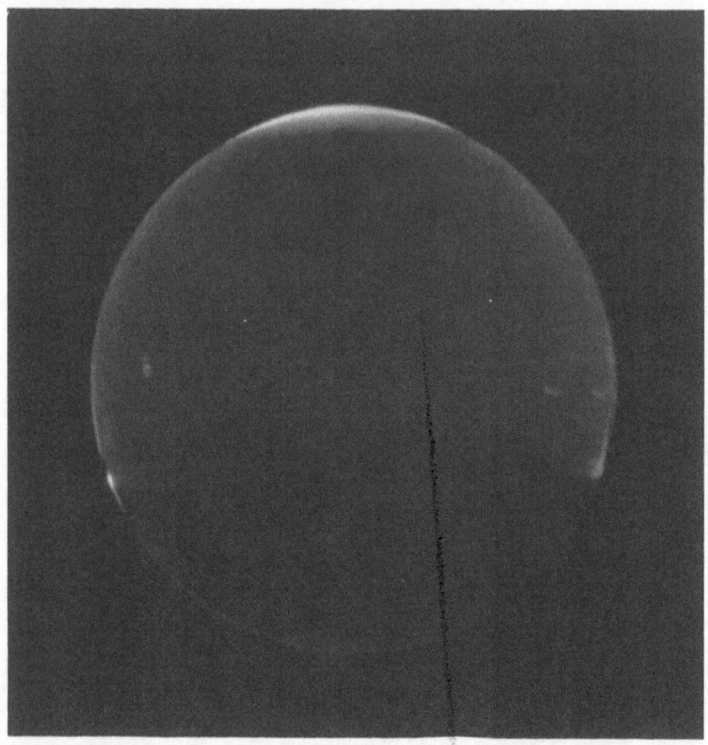

Figure 1b. *Fluorescent spherical membrane (3 mm diameter) in the process of thin-ning to a bilayer. The three regions are an intensely-fluorescent cap, a strongly-flu-orescent upper region, and a weakly-fluorescent lower region. Oil and excess lipid are migrating to the cap*

PLATE III

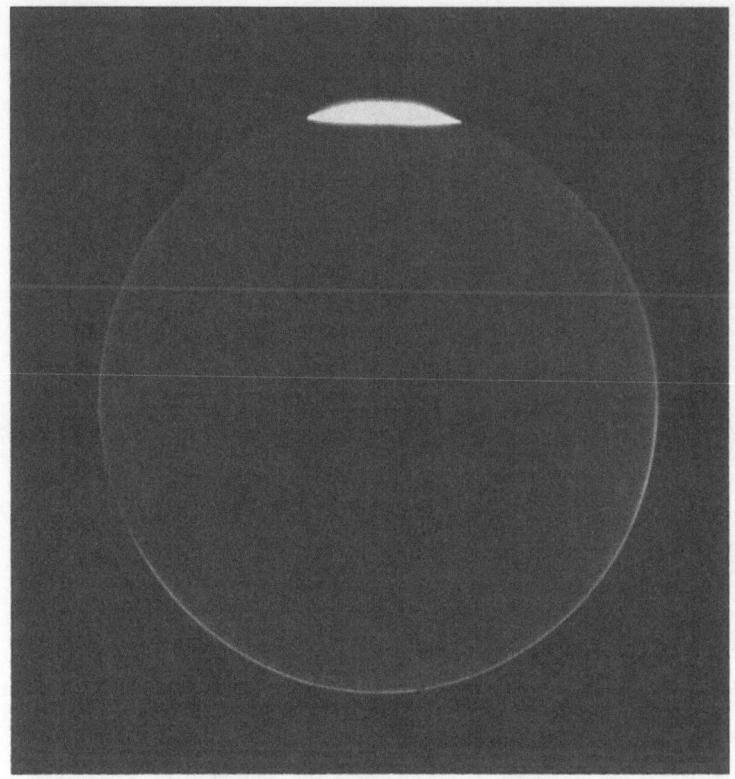

Figure lc. *Fluorescent spherical membrane that has thinned to a bilayer except for the intensely fluorescent cap (Figures 1a – c from Yguerabide and Stryer 1971)*

PLATE IV

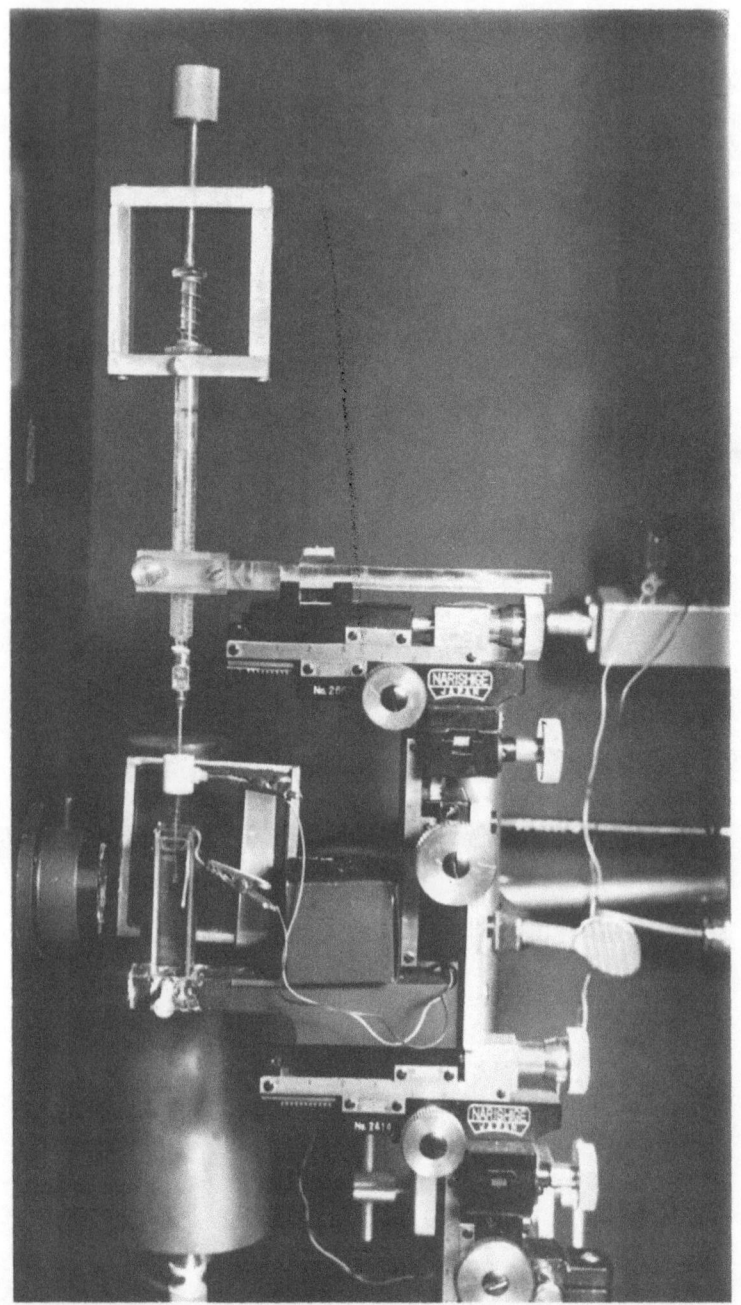

Figure 2. *Experimental arrangement for forming hanging spherical membranes. An electrical potential can be applied to the membrane through a silver chloride electrode on the side of the teflon plug between the needle and glass tube*

When a sphere is illuminated with a beam of white light and viewed against a black background it first shows the specular reflection characteristic of a thick membrane, but as the cap appears, interference colors are transiently observed on the surface of the sphere, this indicates that the membrane thickness is between 1,000 to 5,000 Å thick. The bottom of the sphere then abruptly turns black, concomitant with the appearance of weak fluorescence in this region, a change which denotes a loss of reflectivity characteristic of membranes less than 300 Å thick. When the completely thinned sphere is viewed under ordinary room light, only the caps can be seen floating in the salt gradient. The completely thinned membrane is almost invisible, since it is virtually transparent and reflects little light. Light reflectance and electrical capacitance measurements have shown that the completely thinned oxidized cholesterol membrane is between 40 to 50 Å thick, as expected for a bilayer composed of this lipid.

Spherical membranes can also be made from phospholipids, but these types of membranes are often difficult to form by the wire-loop technique described above. Such membranes, however, can be readily generated with a hypodermic syringe having a small polyethylene or glass tube (2 mm diameter) attached to the end of the syringe needle. To form a membrane, the salt gradient is made first and the syringe is then filled with electrolyte from the middle of the gradient. The tube at the tip of the syringe is next dipped into the membrane-forming solution and some solution is sucked into the tube. The tube is finally immersed into the gradient and a spherical membrane of the desired size is generated by depressing the plunger of the syringe. The bubble is disengaged from the tube by rapidly removing the syringe from the gradient. Phospholipid membranes usually thin more slowly than oxidized cholesterol membranes, and require temperatures above 28° C in order to thin down completely. The thinned phospholipid membranes, however, are very stable, sometimes lasting for six to eight hours.

The spherical membranes suspended in a salt gradient are easy to generate and are very convenient for studying the conformation and dynamics of the membrane lipids. They are not, however, suitable for protein studies where it is necessary to circulate protein solution around the membrane. They are also inconvenient to use in studies where an electric field is to be applied across the membrane during the fluorescence measurements. Both of these difficulties can be eliminated by forming the membrane as described above, but then allowing it to remain attached to the syringe. The electrolyte contained in these hanging membranes is electrically continuous with the electrolyte in the syringe.

Figure 2 (Plate IV) shows the experimental arrangement which we use to form hanging spherical membranes. A salt gradient is not necessary in this case. The syringe is mounted on a micromanipulator which allows the hanging membrane to be moved within the cuvette. For electrical measurements, a hollow teflon cylinder is connected between the needle and the glass or polyethylene tube. A plug transversed by a chloridized silver electrode is inserted into a side hole on the teflon cylinder. Electrical contact with the electrolyte inside of the sphere is made through the chloridized electrode. Solutions of different composition can be flushed around the membrane through a connection at the bottom of the cuvette.

Microspectrofluorometer

Figure 3 (Plate V) shows the optical arrangement used for fluorescence spectroscopic measurements on the spherical membranes suspended in a salt gradient. Collimated light from a high-pressure mercury lamp passes through an interference filter, an infrared filter and a polarizer and is then focused onto a small region of a single spherical membrane. The beam is about 0.1 mm in diameter at its focal point. The quartz cuvette containing the membranes is mounted on a micromanipulator which allows any sphere in the cuvette to be positioned in front of the light beam. The spherical membranes are viewed at right angles to the exciting light beam through a stereoscopic microscope (not shown in Figure 3) whose magnification can be varied from 1 to 100. The intensity of light emitted from any small region of the spherical membrane is detected by a photomultiplier tube that is connected to a side tube on the microscope. A Corning glass filter is placed between the microscope and cuvette to eliminate the scattered exciting light. For fluorescence polarization measurements, a polarizer is also introduced between the cuvette and the microscope.

For measurement of fluorescence spectra, a small Bausch and Lomb monochromator is placed between the photomultiplier and side arm of the microscope. A potentiometer

PLATE V

Figure 3. *Optical arrangement for fluorescence spectroscopic measurements on spherical membranes suspended in salt gradient*

connected to the monochromator grating produces a voltage signal whose magnitude varies linearly with wavelength. This signal goes to the X input of an XY recorder while the photomultiplier tube signal goes to the Y input. A motor rotates the grating during the recording of a fluorescence spectrum. For recording excitation spectra, the mono-chromator is used in conjunction with a xenon lamp to excite the membrane at different wavelengths.

RESULTS

Fluorescence Spectra

The spherical membranes are practically nonfluorescent unless a fluorescent probe is incorporated into them. The fluorescent probe may be either soluble in lipid but highly insoluble in water, or partially soluble in both lipid and water. Chromophores which are partially soluble in water are useful only if the fluorescence efficiency is much lower in water than in lipid. Otherwise, the membrane fluorescence is obscured by emission from the electrolyte. Water-soluble probes such as ANS (1-anilino-8-naph-thalene sulfonate) can be used and are best incorporated into the membrane by dissolv-ing them in the electrolyte. Lipid-soluble probes are incorporated into the bilayer by adding them to the membrane-forming solution.

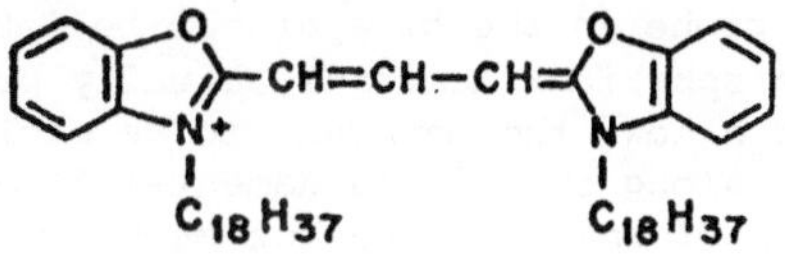

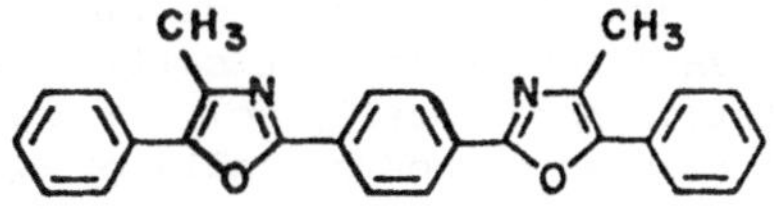

Figure 4. *Lipid-soluble fluorescent probes used in the study of spherical mem-branes. N,N'-di(octadecyl)-oxacarbocyanine (I); 12-(9-anthroyl)-stearic acid (II); p-bis-[2-(4-methyl-5-phenyloxazolyl)]-benzene (III, dimethyl-POPOP)*

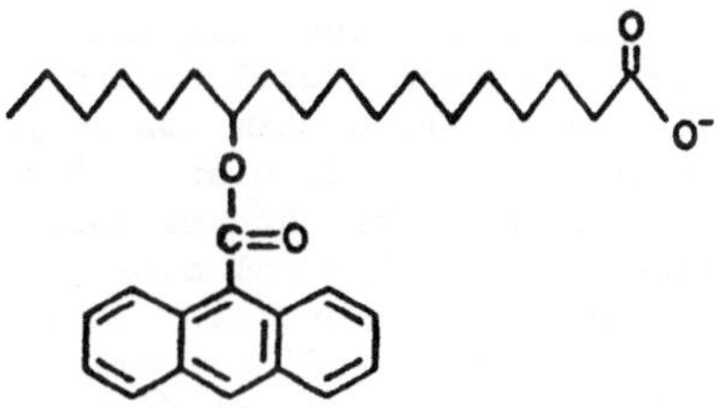

Figure 4 shows examples of the kinds of probes which we have used in the study of model membranes. These probes are insoluble in water. The chromophore of probe I is expected to reside at the membrane surface, whereas the chromophore of probe II is expected to be in the membrane interior. Probe III, commonly known as dimethyl-POPOP, is an example of an ordinary fluorescent chromophore that, because of its insolubility in water, can be used to study the membrane. Figure 5 shows the fluorescence excita-tion spectrum of probe II in both the cap and the bilayer regions of the membrane. The shapes of these spectra agree with the absorption spectrum of the anthroyl chromo-phore. The excitation spectrum from the bilayer, however, is shifted towards longer wavelengths from the spectrum obtained at the cap. Since the cap region is highly non-polar, consisting of lipid and oil, this shift indicates that the inner part of the membrane has an intermediate polarity. The fluorescence spectrum measured from the bilayer is also shifted toward longer wavelengths with respect to the emission spec-trum measured from the cap region.

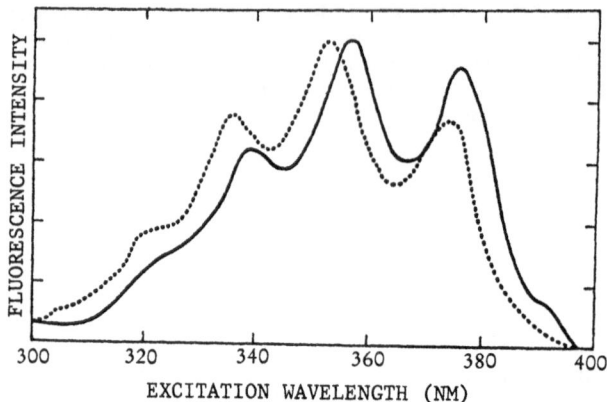

Figure 5. *Excitation spectrum of the bilayer (——) and cap (---) regions of the spherical membrane containing probe II. The emission was detected through a 3-71 Corning filter (Yguerabide and Stryer 1971)*

Molecular Orientation in the Bilayer

The orientation of a probe in the bilayer can be determined by fluorescence polarization measurements. The spherical membrane actually provides an ideal system for such measurements since it allows the membrane to be excited with polarized light and viewed through a polarizer along certain fundamental directions in the bilayer. Figure 6 shows the orientation of small patches of membrane at the side, bottom, and front of the sphere as viewed by the light-detecting system. It should be recalled that in our experimental arrangement, the spherical membrane is viewed along a direction perpendicular to the exciting light beam. Thus when the sphere is positioned so that the exciting beam strikes the side, the beam travels along c and the emitted light is detected along b. Furthermore, the exciting beam can be polarized either in a direction, a (vertical direction, V), which is perpendicular to b and c, or in a direction parallel to b (horizontal direction, H). For each of these directions of polarization, the emitted light can be detected through a polarizer oriented along the vertical direction, a, or along the horizontal direction, c. (It should be noted that in our usage, vertical V refers to a direction extending from the bottom of the sphere to cap, whereas horizontal H refers to a direction perpendicular to V.) From the side of the sphere, one can thus measure four fundamental polarized fluorescence intensities corresponding to the different directions of polarization of the exciting light and the detecting polarizers. These intensities may be designated as: VV (excitation and emission vertically polarized), VH (excitation vertically polarized and emission horizontally polarized), HV (excitation horizontally polarized and emission vertically polarized), and HH (excitation and emission horizontally polarized). Similarly, four basic

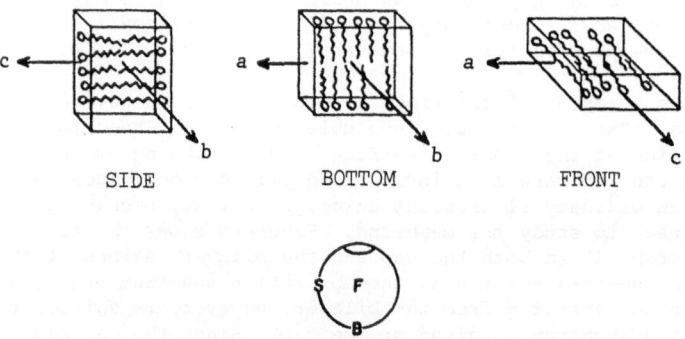

Figure 6. *Orientation of the bilayer membrane in the side (S), front (F), and bottom (B) regions of the spere are viewed by the detection system (Yguerabide and Stryer 1971)*

polarized fluorescence intensities can be measured at the bottom region of the sphere (in which case the exciting light travels along a and the emitted light is detected along b), and four from the front region. A total of twelve fundamental polarized fluorescence intensities can thus be measured. However, some of these measurements are redundant. For a bilayer membrane that is completely symmetric in the __ab__ plane, theory shows that there are only four different basic polarized fluorescence intensities, which we have termed α, β, γ and δ, and of these intensities only three are independent. Table I shows the relation between the fluorescence intensities from the different regions of the spherical membrane and the basic intensities, α, β, γ, and δ.

TABLE I

REGION OF SPHERE

Excitation and Emission Component	Side	Front	Bottom
VV	α	α	δ
VH	γ	β	γ
HV	β	γ	γ
HH	γ	γ	β

The values of the basic intensities depend on the orientation of the absorption and emission transition moments of the chromophores in the bilayer and on their rotational motions during the lifetime of the excited state. It is because of this dependence that one can use fluorescence polarization measurements to measure orientation and mobility in the bilayer. The expressions relating the anisotropic fluorescence parameters to orientation have a simple form when the absorption and emission transition moments are along the same direction in the chromophore and there is no rotational motion during the excited-state lifetime. Thus, if the distribution of the transition moments in the bilayer is $\rho(\phi)$, where ϕ is the angle between the transition moment and the c axis, then in this case, the basic intensities are given by the following expressions:

$$\alpha_0 = \frac{6}{4} \int_0^{\pi/2} \rho(\phi) \, sin^5\phi \, d\phi \tag{1}$$

$$\beta_0 = \frac{1}{2} \int_0^{\pi/2} \rho(\phi) \, sin^5\phi \, d\phi \tag{2}$$

$$\gamma_0 = 2 \int_0^{\pi/2} \rho(\phi) \, cos^2\phi \, sin^3\phi \, d\phi \tag{3}$$

$$\delta_0 = 4 \int_0^{\pi/2} \rho(\phi) \, cos^4\phi \, sin\phi \, d\phi \tag{4}$$

The subscript $_0$ emphasizes that these equations assume that rotational motions are absent during the lifetime of the excited state.

Using equations (1) through (4), we have calculated values of the polarized fluorescence intensities expected from different regions of the membrane for various distributions $\rho(\phi)$. Plots of these intensities are shown in Figure 7. These plots show that the ratio of α_0 to δ_0 is the most sensitive indicator of orientation. For preferential alignment in the ab plane, $\alpha_0 > \delta_0$. For preferential alignment about the c axis, $\delta_0 > \alpha_0$. If there is no preferential alignment, that is, if the distribution of orientations is random, then $\alpha_0 = \delta_0 = 3\beta_0 = 3\gamma_0$. It should be noted that the relation $\alpha_0 = 3\beta_0$ applies even if the distribution is not random. Rotational motion, however, will tend to equalize the values of α and β. Thus the deviation of the ratio α/β from the value of 3 expected for a rigid system can be used as a measure of rotational motion.

Figure 8 shows plots of the polarized fluorescence intensities for bilayer membranes containing the probes I, II and III. The chromophores were excited in their

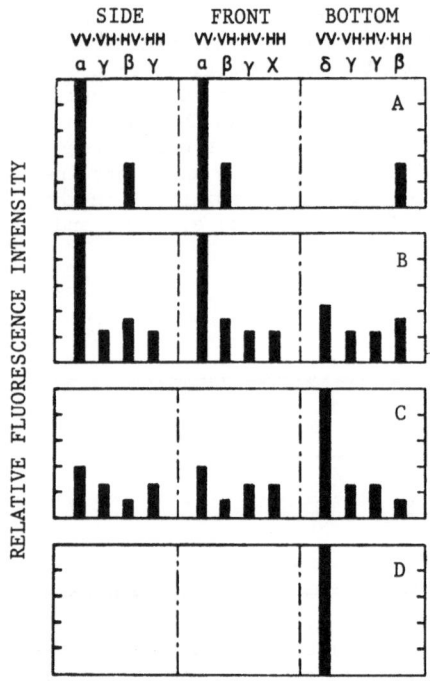

RELATIVE FLUORESCENCE INTENSITY

Figure 7. *Calculated polarized fluores-cence intensities for the side, front, and bottom regions of a spherical bilayer membrane in which (a) the transition moments are per-fectly parallel to the ab plane of the bilayer; (b) the transition moments are partially align-ed parallel to the ab plane of the bilayer; (c) the transition moments are partially aligned parallel to the c axis of the bilayer; and (d) the transition moments are perfectly aligned parallel to the c axis of the bilayer. The in-tensities (b) and (c) were calculated for a Gaussian distribution ρ (φ), in which half of the transition moments are within 43° of the ab plane and c axis, respectively (Yguerabide and Stryer 1971)*

longest-wavelength absorption bands, in which cases the absorption and transition mom-ents are in the same direction. These plots immediately show that the chromophores are oriented, since in all cases α differs from β. The plot for probe I indicates that the probe is highly oriented (α > β), with its transition moment in the ab plane of the bilayer. This orientation is the one expected from the molecular structure of I and from the results of Kuhn and his colleagues on dry, oriented multilayers of fatty acids. The experimental ratio of α/β for this probe is 2.3, which indicates that some rotational motion occurs during the excited-state lifetime.

The plot for probe II indicates that this probe is also preferentially aligned parallel to the ab plane of the bilayer, since α > δ. The emission and longest-wave-length absorption transition moments of anthracene are known to be in the plane of aro-matic ring along its short axis. The results then indicate that this axis is prefer-entially aligned in the ab plane of the membrane. However, the degree of orientation is not so high as for I, as can be seen by comparing the polarized intensity plot of

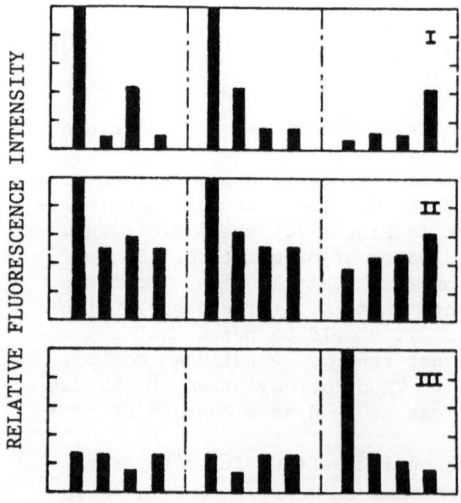

RELATIVE FLUORESCENCE INTENSITY

Figure 8. *Observed polarized fluores-cence intensities for the side, front, and bottom regions of a spherical membrane con-taining the probes I (top), II (middle) and III (bottom). Each chromophore was excited at its longest-wavelength absorption band (Yguerabide and Stryer 1971)*

II with plots similar to those shown in Figure 7. There is then more disorder in the interior of the membrane.

In contrast to probes I and II, the results of Figure 8 indicate that for probe III the transition moment is preferentially aligned parallel to c, since $\delta > \alpha$. A comparison of the observed polarization plot of III (Figure 8) with plots calculated for different distributions (Figure 7c,d) indicates that there is a fairly broad range of orientations for this probe. The emission transition moment of III probably lies along the long axis of the chromophore. The results then indicate that III tends to be oriented in the bilayer with its axis parallel to the c axis of the membrane.

ACKNOWLEDGEMENTS

This work was supported by grants from the National Institute of Health (GM - 16708) and the National Science Foundation (GB - 27408).

REFERENCES

HENN, F.A., and Thompson, T.E., (1969), *Ann. Rev. Biochem.*, **38**, 241.

MUELLER, P., and Rudin, D.O., (1969), *Curr. Top. Bioenerg.*, **3**, 157.

PAGANO, R., and Thompson, T.E., (1967), *Biochim. Biophys. Acta*, **144**, 666.

YGUERABIDE, J., in preparation.

YGUERABIDE, J., and Stryer, L., (1971), *Proc. Nat. Acad. Sci., USA*, **68**, 1217.

ATEBRIN AND RELATED FLUOROCHROMES AS QUANTITATIVE PROBES OF MEMBRANE ENERGIZATION

R. Kraayenhof

Johnson Research Foundation, University of Pennsylvania,
Department of Biophysics and Physical Biochemistry,
*School of Medicine, Philadelphia, Pennsylvania**

INTRODUCTION

The complex structural and functional organization of living organisms depends on the constant availability of chemical energy, mainly in the form of ATP. In both photosynthetic and respiratory organelles, the synthesis of ATP is coupled to electron transport through chains of oxidation-reduction carriers. The mechanism by which the energy set free in the redox reactions is converted into ATP is of fundamental importance and has been the subject of extensive investigations in many laboratories over the past twenty years. These studies have produced many constructive and interesting suggestions, but they do not provide a definite answer to the question: what is the nature of the primary energy-conserving act? An extensive discussion of the several proposed mechanisms of energy conservation can be found in the literature (Greville 1969; Van Dam and Meyer 1971; Slater 1971) and will not be dealt with here.

The primary conservation of energy in both photosynthetic and respiratory organelles probably occurs by fundamentally the same apparatus, which is localized in the osmotically active membranes. The integrity of the membranes is a strict requirement for the proper functioning of this apparatus, a condition which makes it extremely difficult to resolve its detailed composition. It is mainly for this reason that the introduction of fluorescent (Radda 1971) and paramagnetic (McConnell 1970) probes for the study of biological membranes provides a promising new approach to the problem. If the responses of such probes to well-defined chemical and physical changes are known, similar responses in an unknown environment could tell us more about the properties of that environment. Although the extrapolation of knowledge obtained in simple and pure model systems to a complicated biological membrane seems hardly justified, one can derive useful information if one compares the responses of a series of different probes under identical conditions. Additional information can be expected if a probe is actively participating in the mechanism studied; for instance, as a substrate (Velick 1958), an inhibitor (Chance et al. 1969; Berden and Slater 1970), or an uncoupler

The idea that uncoupler molecules may undergo physicochemical changes during their energy-dissipating interaction led to the investigation of the behavior of fluorescent uncouplers (Kraayenhof 1970). Figure 1 depicts a simplified scheme of the pathways along which energy can be generated or utilized in photosynthetic and respiratory organelles. The term, "energized state" (often indicated by a squiggle: $\sim$), illustrates our present lack of knowledge of its nature. It can stand for a high-energy compound or protein conformation (Slater 1971) or for an electrochemical membrane potential (Mitchell 1966). Addition of an uncoupler results in inhibition of energy-utilizing reactions and in stimulation of energy-generating reactions. If the fluorescence of the uncoupler changes as a result of its perturbing action, one should expect concomitant changes of these energy-linked functions.

Abbreviations to be used in this paper include:
    DCMU: 3-(3,4-dichlorophenyl)-1,1-dimethylurea,
    MGA: 9-methylglycine-6-chloro-2-methoxyacridine,
    DMPA: 9-(3-dimethylaminopropylamino)-6-chloro-2-methoxyacridine,
    MBPA: methylamino-*bis*-(9-propylamino-6-chloro-2-methoxyacridine),
    PPA: 9-(propylaminopropylamino)-6-chloro-2-methoxyacridine,
    PEA: 9-(propylaminoethylamino)-6-chloro-2-methoxyacridine,
    HPA: 9-(hexadecylaminopropylamino)-6-chloro-2-methoxyacridine.

* *Present address: Laboratory of Biochemistry, University of Amsterdam, Plantage Muidergracht 12, Amsterdam, The Netherlands.*

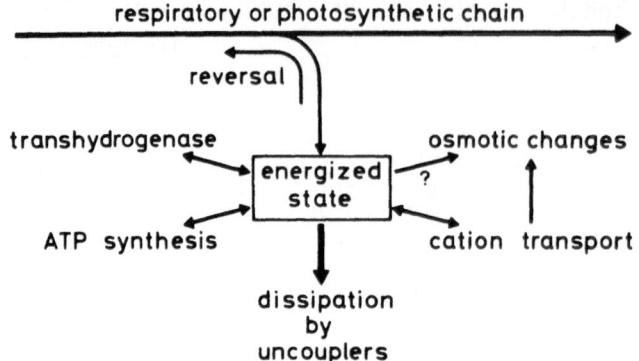

Figure 1. *Pathways of energy generation and utilization in respiratory and photosynthetic organelles. (The energy-linked transhydrogenase and reversal of electron transport apply only to the respiratory organelles)*

In this contribution the results obtained with the antimalarial drug and uncoupler, atebrin (quinacrine; 9-(4-diethylamino-1-methylbutylamino)-6-chloro-2-methoxyacridine), in isolated chloroplasts and other biological membranes will be summarized in order to demonstrate the validity of this compound as a quantitative probe of membrane energization. Some comparative experiments with different substituted acridine dyes serve to illustrate the mechanistic aspects of the energy-linked probe responses.

METHODS AND MATERIALS

Spectrofluorometric measurements were performed with a Perkin Elmer MPF-2A Fluorescence Spectrophotometer. The cuvette compartment was modified to accommodate a thermostated cylindrical glass cuvette fitted with a single flat window on which the excitation light was collimated (angle 60°). Emission was detected at the same window (angle 30°). The cuvette could be illuminated by a cooled quartz-iodine lamp, the light of which was filtered by Corning CS 1-69 (heat) and CS 2-64 (red) filters and passed through a mechanical shutter. The contents of the cuvette were stirred by an overhead Teflon stirrer; additions could be made during the measurements with microliter syringes.

The binding of fluorescent dyes to the organelles was determined (by fluorescence) after rapid centrifugation (Eppendorf 3200 microcentrifuge) or by filtration of the organelles through a glass fiber filter (Whatman GF/B, mounted in a Millipore filter holder) (Brocklehurst 1970). In some cases the kinetics of binding were followed with an apparatus (based on this filtration method) described elsewhere (Kraayenhof 1971).

Intact (Kraayenhof 1969), broken (Izawa and Good 1968) or cyanide-treated chloroplasts (Ouitrakul and Izawa) were prepared from spinach, as previously described. Unless stated otherwise the experiments reported here were performed with broken (Class-II) chloroplasts.

Simultaneous measurements of fluorescence, oxygen, and hydrogen-ion concentrations were performed with an instrument described (Kraayenhof 1969).

The synthesis of substituted acridines is the subject of a separate publication (Kraayenhof et al. 1973).

Commercial sources were for Atebrin Nutr. Biochem. Corp., and for quinacrine mustard, Polysciences, Inc.

RESULTS AND DISCUSSION

Response of Atebrin to Energization of Chloroplasts

When atebrin is added to a suspension of chloroplasts in the dark, there is an initial lowering of the fluorescence (Figure 2A) due to a binding of part of the dye to the chloroplasts. When the chloroplasts are then energized by illumination, the remaining atebrin fluorescence disappears. Figure 2A shows this effect when light-induced electron transport occurs from water to diquat. (For general information on photosynthetic electron transport and associated reactions, see Rabinowitch 1969.) This fluorescence lowering does not occur when electron transport is blocked by DCMU, but is restored after addition of ascorbate *plus* diaminodurene, (these donate electrons to the photosynthetic chain after the DCMU-block). Similar results were obtained with other electron transport systems (Kraayenhof et al. 1972). Under conditions of ATP hydrolysis (Figure 2B), the fluorescence lowering is maintained in the

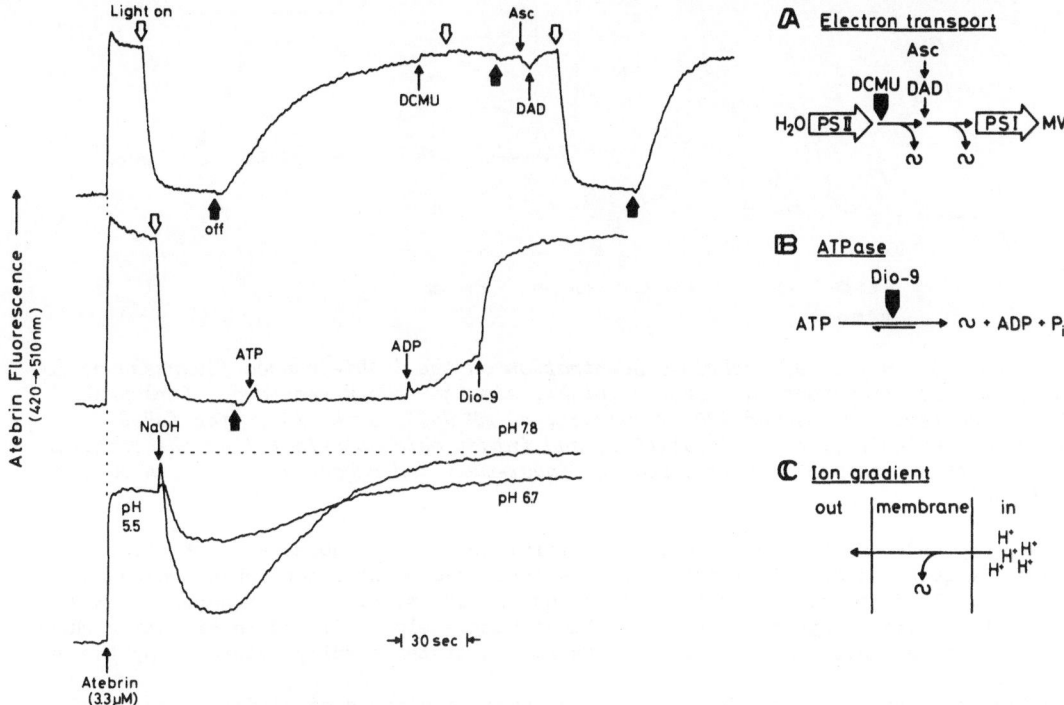

Figure 2. *Atebrin fluorescence lowering in chloroplasts induced by electron transport, ATP hydrolysis, or an ion gradient. The standard reaction medium (1.5 ml, temperature 22°C) contained 250 mM sucrose, 20 mM NaCl 20 mM Tricine buffer (pH 8.0), and 3 mM MgCl₂. The concentrations of additional components were: (A) methylviologen (MV) 47 μM, DCMU 3.3 μM, ascorbate (Asc) 3.3 mM, diaminodurene (DAD) 0.33 mM, and chloroplasts (47 μg chlorophyll/ml); (B) pyocyanine 5 μM, dithioerythritol 3 mM, ATP 3 mM, ADP 1 mM, Dio-9 5 μg/ml, and 72 μg chlorophyll/ml; (C) succinic acid (about 10 mM) to give pH 5.5, NaOH 4 or 10 mM and 61 μg chlorophyll/ml*

dark as long as the ATP/ADP ratio is high enough. A restoration of the atebrin fluorescence is brought about by addition of ADP and, more rapidly, by the inhibitor Dio-9. Chloroplasts are capable of active energy-dependent proton uptake (Neumann and Jagendorf 1964). On the other hand, if one allows internally accumulated protons (by incubation with succinic acid) to rapidly equilibrate with the outside of the organelle, one evokes an energized state that can effectively be used for ATP synthesis (Jagendorf and Uribe 1966). Under conditions of this so-called acid-base jump (Figure 2C), one indeed observes a transient fluorescence lowering, the magnitude of which is dependent on the induced pH gradient.

Thus, three different ways of energizing the chloroplast membrane lead to a lowering of atebrin fluorescence.

Stoichiometry Between Atebrin Response and Energy Level in Chloroplasts

The experiment illustrated in Figure 3 is a simultaneous display of the atebrin fluorescence, electron transport, and proton uptake as a function of atebrin concentration. Illumination initiates electron transport and uptake of protons. Upon subsequent additions of atebrin, the rate of electron transport increases and the extent

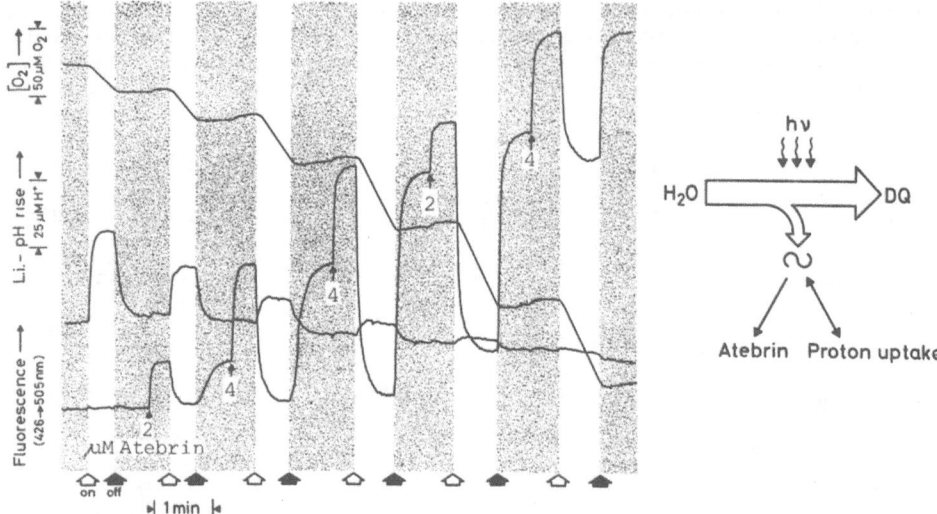

Figure 3. *Effect of atebrin concentration on the light-induced fluorescence low-ering, electron transport, and proton uptake, simultaneously recorded. (Kraayenhof 1970). The medium contained 160 mM sucrose, 40 mM NaCl, 4 mM TES buffer (pH 6.8), 3 mM dithioerythritol, 5 μM diquat (DQ), and intact chloroplasts (71 μg chlorophyll/ ml). Electron transport was measured as diquat-mediated oxygen uptake (with oxygen electrode)*

of the energy-dependent proton uptake is diminished. The fluorescence of atebrin is completely quenched by illumination up to a certain concentration, above which the re-sponse sharply declines. At this atebrin concentration, electron transport is max-imally stimulated and proton uptake is almost completely abolished, a situation which suggests that at this concentration the energized state is fully saturated by the un-coupler.

Similar titrations have been performed under conditions of different *rates* of energy generation. In Figure 4 the rate of electron transport was varied by changing the intensity of the light. The energy-dependent fluorescence lowering is plotted as percentage of the fluorescence signal in the dark at each probe concentration. The inflection point in the titration curve is shifted towards lower concentration when electron transport is slowed down, and this occurs in a proportional way. A similar stoichiometric relation was found when ATP served as the energy donor and the rate of the ATPase was varied by different Dio-9 concentrations (Kraayenhof et al. 1972).

Figure 5 shows a set of titrations performed at identical light intensities, but with different amounts of an additional uncoupler (S-13). S-13 competes with atebrin for interaction with the energized state, but as long as the combination of the two uncouplers does not fully saturate the energized state, there is no effect on the atebrin fluorescence lowering.

If, at the same rate of electron transport, the number of energy-conserving sites is varied, we should expect a proportional change in the atebrin titration. Recent evidence suggests the existence of two energy-conserving sites (instead of one) cou-pled to linear photosynthetic electron transport (Saha et al. 1971; Neumann et al. 1971). Atebrin titrations have been carried out in three different electron transport systems. Figure 6A shows the case of the linear electron transport from water to fer-ricyanide (an identical curve was obtained with diquat or methylviologen). In cyanide-treated chloroplasts (Figure 6B), electron transport is blocked at plastocyanin (Ouitrakul and Izawa in prep.; Izawa et al. 1973), which enables the study of the "first half" of the photosynthetic chain (in the presence of diaminodurene, oxidized by ferricyanide [Saha et al. 1971], as electron acceptor). Similarly, the "second half" of the chain (Figure 6C) can be studied in the presence of DCMU, ascorbate-reduced dia-minodurene as electron donor and diquat as the electron acceptor. The obtained sat-urating atebrin concentrations in the three systems were then related on the basis of equal rates of electron transport to give the "atebrin/e2 ratio". Surprisingly,

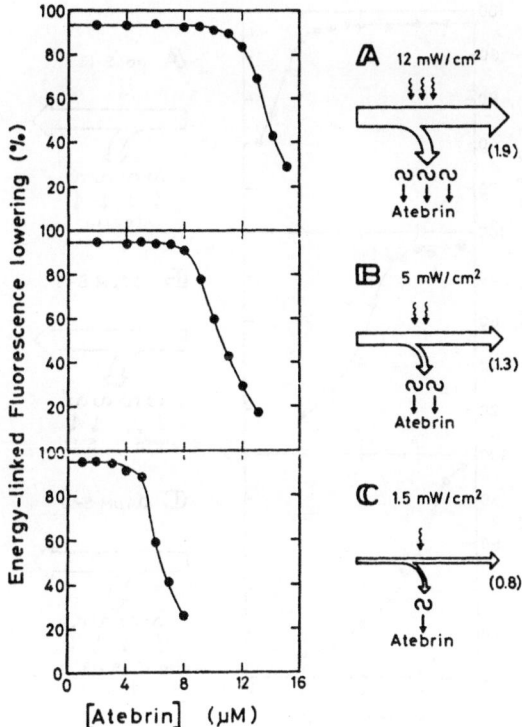

Figure 4. *The atebrin fluorescence lowering as a function of atebrin concentration at different rates of electron transport. The standard medium (Figure 2) contained in addition 5 μM diquat and 48 μg chlorophyll/ml. The different light intensities were as indicated; the rates of electron transport (at saturating atebrin concentrations), indicated between brackets, are expressed as μmoles/min. mg chlorophyll*

the sum of these ratios obtained in the two partial electron-transport systems comes very close to the ratio obtained in the complete system. This lends support to the view that at least two energy-conserving sites are coupled to the linear electron-transport chain (Izawa and Good 1968; Saha et al. 1971; Neumann et al. 1971), and that the different steady-state levels of energy produced by the individual sites or by both together, is quantitatively detected by the fluorescent probe.

Recently, we found that variation of the temperature had a dissimilar effect on different energy-linked processes in chloroplasts (Kraayenhof et al. 1971). Proton uptake (as measured by the H^+ electrode) and osmotic shrinkage (as measured by light-scattering) are almost absent below 10°C, whereas the energy-linked lowering of atebrin fluorescence reaches its full extent at 1°C, with almost the same speed. At lower temperatures, the dark restoration of the probe fluorescence was considerably prolonged. Under similar conditions, ATP synthesis showed a dark overshoot, indicating that at lower temperatures chloroplasts contain a much higher steady-state level of energy, presumably because of the much lower activity of energy-utilizing reactions such as ion transport. This was further verified with atebrin binding titrations. Saturating atebrin concentrations at 26°, 15°, and 4.5°C were: 130, 268, and 495 nmoles/mg chlorophyll, respectively. Another interesting observation was that the dark decay of the energized state as read out by the atebrin fluorescence showed a sharp temperature transition at 19.5°C, and was more temperature sensitive above 19.5° C (Kraayenhof et al. 1971).

Binding of Atebrin to Energized, Nonenergized, and Heat-denatured Chloroplasts

In all cases studied so far, it was found that the observed fluorescence lowering goes hand in hand with a proportional binding of the atebrin to the organelles in

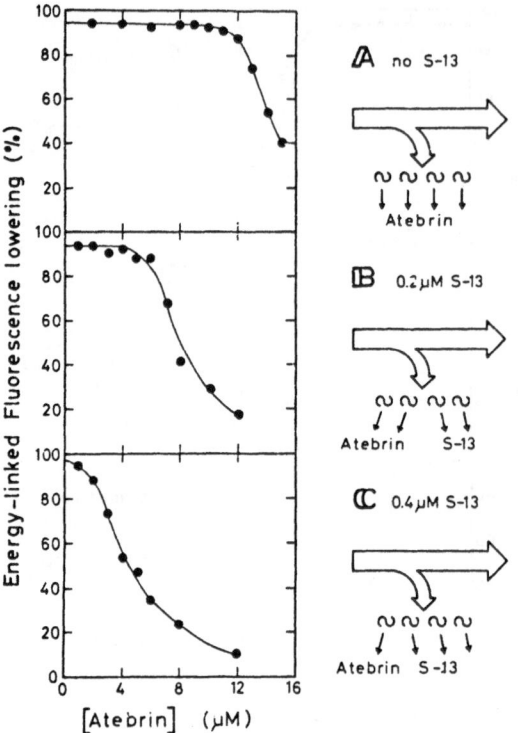

Figure 5. *The atebrin fluorescence lowering as a function of atebrin concentration at different degrees of uncoupling. Conditions as in Figure 4. S-13 concentrations as indicated*

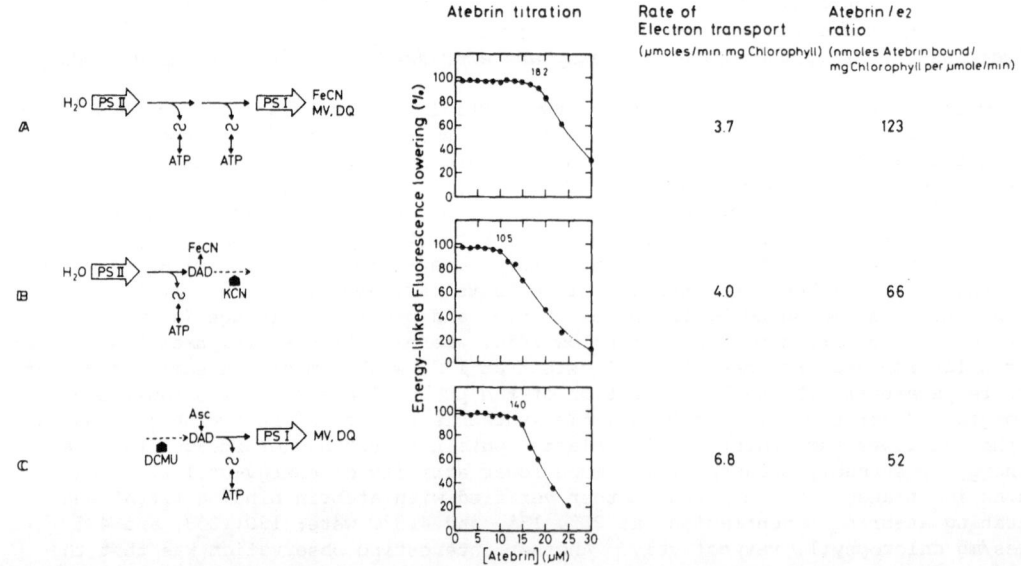

Figure 6. *Atebrin titration in different electron transport systems. The standard reaction medium (Figure 2) contained 40 µg chlorophyll/ml and: (A) 1 mM ferricyanide (FeCN); (B) 0.33 mM diaminodurene (DAD), 1mM ferricyanide and cyanide-treated chloroplasts; (C) 3.3 µM DCMU, 3.3mM ascorbate (Asc), 0.33 mM diaminodurene and 10 µM diquat (DQ). Electron transport was measured as oxygen production (A), (B), or uptake (C). The atebrin/e2 ratio is expressed as the saturating amount of atebrin (nmoles/mg chlorophyll) at a rate of electron transport of 1 µmole-pairs of reducing equivalents (e2)/min per mg chlorophyll*

either dark or light. Also, the kinetics of fluorescence lowering and binding were
identical (Kraayenhof 1971). Control experiments have been performed to see if the
probe responses were absent when the enzymes involved in binding or transport of the
probe were deactivated. Very surprisingly, however, we found that in heat-denatured
chloroplasts the atebrin fluorescence was completely quenched (in a light- and un-
coupler-independent way). In order to see whether this fluorescence lowering was due
to binding to the membrane fragments or to a soluble quencher, binding experiments were
performed under these conditions. Figure 7 compares the fluorescence levels and actual binding levels in dark (b), illuminated (c), and heat-denatured (c') chloroplasts,
as a function of atebrin concentrations. The fluorescence of the free atebrin (a) can
be estimated from the spikes seen after rapid addition of atebrin (see insert). The
fluorescence lowering observed in the dark and in the light coincide very well with
the respective binding levels. The fluorescence and binding data obtained with the
heat-denatured chloroplasts follow those obtained with the energized chloroplasts very
closely. This indicates that the energy-linked binding occurs by the creation of (probably negative) sites on the membrane itself rather than in the aqueous phase inside
the organelle. Presumably, heat denaturation causes an irreversible exposure of part
of these sites so that binding can occur in a nonenergy-linked way.

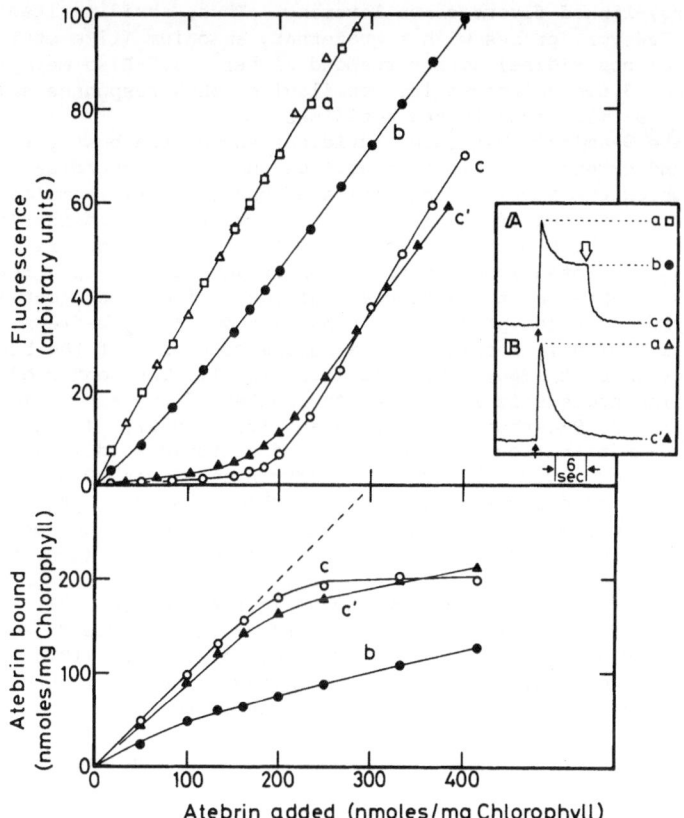

Figure 7. *Atebrin fluorescence and binding titrations in energized, nonenergized
and heat-denatured chloroplasts. The standard medium (Figure 2) contained 5 μM pyocyanine and 44 μg chlorophyll/ml. Fluorescence and binding measurements were done in
separate incubations with the same batch of chloroplasts. Heat treatment: 5 min at 75°.
The insert illustrates examples of the fluorescence lowering (A) in dark and light,
and (B) in heat-denatured chloroplasts (3 μM Atebrin added at arrows)*

Other Biological Membranes

A similar quantitative relationship between energy level and probe response (and
binding) has been observed by Eilermann (1970; 1971) for coupled, uncoupled, and re-

coupled membrane particles of *Azotobacter vinelandii*. The energy-dependent fluorescence lowering of atebrin was further confirmed in chromatophores of *Rhodospirillum
rubrum* by Gromet-Elhanan (1971; 1972) in mitochondrial membrane particles by Lee (1971),
Kraayenhof (1971), and Azzi et al. (1971). Schuldiner et al. (1971; 1972), extended
the work with chloroplasts and subchloroplast particles and also included other dyes
in their studies. Dell'Antone et al. (1972) observed energy-linked absorption changes
with other cationic dyes in mitochondrial membranes that are probably similar to the
interactions described here.

Thus, in a wide variety of biological membranes one observes energy-linked interactions with atebrin and similar cationic dyes. Although interpretations of the results may vary considerably, one can readily recognize the validity of these compounds
as probes of membrane energization. The usefulness of atebrin in quantitative estimation of the energy level in chloroplasts is sufficiently illustrated by the experiments
discussed above.

Comparative Study of Other Acridine Dyes

Investigation of a number of positively and negatively charged dyes (Kraayenhof
and Katan 1972) showed that a net positive charge at the pH of the experiment is a prere-
quisite for an energy-linked fluorescence lowering. Thus, acridine itself (pK$_a$ 5.6)
does not respond. However, probes with a quaternary ammonium (like ethidium bromide
and N-methyl-3,6-diaminoacridine) do not respond either. 3,6-Bisdimethylaminoacridine
(acridine orange) and 3,6-diaminoacridine (proflavine) show responses similar to those
of atebrin and other 9-amino-substituted acridines.

The atebrin-like 9-amino-substituted acridines showed the best performance in
terms of the rate and extent of their fluorescence changes in energized chloroplasts.
In Figure 8, the subsequent steps of ionization of these atebrin homologues are depicted. Comprehensive studies of the ionization behavior of acridines (among other
physicochemical studies) have been carried out by Albert and coworkers (1966) and by
Irvin and Irvin (1950). Potentiometric titration in aqueous solution revealed that
the first proton is accepted by the side-chain nitrogen with a pK$_{a1}$ between 10.0 and
10.5. The second proton is accepted by the ring nitrogen (pK$_{a2}$ between 7.3 and 8.3,
depending on the value of n), resulting in a resonance hybrid of the benzenoid and
quinonoid structures (Albert 1966; Irvin and Irvin 1950). The center of the positive
charge on the acridine nucleus is located at the center of the ring, midway between
the ring nitrogen and the 9-carbon atom (Irvin and Irvin 1950). (A third proton can
be accepted by the 9-amino-nitrogen in concentrated sulfuric acid.) In the case of
atebrin, the pK$_{a1}$ is about 10.2; the pK$_{a2}$, 7.9 (Irvin and Irvin 1950). In the incubation medium, either in the absence or presence of chloroplasts, a pK$_{a2}$ of 7.7 was found
by fluorescence titration, which may indicate that in this case the nucleus pK$_a$ is
not very different in ground and excited states of the molecules (see, however, Loken
et al. 1973).

From different lines of investigation (Schuldiner and Avron 1971; Schuldiner
et al. 1972; Kraayenhof and Katan 1972) it is clear that "ideal" energy-
probes show a certain degree of ionization. Rottenberg et al. (1971; 1972), have
observed an energy-linked binding of ^{14}C-labeled methylamine. This they interpret as
being the result of a redistribution of the amine across the membrane according to the

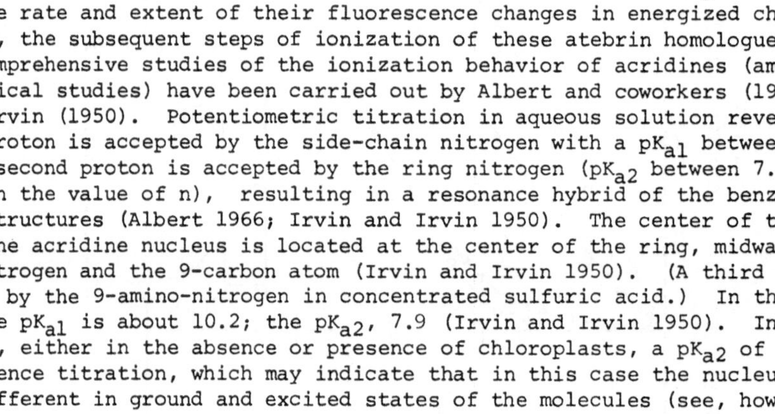

Figure 8. *Ionization steps in alkylamino-substituted acridines.* (Albert 1966;
Irvin and Irvin 1950)

pH gradient (Δ pH) created under energized conditions (Mitchell 1966). Schuldiner et al. (1971; 1972),and Gromet-Elhanan (1971, 1972) extrapolated this (assumed) behavior of amines to their observations with amino-substituted acridines or naphthalenes, and arrived at the far-reaching conclusion that these compounds can be used as ΔpH probes. Also, in the model of Azzi et al. (1971), the binding of acridines to submitochondrial particles is considered a secondary result of preceding anion uptake into the osmotic space. Unfortunately, however, experimental proof for the translocation of acridines (or any amine) across the membranes into the internal water compartment is not available. Yet, before trying to unravel the mechanistic aspects of these responses, it would be highly desirable to know, rather than to assume, in what part of the organelle the energy-linked fluorescence (or absorption) changes take place. The following experiments are part of an approach to this problem.

Figure 9 represents three examples of a series of single- or multiple-charged acridines, synthesized in this laboratory. MGA (pK$_a$ 9.0) is a single-charged acridine that shows a behavior similar to 9-aminoacridine and 9-methylamino-6-chloro-2-methoxy-acridine (not shown). Its response is smaller than that of atebrin at low concentrations (cf. Schuldiner et al. 1972), but increases at higher concentrations, after which it declines again. The unesterified (negatively charged) compound

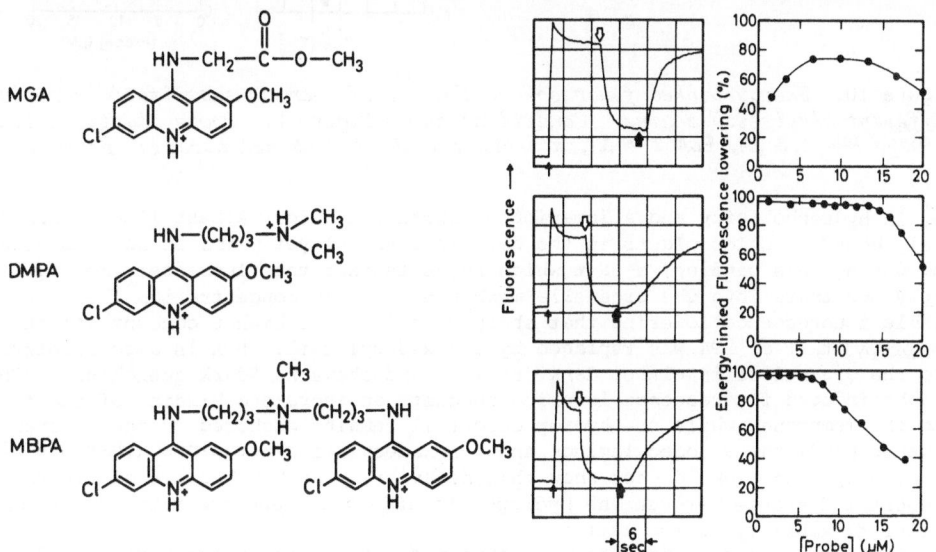

Figure 9. *Energy-linked fluorescence changes of 9-amino-substituted acridines with different number of charges. Conditions as in Figure 7. Probe additions (small arrows) were: MGA 6.7 μM, DMPA 6.7 μM, and MBPA 3.3 μM*

does not respond to illumination of the chloroplasts. The dodecyl and phytyl esters are already completely bound to the chloroplasts in the dark, but show a small (20%) fluorescence lowering upon illumination at higher concentrations (10 to 20 μM). DMPA behaves almost identically to atebrin. Its pK$_{a1}$ is 10.0; the pK$_{a2}$, 7.8 (cf. Irvin and Irvin 1950). MBPA, which is essentially a "double" DMPA, is an interesting compound since it contains two nucleus charges (pK$_{a2}$ 7.8) and one charged nitrogen (pK$_{a1}$ 10.0) in the aminoalkyl bridge. Its titration behavior is similar to two free DMPA molecules, suggesting that the high-pK$_a$ nitrogen does not contribute very much to the binding. Similar results were obtained with other "*bis*" compounds, with varying length of the bridge.

Figure 10 shows another set of homologues synthesized here and tested in chloroplasts. PPA has the same pK$_a$ values as DMPA, and its responses are almost identical to those of DMPA and atebrin. If one now reduces the distance between the two side-chain nitrogens by one methylene group (PEA), both pK$_a$ values are lowered (pK$_{a1}$ 9.7, pK$_{a2}$ 7.3) due to the electrostatic effect one charged group exerts on the ionization of the other group. This small structural change has a large impact on its behavior as a probe. It must be noted that this type of acridine shows both a considerable

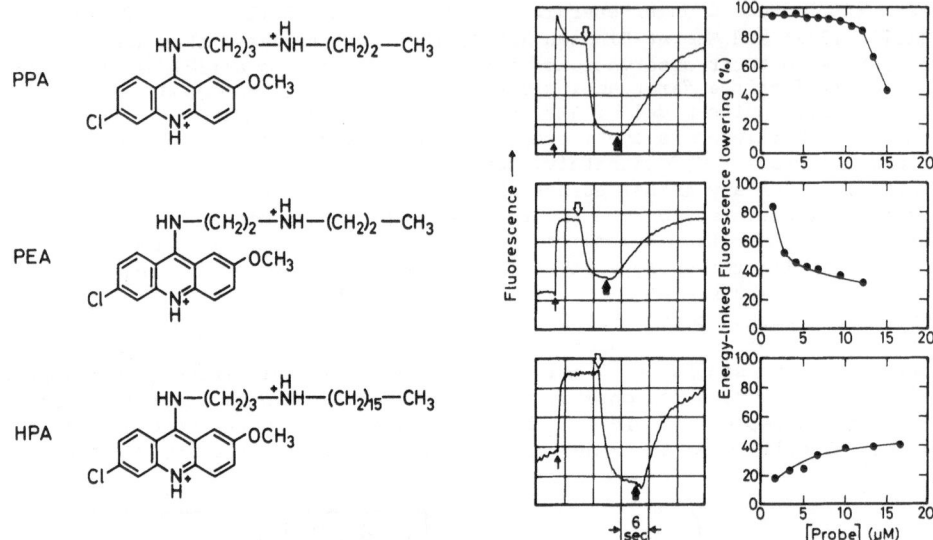

Figure 10. *Energy-linked fluorescence changes of 9-amino-substituted acridines
with different alkyl side-chains. Conditions as in Figure 7. Probe additions (small
arrows) were PPA 1.3 μM, PEA 1.3 μM, and HPA 3.3 μM (6.7 μM HPA already present)*

increase in hydrophobicity and a lowering of surface tension (Albert 1966). PEA is
completely bound to chloroplasts in the dark (not shown),yet there is no fluorescence
lowering during this binding, a fact which suggests that the fluorescent moiety does
not deeply penetrate into the organelle membrane. At low concentration there is an
appreciable fluorescence lowering that sharply declines at higher concentrations. In
HPA, the propyl tail of PPA was replaced by a hexadecyl tail. HPA is also completely
bound to the organelles in either dark or light and shows no "dark quenching". The
small light-induced fluorescence lowering suggests an increased binding of the fluorophore to the membrane, while the hydrophobic tail remains anchored in the membrane.
On the other hand, the 9-dodecylamino- and 9-hexadecylamino-acridines (that are also
bound under any condition) do not show this behavior, possibly because of hindrance
by the shorter distance between the hydrophobic side-chain and the fluorophore and by
the smaller flexibility of the latter.

Qualitatively the same results were obtained with an analogous series of 7-amino-
substituted benz[c]acridines, kindly provided by Dr. E.F.Elslager (Parke, Davis, and
Company) (Elslager et al. 1957). With these bulkier aromatic rings both rate and extent of the responses were reduced by about 30% as compared with the acridines.

Quinacrine mustard (Caspersson et al. 1970; Modest and Sengupta 1973) contains chlo-
rines at the side-chain that might be expected to react covalently with SH or NH_2 groups on
proteins. With bovine serum albumin it formed a tight complex without appreciable change of
its fluorescence. Figure 11 compares the effect of atebrin, quinacrine mustard, and the
complex in intact (A) and sonically disrupted (B) chloroplasts. The atebrin and quinacrine mustard responses are equal in the two preparations since the outer chloroplast
membrane is freely permeable to these compounds. Both outer and grana membranes are imper-
meable to albumin.(Dilley and Rothstein 1967). Hence, the albumin-bound probe cannot reach
the energy-generated binding sites in intact chloroplasts and does not show any light-
induced fluorescence change. This is also a proper control to assure that no free
quinacrine mustard is left behind in the complex. Surprisingly however, in the disrupted chloroplasts, the albumin-bound probe shows a light-induced response similar
to that of unbound quinacrine mustard. This strongly suggests that the energy-generated probe binding sites are peripherally located on the grana membranes. The fluorescence of a complex of quinacrine mustard with rabbit muscle Glyceraldehyde-3-phosphate
dehydrogenase (EC 1.2.1.12) did not respond in this manner. However, the reactive
SH groups seem to be deeply buried in the enzyme molecule (Price 1971) so that bound
quinacrine mustard is in a much more occluded position. Investigations with other

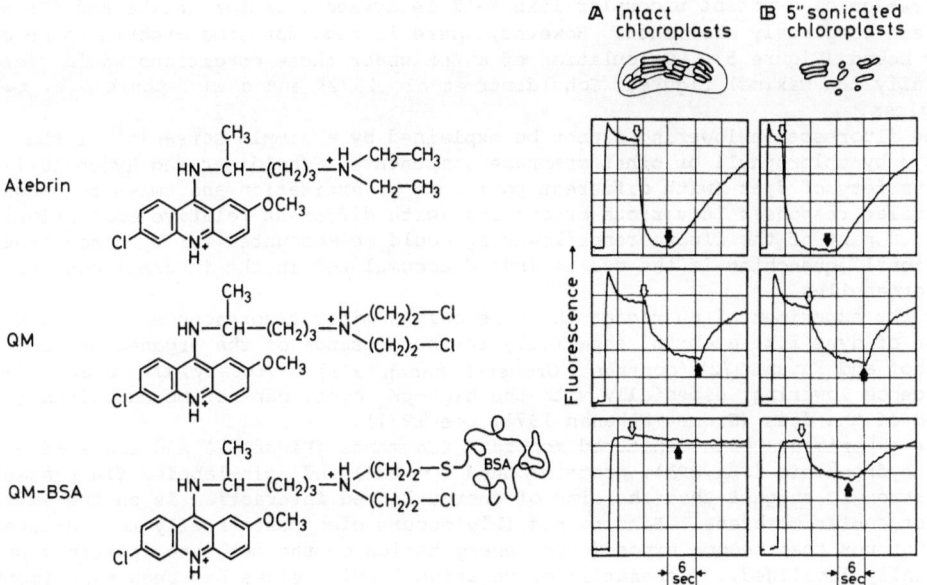

Figure 11. *Comparison of the energy-linked fluorescence changes of atebrin, quinacrine mustard (QM) and quinacrine mustard-bovine serum albumin complex (QM-BSA) in intact and sonically disrupted chloroplasts. Conditions as in Figure 7. The probe concentrations were 3.3 μM (for QM-BSA this refers to the QM-label). Sonication was carried out for 5 sec in a Branson sonifier*

probe-protein complexes are in progress.

In collaboration with Dr. E.K. Ruuge, experiments are being carried out with spin-labeled acridines. Quinacrine mustard, the side-chain of which was attached to a nit-roxyl-free radical (3-amino-2,2,5,5-tetramethyl-1-pyrrolidinyloxy), showed a light-induced response that was intermediate between those of atebrin and quinacrine mustard. However, the EPR spectra taken in the dark and in the light did not reveal any immobil-ization of the probe (at least not of the side-chain). Furthermore, we did not observe any broadening of the spectral lines. If the probe molecules were accumulated inside the organelle in the aqueous phase, one would obtain an internal concentration of several mM (calculated from the fluorescence change). This would give rise to consid-erable broadening of the spectral lines as a result of dipole-dipole interaction.

Mechanistic Considerations

In order to arrive at a satisfactory explanation of the binding or increased bind-ing of acridine dyes to energized organelles (and concomitant fluorescence or absorp-tion changes) the following should be taken into consideration.

The experimental results obtained with different organelles and with different acridines should be accommodated in a general model, although the presence of permeant and lyophilic anions (Lee 1971; Azzi et al. 1971) seems to be a specific requirement for mem-brane particles derived by sonic disruption of mitochondria, but not for other membranes.

All experimental findings, including those with inhibitory effects by ammonium salts and nigericin (Gromet-Elhanan 1971; 1972; Schuldiner and Avron 1971), can be explained with a model in which probe responses and proton uptake (or any ion trans-port) are driven by energy (including a possible electrochemical membrane *potential*) in *parallel*, rather than in *series*. A nice quantitative correlation between the acri-dine fluorescence lowering and the magnitude of the membrane pH gradient (Schuldiner et al. 1972) (as determined from amine distribution) is largely dependent on the proper concentration of the probe relative to that of the organelles. Some dyes, including atebrin and 9-aminoacridine, are already bound to a large extent under nonenergized conditions. With these and other dyes (*i.e.* acridine orange) with one high pK_a one observes an almost complete binding over a large concentration range, es-pecially at lower temperatures, under energized conditions. If a "nearly-saturating" concentration of a potent uncoupler like S-13 is present, proton uptake and ATP synthesis are completely abolished. However, there is room for some atebrin to be completely bound (Figure 5). Calculation of a ΔpH under these conditions would yield essentially the maximal figures (Schuldiner et al. 1972) and seems, therefore, rather meaningless.

The fluorescence lowering cannot be explained by a simple screening of the accumulated dye by chlorophyll or other membrane components (Schuldiner and Avron 1971) since a wide variety of dyes (with different positions of excitation and emission bands) show similar responses in various organelles (with different relative absorption). However, a part of the fluorescence lowering could be accounted for by concentration (collisional) quenching if the dye is indeed accumulated in the internal aqueous phase of the organelles.

In the experimental pH region there is only a minor fluorescence lowering by protonation of dyes like atebrin, especially in the presence of the organelles (Lee 1971; Kraayenhof and Katan 1972) (contrast, Gromet-Elhanan 1971). Obviously, the observed large fluorescence lowering, especially with the high-pK_a dyes, can not be explained by protonation of the dyes (Gromet-Elhanan 1971, Lee 1971).

The experiments with denatured membrane fragments (Figure 7) and those with bulky (MBPA), hydrophobic (PEA, HPA), protein-bound (QM-BSA) and spin-labeled fluorescent probes strongly suggest that the site of energy-linked interaction is on the *outside* of the organelle membrane. Binding possibly occurs electrostatically to negatively charged groups that become exposed upon energization of the membrane (possibly proteins, peripherally localized). Generation of negative binding sites has been experimentally demonstrated by Dilley and Rothstein (1967) for the case of chloroplasts. The fluorescence lowering could occur then by energy transfer to membrane components, by the formation of a nonfluorescent complex between dye and binding site, or by a stacking of the dye similar to its interaction with polyanions (Dell'Antone et al. 1972; Dilley and Rothstein 1967).

It is striking that the acridines that show appreciable energy-linked responses have the physicochemical properties that make them potent antibacterial or antiprotozoal drugs (Albert 1966). Indications are available that these capacities depend on appreciable ionization of the dye and flatness of the aromatic ring system. Interaction with the cell membrane is thought to occur in an aqueous environment at the outer surface of the membrane (Albert 1966). Although the overall perturbation is quite different in the case of the energy-linked dye interactions, it is tempting to suggest that fundamentally the same phenomenon may occur.

ACKNOWLEDGEMENTS

The author is very grateful to Professor E.C. Slater, in whose stimulating environment this work was initiated, and to Professors B. Chance and C.P. Lee for further encouragement and discussions.

This investigation was supported by a U.S. Public Health Service International Postdoctoral Research Fellowship (No. 1-F05-TW-1821-01) and in part, by The Netherlands Organization for the Advancement of Pure Research (Z.W.O.).

REFERENCES

ALBERT, A., (1966), *The Acridines - Their Preparation, Physical, Chemical and Biological Properties and Uses*, Edward Arnold, London.

AZZI, A., Fabbro, A, Santato, M., and Gherardini, P.L., (1971), *Europ. J. Biochem.*, **21**, 404.

BERDEN, J.A., and Slater, E.C., (1970), *Biochim. Biophys. Acta*, **216**, 237.

BROCKLEHURST, J.R., (1970), *Molecular Probes for Biological Functions*, Ph.D. Thesis, Oxford.

CASPERSSON, T., Zech, L., and Modest, E.J., (1970), *Science*, **170**, 762.

CHANCE, B., Azzi, A., Lee, I.Y., Lee, C.P., and Mela, L., (1969), In: *Mitochondria - Structure and Function*, (Ernster, L. and Drahota, Z, eds.), Academic Press, New York, 233.

DELL'ANTONE, P., Colonna, R., and Azzone, G.F., (1972), *Europ. J. Biochem.*, 24, 553.

DILLEY, R. A., and Rothstein, A.,(1967), *Biochim. Biophys. Acta*, 135, 427.

EILERMANN, L.J.M., (1970), *Biochim. Biophys. Acta*, 216, 231.

EILERMANN, L.J.M., (1971), In: *Energy Transduction in Respiration and Photosynthesis*, (Quagliariello, E., Papa, S. and Rossi, C.S., eds.), Adriatica Editrice, Bari, Italy, 659.

ELSLAGER, E.F., Moore, A.M., Short, F.W., Sullivan, M.J., and Tendick, F.H., (1957), *J. Am. Chem. Soc.*, 79, 4699.

GREVILLE, G.D., (1969), In: *Current Topics in Bioenergetics*, Vol.3,(Sanadi,D.R., ed.), Academic Press, New York, 1.

GROMET-ELHANAN, Z. (1971), *FEBS Letters*, 13, 124.

GROMET-ELHANAN, Z. (1972), *Europ. J. Biochem.*, 25, 84.

IRVIN, J.L., and Irvin, E.M., (1950), *J. Am. Chem. Soc.*, 72, 2743.

IZAWA, S. and Good, N.E., (1968), *Biochim. Biophys. Acta*, 162, 380.

IZAWA, S., Ruuge, E.K., Kraayenhof, R., and DeVault, D.,(1973), submitted for publication.

JAGENDORF, A.T., and Uribe, E., (1966), *Proc. Natl. Acad. Sci. U.S.*, 55, 170.

KRAAYENHOF, R., (1969), *Biochim. Biophys. Acta*, 180, 213.

KRAAYENHOF, R., (1970), *FEBS Letters*, 6, 161.

KRAAYENHOF, R., (1971), In: *Energy Transduction in Respiration and Photosynthesis*, (Quagliariello, E., Papa, S., and Rossi, C.S., eds.), Adriatica Editrice, Bari, Italy, 649.

KRAAYENHOF, R., Izawa, S., and Chance, B.,(1972), *Plant Physiol.*, 50, 713.

KRAAYENHOF, R., and Katan, M.B., (1972), *Photosynthesis, Two Centuries after its Discovery by Joseph Priestley*, (Forti, G., Avron, M., and Melandri, A., eds.), Junk Publishers, The Hague, 937.

KRAAYENHOF, R., Katan, M.B., and Grunwald, T., (1971), *FEBS Letters*, 19, 5.

KRAAYENHOF, R., Vanderkooi, J., Ruuge, E.K., Brocklehurst, J.R., Lee, C.P., and Chance, B., (1973), submitted for publication.

LEE, C.P., (1971), *Biochemistry*, 10, 4375.

LOKEN, M.R.,Gohlke, J.R., Brand, L., (1973), This volume.

McCONNELL, H.M., and McFarland, B.G.,(1970), *Quart. Rev. Biophys.*, 3, 91.

MITCHELL, P., (1966), *Biol. Rev.*, 41, 445.

MODEST, E.J., and Sengupta, S.K., (1973), This volume.

MYHR, B.C., (1973), This volume.

NEUMANN, J., Arntzen, C.J., and Dilley, R.A., (1971), *Biochemistry*, 10, 866.

NEUMANN, J., and Jagendorf, A.T., (1964),*Arch. Biochem. Biophys.*, 107, 109.

OUITRAKUL, R., and Izawa, S., in preparation.

PRICE, N.C., (1971),*Chemical Studies of Macromolecular Conformation*, Ph. D. Thesis, Oxford.

RABINOWITCH, E., and Govindjee,(1969), *Photosynthesis*, Wiley, New York.

ROTTENBERG, H., Grunwald, T., and Avron, M., (1971), *FEBS Letters*, 13, 41.

ROTTENBERG, H., Grunwald, T., and Avron, M., (1972), *Europ. J. Biochem.*, 25, 54.

RADDA, G.K., (1971), In: *Current Topics in Bioenergetics*, Vol. 4, (Sanadi, D.R., ed), Academic Press, New York, 81.

SAHA, S., Ouitrakul, R., Izawa, S, and Good, N.E., (1971),*J. Biol. Chem.*, 246, 3204.

SCHULDINER, S., and Avron, M., (1971), *FEBS Letters*, 14, 233.

SCHULDINER, S, Rottenberg, H., and Avron, M.,(1972), *Europ. J. Biochem.*, 25, 64.

SLATER, E.C., (1971), *Quart. Rev. Biophys.*, 4, 35.

VAN DAM, K., and Meyer, A.J., (1971), *Ann. Rev. Biochem.*, 40, 115.

VELICK, S.F, (1958), *J. Biol. Chem.*, 233, 1455.

FLUORESCENT MOLECULAR PROBES IN FLUORESCENCE MICROSPECTROPHOTOMETRY AND MICROSPECTROPOLARIMETRY

Seymour S. West* and Andrew E. Lorincz**

* Department of Engineering Biophysics, University of Alabama in Birmingham, Birmingham, Alabama.
** Professor of Pediatrics and Director, Center for Developmental and Learning Disorders, University of Alabama in Birmingham, Birmingham, Alabama

INTRODUCTION

In recent years the biochemist and the biophysical chemist have made considerable progress in their understanding of the behavior of complex biopolymers as studied in *in vitro* systems. However, it is ordinarily not possible to extrapolate the results of these studies to account with entire satisfaction for the behavior of intact cells. The cytochemist has indeed been devoting considerable attention to this problem, but has been hampered by the nature of his data, which are not comparable with those produced by the biochemist and biophysical chemist. This point is well appreciated and has been explicitly discussed by Spicer, Leppi, and Stoward (1965). As a consequence, it has not been possible for the cytochemist to produce a classification of substances, detected by cytochemical means, which can be considered an accurate classification in biochemical terms. This is particularly true for the carbohydrates and mucosubstances.

More recently, West (1965; 1969; 1970) has developed instrumentation and experimental techniques which produce data that can be compared directly with the results obtained from *in vitro* solution studies. These new approaches to cytochemistry can be grouped under the term, biophysical cytochemistry. Fluorescent molecular probes, which can be dyes such as acridine orange (AO), serve as the means for bridging the gap between the *in situ* and the *in vitro* studies. Biophysical cytochemistry, although not limited to the use of fluorescent molecular probes, has to date been concerned primarily with their development as cytochemical tools.

Fluorescent molecular probes, such as acridine orange, are small planar dye molecules which can serve as vital cytological stains. Such dyes are very sensitive to alterations in environmental conditions (concentration, pH, ionic strength, complexing substrate, etc.). This sensitivity is evidenced by marked alterations in the fluorescence emission spectra, the absorption spectra, in optical rotatory dispersion spectra, and in other physical chemical parameters. Of particular interest to biophysical cytochemistry is fluorescence spectroscopy and optical rotatory dispersion (ORD) the variation of optical activity as a function of wavelength.

Fluorescence microspectrophotometry (FMP) and microspectropolarimetry (MSP)* are nondestructive techniques, which, when used in conjunction with fluorescent vital dyes enable spectroscopic investigation of unfixed or living cells. The freedom to deal with living or unfixed cells that is gained by the use of these methods enables one to bypass the difficulties which attend fixation.

Cytochemical studies on intact cells require methods that are capable of detecting the presence of a single component of a mixture without prior physical or chemical separation. It is the fluorescent molecular probes coupled with spectroscopy which make it possible to satisfy this need. These probes complex with (stain) a limited number of intracellular components, thus making the cytochemical data more explicit.

A further requirement which must be met is extremely high sensitivity to permit the detection of the extremely small quantities of substance present within a single cell. Both FMP and MSP are capable of such sensitivity, particularly when coupled with the use of fluorescent molecular probes. Previous work by the author has indicated that FMP can record the fluorescence spectrum produced by as little as approximately 10^{-17} moles or less of AO. MSP appears to have very nearly the same sensitivity.

---

* *A microspectropolarimeter is a microscope which measures optical activity as a function of wavelength. The resulting data, displayed as a curve of optical rotation versus wavelength, is called an optical rotatory dispersion spectrum.*

Further, the new method which has been developed for measuring optical rotation (West 1970) is insensitive to depolarization scattering and high optical density. Such instrumentation properties are a necessity for dealing with the inhomogeneous interior of the cell. In addition, both FMP and MSP can be applied to *in vitro* solution studies and, hence, provide the means for correlating the biophysical cytochemical investigations with those conducted by the biophysical chemist. This also requires that the spectra be corrected to remove all instrumental distortions. Such corrected spectra enable comparison of cytochemical data not only with chemical data existing in the literature, but also with the results of other cytochemical studies.

Implicit in the above is the requirement that the staining reaction be carried out in a quantitative manner. This necessitates the quantitation of the biopolymer and the dye with which it complexes. Quantitating the total amount of dye present in a given volume of solution is not difficult, but the determination of the amount of a particular biopolymer present in a single cell is a formidable problem. Therefore, as a first approximation, the cells are counted and the total number of cells in a given volume of staining solution is determined. The quantitative dye uptake per cell that is measured, after the staining reaction reaches equilibrium is therefore an average value taken over the entire population of cells. By determining the number of subpopulations and the number of cells in each, and given indices of the amount of biopolymer per cell in each subpopulation, the average dye uptake per cell can be more specifically defined. Alternatively, it is possible to disperse the cells from an inhomogeneous population to obtain samples which contain essentially one cell type. Under these conditions, the average dye uptake per cell may be more representative of every cell in that population, provided the phenomena upon which the separatory means are based are related to the amount of intracellular biopolymer which complexes with the fluorescent molecular probe. To date, purified cell preparation has not been attempted. Nevertheless, marked differences have been observed between normal cells and cells from patients with heritable disorders.

This paper is concerned with application of biophysical cytochemistry to one clinical type from a group of heritable disorders of connective tissue termed the mucopolysaccharidoses. These diseases are characterized by abnormal mucopolysaccharide metabolism which results in the accumulation of these polymers in the tissues throughout the body.

Clinical symptoms of the mucopolysaccharidoses usually become manifest after the first year of life. Growth failure, progressive mental retardation, and gradual development of numerous skeletal deformities generally characterize the clinical course of these disorders. It is not the purpose here to discuss these clinical entities in detail. However, it should be noted that early detection, particularly in the intrauterine condition, of affected individuals, linked with the ability to identify genetic carriers would provide considerable aid in the prevention of these disorders. In addition, if better insights can be gained into the underlying chemistry of the metabolic defect, the possibility of developing effective therapeutic measures should be enhanced. These goals can be served, at least in part, if proper correlations can be established between cytochemical and *in vitro* studies on simplified model systems. It is the purpose of this paper to demonstrate that such possibilities certainly exist in the case of the mucopolysaccharidoses. However, it must be cautioned that the results presented here are to be considered indicative rather than definitive. More research is required to establish firmly the details of the phenomena which have been detected.

MATERIALS AND METHODS

The details of the quantitative staining procedure using acridine orange have been previously described by West, (1969). Purified acridine orange was employed. The importance of using purified dyes in these studies, as in any study involving spectroscopy, cannot be overemphasized. Unfixed or living biological systems impose similar constraints upon the purity of the reagents employed.

The staining reactions were carried out in physiological saline buffered to pH 7 with McIlvaine's citrate-phosphate buffer and backtitrating to pH 7 to provide some buffering capacity with minimum change in ionic strength. The buffered physiological saline was filtered through a 50 millimicron (VM) Millipore filter. Stock solutions of the dye were prepared in deionized water, filtered through a Millipore VM filter, and stored in Desicoted (Beckman) low-actinic glassware at 4° C. The stock solution was determined by micro-Kjeldahl analysis. All solutions were made with deionized water. Solutions used for staining consisted of serial dilutions of the stock dye solution in the buffered physiological saline. The method of determining the dye up-take per cell has been described by West (1969). Fibroblasts, derived from skin cultures obtained from an aborted fetus, were separated for study by standard tissue-culture trypsinization techniques. The mother of the fetus was affected with Type VI mucopolysaccharidosis (Maroteaux-Lamy syndrome). The father was phenotypically normal, and the fetus was, therefore, presumed to be heterozygous for this recessive condition.

Peripheral leukocytes were obtained from freshly-drawn blood collected by veni-puncture into heparinized "Vacutainers" (Becton-Dickinson VBD-4751). Both cultured cells and peripheral leukocytes were washed at least three times in buffered saline prior to staining. Staining reactions were always permitted to reach equilibrium before any measurements were made.

Fluorescence Microspectrophotometry (FMP)

The experimental setup for FMP was the same as that described by West (1969) with the exceptions described below:

The television camera chain was replaced by a rotation linear, circular interference wedge and a multiplier phototube. The plane of the interference wedge was positioned in at the eyepoint of the microscope which then served as the spectroscopic scanning aperture. The multiplier phototube was located after the circular interference wedge. A rotating linear potentiometer was geared to the rotating interference wedge with a gear ratio of 1 to 1. The potentiometer is of the servo type to permit phasing with the rotating interference wedge. A fixed DC voltage placed across the potentiometer enables it to generate the horizontal sweep for the oscilloscope display. The output of the multiplier phototube, after suitable amplification, was fed to the vertical input of the oscilloscope. The oscilloscope display of the fluorescence emission spectrum was photographed on 35mm film and then processed and corrected in the manner described by West (1969). A Leitz epiilluminator (after Ploem) was used for excitation and collection of the fluorescence emission. The microscope fluorometer was used to record emission spectra from cytological preparations, from solutions, and from the precipitates of the dye-mucopolysaccharide complex. The precipitate was placed upon a microscope slide, and sealed under a cover slip with paraffin, prior to obtaining emission spectra. Fluorescence emission spectra were obtained from solutions in one-centimeter polarimeter cells. These preparations were also utilized for the MSP.

The rotating wedge permitted recording an entire fluorescence emission spectrum in one second. The effects of fluorescence fading were negligible when the record of the spectrum was obtained immediately upon beginning of fluorescence excitation. The rotational speed of the interference wedge can easily be increased to take account of more rapid fluorescence fading.

Microspectropolarimetry (MSP)

The instrumentation for microspectropolarimetry has previously been described by West (1970).

RESULTS

Fluorescence emission spectra of AO in 0.9% NaCl solution buffered to pH 7.2 are shown in Figure 1. At concentrations in the neighborhood of 10^{-6}M, a single band is observed with the peak at 540 nm (green fluorescence), characteristic of monomeric AO. As the concentration of AO is increased, a red fluorescence emission (λ_{max} = 660 nm) emerges and becomes the predominant species at about 10^{-3}M. The 660 nm fluorescence emission band is due to the formation of dimers or higher aggregates of the dye molecules. At concentrations of AO intermediate between the extremes shown in Figure 1, both green (540 nm) and red fluorescence (660 nm) coexist. Note that the spectra shown in the Figure are normalized with respect to the most intense peak. As a consequence, it is not possible to tell whether or not the apparent disappearance of the green-fluorescing species observed at the highest AO concentration indicates that the red-fluorescing species is being formed at the expense of the former. This is indeed

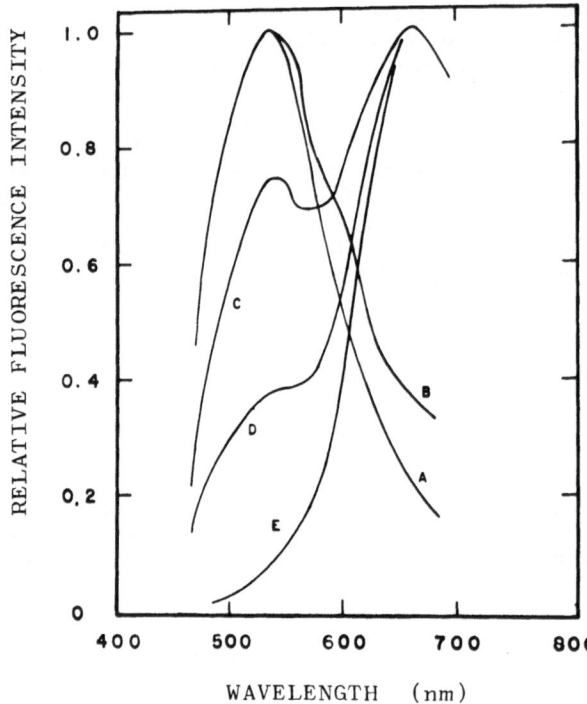

Figure 1. *Fluorescence emission spectra of acridine orange (AO) in 0.9% NaCl solution (pH 7.2, 24 °C) at different AO concentrations (A – E). (A. – 3.5 × 10⁻⁶ M; B. – 7.0 × 10⁻⁵ M; C. – 3.5 × 10⁻⁴ M; D. – 7.0 × 10⁻⁴ M; E. – 3.5 × 10⁻³ M)*

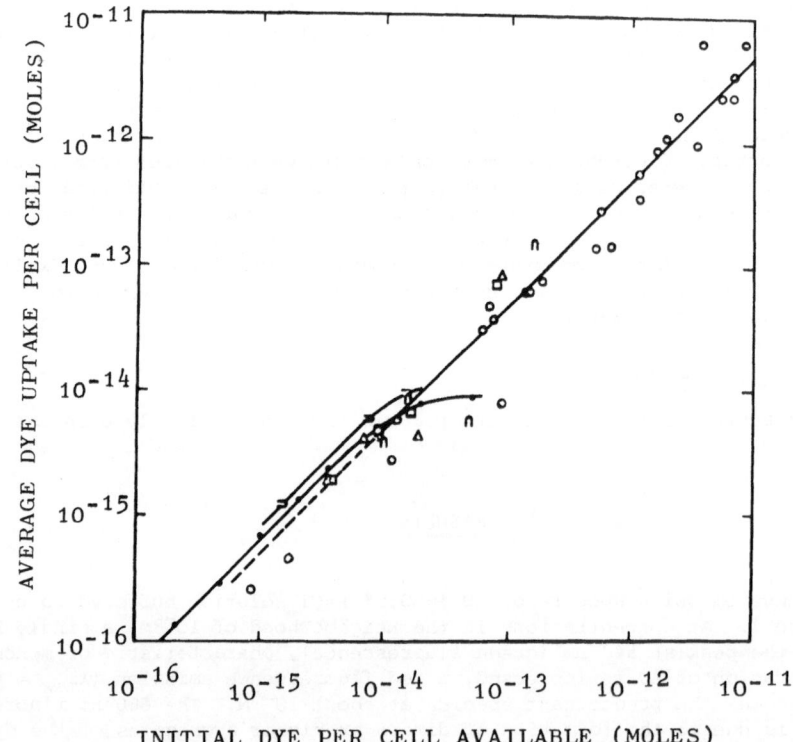

Figure 2. *Average uptake of acridine orange per cell as function of amount of dye per cell initially available. (● = mouse leucocytes; ➔ = normal human leucocytes; △ = MPS-affected leucocytes (B.L.); □ = MPS-affected leucocytes (D.L.); ∩ = MPS-carrier leucocytes (L/M); ●= MPS-affected cultured fibroblasts)*

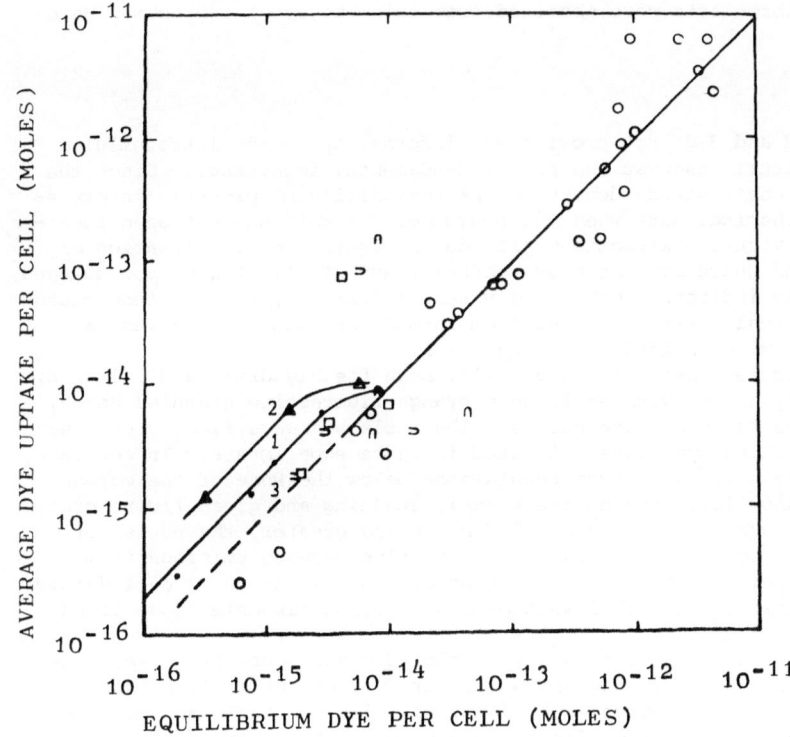

Figure 3. *Average acridine orange uptake per cell as function of the amount of free dye per cell at equilibrium. (1-• = mouse leucocytes; 2-△ = normal human leucocytes; 3-○ = MPS-affected cultured fibroblasts; ⊃ = MPS-affected leucocytes (B.L.); ▫ = MPS-affected leucocytes (D.L.); ⋂ = MPS-carrier leucocytes (L/M)*

the case, however, as is evidenced by the departure from Beer's law as the concentration of the dye approaches 10^{-5} M.

Figures 2 and 3 show the results of quantitatively complexing cells with AO. Measurements of the average dye uptake per cell were made after equilibrium was attained. In Figure 3, the equilibrium dye uptake per cell shown on the abscissa is the free dye left in solution after equilibrium was reached. The free dye uptake per cell was calculated from measurements of the AO concentration in the supernatant fluid at equilibrium.

Although Figures 2 and 3 appear to be quite similar, they represent two different kinds of information. Figure 2 shows the average uptake of dye per cell as a function of the moles of dye *initially available* to the cell. Although it demonstrates that the amount of dye taken up by a cell is linearly proportional to initial dye/cell ratio, as is demonstrated by obtaining a straight line having a slope of 45° on a log-log plot, it does not provide the data needed to determine the binding constant.

Figure 3 shows the average dye uptake per cell as a function of the moles of free dye per cell at *equilibrium*. A binding constant, which is similar to an equilibrium constant, can be calculated from these data.

The most striking feature in Figures 2 and 3 is the marked difference in the amount of dye uptake per cell exhibited by the affected cells as compared with mouse or normal human leukocytes. The affected cells take up at least 1,000 times more dye than the normals and still do not show signs of saturation. The knee of the curve for mouse and normal human leukocytes indicates saturation of the molecular species formed at lower dye-per-cell ratios (< 10^{-15} moles AO/cell). These curves are asymptotic to a dye uptake of approximately 10^{-14} moles AO per cell. The curves for the mouse and normal human leukocytes do not extend beyond the points shown, as cell damage occurs at higher dye-per-cell ratios. The cells from mucopolysaccharidosis (MPS)-affected individuals are resistant to such damage.

The MPS-affected leukocytes B.L., D.L., and L/M were taken from freshly drawn blood and were small samples. Patients B.L. and D.L. are MPS-affected (Maroteaux-Lamy Syndrome). L/M is a phenotypically normal genetic carrier of the Maroteaux-Lamy

Syndrome. The cultured fibroblasts were obtained from skin biopsy of a presumed Maroteaux-Lamy carrier.

Visual Observations

The data in Figures 2 and 3 do not provide any information on the intracellular fate of the dye. Thus, visual observations are of fundamental importance. Since the inhomogeneity of the cell casts strong doubts on the possibility of properly interpreting the biophysical cytochemical data when the source of the data has not been localized within the cell, the visual observations provide information on localization of the fluorescing complex and guard against measurements inadvertantly taken from damaged cells or debris. They also indicate whether the visual appearance, fluorescence microscopy, etc., of other clinical entities differ from normals or among themselves, a determination not possible with unaided instrumentation.

All of the mucopolysaccharidoses-affected cells, both freshly drawn and cultured, exhibit a cytoplasm tightly packed with small deep orange-fluorescing granules which, in some instances, coalesce to form large masses. The nucleus always fluoresces green over the range of intracellular dye content treated in these experiments. In contrast, mouse or normal human leukocytes taken from populations below the knee of the curves shown in Figures 2 and 3 show little or no cytoplasmic staining and green fluorescing nuclei. For intracellular quantities of dye 10^{-15} moles and greater, the nuclei of the unaffected cells show progressive alteration in the fluorescence emission from green to yellow to deep orange. Cytoplasmic staining does not occur until cell damage occurs. Cytoplasmic inclusions are much less common and do not resemble those found in the MPS-affected cells.

Visual examination also reveals differences in photodynamic behavior between unaffected and MPS-affected cells. When the intracellular dye content is less than 10^{-15} moles AO, the green nuclear fluorescence of normal cells is stable and not affected by fluorescence-exciting radiation for periods of eight hours or longer. As the intracellular dye content exceeds 10^{-15} moles, AO fading of fluorescence is observed, which becomes proportionately greater as the intracellular dye content increases. The fading of fluorescence is accompanied by photolysis of the cell. This is initially evidenced by the appearance of red fluorescence in the cytoplasm and the development of an indistinct appearance to the cell boundaries. The rapidity and severity of the photolytic activity is dependent upon the intracellular dye content (only after yellow fluorescence has appeared in the nucleus) and the intensity of the fluorescence-exciting radiation. The fading of fluorescence and the photolytic process can be quite rapid taking only seconds. Biophysical cytochemical data obtained from cells subject to these phenomena must take account of the nonequilibrium state of affairs.

The MPS-affected cells exhibit entirely different behavior as a function of the duration of fluorescence-exciting radiation. Though some fading and also build-up of fluorescence is observed on visual examination, such cells do not exhibit photolysis even when the intracellular dye content is three orders of magnitude greater than that which causes photolysis in unaffected cells. The red-orange fluorescing cytoplasmic material present in the MPS-affected cells does not exhibit appreciable fading and the cells do not lyse even with irradiation times as long as 20 minutes.

Fluorescence Spectroscopy

Although visual observation provides information which cannot be matched by any instrument and, in the case of the MPS-affected cells, even gives strong promise of being an important aid to diagnosis, it does not permit adequate correlation with *in vitro* solution studies. Rapid recording of spectra of fluorescing molecular probes is one means whereby correlations between cytochemistry and solution chemistry may be established.

Figure 4 displays corrected fluorescence emission spectra from individual cultured skin fibroblasts from a fetus heterozygous for the Maroteaux-Lamy syndrome. The spectra were recorded within approximately one second after the cells were exposed to fluorescence-exciting radiation.

The fluorescence emission from the nucleus only was recorded fro the cell containing 4.2×10^{-15} mole AO (solid circles). The spectrum is a single band with a peak at 540 nm which is characteristic of the AO-intranuclear nucleic acid (NA) complex (West, 1969). In the case of the spectra from the cells containing 6.28×10^{-14} and 6.28×10^{-12}

moles AO, (triangles and open circles respectively), both nucleus and cytoplasm were imaged in the photometric aperture. These spectra consist of two overlapping bands. The band with the 540 nm peak is due to the AO-NA complex. The red-fluorescing band with a peak at approximately 660 nm is due to the cytoplasmic AO-complex. Note that these spectra are normalized with respect to the most intense peak. Comparison of relative intensity at a given wavelength is, therefore, not possible from the figure.

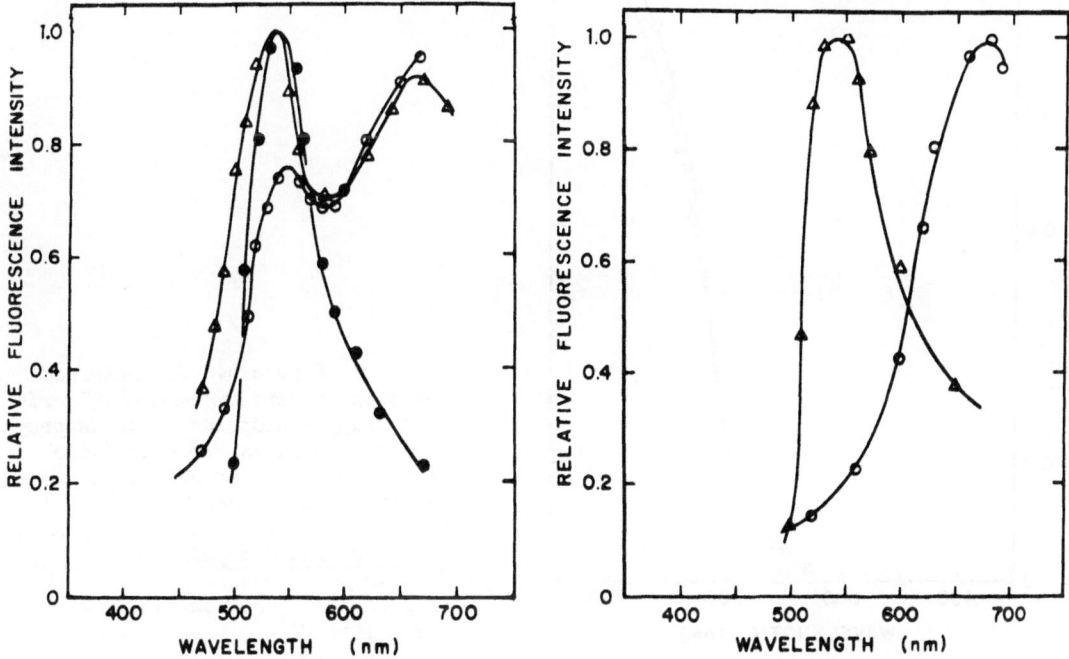

Figure 4. *Fluorescence emission spectra (corrected) of individual fibroblasts derived from skin cultures from an aborted fetus, the mother of which was affected with Type VI mucopolysaccharidosis (Maroteaux-Lamy syndrome) at different acridine orange concentrations. (o - 6.28 x 10^{-12} moles/cell; Δ - 6.28 x 10^{-14} moles/cell; • - 4.21 x 10^{-15} moles/cell)*

Figure 5. *Fluorescence emission spectra (corrected) of normal leucocytes and of those from mucopolysaccharidosis-affected patients. (o - MPS-affected leucocytes 4.16 x 10^{-15} \frac{moles}{cell}; Δ - human normal leucocytes 1.48 x 10^{-15} \frac{moles}{cell})*

Representative corrected spectra from a normal human leukocyte containing 4×10^{-15} moles AO and from an MPS-affected peripheral leukocyte containing 1.5×10^{-15} moles AO are shown in Figure 5. These intracellular quantities of dye are comparable. The spectrum from the normal human leukocyte has an emission peak at approximately 540 nm and fluoresces green. There is no indication of a red-fluorescing spectroscopic species. Visually, there is essentially no cytoplasmic fluorescence. The fluorescence emission spectrum obtained from the MPS-affected leukocyte shows only a red emission band with a peak at approximately 680 nm. This is typical of the spectra obtained from the cytoplasm of MPS-affected leukocytes or cultured affected fibroblasts.

AO complexes with chondroitin sulfates-(CS-A) and (CS-B) in solution have been the subject of preliminary investigations. Except for changes in intensity, little or no alteration in the absorption and fluorescence spectra was observed in solution at low dye/polymer ratios. However, as the dye/polymer ratio approaches 2, the formation of a precipitate is observed.

Figure 6 compares the corrected fluorescence emission spectra of the precipitates that form when AO complexes with CS-A and CS-B at a dye/polymer ratio of approximately 2. These spectra are quite similar in the short-wavelength region of the emission band, but diverge as the peak is approached. The peak of the chondroitin sulfate B complex lies at approximately 680 nm. The peak of the chondroitin sulfate A complex lies beyond the limit of the spectral range over which data was taken.

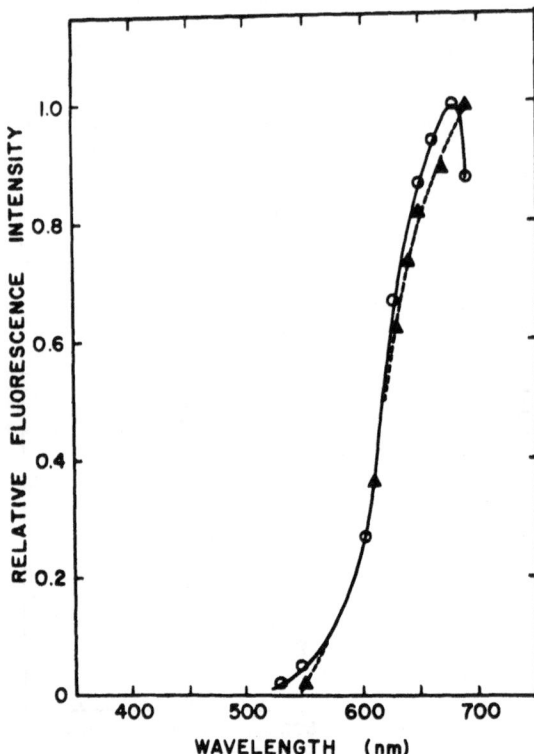

Figure 6. *Fluorescence emission spectra (corrected) of acridine orange-mucopolysaccharide precipitate products. (○ – acridine orange – chondroitin sulfate B; ▲ – acridine orange – chondroitin sulfate A; precipitation product formed at $\frac{(acridine\ orange)}{(mucopolysaccharide)} \cong 2$; mucopolysaccharide concentration = 3.3 x 10⁻⁴ M)*

The fading of fluorescence in the systems under consideration is a relatively slow phenomenon. Consequently, it was possible to follow the course of the fading by recording the entire emission spectrum repeatedly as a function of time. The AO-stained cells were observed visually during the course of these measurements. Dissociation of the cytoplasmic AO-polymer complex with the free dye migrating to other parts of the cell was not observed. The dye remains localized in the granules where it was originally observed throughout the photochemical events which are presumed to underlie the fluorescence fading and build-up phenomena shown in Figures 7 and 8.

The curves shown in Figures 7 and 8 were derived from fluorescence emission spectra recorded at the times indicated after initial exposure to fluorescence-exciting radiation. Excitation was continuous throughout the course of these measurements. The spectra shown are representative of the kinds of behavior that have been observed to date, but do not represent in detail all the variations that have been recorded. Although 660 nm and 540 nm were chosen as the wavelengths at which a measurement of fluorescence emission amplitude would be made as a function of time, their selection is not critical. The fluorescence emission spectra, recorded as a function of time, permit the detection of all fluorescent molecular species which differ from each other on the basis of fluorescence fading.

Figure 7 shows the fluorescence fading behavior obtained from measurements at 660 nm. The fluorescence fading curves obtained from freshly drawn leukocytes from an MPS-affected patient (open circles) show fairly rapid fading of this molecular species, whereas the curve obtained from the cultured fetal heterozygote fibroblast (open triangles) exhibits much slower fading. The fading of the red-fluorescing species in normal human leukocytes (open squares) resembles quite closely the fading behavior of MPS-affected leukocytes.

The curve in the figure represented by the solid circles was produced by fluorescence emission spectra from AO-stained mouse spleen cells obtained from a mouse

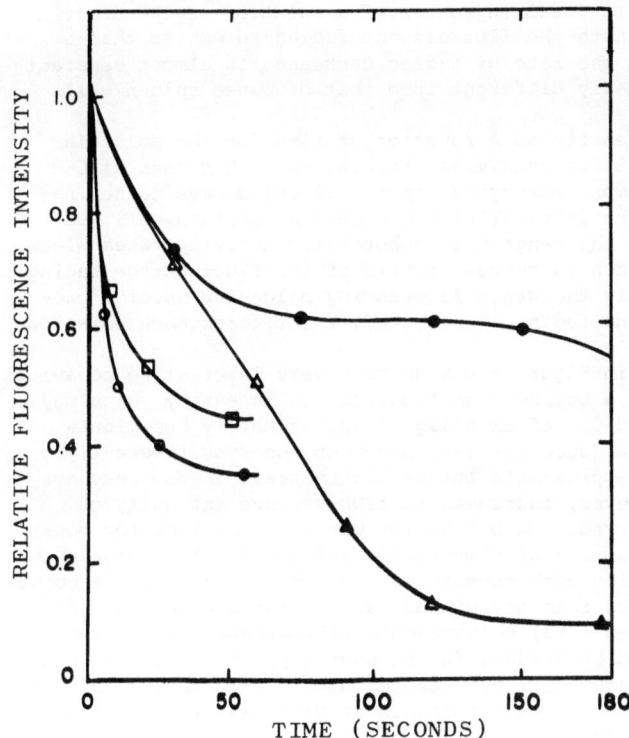

Figure 7. *Fluorescence intensity changes measured at 660 nm on single cells during their irradiation with exciting light. (○ – MPS-affected leucocyte 4.16 x 10^{-15} moles/cell (B.L.); △ – fetal heterozygote cultured fibroblast 3.65 x 10^{-15} moles/cell; □ – normal human leucocyte 1.05 x 10^{-14} moles/ cell; ● – spleen of mouse injected with condroitin sulfate – A)*

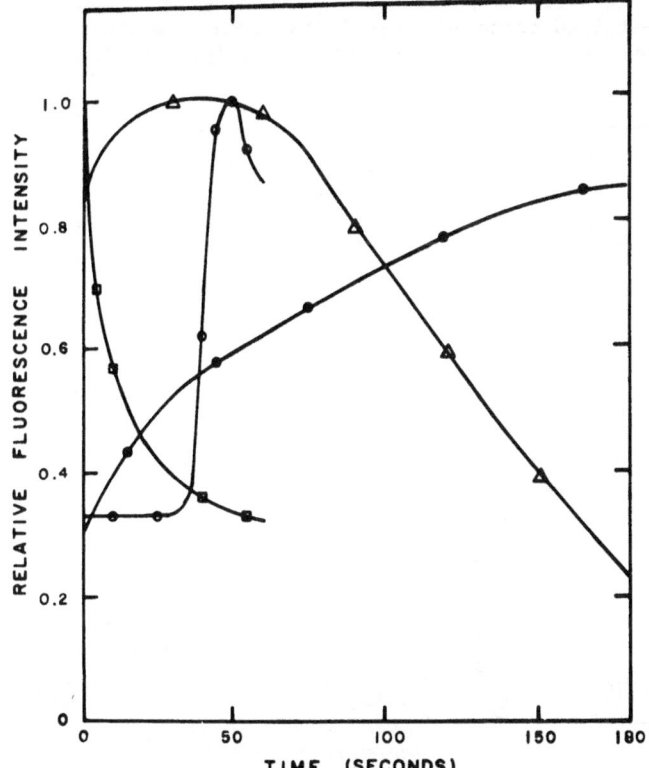

Figure 8. *Fluorescence intensity changes measured at 540 nm on single cells during their irradiation with exciting light. (○ – MPS-affected leucocyte (B.L.) 4.16 x 10^{-15} moles/cell; – fetal heterozygote cultured fibroblast 3.65 x 10^{-15} moles/cell; – normal human leucocyte 1.05 x 10^{-14} moles/cell; ● – spleen of mouse injected with CSA – A)*

proximately one hour after the intraperitoneal injection of a solution of chondroitin sulfonate A. For approximately one minute the fluorescence fading resembles that of the cultured fibroblasts. Thereafter, the rate of fading decreases to almost constant emission. This behavior was also markedly different from that of mouse spleen cells from a control animal.

The variation of fluorescence intensity as a function of time for the molecular species represented at 540 nm (Figure 8) is even more striking than that seen at 660 nm. The curve obtained from normal human leukocytes (open squares) decays rather rapidly. It should be noted here that the intracellular dye content of this cell was approximately 10^{-14} moles, which is in the range where photolytic activity takes place. The decay curve shown should not be taken as representative of the fluorescence fading behavior for normal leukocytes with only the green fluorescing molecular species present. In the latter case, esentially no fading is observed, the fluorescence emission intensity remains constant for hours.

The remaining three curves shown in Figure 8 demonstrate very interesting behavior. Rather than fading, initially there is a build-up in fluorescence intensity which may or may not later fade. The rapid build-up, after a lag of approximately one minute, exhibited by the MPS-affected leukocyte (open circles) has been observed frequently and is very striking. In Figure 8 an approximate three-fold increase in fluorescence intensity at 540 nm is indicated. However, increases in fluorescence intensity of approximately ten times have been observed. Such behavior was also observed for mouse spleen cells after intraperitoneal injection of chondroitin sulfate A. The curve shown for the mouse speen cells after injection with chondroitin sulfate A also shows a build-up of fluorescence, but at a much slower rate and without an initial lag period. Similarly, the curve obtained from cultured fetal heterozygote fibroblasts (open triangles) shows an initial slow build-up followed by gradual fading over a period of 180 seconds. This raises the possibility that the nature of the intracellular dye-mucopolysaccharide complex in freshly drawn leukocytes from affected patients is different from that in cultured fibroblasts from the fetal heterozygote.

Optical Rotatory Dispersion

Figure 9 illustrates the optical rotatory dispersion curves obtained with the microspectropolarimeter both from AO-stained cultured fibroblasts taken from the fetal heterozygote for the Maroteaux-Lamy syndrome and from the AO-CS-B precipitate. Part-

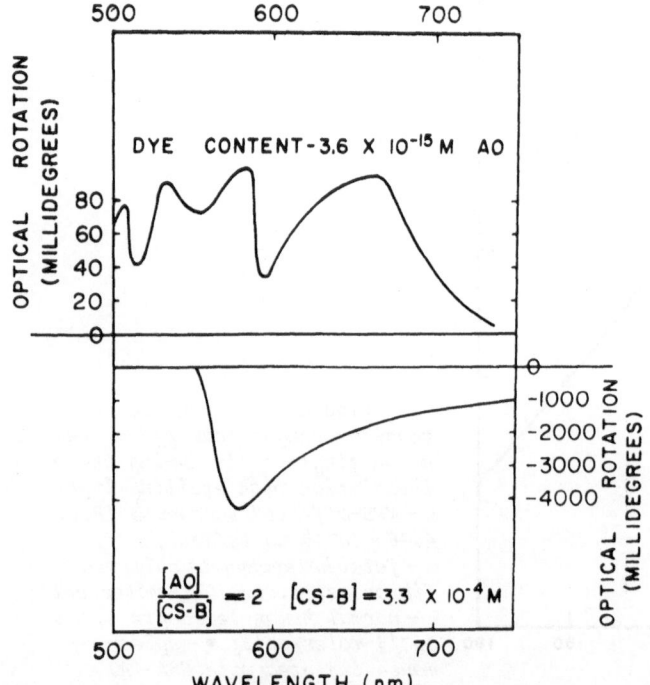

Figure 9. *Optical rotatory dispersion (ORD) spectra obtained from AO-stained cultured fibroblasts from Maroteaux-Lamy patient (upper curve, see also text and Figure 4) and from AO chondroitin sulfate B precipitate (lower curve)*

icular attention should be given to the large trough in the neighborhood of 600 nm in the upper curve, which was obtained from a single cell. When compared with the ORD curve produced by the AO-CS-B precipitate there is apparent similarity, particularly in the position of the trough. This appears to indicate that the intracellular bio-polymer which complexes with AO is the same, or very similar to, CS-B.

DISCUSSION

It is important to realize that a single characteristic parameter such as the fluorescence emission spectrum may not uniquely characterize a particular dye-polymer complex. For example, when acridine orange binds with DNA at a dye-to-polymer ratio less than 0.2, only a green fluorescence band is observed with a peak at 540 nm. This is exactly the same as the fluorescence emission spectrum of the dye alone in dilute solution. Similarly when the dye aggregates on the DNA, forming dimers and higher order polymers, the red fluorescing emission band characteristic of this molecular species has a peak at 660 nm, which is again the same for the dye alone at higher con-centrations (greater than 10^{-5} molar). However, if additional parameters are examined as the basis for a unique identification, then the AO-DNA complex is readily distin-guished from the free dye. There are a fair number of such additional parameters that can readily be employed to make the identification of a given biopolymer unique and unambigious. These include, but are not limited to the following:

a) Enhancement of fluorescence.

b) Formation of one molecular species at the expense of another. For example, this is the case when aggregates form in solutions of the dye alone. It is not the case when the aggregates form on DNA. In the latter instance, the green fluorescing molecular species, which is strongly bound to the DNA, is not affected by the forma-tion of aggregates of dye on the DNA molecule.

c) Behavior as a function of temperature.

d) Behavior as a function of ionic strength.

e) Presence or absence of induced optical activity.

f) Location of the Cotton region in the presence of induced optical activity.

g) Characteristics of fluorescence fading and buildup phenomena.

There are obviously additional parameters that can be utilized. The point to be stressed is that as many independent parameters characterizing a particular dye com-plex be utilized as are found necessary for a unique identification and characteriza-tion. Moreover, this multiplicity of phenomena can possibly be used to characterize a given dye-polymer complex that permits associating data obtained from cells with data obtained from *in vitro* model systems; it can also provide pathways for research on fundamental and applied problems.

This paper deals with only a few of the possible phenomenological approaches, but they demonstrate the principle and emphasize the central role that fluorescent mole-cular probes can play in such developments. This point is more fully elaborated below.

The linear portions of the curves in Figures 2 and 3, provided the slope is unity, delineate the range of equilibrium dye/cell ratios over which the reaction is stoichio-metric and Beer's law holds. The data shown in Figures 2 and 3 display a striking dif-ference in behavior between the cells which contain large amounts of what is presumably a mucopolysaccharide and the cells which do not. Indeed, if a separation of the two cell types were desired, it would be possible to stain a population of cells at suf-ficiently large dye/cell ratios so that cells which do not contain large amounts of mucopolysaccharide suffer photolysis upon irradiation with fluorescence-exciting light. The surviving cells would contain relatively large amounts of mucopolysaccharides.

Figures 2 and 3, however, do not provide any information about the intracellular fate of the dye. Here, visual examination and fluorescence microspectrophotometry and miscrospectropolarimetry must be employed.

Visual examination of the mucopolysaccharide-containing cells shows only the strongly bound form of the AO-NA complex in the nucleus regardless of the equilibrium-free dye-to-cell ratio. This finding is confirmed by the fluorescence emission spectra recorded from the nuclei of such cells. Apparently, the presence of the large quantity of mucopolysaccharide in the cytoplasm completely inhibits aggregation of the dye on the intranuclear nucleic acid, even when the cells contain almost 10^{-11} moles AO. This indicates that the AO-MPS complex has a fairly large free energy of association, although probably not so large as that for the strongly bound AO-NA complex ($\Delta G = -7.4$

Kcal/mole). This impression is strengthened by the fact that the mucopolysaccharide-containing cells resist photolysis. Also, though there are changes in fluorescence intensity of particular molecular species as a function of time, the intracellular morphology of the AO-mucopolysaccharide complex does not change appreciably during the course of these photochemical molecular events. One might perhaps even speculate that this behavior in the presence of AO mimics the behavior of certain normally present metabolites which become strongly bound to the intracellular mucopolysaccharides and are thus prevented from fulfilling their functions in cellular metabolism. Such behavior could perhaps account for the effects of this disease throughout the body. Though this type of thinking is only speculation, it may, nevertheless, be very useful for directing attention to investigations which can lead to a knowledge of the mechanisms underlying the disease. It also helps emphasize the need for correlating cytochemistry with biophysical chemistry. Obviously, the chemist, even if he chooses to work with intact cells, will be severely hampered in seeking to understand the behavior of a particular biopolymer or a system containing two or more biopolymers if his investigations are circumscribed by the physiological needs of the cell. On the other hand, since it is not possible to synthesize a cell in the test tube, it is of prime importance that cytochemical phenomena guide the biophysical chemist, and that the results and concepts of the biophysical chemistry be tested in the cell. The experimental results presented here give strong promise that biophysical cytochemistry can address itself to such goals.

The results presented here also suggest that the cytoplasmic substance which binds AO is a mucopolysaccharide. However, the cytoplasmic AO-biopolymer complex, although it exhibits some similarity with the *in vitro* preparations, also displays some differences, particularly with regard to fluorescence fading and build-up. This may be due to the greater complexity of the intracellular system compared to the *in vitro* model system. It may also be, that more than one biopolymer is involved. It is too early in these studies to reach any firm conclusions on this point.

The photochemical behavior expressed here as a function of time, (Figures 7 and 8) presents some obvious possibilities for differentiating one AO-biopolymer complex from another. Also, investigation of the photochemistry may provide a fruitful means for studying cell physiology. Detailed knowledge of the photochemical phenomena may also serve as a guide to assessing the fidelity with which an *in vitro* model system mimics the intracellular behavior.

Optical rotatory dispersion spectra, such as those shown in Figure 9, add to the certainty with which a given biopolymer can be identified. Beyond this, the optical rotatory dispersion spectrum produced by a macromolecule is related to its conformation. Although it is not reasonable to expect to be able to determine the absolute conformation of an intracellular biopolymer at this time, it is reasonable to expect that *alterations* in the conformation of a particular biopolymer will become evident. Conformational changes are thought to play a fundamental and important role in the metabolism of the cell. Microspectropolarimetry may be able to serve as a means for establishing such a relationship by direct observation.

Finally, important direct clinical applications are possible. Firstly, in the case of heritable disorders, identification of genetic carriers may be possible. It is not yet clear whether the heterozygote for the Maroteaux-Lamy syndrome can be differentiated from the homozygote or phenotypically affected individual. This remains to be more carefully investigated. The possibility of intrauterine diagnosis exists, and a few early trials have already given very promising results. The mucopolysaccharidoses can also serve as a model for investigating other heritable disorders which may be amenable to study by such techniques. Some preliminary investigations have already been devoted to cystic fibrosis. The results so far indicate polymorphonuclear cells from patients affected with cystic fibrosis exhibit behavior which is uniquely different both from cells related to the mucopolysaccharidoses and from normal cells. It would thus appear the biophysical cytochemistry and fluorescent molecular probes have a very broad range of applicability at both the basic and clinical levels; most of this range is still to be explored.

ACKNOWLEDGEMENTS

The authors gratefully acknowledge the unstinting support provided by their colleagues: Dr. Julian Menter, Dr. Wayne Finley, Dr. Sara Finley, Dr. Ralph E. Tiller, and Messrs. Frank Benesh and Lane Nichols, III.

This work was supported, in part by U.S.P.H., National Institute of General Medical Sciences Grant GM 18252; Health Services-Mental Health Administration, Maternal Child Health Project 910; and JPL (NASA) Contract 951925.

REFERENCES

SPICER, S.S., Leppi, T.S., and Stoward, P.J., (1965), *J. Histochem. Cytochem.*, <u>13</u>, 599.

WEST, S.S., (1965), *Acta Histochem.*, Suppl. 6, 135.

WEST. S.S., (1969), In: *Physical Techniques in Biological Research*, (Pollister, A.W., ed.), Vol. 3C, Academic Press, New York, 253-321.

WEST, S.S., (1970), In: *Introduction to Quantitative Cytochemistry-II*, (Wied, G.L., and Bahr, G.F., eds.), Academic Press, New York; 451.

LIST OF ATTENDEES

Dr. E. L. Alpen
Battelle Memorial Institute
Pacific Northwest Laboratories
P.O. Box 999
Richland, Washington 99352

Professor E. P. Benditt
Pathology Department
D511 Health Sciences SK-20
University of Washington
Seattle, Washington 98195

Dr. A. Björklund
Department of Histology
University of Lund
Biskopsgatan 5
Lund, Sweden

Dr. N. Böhm
Institute of Pathology
University of Freiburg
Albertstraße 19, West Germany

Dr. L. Brand
Department of Biochemistry
Mergenthaler Laboratory for Biology
The Johns Hopkins University
Baltimore, Maryland 21218

Dr. T. Caspersson
Institute for Medical Cell Research
and Genetics, Medical Nobel Institute
Karolinska Institute
Stockholm 60, Sweden

Dr. R. F. Chen
Department of Health, Education
and Welfare, Building 10,
National Institute of Health
National Heart and Lung Institute
Bethesda, Maryland 20014

Dr. C. F. Chignell
Laboratory of Chemical Pharmacology
National Heart and Lung Institute
Department of Health, Education
and Welfare, National Institute of Health
Bethesda, Maryland 20014

Dr. J. W. Combs
Department of Pathology
The Milton S. Hershey Medical Center
Pennsylvania State University
Hershey, Pennsylvania 17033

Dr. R. R. Cowden
Department of Anatomy
Albany Medical College
Albany, New York 12208

Mr. L. Scott Cram
Los Alamos Scientific Laboratories
Los Alamos, New Mexico 87544

Professor D. Deranleau
Biochemistry Department
J429A Health Sciences SH-70
University of Washington
Seattle, Washington 98195

Dr. M. Fulwyler
Particle Technology
Los Alamos Scientific Laboratories
Los Alamos, New Mexico 87544

Dr. B. L. Gledhill
Bio-Medical Division
Lawrence Livermore Laboratory
University of California
Livermore, California

Dr. W. Göhde
Institute of Radiation Biology
University of Münster
Münster/Westfalen, West Germany

Dr. G. G. Guilbault
Department of Chemistry
Louisiana State University
Baton Rouge, Louisiana 70803

Mr. A. J. Jesuitis
Division of Biology
California Institute of Technology
1201 East California Boulevard
Pasadena, California 91109

Dr. Gösta Jonsson
Department of Histology
Karolinska Institute
Stockholm 60, Sweden

Dr. T. Jovin
Max-Planck-Institute of
Biophysical Chemistry, Göttingen
Georg-Dehio-Weg 14, Germany

Dr. D. Killander
Institute for Medical Cell Research
and Genetics, Medical Nobel Institute
Karolinska Institute
Stockholm 60, Sweden

Dr. E. Kohen
Papanicolaou Cancer Research Institute
1155 NW 14th Street
Miami, Florida 33136

Dr. R. Kraayenhof
Laboratory of Biochemistry
University of Amsterdam
Plantage Muidergracht 12
Amsterdam, The Netherlands

Dr. J. B. LePecq
Institut Gustave-Roussy
16 bis Avenue Paul Vaillant-Coutirier
94 Villejuif (Val de Marne) France

Dr. W. Leuzinger
Battelle Memorial Institute
7, Route de Drize
1227 Carough-Geneva
Switzerland

Dr. M. Loken
Department of Biochemistry
The Johns Hopkins University
Baltimore, Maryland 21218

Dr. A. E. Lorincz
Biophysics Engineering
Medical Center
University of Alabama in Birmingham
Birmingham, Alabama 35233

Dr. O. Menis
Analytical Chemistry Department
U.S. Department of Commerce
National Bureau of Standards
Washington, DC 20234

Dr. G. Mitchel
Department of Biochemistry
School of Chemical Sciences
University of Illinois
Urbana, Illinois 61801

Dr. E. J. Modest
Children's Cancer Research Foundation, Inc.,
35 Binney Street
Boston, Massachusetts 02115

Dr. B. C. Myhr
Department of Health, Education and Welfare,
National Cancer Institute
Bethesda, Maryland 20014

Professor H. Neurath
Chairman, Biochemistry Department
J405 Health Sciences SH-70
University of Washington
Seattle, Washington 98195

Professor W. W. Parson
Biochemistry Department
J535A Health Sciences SH-70
University of Washington
Seattle, Washington 98195

Dr. R. A. Passwater
American Instrument Company
8030 Georgia Avenue
Silver Springs, Maryland 20910

Dr. G. Prenna
Instituto die Anatomia Comparata
University of Pavia, Piazza Botta 10
Pavia, Italy

Dr. G. K. Radda
Department of Biochemistry
University of Oxford
South Parks Road
Oxford, England

Dr. R. Rigler
Institute for Medical Cell Research
and Genetics, Medical Nobel Institute
Karolinska Institute
Stockholm 60, Sweden

Dr. E. M. Ritzén
Pediatric Endocrinology Unit
Karolinska Institute
Stockholm 60, Sweden

Mr. J. B. Alexander Ross
Biochemistry Department
J429 Health Sciences SH-70
University of Washington
Seattle, Washington 98195

Dr. F. W. D. Rost
Department of Histochemistry
Royal Postgraduate Medical School
University of London, Hammersmith Hospital,
London W. 12, England

Dr. M. B. Rotman
Division of Biological and Medical Science
Brown University
Providence, Rhode Island

Dr. F. Ruch
Department of General Botany
Swiss Federal Institute of Technology
Zurich, Switzerland

Dr. M. Sernetz
Battelle Institut e.V.,
6000 Frankfurt Main/90
Postschliessfach 900160, Germany

Dr. R. D. Spencer
Department of Biochemistry
School of Chemical Sciences
University of Illinois
Urbana, Illinois 61801

Dr. A. Thaer
Battelle Institut e.V.,
6000 Frankfurt Main/90
Postschliessfach 900160, Germany

Dr. Bo Thorell
Pathology Department
Karolinska Institute
Stockholm 60, Sweden

Dr. J. Vanderkooi
Johnson Research Foundation
The School of Medicine
Department of Biophysics and
Physical Biochemistry
University of Pennsylvania
Philadelphia, Pennsylvania 19104

Dr. M. VanDilla
Los Alamos Scientific Laboratories
Los Alamos, New Mexico 87544

Dr. Rance Velapoldi
Analytical Chemistry Department
U.S. Department of Commerce
National Bureau of Standards
Washington, D.C. 20234

Dr. W. Ware
Department of Chemistry
University of Western Ontario
London 72, Ontario, Canada

Dr. G. Weber
Department of Biochemistry
Roger Adams Laboratory
University of Illinois
Urbana, Illinois 61801

Dr. S.S. West
Department of Engineering
Biophysics Chairman, Medical Center
University of Alabama in Birmingham
Birmingham, Alabama 35233

Dr. W. R. Wiley
Battelle Memorial Institute
Pacific Northwest Laboratories
P.O. Box 999
Richland, Washington 99352

Dr. B. Witholt
Department of Chemistry
Revelle College
University of California
P.O. Box 109
La Jolla, California 92037

Dr. D. Wittekind
Institute of Anatomy
University of Freiburg
Freiburg/Breisgau, West Germany

Dr. J. Yguerabide
Department of Biology
University of California at San Diego
La Jolla, California 92037

Edited by V. Neuhoff

Micromethods in Molecular Biology

With contributions
by: G. F. Bahr,
J.-E. Edström,
U. Leemann,
G. M. Lehrer,
V. Neuhoff,
F. Ruch, G. Zimmer

With 275 figs. (2 in color). Approx. 450 pp. 1973
(Molecular Biology, Biochemistry and Bio-
physics, Vol. 14)
Cloth DM 98,—; US $44.10 ISBN 3-540-06319-6
Prices are subject to change without notice

Distribution rights for U.K., Commonwealth, and the Traditional
British Market (excluding Canada): Chapman & Hall, Ltd.

The purpose of this book is to introduce scientists to the use of various highly sensitive micromethods and their application to the broad field
of molecular biology. The methods are described
in great detail, so that any experimenter can adapt
them to his own field of interest. New, unpublished methods and results from the authors' laboratories are included. In essence, this is a "cook
book," giving all the normally unpublished information which is necessary for the successful
application of micromethods.

Detailed "recipes" are given for: micro-disc elec-
trophoresis, micro-electrophoresis on gradient
gels, micro-electrophoresis on SDS gradient
gels, micro-isoelectric focusing, assay of DNA
and RNA polymerase and dehydrogenases in
microgels, micro-determination of amino acids
with dansyl cloride, micro-determination of phospholipids, micro-diffusion techniques, capillary
centrifugation, microphoresis of DNA and RNA
bases, dry-mass determinations, microphotome-
try, cytofluorometry, quantitative autoradiography and several supplementary methods.

Springer-Verlag
Berlin
Heidelberg
New York

München Johannesburg
London New Delhi Paris
Rio de Janeiro Sydney
Tokyo Wien

Edited by
J. K. Koehler

Advanced Techniques in Biological Electron Microscopy

With Contributions
by S. Bullivant,
J. Frank, K. Hama,
T. L. Hayes, J. H. Luft,
F. A. McHenry,
D. C. Pease,
M. M. Salpeter

Contents

With 108 figures. XII, 304 pages. 1973
Cloth DM 40,—; US $18.00   ISBN 3-540-06049-9
Prices are subject to change without notice

The book deals with selected topics of advanced electron optical and preparatory techniques. As such, it is not intended as an elementary guide for beginning workers or students. It contains discussions of new and less well-known embedding media, specimen preparation methods including substitution, and techniques such as inert dehydration, freeze-fracturing and autoradiography, with emphasis on analytical methodology and interpretation. Other chapters deal with image processing using computers, scanning electron microscopy and high-voltage electron microscopy. These papers brought together in one volume are indicative of what constitutes the forefront of research in biological electron microscopy today. No attempt, however, has been made to be exhaustive or all-inclusive with respect to subject matter. In a number of instances, including the chapters on embedding media and computer processing, material not previously available to a general scientific audience is presented in detail. It is hoped that the volume will not only serve as a reference work for scientists already expert in these areas, but will also stimulate biologists to investigate and employ techniques which they may at first consider too exotic or too complex to attempt.

Springer-Verlag Berlin · Heidelberg · New York

München · London · Paris · Sydney · Tokyo · Wien